Astroll
8.9 roy 2

02

Kohl
Cortina
d'Ampezzo
1987

AF854669

Institut für Reine
und Angewandte Kernphysik
Kiel

Long-Term Dynamical Behaviour of Natural and Artificial N-Body Systems

NATO ASI Series

Advanced Science Institutes Series

A Series presenting the results of activities sponsored by the NATO Science Committee, which aims at the dissemination of advanced scientific and technological knowledge, with a view to strengthening links between scientific communities.

The Series is published by an international board of publishers in conjunction with the NATO Scientific Affairs Division

A	Life Sciences	Plenum Publishing Corporation
B	Physics	London and New York
C	Mathematical and Physical Sciences	Kluwer Academic Publishers
D	Behavioural and Social Sciences	Dordrecht, Boston and London
E	Applied Sciences	
F	Computer and Systems Sciences	Springer-Verlag
G	Ecological Sciences	Berlin, Heidelberg, New York, London,
H	Cell Biology	Paris and Tokyo

Series C: Mathematical and Physical Sciences - Vol. 246

1989 K 91

Long-Term Dynamical Behaviour of Natural and Artificial N-Body Systems

edited by

Archie E. Roy
Department of Physics and Astronomy,
University of Glasgow,
Glasgow, U.K.

AUSGESONDERT

Kluwer Academic Publishers

Dordrecht / Boston / London

Published in cooperation with NATO Scientific Affairs Division

Institut für Reine
und Angewandte Kernphysik
Kiel

Proceedings of the NATO Advanced Study Institute on
Long-Term Dynamical Behaviour of Natural and Artificial N-Body Systems
Cortina d'Ampezzo, Italy
August 2–13, 1987

Library of Congress Cataloging in Publication Data

NATO Advanced Study Institute (1987 : Cortina d'Ampezzo, Italy)
Long-term dynamical behaviour of natural and artificial n-body systems : proceedings of the NATO Advanced Study Institute held at Cortina d'Ampezzo, Italy, August 2-13, 1987 / edited by Archie E. Roy.
p. cm. -- (NATO advanced science institutes. Series C, Mathematical and physical sciences ; 246)
Includes indexes.
ISBN 9027728011
1. Many-body problem--Congresses. 2. Mechanics, Celestial--Congresses. 3. Dynamics--Congresses. 4. Planets, Theory of--Congresses. I. Roy, A. E. (Archie E.), 1924- . II. Title. III. Series: NATO ASI series. Series C, Mathematical and physical sciences ; no. 246.
QB362.M3N37 1987
521'.1--dc19 88-8212
CIP

ISBN 90-277-2801-1

Published by Kluwer Academic Publishers,
P.O. Box 17, 3300 AA Dordrecht, The Netherlands.

Kluwer Academic Publishers incorporates the publishing programmes of
D. Reidel, Martinus Nijhoff, Dr W. Junk, and MTP Press.

Sold and distributed in the U.S.A. and Canada
by Kluwer Academic Publishers,
101 Philip Drive, Norwell, MA 02061, U.S.A.

In all other countries, sold and distributed
by Kluwer Academic Publishers Group,
P.O. Box 322, 3300 AH Dordrecht, The Netherlands.

All Rights Reserved
© 1988 by Kluwer Academic Publishers.
No part of the material protected by this copyright notice may be reproduced or utilized in any form or by any means, electronic or mechanical, including photocopying, recording or by any information storage and retrieval system, without written permission from the copyright owner.

Printed in The Netherlands

TABLE OF CONTENTS

PART II: DYNAMICS OF NATURAL AND ARTIFICIAL SATELLITES

PART III: DYNAMICS OF ASTEROIDS

PART IV: DYNAMICS OF STARS AND GALAXIES

PART V : OTHER SPECIFIC TOPICS IN DYNAMICS

PART V. [illegible] IN DYNAMICS

PREFACE

The reader will find in this volume the Proceedings of the NATO Advanced Study Institute held in Cortina d'Ampezzo, Italy between August 3 and August 13, 1987 under the title "Long Term Dynamical Behaviour of Natural and Artificial N-body Systems".

The Institute was the latest in a series held in 1972, 1975, 1978, 1981, 1984 in dynamical astronomy, theoretical mechanics and celestial mechanics under the Directorship of Professor Victor Szebehely. These previous institutes, held in high esteem by the international community of research workers, have resulted in a series of well-received and valuable Proceedings.

In correspondence with Professor Szebehely and in long discussions with him in Colorado in August 1985, I agreed to his request that I undertake the preparation of a new ASI. I was happy to do so knowing I could call upon his vast experience in overseeing such ASI's.

The last quarter century has been a period in which increasingly rapid progress has been made in celestial mechanics and related subjects not only because of the appearance of new problems urgently requiring solution but also because of the advent of new analytical techniques and powerful computer hardware and software. The present Institute's subject, the long-term dynamical behaviour of natural and artificial n-body systems, was shown to be relevant to (i) the older problems of the age and stability of the solar system and its sub-systems of planets, satellites and asteroids, (ii) the dynamics and evolution of planetary rings, (iii) the long-term prediction of the behaviour of artificial satellites, (iv) the dynamics of star clusters and galaxies.

Apart from the series of lectures given by the invited speakers, some 38 presentations on various aspects of the subject were made.

One aspect emphasised by a number of contributions was the importance in the stability of n-body dynamical systems of their hierarchical structure. The progress made in recent years in our understanding of the restricted and general three body problem was also the subject of many papers.

The successful use of the present generation of computers to carry out massive programmes of research into the evolution of the outer solar system over a period of 10^8 years was described as well as their use in allowing the behaviour of cluster and galactic models containing large numbers of stars to be studied. In both fields of computer application, it was emphasised that the efficient use of the computer depended in a highly important way on a careful preliminary selection of the correct model to be integrated, the choosing of suitable analytical and numerical techniques together with the selection of an appropriate data-processing method. A difference of orders of magnitude in computer time, effective capacity, accuracy of results, numbers of bodies processed, real time interval studied could be gained by such careful preliminary studies.

The importance of chaotic behaviour in dynamical systems was the subject of a number of presentations, particularly with respect to the asteroid belt and the stability of planetary and satellite orbits.

Algebraic mapping techniques in the study of n-body dynamical systems, and the study of periodic solutions of analytic Hamiltonian systems with many degrees of freedom were two other approaches shown to be of importance in predicting the long-term behaviour of such dynamical systems.

The wide variety of problems discussed, matched by the number of new techniques, shows clearly that the fields of study embraced by celestial mechanics, theoretical mechanics and dynamical astronomy are in a most healthy state of activity, a view supported by the many researchers at work in these fields.

The high standard of preparation and presentation of invited lectures and other contributions were particularly pleasing: as pleasing were the frequent and sustained discussions and the warm international friendship enjoyed by the participants.

For myself I pay grateful tribute for the help and support I received from the Organising Committee (Professor V. Szbehely, Dr. P. J. Message, Dr. A. Milani). I also thank Dr. G. Volpi and Ms. B. A. Steves for their unstinting efforts before and during the meeting. As always the staff of the Antonelli Institute, where the ASI was held, made our stay a very pleasant one and we extend our thanks to them.

We are also very grateful to the Scientific Affairs Division of NATO for their guidance, counsel and support.

Finally I would like to take this opportunity to thank all those who came to Cortina and helped to make this ASI so enjoyable and productive. Their support and cooperation was very much appreciated.

Cortina d'Ampezzo, Italy
and Glasgow, United
Kingdom.

Archie E. Roy,
Director, NATO Advanced
Study Institute and Editor
of the Proceedings.

PENSIONATO STUDENTESCO
ANTONELLI

LIST OF SPEAKERS AND PARTICIPANTS

Aksnes, K.	(Norwegian)	Norwegian Defence Research Establishment and University of Tromsø, Norway.
Arribas, M.A.	(Spanish)	Dept. Fisica de la Tierra, 7 del Cosmos, U. de Zaragoza, Spain.
Bois, E.	(French)	C.E.R.G. Astronomiques, Grasse, France.
Broucke, R.	(U.S.A.)	Dept. of Aerospace Engineering University of Texas at Austin, U.S.A.
Bryant, J.G.	(U.S.A.)	47 Avenue Felix Faure, Paris, France.
Caranicolas, N.	(Greek)	Department of Astronomy, Univ. of Thessaloniki, Greece.
Carpino, M.	(Italian)	Observatorio Astron. di Brera, Milano, Italy.
Celetti, A.	(Italian)	CNR, Rome, Italy.
Comas, P.	(Spanish)	Dept. of Fisica Teorica, Univ. Autonoma de Barcelona, Spain.
Contopoulos, G.	(Greek)	Dept. of Astronomy, University of Athens, Greece.
Conway, B.	(U.S.A.)	Dept. of Aeronaut. & Astronaut. Eng., Univ. of Illinois, U.S.A.
Delhaise, F.	(Belgium)	Facultes Univ. de Namur, Belgium
Ellis, K.	(U.K.)	Queen Mary College, Univ. of London, U.K.

El-Sabaa, F.	(Kuwaiti)	Dept. of Math., Univ. of Kuwait, Kuwait
Farinella, P.	(Italian)	Dept. of Math., Univ. of Pisa, Pisa, Italy.
Ferrandiz, J.M.	(Spanish)	Dept. of Math. Aplicada, E.T.S. Ingenieros Industriales, Valladolid, Spain.
Ferraz-Mello, S.	(Brazilian)	U.E.R. de Physique, Univ. de Paris, VII, France.
Flammia, A.	(Italian)	Dept. of Math., Univ. of Pisa, Pisa, Italy.
Font, J.	(Spanish)	Fac. de Matematicas, Univ. of Barcelona, Spain.
Froeschlé, Ch.	(French)	Observatoire de Nice, Nice, France.
Froeschlé, Cl.	(French)	Observatoire de Nice, Nice, France.
Hadjidemetriou, J.	(Greek)	Dept. of Theoretical Mechanics, University of Thessaloniki, Greece.
Harper, D.	(U.K.)	Dept. of Applied Math. & Theoret. Physics, Univ. of Liverpool, U.K.
Heggie, D.	(U.K.)	Dept. of Mathematics, University of Edinburgh, U.K.
Henrard, J.	(Belgium)	Facultés Univèrsitaires de Namur, Belgium.
Innanen, K.A.	(Canadian)	Faculty of Sciences, York Univ., North York, Ontario, Canada.
Kaya, D.	(U.S.A.)	Space Command, Peterson Air Force Base, Colorado, U.S.A.
Knezevid, Z.	(Yugoslavia)	Dept. of Math., Univ. of Pisa, Pisa, Italy.

Kovalevsky, J.	(French)	Centre d'Etudes et de Recherches Geodynamiques, Grasse, France.
Leftaki, M.	(Greek)	Dept. of Math., Univ. of Patras, Greece.
Levinson, H.	(U.S.A.)	NASA/AMES Research Centre, Moffett Field, CA 94035.
Liu, J.J.F.	(U.S.A.)	Space Command, Peterson Air Force Base, Colorado, U.S.A.
Marchal, C.	(French)	D.E.S. - Onera, Chatillon, France.
Marzari, F.	(Italian)	Dept. di Fisica, Univ. of Padova, Italy.
Masdemont, J.M.	(Spanish)	Univ. Politecnica de Catalunya, Spain.
McCarter, L.A.	(U.S.A.)	U.S. Space Command, Peterson Air Force Base, Colorado, U.S.A.
Melchioni, E.	(Italian)	Dept. of Physics, Univ. of Pavia, Italy.
Message, P.J.	(U.K.)	Dept. of Applied Maths., & Theoretical Physics, Univ. of Liverpool, U.K.
Milani, A.	(Italian)	Dipartimento di Matematica, Universita di Pisa, Italy.
Nobili, A.	(Italian)	Dept. of Math., Univ. of Pisa, Pisa, Italy.
Ollé,M.O.,	(Spanish)	Dept.of Math., Univ. Politecnica de Catalunya, Spain.
Olvera, A.	(Spanish)	Fac. de Matematicas, Univ. of Barcelona, Spain.

Onargan, G.	(Turkish)	Fac. of Engineering Div., Dokuz Eylül Univ., Turkey.
Pinotsis, A.D.	(Greek)	Dept. of Astronomy, Univ. of Athens, Greece.
Puel, F.	(French)	Observatoire de Bresancon, France.
Roy, A.E.	(U.K.)	Dept. of Physics and Astronomy, Glasgow Univ. Glasgow, G12 8QQ. U.K.
Sansaturio, M.E.	(Spanish)	Dept. of Math., Univ. of Valladolid, Spain.
Snow, D.	(U.S.A.)	Dept. of Aerospace Eng. & Eng. Mechs., Univ. of Texas at Austin, U.S.A.
Simo, C.	(Spanish)	Facultat de Matematiques, Universitat de Barcelona, Spain.
Steves, B.A.	(Canadian)	Dept. of Physics and Astronomy, Univ. of Glasgow, U.K.
Susin, A.S.	(Spanish)	Dept. of Math., U. Politecnica de Catalunya, Spain.
Szebehely, V.	(U.S.A.)	Dept. of Aerospace Eng., University of Texas at Austin, U.S.A.
Varvoglis, H.	(Greek)	Dept. of Astron. Univ. of Thessaloniki, Greece.
Volpi, G.	(Italian)	Via Washington 102, 20146, Milan, Italy.
Waldvogel, J.	(Swiss)	ETH - Zurich, Seminar fur Ange. Mathematik, Zurich, Switzerland.
Winter, K.	(U.S.A.)	Dept. of Applied Mathematics, Univ. of Washington, FS-20. Seattle, Wa 98115, U.S.A.

Yoshikawa, M.	(Japanese)	Dept. of Astronomy, Univ. of Tokyo, Japan.
Yoshida, H.	(Japanese)	Centre de Physique Theorique, Ecole Polytechnique, Palaiseau, France.
Zafiropoulos, B.	(Greek)	Dept. of Physics, Univ. of Patras, Greece.

PART I

PLANETARY DYNAMICS

DYNAMICAL EVOLUTION OF PLANETARY SYSTEMS AND THE SIGNIFICANCE OF BODE'S LAW

Bruce A. Conway
Department of Aeronautical and Astronautical Engineering
University of Illinois
Urbana, Il 61801 USA

Timothy J. Elsner
Lockheed Missiles and Space Co.
Sunnyvale, CA USA

ABSTRACT

We examine the question of whether an arbitrary configuration of planets in orbit about the sun could evolve by point gravitational interaction to become a very regular system such as the actual solar system. The evolution of a number of sample planetary systems with varying original distributions has been examined to determine the degree of "arbitrariness" allowed to the initial conditions which will yet yield the regularities presently observed in the solar system. Previous researchers have identified "quasi-stable" periods in the evolution of such systems but we show that frequently, when the integrations are continued to greater lengths, violent interactions again occur. This study suggests that only a geometric (or Bode-type) initial distribution of planetary semimajor axes is likely to yield the present (geometric) spacing of the planets of the solar system.

INTRODUCTION

The degree of significance which may be given to the present distribution of planets and natural satellites of the solar system, as an indication of the original distribution or of the origin of the solar system, is controversial. Hills[1] and Ovenden[2], from numerical studies of the evolution of sample planetary systems, among other arguments, claim that the "regularity" of the present solar system could result from "dynamic relaxation" i.e. evolution by point-gravitational interaction from an arbitrary initial configuration. However this is only a conjecture, relying principally on the observations of these researchers that during the numerical integration pairs of bodies show commensurabilities in mean motion significantly more often than chance would allow - an observation which these researchers claim is also true of the present

A. E. Roy (ed.), Long-Term Dynamical Behaviour of Natural and Artificial N-Body Systems, 3–12.
© 1988 by Kluwer Academic Publishers.

distribution. No numerical investigations of which we are aware have demonstrated evolution to the very stable configuration evidenced by the present solar system[3] from the quasi-stable configurations found at the end of some of the Hills[1] and Ovenden[2] integrations. (The "quasi-stability" itself may be illusory as we will show here with integrations of greater duration). In fact no theory requires, in the absence of dissipation, this type of evolution.

Ovenden[2] concludes that "the present distribution gives no information concerning the origin of the solar system." Interpreting this to mean that it does not give any information regarding the original distribution of solar system bodies surely this conclusion is not supported by Ovenden's own work. Unexamined by Ovenden and others, is the degree of arbitrariness allowed to the original distribution which will yet yield at least the gross regularity presently observed, that is, the increasing separation between planets, moving outward from the Sun.

OBJECTIVES AND METHOD

In the present work a number of sample planetary systems have been allowed to evolve due to point gravitational interaction. We have varied the number of planets, the distribution of planetary mass, and the initial distribution of planetary semimajor axes. The various cases integrated are shown in Tables I and II. In some cases the integration was allowed to proceed for a much greater length of time (up to 40K years) than the Hills[1] or Ovenden[2] integrations.

The objectives of the study were:

i) To simply continue the evolution of such systems for a greater period than had previously been done. We hoped to then be able to reach some heuristic conclusions regarding the existence or not of a "tendency to stability".

ii) To examine, again with a heuristic analysis, the effects of initial distribution of planetary masses and semimajor axes. We were particularly interested in determining the comparative stability of a system similar to the solar system, with a (Bode's Law) geometric spacing of semimajor axes and with the smallest planets closest to the sun.

iii) To examine (or re-examine) the claim that, during the evolution due to mutual gravitational perturbations, pairs of planets will, more often than not, adopt semimajor axes such that the ratios of their orbit periods will be commensurable.

The results of this study suggest that it is highly unlikely that a planetary system with an initially arbitrary distribution of planets could evolve from a series of dynamical encounters to a distribution which approximated Bode's law.

TABLE 1
SYSTEMS WITH ARBITRARY INITIAL CONDITIONS

System	No. of Planets	Masses of Planets (In Solar Masses)	Init. Distances (A.U.)	Evol. Time (Years)
1	6	4.46×10^{-4}	5.0, 5.75, 6.5, 7.25, 8.0, 8.75	40,000
2	6	2.23×10^{-4}	5.0, 6.0, 7.0, 8.0, 9.0, 10.0	10,000
3	6	4.46×10^{-4}	5.5, 6.0, 6.5 7.0, 7.5, 8.0	10,000
4	6	4.46×10^{-4}	5.0, 6.0, 7.0 8.0, 9.0, 10.0	10,000
5	6	2.23×10^{-4}	2.0, 4.0, 6.0, 8.0, 10.0, 12.0	10,000
6	8	6.69×10^{-4}	1.0, 7.42, 13.83, 20.2, 26.66, 33.08, 39.5, 45.91	20,000
7	8	3.35×10^{-4}	1.0, 7.42, 13.83, 20.2, 26.66, 33.08, 39.5, 45.91	20,000

When planets were originally placed arbitrarily very close together (an unlikely distribution of the planets formed from an early nebula?) the planets underwent an early phase of violent encounters resulting in many of the planets being thrown to more distant orbits where the encounters were less frequent (e.g. System 1 in Figs. 1, 2, and 3). Similar integrations by Hills[1] and Ovenden[2] demonstrated the same pattern. Hills claimed that this showed an increase in stability and then estimated the time required to attain a final stability. However, several of the integrations in this study, which were considerably longer than Hills' or Ovenden's, showed no long term trends of increasing stability beyond that brief initial phase. In the history of System 1, illustrated in Figure 2, for example, there is a period of nonviolent interaction beginning at approximately 13,000 years (immediately following the ejection of Planet 3) analogous to periods interpreted by Hills and Ovenden as "quasi-stable". But at 24,000 years the evolution resumes its former violence and this continues to the end of the integration at 40,000 years. Therefore, estimates of the length of time required to reach a final stability, based on only an early brief trend, are meaningless.

When the initial orbits were spaced further apart, (e.g. System 5 in Fig.4) the orbital distances were briefly maintained but within a short period of time progressively larger perturbations resulted. Finally the system evolved to an unstable state with erratically changing orbits.

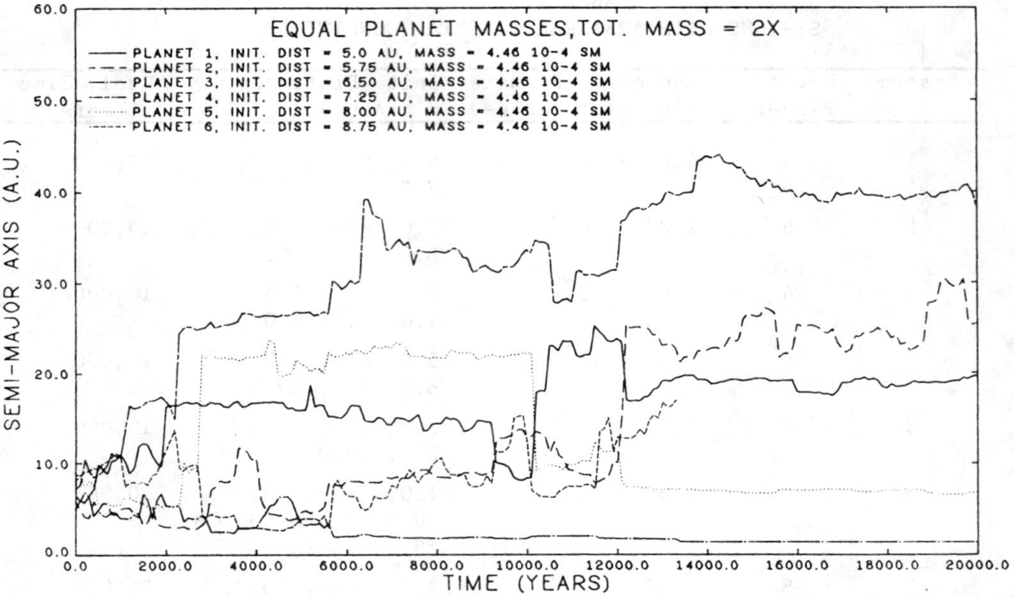

FIGURE 1 HISTORY OF SEMI-MAJOR AXES FOR SYSTEM 1

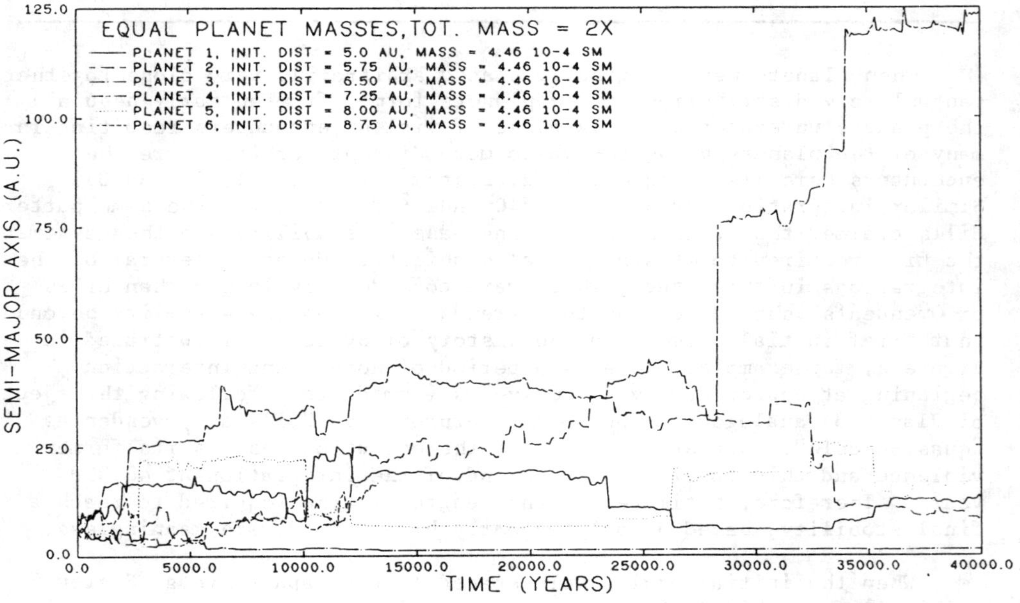

FIGURE 2 HISTORY OF SEMI-MAJOR AXES FOR SYSTEM 1

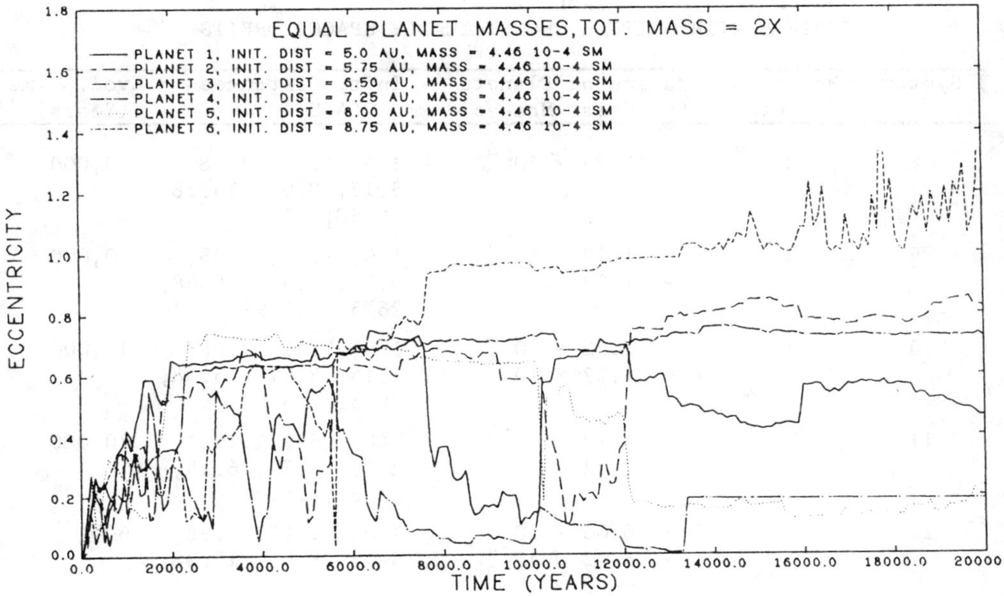

FIGURE 3 HISTORY OF ECCENTRICITIES FOR SYSTEM 1

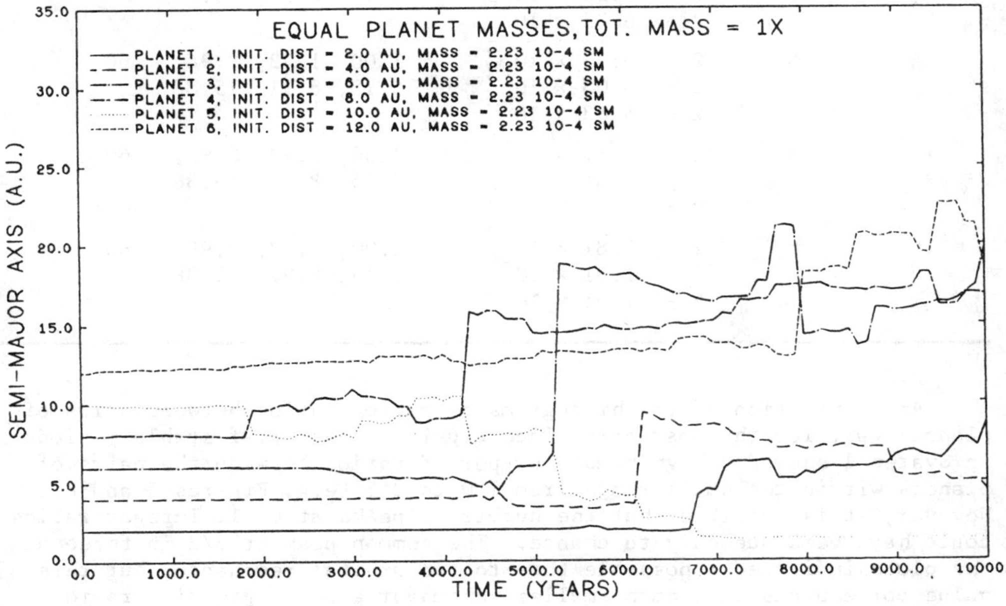

FIGURE 4 HISTORY OF SEMI-MAJOR AXES FOR SYSTEM 5

TABLE 2
SYSTEMS WITH INITIAL GEOMETRICALLY SPACED ORBITS

System	No. of Planets	Masses of Planets (in Solar Masses)	Init. Distances (A.U.)	Evol. Time (Years)
8	8	3.35×10^{-4}	1.0, 1.72, 2.98, 5.15, 8.91, 15.38, 26.58, 45.91	10,000
9	8	4 - 1.22×10^{-4} 4 - 1.22×10^{-3}	1.0, 1.72, 2.98, 5.15, 8.91, 15.38, 26.58, 45.91	10,000
10	8	4 - 1.22×10^{-3} 4 - 1.22×10^{-4}	1.0, 1.72, 2.98, 5.15, 8.91, 15.38, 26.58, 45.91	10,000
11	8	4 - 1.22×10^{-4} 4 - 1.22×10^{-3}	.40, .69, 1.19, 2.06, 3.56, 6.15, 10.63, 18.36	10,000
12	6	2 - 6.03×10^{-5} 2 - 6.03×10^{-4} 2 - 6.03×10^{-3}	1.00, 1.72, 2.98, 5.15, 8.91, 15.38	60
13	6	2 - 6.03×10^{-3} 2 - 6.03×10^{-4} 2 - 6.03×10^{-5}	1.00, 1.72, 2.98, 5.15, 8.91, 15.38	60
14	6	2 - 6.03×10^{-5} 2 - 6.03×10^{-3} 2 - 6.03×10^{-4}	1.00, 1.72, 2.98, 5.15, 8.91, 15.38	60
15	6	2 - 1.81×10^{-4} 2 - 1.81×10^{-3} 2 - 1.81×10^{-2}	1.00, 1.72, 2.98, 5.15, 8.91, 15.38	60
16	6	2 - 1.81×10^{-3} 2 - 1.81×10^{-4} 2 - 1.81×10^{-2}	1.00, 1.72, 2.98, 5.15, 8.91, 15.38	60

An examination of the histograms of period ratios between pairs of planets was, for the most part, inconclusive. The brief stable periods in System 1 showed a high number of period ratios between the pairs of planets within the small range from 1.0 to 2.5 (e.g. Figures 5 and 6). However, it is possible that the number of peaks at small integer ratios could have been due only to chance. The common peak at 3/2 in three of the plots indicated a possible three to two orbital resonance, but this value corresponds to a much smaller semimajor axis progression ratio than that which is found in our solar system. If there was any resonance effect it came across in the semimajor axis histories as a very weak mechanism at best.

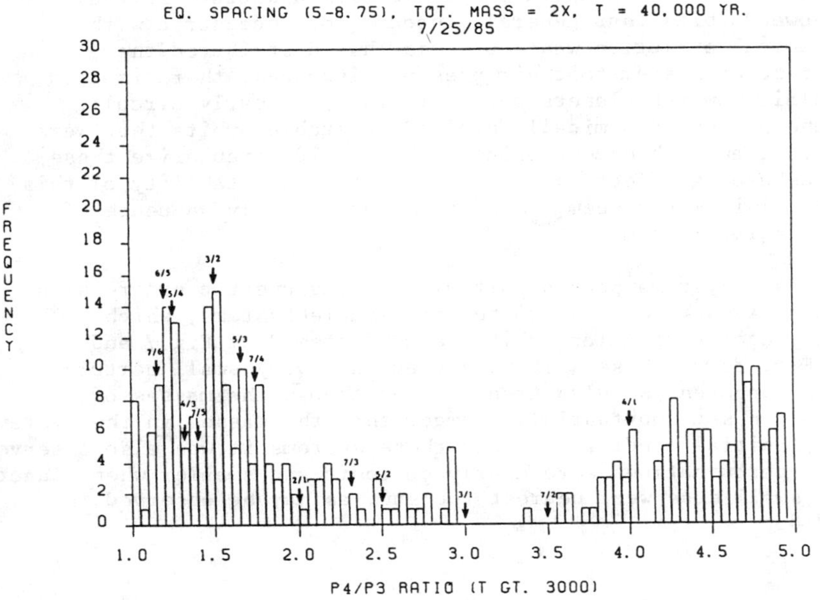

FIGURE 5 A HISTOGRAM OF THE RATIOS OF PERIODS OF PLANET 4 AND PLANET 3

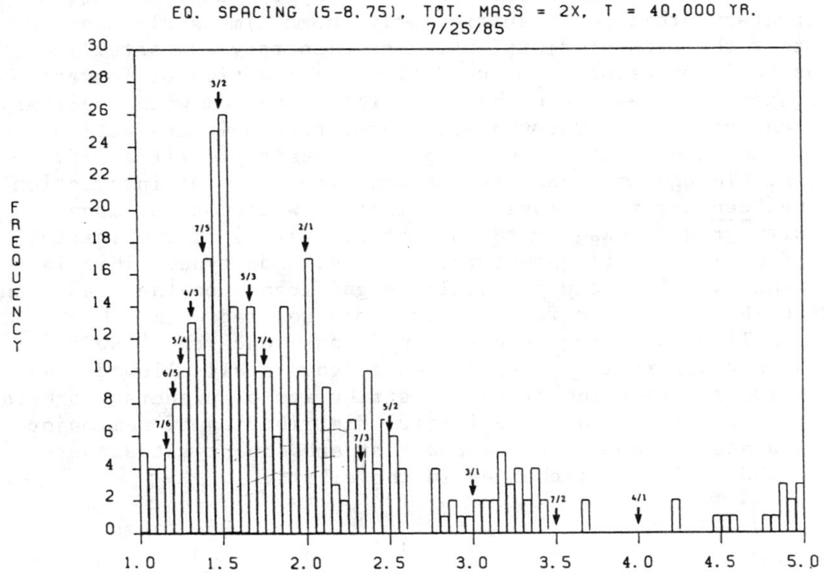

FIGURE 6 A HISTOGRAM OF THE RATIOS OF PERIODS OF PLANET 5 AND PLANET 4

All the evolutions of planetary systems with arbitrary initial conditions showed orbits that generally became more eccentric with time. If dynamical evolution was a process that distributed the planets in our solar system to their present distances, there is then a need to explain how the planets were left in very nearly circular orbits. If the planets dynamically evolved to stable orbits that were highly eccentric, any subsequent process that would circularize these orbits (such as a drag effect) would likely alter the stability of this earlier state. This would seem, then, to be an unlikely sequence of events for our solar system.

In contrast, systems placed initially in a geometric progression in the semimajor axes were shown to be very stable systems, which maintained very nearly circular orbits (e.g. System 9 of Fig.7 and System 10 of Fig. 8). These systems showed only very small perturbations with no apparent secular trends, even though the masses of the planets were often set substantially larger than the masses in the systems with arbitrary initial conditions. For these systems it was also observed that the orbit perturbations were nearly periodic and smaller when planets of relatively small mass were nearest the sun, as may be observed in Figures 7 and 8.

CONCLUSION

We believe Ovenden is incorrect when he says that[2] , "... if it can be shown that processes operate within the solar system that can rearrange the planetary orbits on a sufficiently short time scale then we must conclude that the present distribution of planetary and satellite orbits contains no information about conditions at the time of formation of the solar system." There is no evidence that a system with arbitrary initial conditions can evolve through point gravitational interaction into a system resembling (that is, showing the gross regularities of) the solar system. Though the dramatic and sometimes violent interaction Ovenden describes <u>can</u> occur, we have shown that it will not for many systems (even with greatly exaggerated planetary masses) if the initial semimajor axis distribution is geometric, that is, Bode-type. This is not to suggest that we place any particular significance on the traditional form of the Titus-Bode law; in fact we have done related work[4] which suggests that the Titus-Bode law is not significant, once the "law of increasing planetary separations" is assumed to govern all allowed distributions. We suggest that the present distribution of planetary orbits <u>is</u> informative and indicates that the initial distribution of semimajor axes did follow a nearly geometric law and that subsequent mutual perturbations have made only small changes in the orbits.

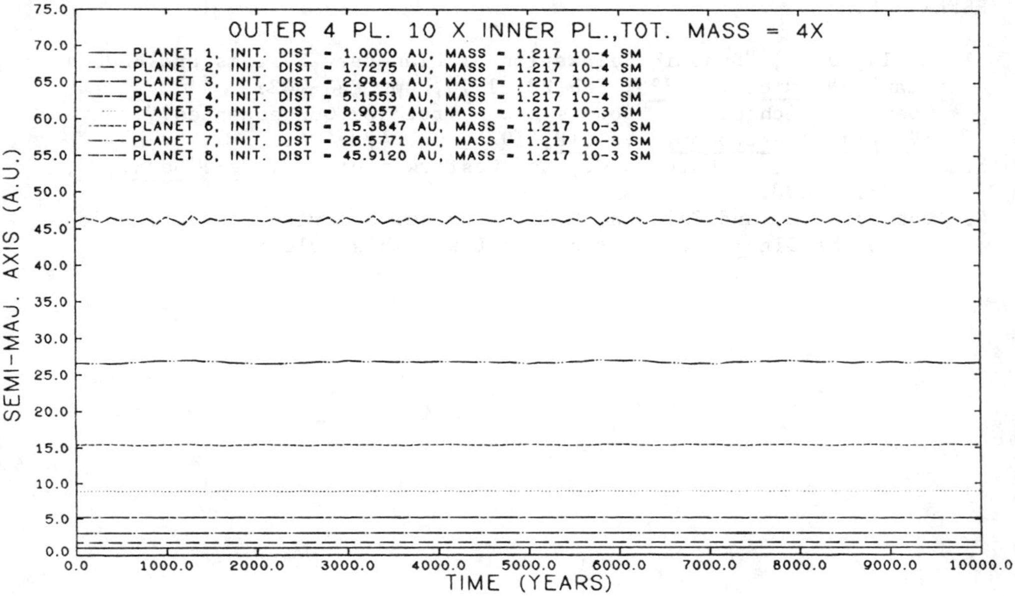

FIGURE 7 HISTORY OF SEMI-MAJOR AXES FOR SYSTEM 9

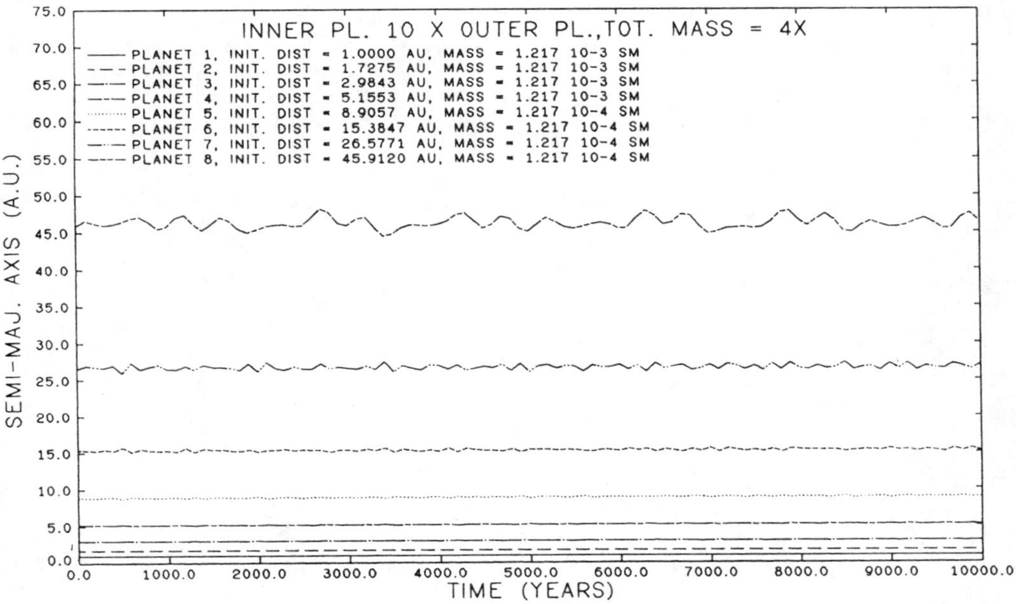

FIGURE 8 HISTORY OF SEMI-MAJOR AXES FOR SYSTEM 10

REFERENCES

1. Hills, J.G., "Dynamic Relaxation of Planetary Systems and Bode's Law", Nature, Vol.225, Feb.28, 1970, pp. 840-842.
2. Ovenden, Michael W., "Bode's Law - Truth or Consequences?", Vistas in Astronomy, Vol.18, 1975, pp. 473-496.
3. Cohen, C.J., Hubbard, E.C., and Oesterwinter, C., Cel. Mech., 7, 1973, p.438.
4. Conway, B.A. and Zelenka, R.E., "Further Numerical Investigations into the Significance of Bode's Law. This Volume.

FURTHER NUMERICAL INVESTIGATIONS INTO THE SIGNIFICANCE OF BODE'S LAW

Bruce A. Conway
Richard E. Zelenka
Department of Aeronautical and Astronautical Engineering, University of Illinois, Urbana, IL 61801

ABSTRACT:

In a 1975 paper Ovenden[1] claimed that "... it is logically impossible to answer the question 'Is Bode's Law due to chance?'." We do not believe this to be correct despite the apparent difficulty cited by Ovenden that we have knowledge only of our own planetary system. In this work the validity of the Titius-Bode law is assessed by comparing the closeness of fit of semi-random, computer - generated distance distributions to that of the actual distribution. The sample distributions are semi-random in that they are drawn from a population which obeys the "law of increasing differences", expressing the fact that the separations between planets increase, moving outward from the Sun. We consider the solar system excluding Pluto, whose orbit is very irregular, and two further abridged systems, lacking Mercury and/or Neptune. We also apply the analysis to the regular satellites of the giant planets. The results are not favourable to Bode's Law. This is partly explained by the fact that the "law" loses much of its unusualness when it is a priori assumed that all stable planetary distributions obey the stronger "law of increasing differencws".

INTRODUCTION

The Titius-Bode Law of planetary distances, or simply Bode's Law, has been the subject of controversy among both historians and scientists. The historical controversy centres on the question of whether it was Johann Titius or Johann Bode who should be credited with the invention of the law. Nieto[2] indicates that it was Titius who first proposed the law in 1766, although a series of events led to Bode unjustly receiving principal credit. The scientific controversy, which will be the focus of this paper, regards the degree of significance which the law should be accorded.

A. E. Roy (ed.), Long-Term Dynamical Behaviour of Natural and Artificial N-Body Systems, 13–20.
© 1988 by Kluwer Academic Publishers.

Bode's Law states that the mean distances from the Sun to each of the planets in the solar system can be written as a geometric progression:

$$d_n = a + b \cdot 2^n \qquad (1)$$

where n is the planet's place number, i.e. n = 1 for Mercury, n = 2 for Venus, etc. The constants a and b are set to 0.4 and 0.075 respectively, in order to provide a distance of 1.0 A.U. from the Sun to the Earth. This generates a good correspondence with the actual planetary distances if the results for Neptune and Pluto are ignored, as shown in Table 1.

Table 1: Bode and Actual Distances

Planet	Bode's Law Distance	Actual Distance
Mercury	0.550	0.387
Venus	0.700	0.723
Earth	1.000	1.000
Mars	1.600	1.524
Asteroids	2.800	2.9
Jupiter	5.100	5.203
Saturn	10.000	9.555
Uranus	19.600	19.218
Neptune	38.80	30.110
Pluto	77.20	39.44

We approach the determination of the validity of the law from the point of view of a statistician. Two previous studies have examined Bode's Law in this way. I. J. Good[2] conducted what he termed a "subjective evaluation" of the law. At the outset, he argued for the removal of those planets whose distances from the Sun were not in good agreement with those predicted by Bode's Law. Pluto was discarded on the basis of its rather irregular orbit, Mercury and Neptune were removed by "conjecturing that the planets condensed from a proto-planetary cloud that was somewhat irregular at its outer and inner edges". Good felt that a penalty factor of 1/5 would generously account for any error induced by the removal of these planets. Good's conclusion was favourable to Bode's Law; a likelihood ratio of 133/1 that the distribution of planets is more significant than chance would allow. Another statistician, Bradley Efron, responding to Good's work two years later (1971) worked with the same subset of the solar system and used the same penalty factor for deleting the non-conforming planets[3]. In a computer simulation, Efron generated semi-random distributions of planetary distances, thus creating systems to which the significance of the Venus - Uranus system could be compared. Instead of fixing the base number in the Bode Law at two, as Good did, he optimized the base number to yield the best fit, as others had done before[2], recognising that the original base number is not even an optimum for the actual solar system.

Efron also differed in his use of non-Bayesian methods as opposed to Good's Bayesian methods. Efron's analysis yielded a likelihood ratio extremely near to one (1.6/1) leading him to conclude that Bode's Law is artifactual.

The work to be presented here principally concerns the assessment of the validity of Bode's Law with only Pluto removed from the solar system. Established arguments, centering on the abnormally high eccentricity and inclination, and the low mass of Pluto, suggest that Pluto had either an unusual origin, e.g. capture by Neptune, or has suffered from cataclysmic event in its history. We do not believe that arguments for removing additional planets, i.e. Mercury and Neptune, are as strong. Abridged systems will, however, be considered for the purpose of determining if Good's penalty factor of 1/5 is really appropriate or sufficient for discarding Mercury and Neptune. We will also extend the same analysis to the case of the regular satellites of the major planets. Finally, a judgement of the validity of Bode's Law will be made.

METHOD

We will employ the same method used by Efron[4], that is, the significance of Bode's Law will be determined by generating many random distributions and comparing those distributions to the actual distribution of the planets. For the system of planets from Mercury to Neptune:

Let $$u_i = \frac{d_{i+1} - d_i}{d_9 - d_1} \quad \text{for } i = 1,2,\ldots\ 8.$$

The u_i represent dimensionless spacings between the planets which are of increasing length, and which alternatively describe the nine distances of the planets (and the asteroid belt) from the Sun.

Suppose that the distances observed were perfectly geometric, that is they fit a generic Bode-type law of the form

$$d_n = a + b\cdot c^n \qquad \text{for } n = 1,2,\ldots\ 9.$$

In this case the spacings u_i, would satisfy

$$u_i = c\, u_{i-1} \qquad \text{for } i = 1,2,3,\ \ldots\ 8,$$

or

$$u_i = \frac{c^{i-1}}{S_c} \qquad \text{for } \quad i = 1,2,\ldots\ 8, \tag{2}$$

With

$$S_c = \frac{c^8 - 1}{c - 1}\ .$$

This is the ideal fit for a given set of distances, if the distances are to adhere to a geometric law. Any observed distribution must be compared to this one. The closeness of fit of sample system distances to a geometric law will be measured with the distance statistic

$$\Delta = \left[\sum_{i=1}^{8} \frac{(u_i - c^{i-1}/S_c)^2}{(c^{-1}/S_c)^q} \right]^{\frac{1}{2}} \quad \text{for } 1 < c < \infty. \tag{3}$$

The value of q is set at one, although some additional analysis with q=0 (essentially a least squares fit) and q=2 was also performed. A computer is then used to find the optimal value of c; that which minimizes the distance statistic and therefore yields the best fit.

A similar approach is employed for the random distributions to be generated. A computer program was written which creates nine random numbers between 0 and 1. These numbers are then arranged in ascending order and spacings between each determined. Upon division by the difference of the largest and smallest numbers, and a final reordering of the spacings in ascending order, a distribution of spacings summing to one will have been created. The code then searches for the minimum Δ for best fit, storing this Δ and then repeating the whole procedure for another distribution. One thousand such distributions were generated for comparison to each actual distribution under considering. It may then be observed how many times out of 1000 trials the minimum Δ for a distribution was found to be less than that for the actual distribution, Δ_{min}, indicating a better geometric fit.

Reordering the randomly generated spacings so that $u_i < u_{i+1}$ for all i greatly reduces the randomness of the distribution but is physically reasonable. Efron[4] assumed that this "law of increasing differences" must be a result of the planet formation mechanism so that it would not be reasonable to compare the goodness of fit of the actual planetary spacings (to a geometric law) with that of completely random distributions of planetary spacings. One of the present authors[5] has conducted research into this hypothesis, examining the stability of sample planetary systems with different initial mass and semimajor axis distributions. Although the number of sample cases examined was limited only distributions which obeyed this "law of increasing differences" exhibited the regularity of the actual solar system as they evolved due to mutual gravitational interaction. There are also the examples of the (major) satellites of the giant planets, whose orbits generally obey this rule, although it cannot be said for certain that the same mechanism is responsible for this. Bode's Law then obviously loses some degree of the significance it gains by predicting increasing distances between planets when we compare it only to these semi-random alternative distributions.

RESULTS

One thousand randomly generated distributions were compared to a few actual distributions for the planetary system, in the manner described above, with results given in Table 2. Efron[4] examined only the Venus-Uranus system but we have considered two others as well.

Table 2: The Planetary System Fitted to a Geometric Law

Planetary System Considered	q	$C_{optimal}$	Δ_{min}	$\Delta < \Delta_{min}$ (per 1000)
	0	2.0515	0.0234	18
Venus → Uranus	1	2.00855	0.0645	14
	2	2.00911	0.224	31
	0	1.625	0.0925	757
Venus → Neptune	1	1.746	0.201	471
	0	1.629	0.0916	799
Mercury → Neptune	1	1.733	0.202	470
	2	1.742	0.690	392

It is evident from Table 2 that whether Neptune can or cannot be neglected is vital in determining the significance of Bode's Law. For q=1, roughly half of the semi-random distributions fit a Bode-type law as well as the actual planetary system (including Neptune), suggesting there is little substance to Bode's Law. However, if an abridged system which excludes Neptune is considered, the results appear to be very favourable toward Bode's Law. The question of removing Mercury seems to be insignificant in any assessment of Bode's Law. As stated earlier, the exclusion of any body in the planetary system aside from Pluto seems unwarranted.

Another simple law is now considered as an alternative to the geometric law that Johann Titius had suggested. The simple law to be considered is that of a quadratic, that is, the distances are described by:

$$d_n = A + Bn + Cn^2 \qquad \text{for } n = 1,2,\ldots\ldots 9. \qquad (4)$$

The spacings satisfy

$$u_i = \frac{d_{i+1} - d_i}{d_9 - d_1} \qquad \text{for } i = 1,2, \ldots\ldots 8. \qquad (5)$$

as before. If the distances were exactly quadratic, then

$$u_i = \frac{1 + C/B(2i + 1)}{8(1 + 10\ C/B)} \qquad \text{for } i = 1,2, \ldots\ldots 8. \qquad (6)$$

A distance statistic, Δ^*, similar to that used for the geometric law will be employed here, i.e.

$$\Delta^* = \left\{ \sum_{i=1}^{8} \frac{[u_i - [(1 + C/B(2i+1))/8(1 + 10\, C/B)]]^2}{[(1 + C/B(2i+1)/8(1 + 10\, C/B)]^q} \right\}^{\frac{1}{2}} \text{ for } -\infty < C/B < \infty. \quad (7)$$

As was done for the geometric law, a computer is allowed to search for the optimal C/B which gives the minimum Δ^* and in turn the best fit. The semi-random distributions are generated just as before. One thousand semi-random distributions were generated and compared to the observed distribution of the 7 planet, Venus to Uranus system and 9 planet, Mercury to Neptune system. Table 3 summarizes the results.

Table 3: The Planetary System Fitted to a Quadratic Law

Planetary System Considered	q	$C/B_{optimal}$	Δ^*_{min}	$\Delta < \Delta^*_{min}$ (per 1000)
Venus → Uranus	0	-0.157	0.192	696
	1	-0.163	0.451	747
Mercury → Neptune	0	-0.226	0.211	772
	1	-0.219	0.542	865

It is clear that when the observed distribution is compared with that generated by chance, using a quadratic law, there is no special significance to the observed distribution. This is the case whether the full Mercury to Neptune system is considered or the abridged, Venus to Neptune system.

The planetary satellite systems of a few of the major planets were also considered. A geometric (Bode-type) law and error statistic with q=1 were exclusively used for these systems. One thousand random distributions were generated and compared with the observed distributions in a manner completely analogous to that previously described. Table 4 describes the results.

Table 4: The Planetary-Satellite System Fitted with a Geometric Law

System	Satellites Included	$C_{optimal}$	Δ_{min}	$\Delta < \Delta_{min}$ (per 1000)
1. Jupiter	J1→J6	3.382	0.422	980
2. Jupiter	J1→J5	1.577	0.151	384
3. Saturn	S1→S8	2.003	0.460	997
4. Saturn	S1→S6	2.213	0.333	893
5. Uranus	U1→U5	1.380	0.225	638

Table 5: Satellites Included in the Systems of Table 4

Jupiter		Saturn		Uranus	
J1	Io	S1	Mimas	U1	Ariel
J2	Europa	S2	Enceladus	U2	Umbriel
J3	Ganymede	S3	Tethys	U3	Titania
J4	Callisto	S4	Dione	U4	Oberon
J5	Amalthea	S5	Rhea	U5	Miranda
J6	Himalia	S6	Titan		
		S7	Hyperion		
		S8	Iapetus		

Table 5 shows which satellites were considered. It is clear that, even for the abridged systems (J1 → J5, S1 → S6) there is nothing special about the goodness of fit of the satellite distances to a Bode-type law.

CONCLUSION

In order to compare our results with those of Good[3] and Efron[4] a likelihood ratio will be determined using the data of Table 2 for the system which they considered, Venus-Uranus. The likelihood ratio is defined as:

$$\text{likelihood ratio} = \gamma \cdot \frac{1}{\text{tail probability}} \qquad (8)$$

For the system Venus-Uranus (with q=1) one can see that the tail probability is 0.014. Good[3] and Efron[4] both note that the significance of Bode's Law must be diminished by the observation that if a geometric law had not fit the distances well perhaps some other simple law, such as the quadratic law examined previously, might have. Good[3] judged that geometric progressions represent 1/3 of all the "simple" laws. Adopting this estimate, as Efron[4] did, we must increase the tail probability by a factor of 1/(2/3), to .021. With Efron's choice of $\gamma = 1/4$ this results in a likelihood ratio slightly less than 12 in support of Bode's Law. Efron then proceeds to apply Good's penalty factor of 1/5 for ignoring Mercury, Neptune, and Pluto. Such a penalty factor, if considered just and applied here, would result in a likelihood ratio of nearly 2 in favour of Bode's Law. Similar analysis of the remaining Venus to Neptune and Mercury to Neptune systems of Table 2 would result in likelihood ratios much less than one (without any type of penalty factor), indicating a distribution due entirely to chance.

The penalty factor of 1/5 that both Good and Efron employed for discarding Mercury, Neptune, and Pluto is considerably too low. Based on the distance statistic with q=1, a factor of 1/33 is appropriate. Should q be set to zero, a penalty factor of 1/44 would be more suitable.

Such a factor, when applied to the Venus to Uranus system discussed, yields a likelihood ratio of 0.30 to 0.23.

The paradox of the Titius-Bode law is that a casual study of the law suggests that it is meaningful and significant. However, we have shown that the law receives much of its apparent significance from predicting increasing separations between planets. When increasing separations are assumed a priori, as a necessary condition for the stability of this and presumably other solar systems, and when the large error of the Bode Law at Neptune is not explained away, the law loses its unusualness. The fact that we have direct evidence of only one planetary system, the difficulty cited by Ovenden[1], does not prevent us from determining the statistical significance of the law.

REFERENCES

(1) Ovenden M.W. "Bode's Law - Truth or Consequences?" Vistas in Astronomy, 18 (1975), pp.473-496.
(2) Nieto, Michael M. The Titius-Bode Law of Planetary Distances: Its History and Theory. Oxford: Pergamon Press, 1972.
(3) Good, I.J. "A Subjective Evaluation of Bode's Law and an 'Objective' Test for Approximate Numerical Rationality". Journal of the American Statistical Association, 64 (March 1969). pp.23-40.
(4) Efron, Bradley. "Does an Observed Sequence of Numbers Follow a Simple Rule? (Another Look at Bode's Law". Journal of the American Statistical Association, 66 (September 1971), pp. 552-559.
(5) Conway, B.A. and Elsner, T.J. "Dynamical Evolution of Planetary Systems and the Significance of Bode's Law". This Volume.

WHERE ARE THE SATURNIAN TROJANS?

K.A. Innanen and Seppo Mikkola*
Department of Physics, York University,
4700 Keele Street, Toronto,
Ontario, M3J 1P3, Canada

*On leave from: Turku University Observatory,
Itäinen Pitkäkatu 1, 20520 Turku, Finland.

ABSTRACT

Numerical N-body integrations show that there may exist stable 1-1 resonance orbits of minor planets for each of the outer planets from Jupiter to Neptune. Such orbits show no sign of instability, for at least a million years. For Saturn the existence of such orbits is surprising because of the periodic proximity of Jupiter. We discuss the results of our numerical study, and consider the possibilities of observational confirmation of our conclusions. There is preliminary evidence for similar stability for the Trojan analogues of the inner planets.

INTRODUCTION

In 1906 astronomers discovered a minor planet, now known as Achilles, which proved to be the first example of a celestial body moving in the neighbourhood of a triangular Lagrangian point. As shown by Lagrange, these points, which are leading or following a planet in its orbit plane by 60 degrees, are stable libration points in the absence of external perturbations. Achilles and about three dozen other Trojan asteroids are now known to be companions of Jupiter. Thus the total absence of external perturbations is not a necessary condition for the stability of these orbits, and the question arises whether Saturn, Uranus and/or Neptune may have analogous companions somewhere along their orbits. However, no other planets are known to have similar bodies in their Lagrangian points. One of the reasons may be that e.g. a Saturnian Trojan would be rather faint. Tombaugh[1], the discoverer of the planet Pluto, mentions that he has checked Saturn's Lagrangian points to a magnitude limit of 17^m. However, Saturn has twice the distance of Jupiter and consequently a Trojan substituted at this distance would be about 16 times fainter. This means that the Saturnian Trojans may be of 18^m or fainter and Tombaugh's limiting magnitude thus was too bright. However, with a larger telescope, these so far hypothetical bodies should be easily observable.

A. E. Roy (ed.), Long-Term Dynamical Behaviour of Natural and Artificial N-Body Systems, 21–26.
© 1988 by Kluwer Academic Publishers.

Kowal[2] actually has photographed two-thirds of the ecliptic region to a limiting magnitude of 20-21. This should be enough to detect a planetoid at Saturn's distance; however, it is not known to us whether these plates have been checked for the possible existence of trans-Jovian Trojans.

The restricted three-body problem has been widely studied (e.g. Szebehely[3]), but no comparable theory is available for the more general problem of particle motion in the solar system. A possible a priori objection against the existence of Saturnian Trojans may be presented by noting the proximity of Jupiter. Jupiter's perturbations on the Saturnian Lagrange-points are indeed rather strong. On the other hand, Jupiter moves fast relative to Saturn. Having a mean distance of about a half, its period is roughly 2/5 of that of Saturn.

Whether Jupiter actually moves rapidly enough to average the perturbations out without destabilizing the Saturnian Lagrangian points can only be checked by numerical integration. This is what we have done.

The motivation to this work originally came from results by K.A. Innanen and S.-P. Zhang[4] when they found evidence for stability of Saturn's Lagrangian points by using an idealized model. In the long run, however, the orbits of the planets experience considerable periodic fluctuations and thus a self-consistent numerical integration seemed to be worth conducting. This paper conveys a simple empirical result: self-consistent computer simulations show that long-term orbital stability persists in the neighbourhoods of the classical Lagrangian triangular points for all of the outer planets, and probably also for the inner planets. Observational searches are thus justified.

NUMERICAL PROCEDURE

We used a simple N-body routine: The Newtonian equations of motion were integrated using a Bulirsch-Stoer[5] second-order integrator. Our model for the solar system consists of the outer major planets from Jupiter to Neptune. As initial values, we used the positions and velocities corresponding to the epoch 1941-Jan-6.0[6]. The initial values for the particles at the Lagrangian points were produced simply by rotating the planet's position and velocity by 60 degrees. The calculations were made in the ecliptic system, and in most cases we made the rotation in the xy-plane. Thus all the other orbital elements for the planet and the particle were approximately the same except the longitudes of the ascending nodes, which differed by 60 degrees. Consequently, the orbits also have a few degrees of mutual inclination. However, in some runs we made the entire solar system exactly planar. The comparison of results showed that the inclinations seem not to have any qualitative effect on the behaviour of the system.

RESULTS

Our main conclusions are as follows: The classical Lagrangian points seem to be stable, despite the perturbations by the other members of the solar system, for all other cases but Saturn. When started at the classical Lagrangian point, the orbits of the Saturnian Trojan show signs of chaotic behaviour in the timescale of one hundred thousand years. This happens for both inclined and purely planar initial conditions. The exact calculation of the orbital motion is apparently impossible, for the results, such as the time of the onset of chaos, depend very sensitively on the details of the computational procedure. This, however, is typical of chaotic systems and it is the physics of the system which is unstable, not the numerical procedure. Are the Lagrangian points of Saturn unstable, and Saturnian Trojans consequently nonexistent? It is something of a surprise to find that this actually is not necessarily the case.

When we increased the initial semi-major axes of the Saturn's companion particle a little (by about 1 percent) we got a stable orbit! Our calculations extend over one million years and at least over this time the system shows no sign of destabilisation. However, a rather wide oscillation around the Lagrangian points develops rapidly. Figures 1 and 2 illustrate the motions of the particles for the two mentioned sets of initial conditions. From these one can see that the particles may experience wide oscillations (amplitude 30 degrees or more) about the theoretical equilibrium points. Even in the case that the particle starts at the most stable position, one can expect it to develop a tadpole orbit, several tens of degrees long, and possibly, in the timescale of the age of the solar system, all the orbits may become more or less horseshoe-like. Thus an observer may have a quite wide area to cover on the sky by long-exposure plates before having a good chance to find a trans-Jovian Trojan planetoid.

Thus far we have concluded that there is no theoretical objection against the existence of Trojan-like companions of the outer planets. What about the inner planets? May Venus, Earth and/or Mars have equivalent particles in their orbits? To give a conclusive answer by using numerical experiments is difficult. The small masses of the inner planets make all orbits evolve slowly, while the short periods of motion require a short stepsize in numerical integration. Thus a convincing check would require much computer time. However, we made some runs for an equal number of planetary periods as in the case of the outer part of the Solar System. The results we obtained show only stable orbital motion.

Thus we conclude that there is no evidence against the existence of stable orbits somewhere in the vicinity of every one of the triangular Lagrangian points in the entire Solar System. If these places are unoccupied by natural objects, then they are potential interesting sites for practical astronautical and astrophysical applications[4].

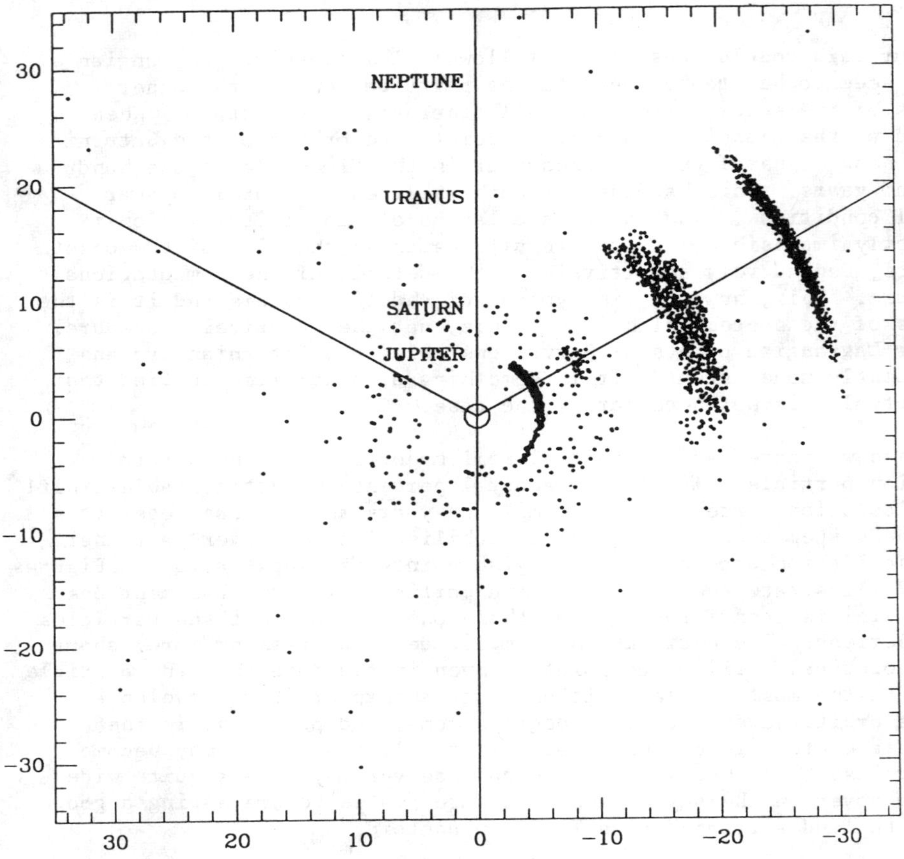

Fig.1 The scatter of positions of Jovian, Saturnian, Uranian and Neptunian Trojans in the rotating coordinate system in which the big planet is at rest. Positions are plotted at intervals of 1200 years over the time span of 10^6 years. The points scattered around are positions of the particle originally at the Saturnian Lagrange-point. All the particles started at the classical equilibrium position. Distance scales are in astronomical units.

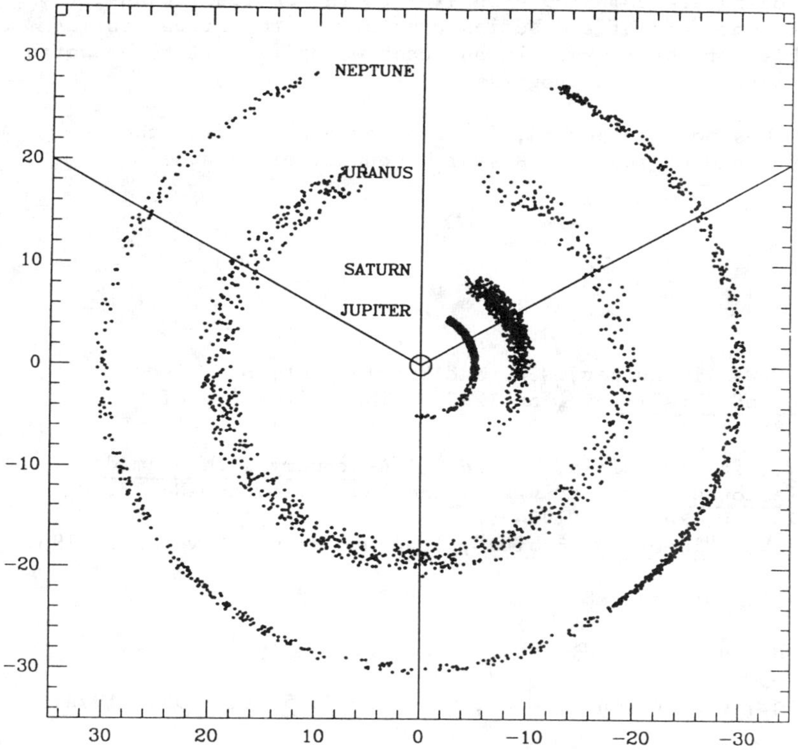

Fig.2 Otherwise the same as Fig.1 but the initial semimajor axes were increased by one percent. Note the stabilizing effect on the orbit of the Saturnian Trojan.

POSSIBILITY OF OBSERVATIONAL CONFIRMATION

Let us estimate the needed limiting magnitudes for minor planets in orbits of the various major planets. From the distances one easily finds that if the bodies are similar to the Jovian Trojans, then their apparent magnitudes are respectively about 3^m, 6^m and 8^m fainter for Saturnian, Uranian and Neptunian Lagrange-points. Recalling that the brightest known Trojan has magnitude of about 15^m, we may conclude that the case of Saturn is not observationally hopeless; however, the possible discovery of Uranian and Neptunian Trojans may require a very large telescope. Large telescopes, however, have small fields which make the discovery unlikely. On the other hand, the estimates of magnitude (18^m, 21^m and 23^m) following from the above figures do apply only if the bodies are comparable in size to the Jovian Trojans. If it happens that some of the bodies are larger, the situation is more favourable for observers. In any case we believe it to be worth trying; one proposal is in progress.

This research has been supported, in part, by a grant from the Natural Science and Engineering Research Council of Canada.

REFERENCES

1. Tombaugh, C.W., in Kuiper, G.P. and Middlehurst, B.M. (ed. "Planets and Satellites", pp.12-30. The University of Chicago Press, (1961).
2. Kowal, C.T., in Capacciolo, M. (ed.) "Astronomy with Schmidt-type telescopes", Proceedings of the IAU Colloquium No.78, pp.229-232. D. Reidel, (1984).
3. Szebehely, V., "Theory of ôrbits", Academic Press, New York, (1967).
4. Innanen, K.A. and Zhang S.-P., submitted, (1987).
5. Bulirsch, R. and Stoer, J., Num. Math., 8, 1, (1966).
6. Oesterwinter, C. and Cohen, C.J., Cel. Mech., 5, 317-395, (1972).

ORBITAL EVOLUTION

J. Kovalevsky
CERGA
Avenue Copernic
06130 Grasse, France.

ABSTRACT: In the first part of the paper, the notion of evolution is introduced, as opposed to that of stability and instability. A definition is proposed that involves the variations of the energy and angular momentum integrals of a subgroup of an isolated N body system.

When there are only gravitational interactions between the bodies, the conditions for evolution are outlined and some examples are given. This is in particular applied to the case when there exists a massive central body.

Several examples of evolution when non-conservative forces are also acting are proposed. It is shown that evolution takes place because generally there exists a non-vanishing component of the disturbing force perpendicular to the radius vector in the orbital plane. This is the case of the atmospheric drag, of the Poynting-Robertson effect and of tidal effects. Some details are presented for the latter and the results are applied to the evolution of the Earth-Moon system.

1. INTRODUCTORY REMARKS

When speaking of long-term evolution of a dynamical system, it seems natural to use the concepts of stability of the system or its opposite, instability. We know how difficult it is to define stability and how many different concepts may be used to study it. In preceding Cortina advanced study institutes, Szebehely (1979 and 1985) has discussed and compared some of these approaches, and it is clear that in most of the cases the various stability criteria are practically much too stringent and that there are many systems that we call by experience stable and for which the stability criteria give negative results. This is the case for instance for the Moon using elliptic restricted 3 body problem approach and even the two body problem in Liapounov's sense. Would one turn the problem and try to define criteria for instability, the difficulties will remain the same. There will be cases when the result will depend on the definitions and on the criteria applied, even if one uses approaches which are not simply the non-veri-

A. E. Roy (ed.), Long-Term Dynamical Behaviour of Natural and Artificial N-Body Systems, 27–46.
© 1988 by Kluwer Academic Publishers.

fication of stability criteria, like, for instance, the tests of escape studied by Marchal (1985).

It is enlightening to turn to real systems and analyse more deeply what one is really trying to learn when studying the stability. Let us consider the major bodies of the solar system: the planets, the Moon and some other large satellites. For many physical reasons, we know that they have existed for more than 4 billion years and that at that time, they were already members of the solar system. We do not know how long they will remain, but it is clear that one can study the system backwards in time and try to apply general criteria that agree with the behaviour that would be found in the past.

Already from the very simple fact of the existence of the solar system for so long, one can derive the following conclusions:

1.1 The stability or instability studies for the solar system should take into account much larger times than those presently accessible to the most efficient numerical methods. In particular, none of the presently used stability criteria can be in practice applied.

1.2 Although 4.10^9 years is physically a very long time, it is not the mathematical infinity. So the results on the behaviour of the system that one wishes to obtain is not the limiting behaviour for $t = \pm\infty$, but what happens between two finite times.

1.3 The reduction of the problem to the Newtonian gravitational interactions is no more valid for such physically long times and other forces will have to be taken into account.

1.4 During some shorter but still long time, we know that the set of elements belonging to the system (Sun, major planets and some satellites) has not been very much modified. One can even say that for a similar time the subsets Sun, planets and satellites were also rather "stable". This was also the case of their hierarchy.

The last properties are not theoretically proven, but Nature is somewhat similar to an analog computer that shows what result one should obtain. It is important to keep this in mind as an indicator of what one wishes to prove.

2. EVOLUTION

The invariance of the hierarchical structure of the solar system for a finite time does not mean that the dynamical appearance of the system has remained unchanged. The comparison of the aspects A_1 and A_2 of the system at times T_1 and T_2 would possibly show large differences and if various stability criteria would be applied to the transformation from one aspect to another, one would infer the instability of the system even if for short periods of time around T_1 and T_2, stability criteria like the Poincaré orbital stability may be verified.

We shall tell that the system has evolved from the aspect A_1 to aspect A_2 between times T_1 and T_2. Before attempting to somewhat formalize a definition, let us give some qualitative examples showing that the concept of orbital evolution does not coincide with that of orbital stability. This is essentially due to the fact that the stability definition implies the study of a set of orbits differing by small variations of the initial conditions, while the evolution is an intrinsic property of a given orbit.

2.1 The two body problem is stable and is not evolving.

2.2 All periodic orbits are not evolving, although some are unstable.

2.3 Capture or escape orbits are unstable and evolving.

2.4 Orbits under tidal forces are evolving, but can be found to be practically stable when stability criteria are applied.

2.5 An ergodic solution in a finite region of phase space is unstable. In some way, it does not evolve.

In order to give a practical definition of the orbital evolution, let us restrict ourselves to cases applicable to the solar system, that is a set of massive bodies that we shall in a first approximation consider as point-like. If only Newtonian forces are present, in such a system, it is well known that it has an energy integral and a vectorial integral of angular momentum. If $P_i(x_i, y_i, z_i$ mass $m_i)$ are the N bodies and 0 the origin of coordinates coinciding with the centre of mass of the system, one has:

$$H = \frac{1}{2} \sum_{i=1}^{N} m_i \left(\frac{d\overrightarrow{OP}_i}{dt} \right)^2 - G \sum_{i=1}^{N} \sum_{j=i+1}^{N} \frac{m_i m_j}{P_i P_j}$$

$$\vec{C} = \sum_{i=1}^{N} m_i \, \overrightarrow{OP}_i \wedge \frac{d\overrightarrow{OP}_i}{dt}$$

where G is the constant of gravitation.

We shall say that the system of N bodies evolves if at least one of these integrals changes with time within bounds that have secular variations or if this happens to one of its subsystems. This evolutionary situation can happen at least in two cases.

1. When one considers a subsystem S of the whole system for which H_s and $\vec{C}_s$ are strongly modified. It could be called gravitational evolution.
2. When the equations of motion are not purely Newtonian and that there are additional terms. This does not necessarily imply evolution, but if it is the case, it is generally because there are non-conservative forces present in the dynamics of the system. This is

why we shall call it non-conservative evolution.

If now, we come back to what one is really looking for in studying real systems like the solar system or some subsystem of it, we can say that there are two different types of results one wishes to achieve.

1. We want to know whether the system is evolving or not, using the kind of definition we propose together with some decision to be taken on setting the criteria for the bounds within which H and $\vec{C}$ should remain, one will compute H and $\vec{C}$ and see if they come out of bounds - this meaning evolution - or not - this meaning non-evolution. The latter could also be called stationarity.
2. We also want to know whether this remains true for a set of initial conditions and masses that include certainly the actual ones, that is to say the best known initial conditions and masses surrounded by error bars. This type of question is akin to a stability problem, however not to any type of stability, but to a definition that would imply that in the manifold of varying parameters, the evolutionary or stationarity properties are conserved. One could call this "evolutionary stability".

3. GRAVITATIONAL EVOLUTION

An isolated gravitational system of N bodies has an energy integral and an angular momentum integral. Therefore, globally there is no evolution. But one has also to consider subsystems that are almost isolated in the sense that they hardly interact between them, so that one may wish to consider them independently. In this case, one would construct quantities corresponding to the energy and to the angular momentum with respect to the barycentres of the subsystems. If most of these quantities are either constant or undergo only periodic variations, there would be no evolution either in the system, or in the subsystems. But if, at some moment, the two subsystems interact strongly, there may be some irreversible exchange of energy and angular momentum. It would mean an evolution of the system. Let us consider two possible approaches of this definition.

3.1 Separate subsystems

We shall discuss first the case of the energy integral. Let us consider the N body problem and divide it into two subsystems:
- Subsystem 1: n bodies $P_1, \ldots P_n$ with masses $m_1 \ldots m_n$. B_1 is its centre of mass and

$$M_1 = \sum_{i=1}^{n} m_i$$

- Subsystem 2: N-n bodies $P_{n+1} \ldots P_N$ with masses $m_{n+1} \ldots m_N$. B_2 its centre of mass and

$$M_2 = \sum_{i=n+1}^{N} m_i$$

The energy integral of the whole system is:

$$H = \frac{1}{2} \sum_{i=1}^{N} m_i \left(\frac{d\overrightarrow{OP_i}}{dt}\right)^2 - G \sum_{i=1}^{N} \sum_{j=i+1}^{N} \frac{m_i m_j}{|P_i P_j|} \qquad (1)$$

If we consider each of the subsystems independently as if they were isolated, one would have the following expressions of the energy:

$$H_1 = \frac{1}{2} \sum_{i=1}^{n} m_i \left(\frac{d\overrightarrow{B_1P_i}}{dt}\right)^2 - G \sum_{i=1}^{n} \sum_{j=i+1}^{n} \frac{m_i m_j}{|P_i P_j|} \qquad (2)$$

$$H_2 = \frac{1}{2} \sum_{i=n+1}^{N} m_i \left(\frac{d\overrightarrow{B_2P_1}}{dt}\right)^2 - G \sum_{i=n+1}^{N} \sum_{j=i+1}^{N} \frac{m_i m_j}{|P_i P_j|} \qquad (3)$$

Evidently, H is not the sum of H_1 and H_2; the missing quantities may be written as the sum of H_3, H_4 and H_5 defined as follows:

$$H_3 = -G \sum_{i=1}^{n} \sum_{j=n+1}^{N} \frac{m_I m_j}{P_i P_j} \qquad (4)$$

$$H_4 = \frac{1}{2} \sum_{i=1}^{n} m_i \left[\left(\frac{d\overrightarrow{OP_i}}{dt}\right)^2 - \left(\frac{d\overrightarrow{B_1P_i}}{dt}\right)^2\right]$$

$$H_5 = \frac{1}{2} \sum_{j=n+1}^{N} m_i \left[\left(\frac{d\overrightarrow{OP_i}}{dt}\right)^2 - \left(\frac{d\overrightarrow{B_2P_1}}{dt}\right)^2\right]$$

and since $\sum_{i=1}^{n} m_i \overrightarrow{B_iP_i} = 0$, one has

$$H_4 = \frac{1}{2} M_1 \left(\frac{d\overrightarrow{OB_1}}{dt}\right)^2 \qquad (5)$$

and similarly:

$$H_5 = \frac{1}{2} M_2 \left(\frac{d\overrightarrow{OB_2}}{dt}\right)^2 \qquad (6)$$

One has finally:

$$H = H_1 + H_2 + H_3 + H_4 + H_5$$

If the two subsystems are almost isolated, their distance is large. It results that H_3 is small and it produces only a small disturbing effect on the motion of each member of the subsystems and consequently also on B_1 and B_2 so that H_4 and H_5 are almost constant. In this case H_1 and H_2 are also almost constant and their variation is limited: there is no evolution.

Similarly, the angular momentum integral:

$$\vec{C} = \sum_{i=1}^{N} m_i \, \overrightarrow{OP}_i \wedge \frac{d\overrightarrow{OP}_i}{dt} \tag{7}$$

can be written in terms of the angular momentum of each subsystem:

$$C_1 = \sum_{i=1}^{n} m_i \, \overrightarrow{B_1P}_i \wedge \frac{d\overrightarrow{B_1P}_i}{dt} \tag{8}$$

$$\vec{C_2} = \sum_{i=n+1}^{N} m_i \, \overrightarrow{B_2P}_i \wedge \frac{d\overrightarrow{B_2P}_i}{dt} \tag{9}$$

with the addition of two corrective terms:

$$\vec{C_3} = M_1 \, \overrightarrow{OB}_1 \wedge \frac{d\overrightarrow{OB}_1}{dt} \tag{10}$$

and

$$\vec{C_4} = M_2 \, \overrightarrow{OB}_2 \wedge \frac{d\overrightarrow{OB}_2}{dt} \tag{11}$$

so that

$$\vec{C} = \vec{C}_1 + \vec{C}_2 + \vec{C}_3 + \vec{C}_4 \tag{12}$$

As in the case of the energy integral, if the subsystems are well separated, so that the motion of the centres of mass are almost linear and $\vec{C}_3 + \vec{C}_4$ is almost constant, one can again derive the same conclusions.

In contrast, when the two subsystems are mixed or very close one to another, H_3 is no longer negligible but is large with rapid and important variations due to the motion of the masses. Not much can be said a priori about H_4, H_5, $\vec{C}_3$ and $\vec{C}_4$, but the conclusion that they are almost constant is no longer valid in general. Let us consider only the integral of energy. Since H is constant and H_3 is strongly varying and large, $H_1 + H_2$ cannot any more be almost constant. There exist two times T_1 and T_2 for which at least H_1 or H_2 has had a large variation. Hence, the corresponding subsystem has undergone a gravitational evolution. Let us illustrate this by a few examples.

In the general three body problem, it is well known that there exist many cases when one of the three bodies escapes. A sufficient condition for this was computed by Standish (1971). See also Marchal (1974). In a more general case, Pollard and Saari (1970) have proven that with very general conditions of an N body problem, at least two bodies must escape. The escapes are usually produced when three bodies have a close triple approach (Szebehely, 1973). One can generalize these situations by replacing the escaping body by a close binary whose separation is small with respect to the minimum distance between bodies at the close approach. While all the bodies are together, before the escape takes place, the second situation given above takes place. Some time after escape, the subsystems are widely separated and one has the first situation, providing examples of gravitational evolution.

Harrington (1974) has made a numerical study of the dynamics of 200 randomly chosen four body systems and later (1975), of 120 five body systems with unequal masses. In the later tests, he found that in 30% of cases, there was a disruption with production of a stable triple system and/or a stable double system was formed within the same percentage of cases. This shows that the gravitation evolution as defined above is a common feature.

A similar discussion could be made in case of capture instead of disruption. Then, the two subsystems are well defined in the beginning. An exchange mode of the hyperbolic-elliptic class of motion also exists in the three body problem. Extended to more bodies, it corresponds also to the disruption of the two initial subsystems and the creation of new ones.

3.2 Case of a massive central body

The preceding model cannot be applied to the solar system if one looks for an evolution without escape from the system. A slight modification of the definition of subsystems is necessary, using the particularity that the central body, the Sun, has a mass much larger than that of the other bodies of the system: we shall include the Sun in each of the sub-systems.

Let us consider the N body system where $P_1(m_1)$ is the Sun. The centre of mass, O is very close ot P_1.

Subsystem 1 includes P_1 and n-1 bodies $P_2 \ldots P_n(m_2 \ldots m_n)$. The centre of mass B_1 is close to P_1. The mass is m_1+M_1.

Subsystem 2 includes P_1 and N-n bodies $P_{n+1} \ldots P_n(m_{n+1} \ldots m_N)$. The centre of mass B_2 is also close to P_1. The mass is $m_1 + M_2$.

One can remark at this stage (Milani, private communication), that this decomposition can be generalized and several bodies instead of one may be put in both subsystems. An example would be to include the Sun and the inner planets in both subsystems when studying the behaviour of the outer planets.

The energy integral of the whole system is, in the case described above:

$$H = \frac{1}{2} m_1 \left(\frac{d\overrightarrow{OP_1}}{dt}\right)^2 + \frac{1}{2} \sum_{i=2}^{N} m_i \left(\frac{d\overrightarrow{OP_i}}{dt}\right)^2 - G \sum_{j=2}^{N} \frac{m_1 m_j}{|P_1 P_j|} - G \sum_{i=2}^{N} \sum_{j=i+1}^{N} \frac{m_i m_j}{|P_i P_j|}$$

For the first subsystem, it would be:

$$H_1 = \frac{1}{2} m_1 \left(\frac{d\overrightarrow{B_1 P_1}}{dt}\right)^2 + \frac{1}{2} \sum_{i=2}^{n} m_i \left(\frac{d\overrightarrow{B_1 P_i}}{dt}\right)^2 - G \sum_{j=2}^{n} \frac{m_1 m_j}{|P_1 P_j|} - G \sum_{i=2}^{n} \sum_{j=1+i}^{n} \frac{m_i m_j}{|P_i P_j|} \qquad (13)$$

and, for the second subsystem:

$$H_2 = \frac{1}{2} m_1 \left(\frac{d\overrightarrow{B_2 P_1}}{dt}\right)^2 + \frac{1}{2} \sum_{i=n+1}^{N} m_i \left(\frac{d\overrightarrow{B_i P_i}}{dt}\right)^2 - G \sum_{j=n+1}^{N} \frac{m_1 m_j}{|P_1 P_j|} - G \sum_{i=n+1}^{N} \sum_{j=1+i}^{N} \frac{m_i m_j}{|P_i P_j|} \qquad (14)$$

As in the case of separate subsystems, one can set:

$$H_3 = - G \sum_{i=2}^{n} \sum_{j=n+1}^{N} \frac{m_i m_j}{|P_i P_j|} \qquad (15)$$

and, taking into account that the term in P_1 is present in both H_1 and H_2, the remainder is:

$$H_4 = \frac{1}{2} m_1 \left(\left(\frac{d\overrightarrow{OP_1}}{dt}\right)^2 - \left(\frac{d\overrightarrow{B_1 P_1}}{dt}\right)^2 - \left(\frac{d\overrightarrow{B_2 P_1}}{dt}\right)^2\right) + \frac{1}{2} M_1 \left(\frac{d\overrightarrow{OB_1}}{dt}\right)^2 + \frac{1}{2} M_2 \left(\frac{d\overrightarrow{OB_2}}{dt}\right) \qquad (16)$$

so that one has:

$$H = H_1 + H_2 + H_3 + H_4 \qquad (17)$$

Similarly, the angular momentum integral of the total system can be written as:

$$\vec{C} = \vec{C}_1 + \vec{C}_2 + \vec{C}_3 \qquad (18)$$

where $\vec{C}_1$ and $\vec{C}_2$ are the angular momenta of the two subsystems and:

$$\vec{C}_3 = M_1 \overrightarrow{OB_1} \wedge \frac{d\overrightarrow{OB_1}}{dt} + M_2 \overrightarrow{OB_2} \wedge \frac{d\overrightarrow{OB_2}}{dt} - m_1 \overrightarrow{OP_1} \wedge \frac{d\overrightarrow{OP_1}}{dt} \qquad (19)$$

If we consider only the energy integral, it is clear that the behaviour of the sub-systems will depend essentially on the properties and the order of magnitude of H_3. To consider the variation of $(\vec{C}_1)^2 H_1$ and $(\vec{C}_2)^2 H_2$ is another simplified technique to check the evolution that has been used in particular by Milani and Nobili (1985). However the correct criterion is to consider the four scalar parameters.

This can be applied to any group of planets acting on any other group, each one being treated as a subsystem. Let us, for simplicity sake, consider only the energy integral. Since H is constant, formula (17) shows that the maximum variation allowed for H_1 and H_2 is:

$$V = \max \left(|H_3| + |H_4| \right)$$

If the excursions of H_1 and H_2 are smaller than V, no conclusion can be drawn on a possible evolution of the system. On the contrary, if between times T_1 and T_2, H_1 and/or H_2 have been modified by significantly more than V, then one can conclude for evolution. Let us give examples.

If we take subsystem 1 as consisting of Sun, Jupiter and Saturn, while subsystem 2 consists of Sun, Uranus and Neptune, as studied by Milani and Nobili (1985), then a rough calculation gives the following upper bounds:

for H_3 : 2.10^{-7}

for H_4 : 3.10^{-6}

where the units are the year, the solar mass and the astronomical unit.

Since the total energy of the four planets is of the order of 10^{-2}, one can see that only excursions of H_1 and H_2 larger than 3.10^{-4} times the total energy would be significant for an evolution. The results found in the article cannot be interpreted as an evolution of this part of the solar system.

Similarly, Milani et al. (1987) have studied the long term interactions between Uranus and Neptune. The variation found by the authors is about 60 times smaller than a variation that might be significant for an evolution.

Let us now consider the case of the gravitational evolution of comets. The first subsystem will consist of the Sun and all major planets. The second subsystem would consist only of the Sun and the comet. Since the mass m_N of the comet is very small, one can identify B_2 with P_1. One gets:

$$H_1 = \frac{1}{2} m_1 \left(\frac{d\overrightarrow{OP_1}}{dt} \right)^2 + \frac{1}{2} \sum_{i=2}^{N-1} m_i \left(\frac{d\overrightarrow{OP_i}}{dt} \right)^2 - G \sum_{j=2}^{N-1} \frac{m_1 m_j}{|P_1 P_J|} - G \sum_{i=2}^{N-1} \sum_{j=i+1}^{N-1} \frac{m_i m_j}{|P_i P_j|}$$

and

$$H_2 = \frac{1}{2} m_N \left(\frac{d\overrightarrow{P_1 P_N}}{dt} \right)^2 - G \frac{m_1 m_N}{|P_1 P_N|}$$

H_2 is the two body problem integral of energy for the comet.

$$H_3 = - G \sum_{i=2}^{N} \frac{m_i m_N}{|P_i P_N|}$$

$$H_4 = \frac{1}{2} m_N \left(\frac{d\overrightarrow{OP_1}}{dt} \right)^2$$

While H_4 is a limited variation quantity, defined by the motion of all planets, H_3 may undergo large changes when P_iP_N becomes small that is at the occasion of close approaches. While, because of the small size of m_N, H_1 will have no relative change, the effect on H_2 can be very large and therefore produce drastic changes in the orbit of the comet even if only a minute relative amount of energy has been taken from H_1 and transferred to H_2 (or vice-versa).

This gravitational evolution can be produced only by those planets whose distance to the Sun is included in the comet's perihelion-aphelion interval. The example of comet Oterma is enlightening in this respect (see Carusi et al., 1985) since before the 19th century, close approaches were possible with Saturn. The evolution was such that the aphelion distance regressed as well as the perihelion distance so that it could not be any more strongly perturbed by Saturn and its later evolution was entirely governed by Jupiter. Many more examples of various cometary evolutions are given in the catalogue of short period comets by Belyaev et al. (1986).

4. NON CONSERVATIVE EVOLUTION

Let us now assume that in the interactions between the N bodies, in addition to the Newtonian attraction, there exists another type of forces. These forces may act on all or only some of the bodies, they may depend on the coordinates and velocity of the body on which it is applied or on any other parameter present in the kinematic or dynamical description of the system.

In practice, for the planets and major satellites of the solar system, there is only one cause of such additional forces: the tidal bulge, caused by the gravitational deformation of bodies, acts on surrounding bodies and, in particular, on the one that causes the tidal deformation. Vice versa this bulge is attracted and a torque is created that modifies the dynamical behaviour of the corresponding body. Another effect might be described similarly if its existence was proved: a possible secular change of the constant of gravitation or, more generally, the variability of masses of some of the bodies.

Smaller objects like artificial satellites or small particles in rings and, on a long run even large satellites undergo in addition gas drag if they happen to orbit in some non-zero density medium, radiation pressure and possibly some electromagnetic effects if they are charged or magnetic objects. They can also undergo discontinuous drag due to collisions with smaller objects. Finally, very small objects and dust are sensitive not only to direct radiation pressure, but also to a transversal component called the Poynting-Robertson effect.

All these effects have been very much studied. For instance the effect of atmospheric drag is studied in detail by King-Hele (1964) and the perturbations due to radiation pressure by Burns et al. (1979). As for the tidal effects, a large number of papers were written on the subject in the last 25 years since the complete development of the tidal potential by Kaula (1964). Among recent treatments, one should quote Mignard (1979 and 1980). In all these cases, in the Newtonian equations of motion of a planet or of a satellite, there is an additional term that depends upon a small parameter.

Let us consider a certain configuration of N bodies ($N \geqslant 2$) that is typical of the situation of major bodies in the solar system, namely that the motions are perturbed two body motions. We shall in addition assume that two of the bodies are a planet with its satellite and we are proposing to study what is involved in the theory of motion of this satellite if there exists one of the effects listed above acting on this particular satellite. We also assume that no (N+1)th body approaches one of the N bodies sufficiently close so as to produce a gravitational evolution of the system.

Under these conditions, the motion of the satellite around the planet can be expressed formally as a sum of multi-periodic functions of time (see Kovalevsky, 1982, in the case of the Moon or any other satellite in a planet-satellite-Sun system). This means that the system is not subject to gravitational evolution. This can be in particular expressed by stating that while the three angular variables of the theory of motion of the satellite can take any value, the three action variables (or their equivalent in orbital elements: semi-major axis a, eccentricity e and inclination i) have bounded variations. For all the times of the motion, one has:

$$\begin{aligned} a_o - \Delta a \lesssim\ & a \lesssim a_o + \Delta a \\ 0 \text{ or } e_o - \Delta e \lesssim\ & e \lesssim e_o + \Delta e \\ 0 \text{ or } i_o - \Delta i \lesssim\ & i \lesssim i_o + \Delta i \end{aligned} \tag{20}$$

which give similar bounds in the energy and angular momentum integrals since:

$$\begin{aligned} C^2 &= G(M+m)a(1-e^2) \\ &= n^2a^2(1-e^2) \\ k &= -\,G(M+m)/a \\ h &= n^2a^2 \end{aligned}$$

Using the osculating elements, and calling F the disturbing function of the gravitational problem considered, these three metric elements are solution of the first three equations of the Lagrange system of equations. If Ω, ω and M are respectively the longitude of the node, the argument of pericentre and the mean anomaly, the Lagrange equations are:

$$\frac{da}{dt} = \frac{2}{n\,a}\ \frac{\partial F}{\partial M}$$

$$\frac{de}{dt} = \frac{-\sqrt{1-e^2}}{n\,a^2 e}\ \frac{\partial F}{\partial \omega} + \frac{1-e^2}{n\,a^2 e}\ \frac{\partial F}{\partial M}$$

$$\frac{di}{dt} = \frac{-1}{n\,a^2\sqrt{1-e^2}\sin i}\ \frac{\partial F}{\partial \Omega} + \frac{\cos i}{n\,a^2\sqrt{1-e^2}\sin i}\ \frac{\partial F}{\partial \omega}$$

$$\frac{d\Omega}{dt} = \frac{1}{n\,a^2\sqrt{1-e^2}\sin i}\ \frac{\partial F}{\partial i} \qquad (21)$$

$$\frac{d\omega}{dt} = \frac{\sqrt{1-e^2}}{n\,a^2 e}\ \frac{\partial F}{\partial e} - \frac{\cos i}{n\,a^2\sqrt{1-e^2}\sin i}\ \frac{\partial F}{\partial \Omega}$$

$$\frac{dM}{dt} = n - \frac{2}{n\,a}\ \frac{\partial F}{\partial a} - \frac{1-e^2}{n\,a^2 e}\ \frac{\partial F}{\partial e}$$

where $n^2a^3 = G(M+m)$, M being the mass of the planet and m the mass of the satellite. The Lagrange equations may also be replaced by their equivalent for near zero eccentricities and/or inclinations. We shall assume in what follows that this is not the case, though with some more algebra one can use them and obtain similar results.

The additional forces that we are considering here can usually be expressed in terms of rectangular components depending on time, on satellite elements and other external parameters. If they derive from a force function F', their effects can be computed using equations (21). If not, let us call R, S and T respectively the radial component (OX axis), S the perpendicular component in the instantaneous orbital plane (OY axis) and T the component perpendicular to R and S (OZ axis). The Gauss system of equations can be used to take into account these forces (see for instance, Kovalevsky, 1967).

$$\frac{da}{dt} = \frac{2e\sin v}{n\sqrt{1-e^2}}\,R + \frac{2a\sqrt{1-e^2}}{n\,r}\,S$$

$$\frac{de}{dt} = \frac{\sqrt{1-e^2}\sin v}{n\,a}\,R + \frac{\sqrt{1-e^2}}{n\,a\,e}\left[\frac{a(1-e^2)}{r} - \frac{r}{a}\right]S$$

$$\frac{di}{dt} = \frac{r\cos(\omega+v)}{n\,a^2\sqrt{1-e^2}}\,T \qquad (22)$$

$$\frac{d\Omega}{dt} = \frac{r\sin(\omega+v)}{n\,a^2\sqrt{1-e^2}\sin i}\,T$$

$$\frac{d\omega}{dt} = \frac{-\sqrt{1-e^2}\cos v}{n\,a\,e} R + \frac{\sqrt{1-e^2}\sin v}{n\,a\,e}\left[1 + \frac{r}{a(1-e^2)}\right]S - \frac{r\sin(\omega+v)\cot g\, i}{na^2\sqrt{1-e^2}} T$$

$$\frac{dM}{dt} = n - \frac{1}{n\,a}\left[2\frac{r}{a} - \frac{(1-e^2)\cos v}{e}\right] R - \frac{(1-e^2)\sin v}{n\,a\,e}\left[1 + \frac{r}{a(1-e^2)}\right]S$$

where, in addition to the notations already defined, r and v are the radius vector and the true anomaly of the satellite.

The actual motion including all the effects is represented by a system of differential equations where all the second members of type (21) and (22) are combined. Let us consider successively the three examples proposed in the beginning.

4.1 Atmospheric drag

It produces a force opposite to the velocity vector $\vec{V}$ of the satellite proportional to the square of this velocity and to the density ρ of the atmosphere:

$$\vec{F} = - K\rho V.\ \vec{V}$$

where K is a coefficient that characterizes the satellite. Strictly speaking, K depends on the attitude of the satellite, but in practice, one can average it and consider a mean value of K. The density ρ depends upon the position of the satellite in a planetocentric system of reference, identical to the one chosen to express the orbital elements.

The velocity vector components in the orbital plane are:

$$\frac{dr}{dt} = \frac{n\,a\,e\,\sin v}{\sqrt{1-e^2}}$$

$$r\frac{dv}{dt} = \frac{n\,a^2\sqrt{1-e^2}}{r}$$

Hence setting $V^2 = (\frac{dr}{dt})^2 + r^2(\frac{dv}{dt})^2 = \frac{n^2a^2(1+e^2+2e\cos v)}{1-e^2}$, one has:

$$\left.\begin{aligned} r &= - K\rho(x,y,z)\, V \frac{n\,a\,e\,\sin v}{\sqrt{1-e^2}} \\ S &= - K\rho(x,y,z)\, V \frac{n\,a^2\sqrt{1-e^2}}{r} \end{aligned}\right\} \quad (23)$$

4.2 Radiation pressure

The solar radiation pressure may be represented as a constant force along the direction Sun-satellite. The proportionality coefficient K' depends on the dimensions of the satellite and has to be averaged for all orientations. The computation of R, S and T for a constant force with a slowly varying direction is very simple. Each of the three components is proportional to K and is also a periodic function of time, that depend also on the coordinates of the satellites. The same is true for the effects of the radiation pressure from the planet. In contrast, the Poynting-Robertson effect has a clear systematic effect along the S component.

4.3 Tidal effects

The satellite attracts differently various points of the planet, and deforms it. The difference with the undisturbed planet is the "tidal bulge". This bulge produces a "tidal potential" that has to be added to the normal potential produced by the undisturbed planet and that is used to construct the theory of the motion of the satellite. This potential ΔV is expressed in terms of Legendre polynomials of which we shall keep only the second order term, although general formulae in the forces are available (see for instance Mignard, 1980).

Would this picture be complete, ΔV would follow the direction of the satellite and, assuming that the planet has a rotational symmetry, ΔV would have a symmetry with respect to the OXZ plane. One will have, for R,S and T:

$$R_0 = \frac{\partial \Delta V}{\partial X} : \quad S_0 = \frac{\partial \Delta V}{\partial Y} ; \quad T_0 = \frac{\partial \Delta V}{\partial Z}$$

Because of the symmetry, $S_0 = 0$. More generally, it can be shown that this particular tidal perturbation does not produce evolution of the system.

But in reality, the planet is not a perfect tidal perfect solid, and the bulge gets its shape only after a certain delay Δt during which the planet has rotated by an angle:

$$\Delta\lambda = \nu \, \Delta t$$

called the tidal phase lag, ν being the rate of rotation of the planet.

But, in the same time, the satellite has moved on its orbit by the angle:

$$\Delta v = \Delta t \, \frac{dv}{dt}$$

Neglecting the periodic terms, this is equal to $\bar{n}\,\Delta t$ where $\bar{n}$ is the mean motion of the satellite. So actually, the plane of symmetry of the tidal potential makes, with the direction of the satellite, an angle that is the difference between the elementary rotation of the planet $\nu\Delta t$ and the elementary motion of the satellite on its orbit $\bar{n}\Delta t$. In particular, one can see that S has a systematic component always directed towards the direction of the planet-satellite radius vector. Let us take Δt as the small quantity; one can develop the force as:

$$S = S_0 + S_1 \Delta t + S_2 \Delta t^2 + \dots \text{ with } S_1 \neq 0$$

and similar expressions in R and T.

R_o, S_o and T_o cannot produce evolution, their effects can be associated with the basic non-evolutive theory of the motion of the satellite. The remainder forces are multiplied by Δt.

4.4 Consequences

From these examples, there appears a general feature: the additional force is present in the S component that does not vanish and has consequently a non-zero mean value $\bar{S}$. Let us consider what this means in the solution of the system (22). The coefficients of S in da/dt and de/dt can be developed in trigonometric series of the mean anomaly M. One has:

$$\frac{de}{dt} = \frac{\sqrt{1-e^2}\sin v}{n\,a} R - \frac{3}{2}\frac{\sqrt{1-e^2}}{n\,a} e\, S\, [1 + \Sigma \beta_i \cos i M]$$

$$\frac{da}{dt} = \frac{2\, e \sin v}{n\sqrt{1-e^2}} R + \frac{2\sqrt{1-e^2}}{n} S\, [1 + \Sigma \alpha_i \cos i M] \qquad (24)$$

The first part of these equations, as well as the equation in di/dt would produce a constant term only if the development of R would contain erms in kM (k non-zero integer) or if the development of T would include terms in $\omega + \overline{kM}$. On the contrary in the second part there exists a non-periodic term:

$$\frac{2\sqrt{1-e^2}}{n}\bar{S} \quad \text{for a and} \quad -\frac{3\sqrt{1-e^2}}{2n\,a} e\,\bar{S} \quad \text{for e}$$

The solution with this periodic term depends of course on the way that $\bar{S}$ itself depends upon a, e and possible i. In any case, in a first approximation, with the initial conditions a_o, e_o, i_o, one would have the following solution:

$$a = a_o + \frac{2\sqrt{1-e_0^2}}{n_0}\bar{S}(a_0, e_0, i_0)(t-t_0) + \dots$$

$$e = e_0 - \frac{3\sqrt{1-e_0^2}}{2\, a_0 n_0}\bar{S}(a_0, e_0, i_0)(t-t_0) + \dots$$

$$i = i_0$$

where the remainder is a development in power series of $(t-t_o)$.

After a certain time the developments make a and e exceed the limits (20) and, as a consequence, the energy and C also so that one has non-gravitational evolution. Let us only remark that in the case of tidal evolution, although the tidal bulge is indeed a purely gravitational effect, the existence of the tidal phase lag is a non-gravitational effect and is produced by the rheological properties of the planet.

5. APPLICATION TO THE EARTH-MOON SYSTEM

The actual situation in the Earth-Moon system is the one described above for the tidal effects. The equations for a, e and i corresponding to the system (22) as given explicitly in Kovalevsky (1985) are the following, for the second order harmonic of the tide:

$$\frac{da}{dt} = -\frac{2k}{a^7}\cdot\frac{1}{1-e^2}\left(\frac{a}{r}\right)^8 (1 + 2e^2 + 2\,e\cos v - e^2\cos 2v) \qquad (25)$$

$$+ \frac{2\nu}{n}\frac{k}{a^7}\sqrt{1-e^2}\left(\frac{a}{r}\right)^8 (\cos\varepsilon\cos i + \sin\varepsilon\sin i\cos\Omega)$$

$$\frac{de}{dt} = \frac{-k}{a^8}\left(\frac{a}{r}\right)^8 (3\,e + 2\cos v - e\cos 2v) \qquad (26)$$

$$+ \frac{\nu k}{n\,a^8}\frac{1}{\sqrt{1-e^2}}\left(\frac{a}{r}\right)^6\left(\frac{3e}{2} + 2\cos 2v + \frac{e}{2}\cos v\right)(\cos\varepsilon\cos i + \sin\varepsilon\sin i\cos\Omega)$$

$$\frac{di}{dt} = -\frac{1}{4}\frac{\nu}{n}\frac{k}{a^8}\frac{1}{\sqrt{1-e^2}}\left(\frac{a}{r}\right)^6\begin{bmatrix} 2\sin i(1+2\cos(2\omega+2v)) + \\ +2\sin\varepsilon\cos i\cos\Omega + \sin\varepsilon(1+\cos i)\cos(2\omega+2v+\Omega) \\ -\sin\varepsilon(1-\cos i)\cos(2\omega+2v-\Omega) \end{bmatrix} \qquad (27)$$

where, in addition to the already defined symbols, one has:

$$k = \frac{3\,G\,m\,(M+m)}{M}\,k_2\,R^5\,\Delta t$$

- k_2 is the Love number of order 2,
- R is the radius of the Earth,
- ε is the obliquity of the ecliptic.

The first term in equations (25) and (26) is due to the motion of the Moon during the time lag Δt in the ecliptic plane and the second term corresponds to the rotation of the Earth during the same time in the equatorial plane.

Clearly, a complete rigorous solution should be obtained by adding the terms of (25) and to (27) as well as the terms produced in the equations for angular variables to the general equation of the motion of the Moon as described for instance in Kovalevsky (1982), and we should keep this in mind. However, since we are looking for very long term evolution, a good approximation can be obtained by averaging the equations with respect to the angular variables. Since the eccentricity is a small quantity, the resulting equations can be developed in powers of e and one gets the expression obtained by Mignard (1980) in which, and from now on a, e and i designate the secular part of the variation of the metric variables:

$$\frac{da}{dt} = \frac{2\,k}{a^7}\left[\frac{-1}{(1-e^2)^{15/2}}\left(1+\frac{31e^2}{2}+\frac{255e^4}{8}+\ldots\right) + \frac{\nu}{n}\frac{\cos i}{(1-e^2)^6}\left(1+\frac{15e^2}{2}+\frac{45e^4}{8}+\ldots\right)\right] \tag{28}$$

$$\frac{de}{dt} = \frac{k}{a^8}\left[\frac{-1}{(1-e^2)^{13/2}}\left(9e+\frac{135e^8}{4}+\frac{135e^5}{8}+\ldots\right) + \frac{v}{n}\frac{\cos i}{(1-e^2)^5}\left(\frac{11e}{2}+\frac{33e^3}{4}+\frac{11e^5}{16}\right)\right]$$

$$\frac{di}{dt} = -\frac{1}{2}\,\frac{k}{a^8}\,\frac{\nu}{n}\,\frac{\sin i}{(1-e^2)^5}\left(1+3e^2+\frac{3e^4}{8}\right)$$

Although the expressions in e are polynomials and not series, the solution of such a system is obviously not straightforward and one has to proceed with numerical integration. There are several difficulties in this approach.

5.1 The parameter k includes two quantities K_2 and Δt that depend upon rheological properties of the Earth, and nothing proves that they have been the same throughout the four and a half billion years of the existence of the Earth-Moon system. Actually there are many reasons to believe that the contrary was true (see Lambeck, 1977).

Indeed, if we keep the values of k_2 and Δt (0.25 and 10 mins. respectively) presently known, one obtains unduly small values for the time for which the semimajor axis of the lunar orbit was zero. A very crude way in obtaining this number is to neglect e and i in the first equation (28), and to keep only the second term, since ν/n is of the order of 27. Since:

$$n = \sqrt{\frac{G(M+m)}{a^3}}\,,\ \text{one gets:}$$

$$\frac{da}{dt} = \frac{2\,k\,\nu}{\sqrt{G(M+m)}\,a^{11/2}} = A\,a^{-11/2}$$

that resolves in:

$$a = a_0 \left[1 + \frac{13}{2} A t \right]^{2/13} \qquad (29)$$

and gives a = 0 for t of the order of 2 billion years.

A way of overcoming this difficulty is to eliminate the time betweek the three equations (28) and to consider a as the independent variable. This gives the variations of e and i independently of any geophysical assumption, since this also eliminates k from the equations.

5.2 Another difficulty is that (28) is only a subset of the whole system of equations and that other dynamical effects are overlooked. One should consider the whole Earth-Moon-Sun system and all the gravitational and tidal interactions between these three bodies with their physical properties. In some cases, even planets may have an action. Let us give three examples.

5.2.1 Interaction with Earth's rotation

If we assume that the Earth-Moon system is isolated, the total angular momentum should be constant. Then one would have:

$$I \frac{d\nu}{dt} + \frac{d}{dt} (ma^2 n) = 0$$

where I is the moment of inertia of the Earth with respect to its axis of rotation. This can be written as:

$$\frac{d\nu}{dt} = \frac{1}{3} \frac{MR^2}{I} \frac{m}{M} \left(\frac{a}{R}\right)^2 \frac{dn}{dt} \qquad (30)$$

and gives the relation between the evolution of the lunar orbit and the retardation of the rate of rotation of the Earth. The relative deceleration due to the Moon:

$$\frac{I}{\nu} \frac{d\nu}{dt}$$

is estimated presently to be 21.10^{-9} per century. If one takes into account the effect of the solar tides, it increases by 10% inducing a corresponding deceleration of the orbit of the Earth around the Sun.

5.2.2. Resonances with planets

It has been shown (Kovalevsky, 1985) that in some cases, some planetary terms in the lunar theory may become resonant while the semi-major axis is evolving. A temporary capture into resonance may occur so that during a certain interval of time, the mean values of a and n do not change. Since there is still a tidal torque acting on the Earth, the resulting change in the Earth's angular momentum is transferred to the orbits of the Earth and of the disturbing planet.

5.2.3 Lunar tidal bulge

Conversely the Earth gravitational field exerts a torque on the Moon and this produces some tidal dissipation in the Moon, contributing to the general evolution of the system, in particular in the lunar orbit. However, presently our knowledge of the lunar interior is not sufficient to evaluate the magnitude of this effects. Mignard (1980) has studied the past evolution of the lunar orbit as a function of the ratio of the tidal effects in the Moon and in the Earth. The main consequence is on the variations of the inclination and the eccentricities of the lunar orbit in the early phases of the evolution.

5.3 Another consequence of this tidal evolution of the Earth-Moon system is a tilt of the Earth's axis of rotation. The projection of the angular momentum of the Earth-Moon system on an axis perpendicular to the ecliptic is:

$$I \cos \varepsilon + \sqrt{G(M+m)}\ a^{\frac{1}{2}} \sqrt{1-e^2} \cos i$$

It should be constant and cos ε cannot be constant. There is an evolution of the obliquity of the ecliptic. And since the ecliptic plane is not affected, this means that the axis of rotation of the Earth has a secular motion.

5.4 In conclusion, many different effects are simultaneously present in the dynamics of the Earth-Moon system. The general trend is governed by the tidal interaction that imposes a general evolution as sketched in this section. However, the exact behaviour during the past four and a half billion years is still far from being known, since even the time-scale problem is not yet solved. In particular, the problem of the origin of the Moon - separation from the Earth, double planet from the origin, capture - is still strongly debated. Let us only note that, if the origin was a capture, it involves either a very close approach by a third body in the case of a gravitational capture as described in section 3 or a very large dissipation of energy by friction or collision.

BIBLIOGRAPHY

Belyaev, N.A., Kresàk, L., Pittich, E.M. and Pushkarev, A.N., 1986, "Catalogue of short period comets", Astronomical Institute of the Slovak Academy of Sciences publisher, Bratislava.

Burns, J.A., Lamy, P.L. and Soter, S., 1979, Icarus, 40, p.1

Carusi, A., Kresak, L., Perozzi, E. and Valsecchi, G.B., 1985, in "Dynamics of Comets: their origin and evolution", A. Carusi and G.B. Valsecchi eds., IAU Colloquium 83, Rome, D. Reidel Publ. Co. Dordrecht, p.319.

Harrington, R.S., 1974, Celestial Mechanics, 9, p.465

Harrington, R.S., 1975, Astronomical Journal, 80, p.1081

Kaula, W.M., 1964, Reviews of geophysics and Space Physics, 2, p.661.
King-Hele, D.G., 1964,"Theory of satellite orbits in an atmosphere", Butterworths Publ. Co., London.
Kovalevsky, J., 1967, "Introduction to Celestial Mechanics", Reidel Publ. Co., Dordrecht, p.116.
Kovalevsky, J., 1982, in "Applications of Modern Dynamics to Celestial Mechanics and Astrodynamics", V. Szebehely ed., Reidel Publ. Co., Dordrecht, p.59.
Kovalevsky, J., 1985, in "Stability of the Solar System and its Minor Natural and Artificial Bodies", V.G. Szebehely ed., Reidel Publ. Co., Dordrecht, p.39.
Lambeck, K., 1977, Phil. Transactions, Royal Society, London, 287, p.545.
Marchal, C., 1974, Celestial Mechanics, 9, p.381.
Marchal, C., 1985, in "Stability of the Solar System and its Minor Natural and Artificial Bodies", V.G. Szebehely ed., Reidel Publ. Co., Dordrecht, p.115.
Mignard, F., 1979, The Moon and the Planets, 20, p.301.
Mignard, F., 1980, The Moon and the Planets, 23, p.185.
Milani, A., and Nobili, A.M., 1985, in "Stability of the Solar System and its Minor Natural and Artificial Bodies", V.G. Szebehely ed., Reidel Publ. Co., Dordrecht, p.139.
Milani, A., Nobili, A.M. and Carpino, M., 1987, Astronomy and Astrophys. 172, p.265.
Pollard, H. and Saari, D.G., 1970, Celestial Mechanics, 1, p.347.
Standish, E.M., 1971, Celestial Mechanics, 4, p.44.
Szebehely, V.G., 1973, Celestial Mechanics, 8, p.163.
Szebehely, V.G., 1979, in "Instabilities in Dynamical Systems", V.G. Szebehely ed., Reidel Publ. Co., Dordrecht, p.61.
Szebehely, V.G., 1985, in "Stability of the Solar System and its Minor Natural and Artificial bodies", V.G. Szebehely ed., Reidel Publ. Co., Dordrecht, p.175.

PLANETARY PERTURBATION THEORY FROM LIE SERIES, INCLUDING RESONANCE AND CRITICAL ARGUMENTS

P.J. Message,
Department of Applied Mathematics & Theoretical Physics,
Liverpool University,
Liverpool, U.K.

ABSTRACT: In these lectures the Lie series transformation method will be used (after a brief description of the concepts on which it is based) to derive the main properties of the first- and second-order periodic terms in planetary perturbation theory, indicating how periodic terms of further orders may be derived. Then the formal solution of the complete problem, including long-term changes, is derived, by use of further transformations, including a further Lie series transformation to derive the complete expressions for the long-period variations. The solution thus obtained expresses the orbital elements in series entirely of periodic terms, some of the frequencies being those of the long-period "secular" variations. It is seen how near-resonance conditions present difficulties, and it is shown how, in the case of a single close resonance, they may be surmounted by an appropriate modification of the method.

1. THE LIE SERIES TRANSFORMATION

Consider a dynamical system in which the coordinates are $q_1, q_2, q_3, \ldots, q_n$ (sometimes they will be collectively referred to as "q"), and the momenta conjugate to them are, respectively, $p_1, p_2, p_3 \ldots \ldots p_n$ (sometimes denoted by "p"), and the Hamiltonian function is $H(q,p)$, so that the equations of motion are

$$\dot{q}_i = \frac{\partial H}{\partial p_i} \quad , \quad \dot{p}_i = - \frac{\partial H}{\partial q_i} \quad , \quad (i=1,2,3,\ldots,n) \tag{1.1}$$

Then for any function f of the q_i and p_i, we have

$$\frac{df}{dt} = \sum_{i=1}^{n} \left(\frac{\partial f}{\partial q_i} \dot{q}_i + \frac{\partial f}{\partial p_i} \dot{p}_i\right) = \{f,H\} \ , \tag{1.2}$$

where for any two functions f and g, the Poisson bracket $\{f,H\}$ is defined to be

A. E. Roy (ed.), Long-Term Dynamical Behaviour of Natural and Artificial N-Body Systems, 47–72.
© 1988 by Kluwer Academic Publishers.

$$\sum_{i=1}^{n} \left(\frac{\partial f}{\partial q_i}\frac{\partial g}{\partial p_i} - \frac{\partial f}{\partial p_i}\frac{\partial g}{\partial q_i}\right) . \tag{1.3}$$

Also

$$\frac{d^2 f}{dt^2} = \frac{d}{dt}\{f,H\} = \{\{f, H\} , H \}, \tag{1.4}$$

and so on for higher derivatives. If we define the operator $\mathcal{L}_H$ by

$$\mathcal{L}_H f = \{f,H\}, \tag{1.5}$$

$$\mathcal{L}^2_H \text{ by } \mathcal{L}^2_H f = \mathcal{L}_H (\mathcal{L}_H f) , \tag{1.6}$$

$$\text{and } \mathcal{L}^k_H \text{ by } \mathcal{L}_H^k f = \mathcal{L}_H (\mathcal{L}^{k-1}_H f) \qquad (k=2,3,\ldots) \tag{1.7}$$

then

$$\frac{d^2 f}{dt^2} = \mathcal{L}^2_H f , \quad \frac{d^k f}{dt^k} = \mathcal{L}^k_H f. \tag{1.8}$$

The value of f at $t=t_1$ can then be expressed in terms of values at $t = t_o$ using Taylor's theorem in the form

$$f(q(t_1),p(t_1))=f(q(t_o),p(t_o))+ \sum_{k=1}^{\infty} \frac{(t_1-t_o)^k}{k!} \mathcal{L}^k_H f \Big|_{q=q(t_o),p=p(t_o)} \tag{1.9}$$

including the special cases, first, putting $f = q_i$, and noting that $\mathcal{L}_H q_i = \frac{\partial H}{\partial p_i}$, we obtain

$$q(t_1) = q(t_o) + (t_1-t_o)\frac{\partial H}{\partial p_i}\Big|_{t_o} + \sum_{k=2}^{\infty} \frac{(t_1-t_o)^k}{k!} \mathcal{L}^{k-1}_H \frac{\partial H}{\partial p_i}\Big|_{t=t_o} \tag{1.10}$$

and second, putting $f = p_i$, and noting that $\mathcal{L}_H p_i = -\frac{\partial H}{\partial q_i}$, we obtain

$$p(t_1) = p(t_o) - t_i-t_o)\frac{\partial H}{\partial q_i}\Big|_{t_o} - \sum_{k=2}^{\infty} \frac{(t_1-t_o)k}{k!} \mathcal{L}^{k-1}_H \frac{\partial H}{\partial q_i}\Big|_{t_o} . \tag{1.11}$$

In what follows we will use the concept of a canonical transformation, that is, a transformation in which the equations of motion retain the Hamiltonian form (1.1), so that if the new coordinates are q^*_i, and the new momenta are p^*_i, then there is a Hamiltonian function $H^*(q^*,p^*)$ corresponding to them so that the equations of motion become after transformation

$$\dot{q}^*_i = \frac{\partial H^*}{\partial p^*_i} , \quad \dot{p}^*_i = - \frac{\partial H^*}{\partial q^*_i} , \qquad (i=1,2,3,\ldots,n). \tag{1.12}$$

If the time does not enter explicitly into the relations between the oroginal variables q,p, and the new variables q*,p*, then the new Hamiltonian function H* is numerically equal to the original one H. Now it is a property of Hamiltonian dynamical systems that the relations between the dynamical variables $q(t_1)$, $p(t_1)$ at any time t_1 and those $q(t_o)$, $p(t_o)$ at any other time t_o have the correct mathematical properties to make the transformation

$$(q(t_o),p(t_o)) \longrightarrow (q(t_1), p(t_1))$$

a canonical transformation, that is to say, if the transformation has the mathematical form given by equations (1.10) and (1.11), it is a canonical transformation. This mathematical form will still be possessed by the transformation if, in (1.10) and (1.11), we replace H by another function, and choose any value for t_1-t_o. Thus if we choose a suitable function W(q,p), and take $t_1-t_o = 1$, and define a transformation

$$(q,p) \rightarrow (q^*,p^*) \tag{1.13}$$

by the equations, taken from (1.10) and (1.11):

$$q_i^* = q_i + \frac{\partial W}{\partial p_i} + \sum_{k=2}^{\infty} \frac{1}{k!} \mathcal{L}_W^{k-1} \frac{\partial W}{\partial p_i}$$

$$\text{and} \quad p_i^* = p_i - \frac{\partial W}{\partial q_i} - \sum_{k=2}^{\infty} \frac{1}{k!} \mathcal{L}_W^{k-1} \frac{\partial W}{\partial q_i} \qquad (i=1,2,3,\ldots,n) \tag{1.14}$$

then this is a canonical transformation. Then for any function f(q,p), (1.9) shows that

$$f(q^*,p^*) = f(q,p) + \sum_{k=1}^{\infty} \frac{1}{k!} \mathcal{L}_W^k f \Big|_{q,p} \tag{1.15}$$

Then the transformation (1.13) is time independant, so that the Hamiltonian function H*(q*,p*) for the new variables is given by

$$H^*(q^*,p^*) \equiv H(q,p). \tag{1.16}$$

A transformation defined by equations of the type (1.14) is called a Lie series transformation.

To obtain the inverse of a Lie series transformation, we observe that changing the sign of the Hamiltonian function simply reverses the dynamical motion, so that the inverse transformation is obtained by putting -W in place of W:

$$q_i = q_i^* - \frac{\partial W}{\partial p_i^*}(q^*,p^*) + \sum_{k=2}^{\infty} \frac{(-1)^k}{k!} \mathcal{L}_W^{k-1} \frac{\partial W}{\partial p_i^*}\Big|_{q^*,p^*}$$

$$\text{and} \quad p_i = p_i^* + \frac{\partial W}{\partial q_i^*}(q^*,p^*) - \sum_{k=2}^{\infty} \frac{(-1)^k}{k!} \mathcal{L}_W^{k-1} \frac{\partial W}{\partial q_i^*}\Big|_{q^*,p^*} \tag{1.17}$$

and we also have

$$f(q,p) = f(q^*,p^*) + \sum_{k=1}^{\infty} \frac{(-1)^k}{k!} \mathcal{L}_W^k f \Big|_{q^*,p^*}. \tag{1.18}$$

Hence from (1.16) we have

$$H^*(q^*,p^*) \equiv H(q,p) = H(q^*,p^*) + \sum_{k=1}^{\infty} \frac{(-1)^k}{k!} \mathcal{L}_W^k H \Big|_{q^*,p^*} \tag{1.19}$$

In the use we will make of the Lie series transformation, we will use this equation to form our choice of W to arrange for H* to have the form desired, which in the first application will be for it to be without short-period terms.

2. THE PLANETARY (n+1)-BODY PROBLEM IN HAMILTONIAN FORM

Consider a system consisting of a primary body P_o, of mass m_o, and n planets (considered as particles) P_1, P_2, P_3,... P_n (the mass of P_i being m_i), and each body attracting each of the others according to Newton's gravitational law. We will use Jacobi's system of position vectors to describe the relative motions, that is,

$$\underset{\sim}{x}_i = \overrightarrow{P_oP_1} \quad \text{and,} \quad \text{for} \quad 2 \leq j < n, \quad \underset{\sim}{x}_j = \overrightarrow{G_{j-1}P_i} \tag{2.1}$$

where G_j is the mass-centre of P_o, $P_1, P_2, \ldots P_j$. Then we find (see, e.g., Message (1982a) section 2) that the kinetic energy of the system may be written as

$$T = \frac{1}{2} M \dot{\underset{\sim}{\bar{r}}}^2 + \frac{1}{2} \sum_{j=1}^{n} \varepsilon\beta_j \dot{\underset{\sim}{x}}_j^2 \tag{2.2}$$

where $M = \sum_{j=0}^{n} m_j$,

$\underset{\sim}{\bar{r}}$ is the position vector of the mass-centre of the entire system, referred to an inertial frame,

$$\varepsilon\beta_j = m_{j_k} M_{j-1}/M_j , \tag{2.3}$$

and $M_j = \sum_{k=0} m_k$.

The factor ε may be chosen as a small parameter for the construction of perturbation theory, for example it may be taken to be the largest of the m_j/m_o $(1 \leq j \leq n)$. Note that

$$m_j = \varepsilon\beta_j/\{1 - \varepsilon\beta_j/M_{j-1}\} , \tag{2.4}$$

so that the relative sizes of the β_j are approximately those of the m_j. The potential energy of the system is

$$V = - G \sum_{(i,j)} \frac{m_i m_j}{P_i P_j} \tag{2.5}$$

where G is the Newtonian gravitational constant, the summation being over all pairs (i,j) for which $0 \leq i < j \leq n$. The equations of motion then take the form, after cancelling a factor ε ,

$$\beta_j \ddot{\underset{\sim}{x}}_j = - \mu_j \beta_j \underset{\sim}{x}_j/|\underset{\sim}{x}_j|^3 + \frac{\partial R}{\partial \underset{\sim}{x}_j} \tag{2.6}$$

where $\mu_j = G\, m_o m_j/(\varepsilon\beta_j)$ (2.7)

from which $\mu_j = G\,\{m_o + \varepsilon\beta_j + O(\varepsilon^2)\}$

and R is such that

$$V = - \sum_{j=1}^{n} \mu_j \varepsilon\beta_j/|\underset{\sim}{x}_j| - \varepsilon R \tag{2.8}$$

and is a function of the positions only through the $|\underset{\sim}{x}_j|$ and $\underset{\sim}{x}_j . \underset{\sim}{x}_k$, and also depends on the β_j and on ε. Now note that, for $0 < j \leq n$,

$$\overrightarrow{P_o P_j} = \sum_{k=1}^{j-1} \frac{m_k}{M_k} \underset{\sim}{x}_k + \underset{\sim}{x}_j$$

so that $P_o P_j^2 = |\underset{\sim}{x}_j|^2 + 2 \sum_{k=1}^{j-1} \frac{m_k}{M_k} \underset{\sim}{x}_j . \underset{\sim}{x}_k + O(\varepsilon^2)$

and so

$$\frac{G\, m_o m_j}{P_o P_j} = \frac{\mu_j \varepsilon\beta_j}{|\underset{\sim}{x}_j|} - G \sum_{k=1}^{j-1} m_j m_k \left(\frac{m_o}{M_k}\right) \frac{\underset{\sim}{x}_j . \underset{\sim}{x}_k}{|\underset{\sim}{x}_j|^3} + O(\varepsilon^3) \tag{2.9}$$

and, for $0 < i < j \leq n$,

$$P_i P_j = - \frac{M_{i-1}}{M_i} \underset{\sim}{x}_i + \sum_{k=i+1}^{j-1} \frac{m_k}{M_k} \underset{\sim}{x}_k + \underset{\sim}{x}_j$$

and thus

$$P_iP_j^2 = (\underset{\sim}{x}_j - \underset{\sim}{x}_i)^2 + O(\varepsilon).$$

From these we see that

$$R = G \sum_{o<i<j\leq n} 2\beta_i\beta_j \left\{ \frac{1}{|\underset{\sim}{x}_i - \underset{\sim}{x}_j|} - \frac{\underset{\sim}{x}_i \cdot \underset{\sim}{x}_j}{|\underset{\sim}{x}_j|^3} \right\} + O(\varepsilon^2) \qquad (2.10)$$

so that R contains all the usual direct terms, and all the indirect terms for the action of an exterior on an interior planet, which arise in the disturbing functions for conventional planetary perturbation theory using heliocentric position vectors.

Now the equations of motion (2.6) may be put into Hamiltonian form, taking

$$\underset{\sim}{p}_j = \beta_j \underset{\sim}{x}_j \qquad (2.11)$$

as the momentum conjugate to $\underset{\sim}{x}_j$, with the Hamiltonian function

$$H(\underset{\sim}{x},\underset{\sim}{p}) = \sum_{j=1}^{n} \{ \underset{\sim}{p}_j^2/(2\beta_j) - \mu_j\beta_j/|\underset{\sim}{x}_j| \} - R \qquad (2.12)$$

Then taking $\varepsilon = 0$ (so that R=0) now results in n separate Kepler inverse-square central force problems. At any time each pair $(\underset{\sim}{x}_j,\underset{\sim}{p}_j)$ define an instantaneous Kepler orbit in the central force problem corresponding to $\varepsilon = 0$, and we may use the parameters of this orbit to describe the state of relative motion at that time. Suppose that the Kepler ellipse corresponding to $(\underset{\sim}{x}_j,\underset{\sim}{p}_k)$ has major semi-axis a_i, eccentricity e_i, and inclination I_j, referred to a plane of inertial orientation, and let the longitude of the ascending node on that plane, measured from an inertially fixed direction, be Ω_i, the longitude of the near apse $\tilde{\omega}_j$, and the mean longitude of P_j in this orbit be λ_i. Then a set of canonical coordinates for the relative motion of the system is

$$(\lambda,\tilde{\omega},\Omega) = (\lambda_1,\lambda_2,\ldots,\lambda_n,\tilde{\omega}_1,\tilde{\omega}_2,\ldots,\tilde{\omega}_n, \Omega_1,\Omega_2,\ldots,\Omega_n)$$

to which the conjugate momenta are

$$(\Lambda,\Pi,N) = (\Lambda_1,\Lambda_2,\ldots,\Lambda_n,\Pi_1,\Pi_2,\ldots,\Pi_n,N_1,N_2,\ldots,N_n)$$

where

$$\Lambda_j = \beta_j\sqrt{(\mu_j a_j)},$$

$$\Pi_j = \Lambda_j\{\sqrt{(1-e_j^2)} - 1\} = -\frac{1}{2}\Lambda_j \varepsilon_j^2,$$

$$N_j = -2\Lambda_j \sqrt{(1-e_j^2)} \sin^2(\tfrac{1}{2} I_j), \qquad (2.13)$$

in which we have put

$$\varepsilon_j^2 = 2\{1 - \sqrt{(1 - e_j^2)}\} = e_j^2 + \frac{1}{4} e_j^4 + O(e_j^6). \tag{2.14}$$

From the formulae of Keplerian motion we have

$$p_j^2/(2\beta_j) - \mu_j\beta_j/|x_j| = -\beta_j^3\mu_j^2/(2\Lambda_j^2), \tag{2.15}$$

so that the Hamiltonian function is

$$H^+(\lambda,\tilde{\omega},\Omega;\Lambda,\Pi,N) = -\frac{1}{2}\sum_{j=1}^{n}\beta_j^3\mu_j^2/\Lambda_j^2 - R$$

$$= H_o - R, \text{ say}. \tag{2.16}$$

We recall that R is a function of the coordinates only through the $|x_j|$ and $x_j.x_k$, and note that these have period 2Π in each of the angles λ_i, $\tilde{\omega}_i$, and Ω_i, thus we see that we may express R as a multiple Fourier series in the form

$$R = \sum_j K_j \cos N_j \tag{2.17}$$

where

$$N_j = \sum_{i=1}^{n} (j_{1i}\lambda_i + j_{2i}\tilde{\omega}_i + j_{3i}\Omega_i), \tag{2.18}$$

and the summation is over all sets of integers

$$j = (j_{11}, j_{12}, \ldots, j_{1n}, j_{21}, j_{22}, \ldots, j_{2n}, j_{31}, j_{32}, \ldots, j_{3n})$$

with $j_{11} \geq 0$. Since the value of R does not depend on the position of the fixed reference line, only those terms appear which have

$$\sum_{i=1}^{n} (j_{1i} + j_{2i} + j_{3i}) = 0 \tag{2.19}$$

Now the expressions for the $|x_j|$ and $x_j.x_k$ in terms of the λ_i and $\tilde{\omega}_i$ possess the d'Alembert property, that is, a term in the Fourier series whose angular argument contains the multiple j_{2i} of $\tilde{\omega}_i$ will have as coefficient $e_i^{|j_{2i}|}$ times a power series in e_i^2. This is equivalent to saying that the dependence of $|x_i|$ and $x_i.x_j$ on e_i and $\tilde{\omega}_i$ can be expressed by power series in $e_i \cos \tilde{\omega}_i$ and $e_i \sin \tilde{\omega}_i$. This property is preserved on adding or multiplying expressions which possess it, from which it follows that it is possessed also by R. In a similar manner the dependence of R on I_i and Ω_i shows a corresponding property, so that K_j has the factors $e_i^{|j_{2i}|}$ and $\sin^{|j_{3i}|}(\frac{1}{2} I_i)$, and other wise only even positive powers of e_i and $\sin(\frac{1}{2} I_i)$. We will see that this is important in what follows.

Institut für Reine und Angewandte Kernphysik Kiel

3. THE USE OF LIE SERIES TO DERIVE THE SHORT-PERIOD PERTURBATIONS.

A term in the expansion (2.17) of R in which not all of the j_{1i} are zero will usually be called a "short-period" term, in recognition of the fact that it will be found to lead to perturbations in the orbital elements with the frequency

$$\nu_j = \sum_{i=1}^{n} j_{1i}\, n_i \,, \quad \text{where}$$

$$n_i = \frac{\partial H_o}{\partial \Lambda_i} \qquad (i=1,2,\ldots,n) \tag{3.1}$$

is the mean orbital motion of P_i, so that the period is of the same order of magnitude as the periods of the planets in their orbits. In cases where there is an important near-commensurability of the mean motions, so that for some terms ν_i is of smaller order of magnitude than the n_i, we will call such terms "critical", and the argument N_i a "critical argument". We will return to the consideration of such terms later.

We will use a Lie series transformation

$$(\lambda,\tilde{\omega},\Omega;\Lambda,\Pi,N) \longrightarrow (\lambda^*,\tilde{\omega}^*,\Omega^*,\Lambda^*,\Pi^*,N^*) \tag{3.2}$$

defined explicitly, following (1.14), by

$$\lambda_i^* = \lambda_i + \frac{\partial W}{\partial \Lambda_i} + \sum_{k=2}^{\infty} \frac{1}{k!} \mathcal{L}_W^{k-1} \frac{\partial W}{\partial \Lambda_i}$$

$$\tilde{\omega}_i^* = \tilde{\omega}_i + \frac{\partial W}{\partial \Pi_i} + \sum_{k=2}^{\infty} \frac{1}{k!} \mathcal{L}_W^{k-1} \frac{\partial W}{\partial \Pi_i}$$

$$\Omega_i^* = \Omega_i + \frac{\partial W}{\partial N_i} + \sum_{k=2}^{\infty} \frac{1}{k!} \mathcal{L}_W^{k-1} \frac{\partial W}{\partial N_i}$$

$$\Lambda_i^* = \Lambda_i - \frac{\partial W}{\partial \lambda_i} - \sum_{k=2}^{\infty} \frac{1}{k!} \mathcal{L}_W^{k-1} \frac{\partial W}{\partial \lambda_i}$$

$$\Pi_i^* = \Pi_i - \frac{\partial W}{\partial \tilde{\omega}_i} - \sum_{k=2}^{\infty} \frac{1}{k!} \mathcal{L}_W^{k-1} \frac{\partial W}{\partial \tilde{\omega}_i}$$

$$N_i^* = N_i - \frac{\partial W}{\partial \Omega_i} - \sum_{k=2}^{\infty} \frac{1}{k!} \mathcal{L}_W^{k-1} \frac{\partial W}{\partial \Omega_i} \qquad (i=1,2,3,\ldots,n) \tag{3.3}$$

the inverse being defined by (following (1.17))

Institut für Reine und Angewandte Kernphysik Kiel

$$\lambda_i = \lambda_i^* - \frac{\partial W}{\partial \Lambda_i^*} + \sum_{k=2}^{\infty} \frac{(-1)^k}{k!} \mathcal{L}_W^{k-1} \frac{\partial W}{\partial \Lambda_i^*}$$

$$\tilde{\omega}_i = \tilde{\omega}_i^* - \frac{\partial W}{\partial \Pi_i^*} + \sum_{k=2}^{\infty} \frac{(-1)^k}{k!} \mathcal{L}_W^{k-1} \frac{\partial W}{\partial \Pi_i^*}$$

$$\Omega_i = \Omega_i^* - \frac{\partial W}{\partial N_i^*} + \sum_{k=2}^{\infty} \frac{(-1)^k}{k!} \mathcal{L}_W^{k-1} \frac{\partial W}{\partial N_i^*}$$

$$\Lambda_i = \Lambda_i^* + \frac{\partial W}{\partial \lambda_i^*} - \sum_{k=2}^{\infty} \frac{(-1)^k}{k!} \mathcal{L}_W^{k-1} \frac{\partial W}{\partial \lambda_i^*}$$

$$\Pi_i = \Pi_i^* + \frac{\partial W}{\partial \tilde{\omega}_i^*} - \sum_{k=2}^{\infty} \frac{(-1)^k}{k!} \mathcal{L}_W^{k-1} \frac{\partial W}{\partial \tilde{\omega}_i^*}$$

$$N_i = N_i^* + \frac{\partial W}{\partial \Omega_i^*} - \sum_{k=2}^{\infty} \frac{(-1)^k}{k!} \mathcal{L}_W^{k-1} \frac{\partial W}{\partial \Omega_i^*}$$

$$(i=1,2,3,\ldots,n) \qquad (3.4)$$

where, in (3.4), the right-hand sides are evaluated by substituting the transformed variables into W. We shall seek to choose W so that the new Hamiltonian function H* contains no short-period terms. We recall that

$$H = H_o - R, \qquad (3.5)$$

and that R contains parts of order ε, and some of higher degree in ε, so that we may write

$$R = R_1 + R_2 + R_3 + \ldots. \qquad (3.6)$$

where by R_i we mean that part of R which is of explicit degree i in ε. We may expect that H* also will be found to contain parts of each degree explicitly in ε:

$$H^* = H_o^* + H_1^* + H_2^* + \ldots. \qquad (3.7)$$

and likewise W:

$$W = W_1 + W_2 + W_3 + \ldots \qquad (3.8)$$

where H_i^* and W_i are each of explicit degree i and ε. The equation (1.19), giving the Hamiltonian function of the transformed system in terms of the original one, showing its first few terms, is

$$H^* \equiv H + \sum_{k=1}^{\infty} \frac{(-1)^k}{k!} \mathcal{L}_W^k H$$

$$= H - \{H,W\} + \frac{1}{2} \{\{H,W\}, W\} - \frac{1}{3!} \{\{\{H,W\}, W\}, W\} \ldots \tag{3.9}$$

which becomes, on inserting the expansion (3.6), (3.7), and (3.8),

$$\begin{aligned} H^*_o + H^*_1 + H^*_2 + \ldots \equiv H_o &- R_1 - R_2 - \ldots \\ &-\{H_o,W_1\} - \{H_o,W_2\} - \ldots \\ &+\{R_1,W_1\} + \{R_2,W_1\} + \ldots \\ &+\{R_1,W_2\} + \ldots \\ &+ \frac{1}{2} \{\{H_o,W_1\}, W_1\} + \ldots \end{aligned} \tag{3.10}$$

In this we equate the terms of each explicit degree in ε. First those independent of ε :

$$H^*_o \equiv H_o \tag{3.11}$$

showing that the transformation (3.2) leaves the dynamical system unchanged in the unperturbed case $\varepsilon = 0$, in which case, of course, the transformation (3.2) would be the identity, since then $W = 0$. Equating terms of degree one explicitly in ε :

$$H^*_1 \equiv - R_1 - \{H_o,W_1\}, \tag{3.12}$$

and equating terms of degree two:

$$H^* \equiv - R_2 - \{H_o,W_2\} + \{R_1,W_1\} + \frac{1}{2} \{\{H_o,W_1\},W_1\}, \tag{3.13}$$

and similarly for each successive order in ε . Denote by $\tilde{R}_1$ the short-period terms of R_1 , and by $\bar{R}_1$ the remainder of R_1 , so that

$$R_1 = \bar{R}_1 + \tilde{R}_1. \tag{3.14}$$

Then take

$$H^*_1 \equiv - R_1, \tag{3.15}$$

so that H^*_1 has no short-period terms, as desired. This leaves

$$\{H_o,W_1\} \equiv - \tilde{R}_1 \tag{3.16}$$

Now H_o depends only on the Λ_i , and $\frac{\partial H_o}{\partial \Lambda_i} = n_i$, so that

$$\{H_o, W_1\} \equiv \sum_j n_j \frac{\partial W_1}{\partial \lambda_j} \tag{3.17}$$

which would be the rate of change of W_1 with time in the undisturbed motion. Suppose that

$$\tilde{R}_1 = \sum_{(s.p)} K_j \cos N_j \tag{3.18}$$

the summation being over only short-period terms. Then (3.17) is satisfied by

$$W_1 \equiv - \sum_{(s.p)} K_j \sin N_j / \nu_j \tag{3.19}$$

where

$$\nu_j = \sum_{i=1}^{n} j_{1i} n_i . \tag{3.20}$$

Now that W_1 has been chosen, the quantities in (3.13) which remain undetermined are H_2^* and W_2. Put

$$\begin{aligned} \Phi_2 &= - R_2 + \{R_1, W_1\} + \frac{1}{2} \{\{H_o, W_1\}, W_1\} \\ &= - R_2 + \{R_1, W_1\} - \frac{1}{2} \{\tilde{R}_1, W_1\} \end{aligned} \tag{3.21}$$

(by virtue of (3.16)), which is therefore now known. Thus (3.13) becomes

$$H_2^* \equiv \Phi_2 - \{H_o, W_2\} \tag{3.22}$$

Denote by $\tilde{\Phi}_2$ the short-period terms of Φ_2, and by $\overline{\Phi}_2$ the remainder. We now may choose

$$H_2^* \equiv \overline{\Phi}_2 \tag{3.23}$$

which then has no short-period terms, again as desired. This leaves

$$\{H_o, W_2\} \equiv \tilde{\Phi}_2 , \tag{3.24}$$

and we determine W_2 from $\tilde{\Phi}_2$ in the same way that W_1 was determined from $- \tilde{R}_1$, that is, if

$$\tilde{\Phi}_2 = \sum_{s.p.} L_j^{(2)} \cos N_j , \tag{3.25}$$

then

$$W_2 = \sum_{s.p.} L_j^{(2)} \sin N_j / \nu_j \tag{3.26}$$

In the same way the terms in (3.9) of any order may be dealt with. Thus the terms of order r will take the form

$$H^*_r \equiv \Phi_r - \{H_o, W_r\} \tag{3.27}$$

where Φ_r denotes the sum of terms, all of which will be known by the time the terms of order r-1 have been dealt with. So we choose

$$H^*_r = \overline{\Phi}_r \tag{3.28}$$

the part of Φ_r free of short-period terms, and this leaves

$$\{H_o, W_r\} \equiv \tilde{\Phi}_r \tag{3.29}$$

where $\tilde{\Phi}_r$, the short-period part of Φ_r, may be written as

$$\tilde{\Phi}_r = \sum_{s.p.} L_j^{(r)} \cos N_j, \tag{3.30}$$

which is satisfied by putting

$$W_r = \sum_{s.p.} L_j^{(r)} \sin N_j / \nu_j. \tag{3.31}$$

4. THE MAIN PROPERTIES OF THE SHORT-PERIOD PERTURBATIONS

The function W_1 contains all the information necessary to construct the first-order perturbations, as usually defined, in all the elements, since these are given by the first-order parts of the right-hand sides in equations (3.3), and the interpretation as in Poisson's successive approximation method is achieved by regarding the elements as unperturbed in the right-hand sides, that is, as all constant except the mean longitudes, which are taken as linear functions of time. For the mean longitudes, λ_i, the expressions are

$$\begin{aligned} -\frac{\partial W_1}{\partial \Lambda_i} &= \sum_{s.p.} \frac{\partial}{\partial \Lambda_i}(K_j/\nu_j)\sin N_j \\ &= \frac{1}{\beta_i n_i a_i^2} \sum_{s.p.} \frac{1}{\nu_j}\{2a_i \frac{\partial K_j}{\partial a_i} + 3j_{1i}(\frac{n_i}{\nu_j})K_j \\ &\quad - \frac{b_i(a_i - b_i)}{a_i^2 e_i} \frac{\partial K_j}{\partial e_i} - \tan(\frac{1}{2} I_i)(\frac{a_i}{b_i}) \frac{\partial K_j}{\partial I_i}\} \sin N_j, \end{aligned} \tag{4.1}$$

in the apse longitudes, $\tilde{\omega}_i$, they are

$$-\frac{\partial W_1}{\partial \pi_i} = \sum_{s.p.} \frac{\partial}{\partial \pi_i}(K_j/\nu_j)\sin N_j$$

$$= \frac{1}{\beta_i n_i a_i^2} \sum_{s.p.} \frac{1}{\nu_j}\{-\frac{b_i}{a_i e_i}\frac{\partial K_j}{\partial e_i} - \tan(\tfrac{1}{2} I_i)\frac{a_i}{b_i}\frac{\partial K_j}{\partial I_i}\}\sin N_j, \quad (4.2)$$

in the node longitudes, Ω_i, they are

$$-\frac{\partial W_1}{\partial N_i} = \sum_{s.p.} \frac{\partial}{\partial N_i}(K_j/\nu_j)\sin N_j$$

$$= \frac{-b_i}{\beta_i n_i a_i^2 \sin I_i} \sum_{s.p.} \frac{1}{\nu_j}\frac{\partial K_j}{\partial I_i}\sin N_j \quad (4.3)$$

in the canonical momenta Λ_i they are

$$\frac{\partial W_1}{\partial \lambda_i} = \sum_{s.p.} (K_j/\nu_j)\, j_{1i} \cos N_j \quad (4.4)$$

from which in the major semi-axes, a_i, they are

$$\frac{2}{\beta_i n_i a_i} = \sum_{s.p.} (K_j j_{1i}/\nu_j)\cos N_j. \quad (4.5)$$

In the canonical momenta π_i they are

$$\frac{\partial W_1}{\partial \tilde{\omega}_i} = \sum_{s.p.} (K_j/\nu_j)\, j_{2i} \cos N_j \quad (4.6)$$

from which, in the eccentricities, e_i, they are

$$\frac{1}{\beta_i n_i a_i^2} \sum_{s.p.} \{-\frac{b_i(a_i-b_i)}{a_i^2 e_i} j_{1i} - \frac{b_i}{a_i e_i} j_{ai}\}(K_j/\nu_j)\cos N_j. \quad (4.7)$$

In the momenta N_i they are

$$\frac{\partial W_1}{\partial \Omega_i} = \sum_{s.p.} (K_j/\nu_j)\, j_{3i} \cos N_j \quad (4.8)$$

from which, in the inclinations, I_i, they are

$$\frac{1}{\beta_i n_i a_i b_i} \sum_{s.p.} \{-\tan(\tfrac{1}{2} I_i)(j_{1i}+j_{2i}) - \frac{j_{3i}}{\sin I_i}\}(K_j/\nu_j)\cos N_j \quad (4.9)$$

We note that, in the first-order perturbations in $\tilde{\omega}_i$, some terms will have an uncancelled factor e_i in the denominator, such being terms with $j_{2i} = 1$, so that, from the d'Alembert property, K_j has e_i to first power only as a factor. (In the expression for the perturbation in λ_i, the e_i in the denominator is cancelled by the e_i^2 implicit in the factor $a_i - b_i$ in the numerator). This singularity at $e_i = 0$ corresponds of course to the fact that the apse longitude $\tilde{\omega}_i$ is not defined for a circular orbit. When we come to study the long-period motion, in which the occurrence of zero values of the eccentricities cannot be excluded, we will use a modified set of parameters. A corresponding situation arises for zero values of the inclinations, and modified parameters will be used here also.

Linear combinations ν_j of the n_i which are small correspond to terms of long period (since the period of the term is $2\pi/\nu_i$) and the amplitude has the factor $1/\nu_j$, and so is large unless there is a compensating small factor in the numerator. Usually such small combinations only arise in terms with large values of the $|j_{1i}|$, and this implies, because of equation (2.19), that some at least of the $|j_{2i}|$ and $|j_{3i}|$ must be large also, and so, because of the d'Alembert property, large powers of some at least of the small quantities e_i and $\sin(\frac{1}{2} I_i)$ must occur as factors in K_j, so that the required small factor is indeed present. However, if a small-integer combination of the n_i is small, as for the planets Jupiter and Saturn, where twice the mean motion of Jupiter is close to five times that of Saturn, we do indeed find a long-period term in the perturbations (of period close to 890 years) of large amplitude (called the "great inequality").

The second-order short-period perturbations in the elements can be derived from the function W_2, since, for example for λ_i they are $-\frac{\partial W_2}{\partial \Lambda_i}$, together with the short-period part of $\{\partial W_1/\partial \Lambda_i, W_1\}$. Similarly, for the other elements.

5. TREATMENT OF THE LONG-PERIOD PROBLEM, TO DERIVE ASYMPTOTIC SERIES

In discussing the long-period evolution of the planetary system, we must choose canonical parameters to avoid the singularities at zeroes of the eccentricities and inclinations of the orbits, which were described towards the end of the previous section. Circular orbits are simple cases, not singular ones, and motion in which all the planets move in the reference plane is also a simple one. Thus, in place of the canonical pairs $(\tilde{\omega}_i\ ;\ \pi_i)$, and $(\Omega_i\ ;\ N_i)$, we will use

$$\xi_i = \sqrt{(-2\pi_i)}\cos\tilde{\omega}_i = \sqrt{(\Lambda_i)}\varepsilon_i \cos\tilde{\omega}_i\ ,$$

$$\eta_i = \sqrt{(-2\pi_i)}\sin\tilde{\omega}_i = \sqrt{(\Lambda_i)}\varepsilon_i \sin\tilde{\omega}_i\ , \qquad (5.1)$$

$$q_i = \sqrt{(-2N_i)}\cos\Omega_i = \sqrt{(\Lambda_i+\pi_i)}\sin(\tfrac{1}{2}I_i)\cos\Omega_i,$$

and

$$p_i = \sqrt{(-2N_i)}\sin\Omega_i = \sqrt{(\Lambda_i+\pi_i)}\sin(\tfrac{1}{2}I_i)\sin\Omega_i,\ \text{for } i=1,2,3,\ldots,n).$$

(These are the canonical counterparts of the rectangular-type variables used by Laplace and Lagrange in their secular variation theory). The equations of motion are still of Hamiltonian form, with the λ_i, ξ_i, and q_i being the coordinates, and Λ_i, η_i, and p_i, respectively, as their conjugate momenta, the Hamiltonian function being still

$$H(\lambda,\xi, q; \Lambda,\eta, p) = H_o - R \tag{5.2}$$

though now expressed in terms of the new variables. The disturbing function may be expressed as

$$R = \sum_{j_1} (B_{j_1} \cos N_{j_1} + C_{j_1} \sin N_{j_1}), \tag{5.3}$$

where

$$N_{j_1} = \sum_{i=1}^{n} j_{1i}\, \lambda_i , \tag{5.4}$$

and B_{j_1} and C_{j_1} are functions of the Λ_i, ξ_i, η_i, q_i, and p_i. (see Message (1982a) page 30 for more details).

The Lie series transformation used to separate the short-period from the long-period part of the problem, analogous to (3.2), is

$$(\lambda,\xi,q; \Lambda,\eta, p) \longrightarrow (\lambda',\xi',q'; \Lambda',\eta',p') \tag{5.5}$$

given by

$$\begin{aligned}
\lambda'_i &= \lambda_i + \frac{\partial W'}{\partial \Lambda_i} + \sum_{k=2}^{\infty} \frac{1}{k!} \mathcal{L}^{k-1}_{W'} \frac{\partial W'}{\partial \Lambda_i} \\
\xi'_i &= \xi_i + \frac{\partial W'}{\partial \eta_i} + \sum_{k=2}^{\infty} \frac{1}{k!} \mathcal{L}^{k-1}_{W'} \frac{\partial W'}{\partial \eta_i} \\
q'_i &= q_i + \frac{\partial W'}{\partial p_i} + \sum_{k=2}^{\infty} \frac{1}{k!} \mathcal{L}^{k-1}_{W'} \frac{\partial W'}{\partial p_i} \\
\Lambda'_i &= \Lambda_i - \frac{\partial W'}{\partial \lambda_i} - \sum_{k=2}^{\infty} \frac{1}{k!} \mathcal{L}^{k-1}_{W'} \frac{\partial W'}{\partial \lambda_i} \\
\eta'_i &= \eta_i - \frac{\partial W'}{\partial \xi_i} - \sum_{k=2}^{\infty} \frac{1}{k!} \mathcal{L}^{k-1}_{W'} \frac{\partial W'}{\partial \xi_i} \\
p'_i &= p_i - \frac{\partial W'}{\partial q_i} - \sum_{k=2}^{\infty} \frac{1}{k!} \mathcal{L}^{k-1}_{W'} \frac{\partial W'}{\partial q_i}
\end{aligned}$$

$$(i=1,2,3,\ldots,n) \tag{5.6}$$

with inverse constructed as (3.4) is from (3.3). The generating function

$$W' = \sum_{k=1}^{\infty} W'_k \tag{5.7}$$

is chosen to make the Hamiltonian function for the transformed system,

$$H' = \sum_{k=0}^{\infty} H'_k \tag{5.8}$$

independent of λ'_i . Following the same procedure as in equations (3.9) onwards, we obtain

$$H'_o \equiv H_o , \tag{5.9}$$

and

$$H'_1 \equiv - \bar{R}_1 , \tag{5.10}$$

and

$$\{H_o, W'_1\} \equiv - \tilde{R}_1 . \tag{5.11}$$

Thus if

$$\tilde{R}_1 = \sum_{s.p.} (B_{j_1} \cos N_{j_1} + C_{j_1} \sin N_{j_1}) , \tag{5.12}$$

then

$$W'_1 = \sum_{s.p.} ((- B_{j_1}/\nu_{j_1}) \sin N_{j_1} + (C_{j_1}/\nu_{j_1}) \cos N_{j_1}) , \tag{5.13}$$

and W'_2 , W'_3 , etc. are determined in like manner, as before.

The d'Alembert property will have the consequence that H' will depend on the ξ'_i , η'_i, q'_i, and p'_i through the quantities

$$\xi'^2_i + \eta'^2_i , \quad \xi'_i \xi'_j + \eta'_i \eta'_j, \quad q'^2_i + p'^2_i , \quad q'_i q'_j + p'_i p'_j .$$

Thus the terms of lowest degree will be quadratic, of the form

$$H'_Q = \sum_{ij} \{K_{ij}(\xi'_i \xi'_j + \eta'_i \eta'_j) + K^*_{ij}(q'_i q'_j + p'_i p'_j)\} , \tag{5.14}$$

where the K_{ij} and K^*_{ij} are functions of the Λ'_i only.

The approximation in which all terms of higher degree than these are neglected leads to the "secular variation" theory of Laplace and Lagrange. We now use for the complete problem the linear transformation which gives the solution of that simplified problem, that is, we make a transformation

$$(\lambda', \xi', q'; \Lambda', \eta', p') \longrightarrow (\lambda'', \xi'', q''; \Lambda'', \eta'', p'') \tag{5.15}$$

with

$$\xi'_i = \sum_{j=1}^{n} a_{ij}\, \xi''_j ,$$

$$\eta'_i = \sum_{j=1}^{n} a_{ij}\, \eta''_j ,$$

$$q'_i = \sum_{j=1}^{n} b_{ij}\, q''_j ,$$

and

$$p'_i = \sum_{j=1}^{n} b_{ij}\, p''_j \quad . \qquad (i=1,2,\ldots,n). \tag{5.16}$$

We find that this is a canonical transformation provided that

$$\sum_{j=1}^{n} a_{ij}\, a_{jk} = \delta_{ik} ,$$

and

$$\sum_{j=1}^{n} b_{ij}\, b_{jk} = \delta_{ik} , \tag{5.17}$$

where δ_{ik} is zero if $i \neq k$, and 1 if $i = k$. That is, the matrices (a_{ij}) and (b_{ij}) must be orthogonal. Additional requirements are that

$$\Lambda'_i = \Lambda''_i \quad \text{and} \quad \lambda'_i = \lambda''_i - \sum_{j,k,\ell} \{\xi''_j \frac{\partial a_{jk}}{\partial \Lambda''_i} a_{\ell k}\, \eta''_\ell + q''_j \frac{\partial b_{jk}}{\partial \Lambda''_i} b_{\ell k}\, p''_\ell \} . \tag{5.18}$$

We may choose the matrices (a_{ij}) and (b_{ij}) to diagonalise the quadratic forms, that is, so that

$$\sum_{j,k} a_{r_j}\, K_{jk}\, a_{sk} = \frac{1}{2} A_r\, \delta_{rs}, \quad \text{and} \quad \sum_{j,k} b_{r_j}\, K^*_{jk}\, b_{sk} = \frac{1}{2} B_r\, \delta_{rs} . \tag{5.19}$$

The matrices (a_{ij}) and (b_{ij}) must therefore be functions of the Λ'_i. The Hamiltonian function of the transformed system is

$$H''(\xi'', q''; \Lambda'', \eta'', p'') = H''_o + H''_p \tag{5.20}$$

where

$$H''_o = -\frac{1}{2} \sum_{i=1}^{n} \mu_i^2\, \beta_i^3 / \Lambda''^2_i \tag{5.21}$$

and

$$H''_P \equiv H'_1 + H'_2 + H'_3 + \ldots. \tag{5.22}$$

is a function of the Λ''_i , and, because of the orthogonality of the matrices (a_{ij}) and (b_{ij}), also of the $\xi''_i \xi''_j + \eta''_i \eta''_j$ and $q''_i q''_j + p''_i p''_j$. Its terms of lowest degree are

$$H''_Q = \frac{1}{2} \sum_{r=1}^{n} \{A_r(\xi''^2_r + \eta''^2_r) + B_r(q''^2_r + p''^2_r)\}. \tag{5.23}$$

The A_r are, to first order of ε, the frequencies of the secular variations in the eccentricities and apse motions, and the B_r are, also to first order in ε , the frequencies of the secular variations in the nodes and inclinations.

The next step is to return to action and angle type variables, with a transformation

$$(\lambda''.\xi''.q''; \Lambda'',\eta'',p'') \longrightarrow (\lambda''',\omega; \Lambda''',\Omega) \tag{5.24}$$

where $\omega = (\omega_1,\omega_2,\ldots,\omega_{2n})$ and $\Omega = (\Omega_1,\Omega_2,\ldots,\Omega_{2n})$, putting

$$\begin{aligned}
\xi''_r &= \sqrt{(-2\Omega_r)}\cos\omega_r, \\
\eta''_r &= \sqrt{(-2\Omega_r)}\sin\omega_r, \\
q''_r &= \sqrt{(-2\Omega_{n+r})}\cos\,\omega_r, \\
p''_r &= \sqrt{(-2\Omega_{n+r})}\sin\,\omega_r, \qquad (r=1,2,\ldots,n) \\
\lambda''_i &= \lambda'''_i \ , \quad \text{and} \quad \Lambda''_i = \Lambda'''_i \ (i=1,2,\ldots,n).
\end{aligned} \tag{5.25}$$

The Hamiltonian function for these new variables is

$$H'''(\omega;\Lambda''',\Omega) \equiv H''(\xi'',q'';\Lambda'',\eta'',p''), \tag{5.26}$$

and since

$$\xi''_r \xi''_s + \eta''_r \eta''_s = 2\sqrt{(\Omega_r\Omega_s)}\cos(\omega_r - \omega_s), \tag{5.27}$$

we see that

$$H''' = \sum_\nu K^{**}_\nu \cos N_\nu \ , \tag{5.28}$$

where $\nu = (\nu_1,\nu_2,\ldots,\nu_{2n})$
and

$$N_\nu = \sum_{r=1}^{2n} \nu_r \omega_r \tag{5.29}$$

the coefficients K^{**}_ν being functions of the Λ'''_i and Ω_r, having, because of the d'Alembert property, the factor $\Omega_r^{\frac{1}{2}|\nu_r|}$, and the summation is over all sets ν of 2n integers ν_r with

$$\sum_{r=1}^{2n} \nu_r = 0. \tag{5.30}$$

The part of H''' of lowest order in the Ω_r is

$$H_Q''' = - \sum_{r=1}^{n} (A_r\Omega_r + B_r\Omega_{n+r}) \tag{5.31}$$

and, since this is independent of the angle variables ω_r, we may carry out a further Lie series transformation

$$(\lambda''',\omega;\Lambda''',\Omega) \longrightarrow (\lambda^*,\omega^*;\Lambda^*,\Omega^*), \tag{5.32}$$

defined by

$$\left.\begin{aligned} \lambda_i^* &= \lambda_i''' + \frac{\partial T}{\partial \Lambda_i'''} + \sum_{k=2}^{\infty} \frac{1}{k!} \mathcal{L}_T^{k-1} \frac{\partial T}{\partial \Lambda_i'''} \\ \Lambda_i^* &= \Lambda_i''' - \frac{\partial T}{\partial \lambda_i'''} - \sum_{k=2}^{\infty} \frac{1}{k!} \mathcal{L}_T^{k-1} \frac{\partial T}{\partial \lambda_i'''} \end{aligned}\right\} \quad (i=1,2,\ldots,n)$$

and

$$\left.\begin{aligned} \omega_r^* &= \omega_r + \frac{\partial T}{\partial \Omega_r} + \sum_{k=2}^{\infty} \frac{1}{k!} \mathcal{L}_T^{k-1} \frac{\partial T}{\partial \Omega_r} \\ \Omega^* &= \Omega_r - \frac{\partial T}{\partial \omega_r} + \sum_{k=2}^{\infty} \frac{1}{k!} \mathcal{L}_T^{k-1} \frac{\partial T}{\partial \omega_r} \end{aligned}\right\} \quad (r=1,2,\ldots,n) \tag{5.33}$$

where T is chosen to make the new Hamiltonian function H* independent of the angles ω^* as well as of the angles λ^*. (The small parameter in powers of which T is expanded is in fact chosen to be of the size of the largest of the Ω_r, and so is of the size of the largest of the squares of the eccentricities and of the $\sin^2(\frac{1}{2} I_i)$. For more details of the construction of this transformation see Message (1982a) pages 35-38). In this final form of the problem, all the coordinates λ^* and ω^* are ignorable, so that their conjugate momenta Λ^* and Ω^* are all constants of the motion, and the angles λ^* and ω^* are linear functions of the time. Then, by proceeding back through the inverses of the successive transformations (5.32), (5.24), (5.15), and (5.5), we may obtain explicit formal series expressions for the original Kepler orbital elements entirely in terms of periodic terms, some of the frequencies being those of the orbital motions, others those of the secular variations, and the remainder are linear combinations of these.

At first sight this result appears to promise a means of establishing the continuation for an indefinite time of regular motion in a planetary system. However, the expression (3.19) for W_1, and the corresponding ones for W_2, W_3, etc., consists of terms each of which has a denominator of the form

$\nu_j = \sum_{j=1}^{n} j_{1i} \; n_i$, where the j_{1i} are positive or negative integers, and the n_i are the mean orbital motions of the planets, and the sum may in principle contain corresponding to any set of integers $j = (j_{1i};\, j_{2i};\, j_{3i})$ for which the condition (2.19), i.e.

$$\sum_{i=1}^{n} (j_{1i} + j_{2i} + j_{3i}) = 0 \quad ,$$

is satisfied. Now, whatever the values of the n_i, clearly integers (j_{1i}) may always be chosen to make ν_j as close to zero as we wish. Consequently the series (3.19) cannot be convergent uniformly in the n_i, and so in the Λ_i, which prevents any attempt to provide rigorous analytical proof of statements about the long-term evolution of the planetary system on the basis of this method. Nevertheless, it is found that planetary theories constructed in this and similar ways are much more successful in predicting planetary positions, and in facing test by numerical integration of the equations of motion, than we would be entitled to claim on the basis of any error bounds constructed on the basis of strict analysis. The practical utility of the expressions is in fact saved by the d'Alembert property, for small enough values of the orbital eccentricities and inclinations, as explained at the end of section 4, these quantities being in fact small for the orbits of the eight major planets. Difficulties do in fact arise if we have a close small-integer commensurability or orbital periods, as happens for some minor planets, and we now turn to this case.

6. THE CASE OF CLOSE SMALL-INTEGER COMMENSURABILITY

Consider then the case where there is a small-integer commensurability of orbital periods, close enough so that the expressions (4.1) to (4.8) do not even approximate to the actual perturbations in the orbital elements. This is sometimes called the case of "deep resonance". Let us consider in some detail one of the simpler types of situation in which this occurs, by restricting our attention to a system with two planets only (so that n = 2), in which the motion relative to the primary is confined to a single plane. (Thus we are studying the general problem of three bodies in the plane). If that plane is used as reference plane, then the orbital inclinations are always zero, and the mutual distance P_1P_2 and the disturbing function R, and the Hamiltonian function H, are functions only of the coordinates $\lambda_1, \lambda_2, \tilde{\omega}_1$, and $\tilde{\omega}_2$, and of their conjugate momenta $\Lambda_1, \Lambda_2, \pi_1$, and π_2, respectively, so that we have a dynamical system of four degrees of freedom. (The node longitudes are in fact undefined). Suppose then that there is a near-commensurability relation of the form

$$(p+q)n_2 - p\, n_1 \approx 0, \tag{6.1}$$

for some small integers p and q. Then any term of the disturbing function expansion (2.17), (5.3) which has

$$j_{12} = s(p+q) \quad \text{and} \quad j_{11} = -s\,p \tag{6.2}$$

for any non-zero integer s is called a "critical" term, and such a term must be excluded from the process of forming W_1, because of the small denominator $(p+q)\,n_2 - p\,n_1$. A modification of the procedures of sections 3 and 5 is therefore necessary.

In order to focus attention on the critical terms, let us use the set of coordinates (as used in Message (1982b))

$$\phi = \lambda_2 - \lambda_1$$

$$\theta_1 = (p+q)\,\lambda_2/q - p\,\lambda_1/q - \tilde{\omega}_1,$$

$$\theta_2 = (p+q)\,\lambda_2/q - p\,\lambda_1/q - \tilde{\omega}_2,$$

$$\text{and} \quad \chi = \frac{1}{2}\,(\lambda_1 + \lambda_2)\,, \tag{6.3}$$

to which the conjugate momenta are, respectively,

$$\Phi = \frac{1}{2}\,(\Lambda_2 - \Lambda_1) + (2p+q)(\pi_1 + \pi_2)/(2q),$$

$$\Theta_1 = -\,\pi_1\,,$$

$$\Theta_2 = -\,\pi_2\,,$$

$$\text{and} \quad X = \Lambda_1 + \Lambda_2 + \pi_1 + \pi_2\,. \tag{6.4}$$

The general argument of the distrubing function is

$$\begin{aligned} N_j &= j_{11}\,\lambda_1 + j_{12}\,\lambda_2 + j_{21}\,\tilde{\omega}_1 + j_{22}\,\tilde{\omega}_2 \\ &= -j_{21}\,\theta_1 - j_{22}\,\theta_2 - \{\,p j_{12} + (p+q) j_{11}\}\,\phi/q, \end{aligned} \tag{6.5}$$

using the fact that only those terms appear which have

$$j_{11} + j_{12} + j_{21} + j_{22} = 0.$$

Therefore χ is an ignorable coordinate, so that its conjugate momentum X is constant. This is equivalent to the conservation of angular momentum. The argument of a critical term (called a "critical argument") may, because of (6.2), be expressed as

$$N_j = -j_{21}\,\theta_1 - j_{22}\,\theta_2 \tag{6.6}$$

with

$$j_{21} + j_{22} = -\,j_{11} - j_{22} = q\,s. \tag{6.7}$$

The secular terms in this problem are those whose arguments are multiples of

$$\tilde{\omega}_2 - \tilde{\omega}_1 = \theta_1 - \theta_2$$

so that all the secular and critical arguments are expressible in terms of the θ_i only.

As in section 5, to avoid singularities at zero values of the eccentricities, we use the rectangular-type coordinates

$$\xi_i = \surd(2\Theta_i)\cos\theta_i = \surd(\Lambda_i)\varepsilon_i \cos\theta_i , \ (i=1,2), \tag{6.8}$$

to which the conjugate momenta are

$$\eta_i = \surd(2\Theta_i)\sin\theta_i = \surd(\Lambda_i)\varepsilon_i \sin\theta_i , \ (i=1,2), \tag{6.9}$$

respectively. Thus the non-ignorable coordinates are now ϕ, ξ_1, and ξ_2, and their conjugate momenta are, respectively, Φ, η_1, and η_2. In terms of these

$$R = \sum_{j_1} (B_{j1} \cos N_{j1} + C_{j1} \sin N_{j1}) ,$$

where

$$N_{j1} = \{(p+q)j_{11} + p\, j_{12}\}\, \phi/q, \tag{6.10}$$

the summation being over all pairs of integers $j_1 = (j_{11}, j_{12})$ with $j_{11} \geq 0$. The expressions for the B_{j1} and C_{j1} contain only positive powers of the ξ_i and η_i, because of the d'Alembert property. The secular terms are those with $j_{11} = j_{12} = 0$, and they are expressible entirely in terms of $\xi_1\xi_2 + \eta_1\eta_2$ and the $\xi_i^2 + \eta_i^2$ (which is as in the non-resonance case, though the ξ_i and η_i are defined differently there). The critical terms are those with j_{11} and j_{12} different from zero, but with $N_{j1} = 0$, and (6.7), with the d'Alembert property, shows that the coefficient B_{j1} of a critical term, when expressed in terms of the ξ_i and η_i, has as its term of lowest degree, a term of total degree in the ξ_i and η_i which is a multiple of q.

To remove the non-critical short-period terms, we use the transformation

$$(\phi,\xi_1,\xi_2;\ \Phi,\ \eta_1,\ \eta_2) \longrightarrow (\phi',\xi_1',\ \xi_2'\ ;\ \Phi_1',\ \eta_1',\ \eta_2') \tag{6.11}$$

constructed similarly to (5.5), and so defined by

$$\phi' = \phi + \frac{\partial W'}{\partial \Phi} + \sum_{k=2}^{\infty} \frac{1}{k!} \mathcal{L}_{W'}^{k-1} \frac{\partial W'}{\partial \Phi}$$

$$\xi_i' = \xi_i + \frac{\partial W'}{\partial \eta_i} + \sum_{k=2}^{\infty} \frac{1}{k!} \mathcal{L}_{W'}^{k-1} \frac{\partial W'}{\partial \eta_i} \qquad (i=1,2)$$

$$\Phi' = \Phi + \frac{\partial W'}{\partial \phi} - \sum_{k=2}^{\infty} \frac{1}{k!} \mathcal{L}_{W'}^{k-1} \frac{\partial W'}{\partial \phi}$$

$$\eta_i = \eta_i - \frac{\partial W'}{\partial \xi_i} - \sum_{k=2}^{\infty} \frac{1}{k!} \mathcal{L}_{W'}^{k-1} \frac{\partial W'}{\partial \xi_i} \qquad (i=1,2) \qquad (6.12)$$

with inverse given by

$$\phi = \phi' - \frac{\partial W'}{\partial \Phi'} + \sum_{k=2}^{\infty} \frac{(-1)^k}{k!} \mathcal{L}_{W'}^{k-1} \frac{\partial W'}{\partial \Phi'}$$

$$\xi_i = \xi_i' - \frac{\partial W'}{\partial \eta_i'} + \sum_{k=2}^{\infty} \frac{(-1)^k}{k!} \mathcal{L}_{W'}^{k-1} \frac{\partial W'}{\partial \eta_i'} \qquad (i=1,2)$$

$$\Phi = \Phi' + \frac{\partial W'}{\partial \phi'} - \sum_{k=2}^{\infty} \frac{(-1)^k}{k!} \mathcal{L}_{W'}^{k-1} \frac{\partial W'}{\partial \phi'}$$

and

$$\eta_i = \eta_i' + \frac{\partial W'}{\partial \xi_i'} - \sum_{k=2}^{\infty} \frac{(-1)^k}{k!} \mathcal{L}_{W'}^{k-1} \frac{\partial W'}{\partial \xi_i'} \qquad (i=1,2) \qquad (6.13)$$

with W' and the new Hamiltonian function H' constructed as in (5.7), (5.8), (5.9), (5.10), (5.11), (5.12), and (5.13), with the difference that $\tilde{R}_i$ consists of the non-critical short-period terms of R_i, and, in (5.13), only non-critical short-period terms are taken. For such terms

$$\nu_{j1} = \{(p+q)j_{11} + pj_{12}\} (n_1 - n_2)/q$$

is not small until larger values of $|j_{11}|$ and $|j_{12}|$, and for these the coefficients B_{j1} and C_{j1} have the small factors described at the end of section 4.

Consider first commensurabilities with $q \geq 3$. Since the critical terms are of degree at least q in the ξ_i and η_i , the terms in H' of lowest degree in the ξ_i and η_i are, as in (5.14), the secular terms of degree two:

$$H_Q' = \sum_{(j,k)} K_{jk} (\xi_j' \xi_k' + \eta_i' \eta_k') \qquad (6.14)$$

the summation being over the pairs (1,1), (1,2), and (2,2). We therefore use a transformation corresponding to (5.8):

$$(\xi_1', \xi_2' ; \eta_1', \eta_2') \longrightarrow (\xi_1'', \xi_2'' ; \eta_2'', \eta_2'') \qquad (6.15)$$

defined by

$$\xi_i' = \sum_{j=1}^{2} a_{ij} \xi_j'' \quad ; \quad \eta_i' = \sum_{j=1}^{2} a_{ij} \eta_j'' \quad (i=1,2) \qquad (6.16)$$

so as to put the quadratic part of the Hamiltonian function into the form

$$H''_Q = \sum_{j=1}^{2} \frac{1}{2} A_j(\xi''^2_j + \eta''^2_j) , \tag{6.17}$$

and then we proceed as in (5.24) and the following part of section 5, to complete the formal solution of the problem, as series in periodic terms only.

If however q = 1 or 2, other preparatory transformations are necessary, since there will be the critical terms involving

$$\varepsilon_i^2 \cos 2\theta_i = (\xi_i^2 - \eta_i^2)/\Lambda_i \qquad (i=1,2)$$

and $$\varepsilon_1\varepsilon_2 \cos(\theta_1 + \theta_2) = (\xi_1\xi_2 - \eta_1\eta_2)/\sqrt{(\Lambda_1\Lambda_2)}, \tag{6.18}$$

and, for q = 1, also the terms in the

$$\varepsilon_i \cos \theta_i = \xi_i/\sqrt{(\Lambda_i)}. \tag{6.19}$$

In the latter case, we first remove the terms linear in the ξ'_i . Poincaré (1893) Chapter XII , section 145, shows how we may proceed. (Note that the "Reducing Transformation" (Henrard et al. 1986) removes one of the two linear terms). Since H' has terms linear in the ξ'_i , the origin $\xi'_i = \eta'_i = 0$ is not a stationary point. Suppose

$$\xi'_i = \xi'_{io} , \quad \eta'_i = 0 \qquad (i=1,2) \tag{6.20}$$

is stationary point of H'. It corresponds to an equilibrium point in the long-period problem, and in the full problem to a periodic solution which is of Poincaré's second sort in the case of very close commensurability, on a continuous family of periodic solutions whose members of only moderately close commensurability are of Poincaré's first sort. The values of the ξ'_{1o} will depend on the adelphic constant Φ' . We carry out a transformation

$$(\phi'; \xi'_1,\xi'_2; \Phi'; \eta'_1,\eta'_2) \longrightarrow (\hat{\phi} , \hat{\xi}_1, \hat{\xi}_2; \hat{\Phi}, \hat{\eta}_1,\hat{\eta}_2) \tag{6.21}$$

such that

$$\hat{\xi}_i = \xi'_i - \xi'_{io} , \quad \hat{\eta}_i = \eta'_i \qquad (i=1,2) \tag{6.22}$$

We find that the transformation is canonical provided that

$$\hat{\Phi} = \Phi'$$

and $$\hat{\phi} = \phi' - \sum_{i=1}^{2} \hat{\eta}_i \frac{d\xi'_{io}}{d\Phi'} . \tag{6.23}$$

Then $\hat{\xi}_i = \hat{\eta}_i = 0$ is a stationary point of the Hamiltonian function, which therefore will have no terms linear in the $\hat{\xi}_i$; the case q = 1 is thus reduced by this transformation to the same form as the case q=2, in which the critical terms of lowest degree in the ξ'_i and η'_i are of degree two. The case q=1 may then be treated in the same way as the case q=2, to which we now pass.

In the case q=2, the presence of critical terms of the type (6.18) implies that the terms in H' of lowest degree in the ξ'_i and η'_i will be of the form

$$H'_Q = \sum_{(j,k)} (K_{jk}\xi'_j\,\xi'_k + L_{jk}\,\eta'_j\,\eta'_k) \tag{6.24}$$

where the matrices (K_{jk}) and (L_{jk}), will now be different, so we cannot proceed as in (6.15) to put H'_Q into the form (6.17). A sequence of transformations is now necessary. We may choose the matrix (K_{jk}) to be symmetric, so that an orthogonal matrix (p_{ij}) can be found to reduce it to diagonal form, that is, a matrix (p_{ij}) so that

$$\sum_{j=1}^{2}\sum_{k=1}^{2} p_{ji}\,K_{jk}\,p_{k\ell} = \delta_{i\ell}\,\lambda_i \qquad (i=1,2;\ \ell = 1,2)$$

and

$$\sum_{j=1}^{2} p_{ji}\,p_{j\ell} = \delta_{i\ell} \tag{6.25}$$

Then the λ_i are the eigenvalues of (K_{jk}), and we suppose them to be positive. Then put

$$\xi''_i = \sqrt{\lambda_i}\sum_{j=1}^{2} p_{ji}\,\xi'_j \qquad \text{and } \eta''_i = (1/\sqrt{\lambda_i})\sum_{j=1}^{2} p_{ji}\,\eta'_j \qquad (i=1,2) \tag{6.26}$$

so that

$$(\xi'_1,\xi'_2;\eta'_1,\eta'_2) \longrightarrow (\xi''_1,\xi''_2;\eta''_1,\eta''_2) \tag{6.27}$$

is a canonical transformation. The new Hamiltonian function H" has as its terms of lowest degree

$$H''_Q = \sum_{j=1}^{2} \xi''^2_j + \sum_{j=1}^{2}\sum_{k=1}^{2} C_{jk}\,\eta''_j\,\eta''_k \tag{6.28}$$

where

$$C_{jk} = \sqrt{(\lambda_j\lambda_k)}\sum_{\ell=1}^{2}\sum_{m=1}^{2} p_{\ell j}\,L_{\ell m}\,p_{mk}\,. \tag{6.29}$$

Since (C_{jk}) is symmetric, it also may be reduced by an orthogonal transformation to diagonal form, that is, there is a matrix (q_{ij}) so that

$$\sum_{j=1}^{2}\sum_{k=1}^{2} q_{ji}\,C_{jk}\,q_{k\ell} = \delta_{i\ell}\mu_i$$

and

$$\sum_{j=1}^{2} q_{ji}\, q_{j\ell} = \delta_{i\ell} \, . \qquad (i=1,2; \quad \ell=1,2) \qquad (6.30)$$

Then carry out the further canonical transformation

$$(\xi_1'' , \xi_2'' ; \eta_1'' , \eta_2'') \longrightarrow (\xi_1''' , \xi_2''' ; \eta_1''' , \eta_2''') \qquad (6.31)$$

given by

$$\xi_i''' = \sum_{j=1}^{2} q_{ji}\, \xi_j''/(\mu_i)^{\frac{1}{4}} \quad ; \quad \eta_i''' = \sum_{j=1}^{2} q_{ji}\, \eta_j''\, (\mu_i)^{\frac{1}{4}} \qquad (6.32)$$

so that the new Hamiltonian function H''' has as its terms of lowest degree

$$H'''_Q = \sum_{j=1}^{2} \surd(\mu_j)\, (\xi_j'''^2 + \eta_j'''^2) \, . \qquad (6.33)$$

This is now of the form from which a transformation of the type (5.22) may be carried out, so that the formal solution of the problem in series entirely of periodic terms may be carried out as in the latter part of section 5. (Note that the condition that the $\surd(\mu_i)$ be real is the same as the condition that $\xi_i' = \eta_i' = 0$ be a stable equilibrium point, that is, that the periodic solution about which we are in fact developing the solution is a stable one.

REFERENCES

Henrard, J., Lemaitre, A., Milani, A., and Murray, C.D., 1986: "Celestial Mechanics", Vol.38, pp. 335-344.

Message, P.J., 1982a: Celestial Mechanics, Vol.26, pp.25-39.

Message, P.J., 1982b: in Application of Modern Dynamics to Celestial Mechanics and Astrodynamics, (Proceedings of the N.A.T.O. Advanced Study Institute, 1981) ed. V. Szebehely, (Reidel) pp.77-101.

Poincaré, H. 1893 Les Methodes Nouvelles de la Mécanique Célèste, (Gauthier-Villers).

SECULAR PERTURBATIONS OF PLANETARY ORBITS AND THEIR REPRESENTATION AS SERIES

Andrea Milani
Space Mechanics group
Department of Mathematics
University of Pisa
Via Buonarroti 2
I-56100 Pisa, Italy

ABSTRACT The long term changes of the orbital elements of the planets are described by secular perturbation theories. After a short historical discussion, the secular perturbation equations are derived by means of the formalism of the Lie series transformations. To solve the classical problem of the long term changes in the major semiaxes second order effects have to be computed. Such a computation is feasible within a modern formalism and succeeds in matching the data from the numerical integrations over time spans of a few million years. As for the long term changes in the eccentricities and inclinations, they can be computed by means of higher degree theories. However the time span over which the latter apply cannot be increased at will. This because of the divergence of the perturbative series, a fundamental property of a non-integrable system such as the N-body problem and not only a technical difficulty. Numerical integrations are therefore an essential tool both to assess the reliability of any analytic theory and to provide data on the fundamental frequencies of the secular system and on the occurence of secular resonances. Examples of this use are taken from the LONGSTOP integrations of the outer planets for 100 million years.

1 Introduction

The long term evolution of the orbital elements of the planets being one of the oldest subjects of investigation in Celestial Mechanics, it is better to look at the current research efforts in the appropriate historical perspective.

As for the major semiaxes of the planetary orbits, it was already shown by the italian mathematician of the XVIII century Lagrange that they undergo no secular perturbations, provided the terms in the differential equations of motion of the second order in the planetary masses can be neglected. In the XIX century it was shown that no secular perturbation on the major semiaxes occurs even taking into account the second order terms (Tisserand, 1876) but the same result does not extend to the third order (Eginitis, 1889). The masses of the planets m_i being smaller than 1/1000 (having set the mass of the Sun $m_\odot = 1$), the effect of these third order terms is either very small or very slow, or both, and for computations of the orbits of the planets over the time span of the astronomical observations these terms can be entirely

A. E. Roy (ed.), Long-Term Dynamical Behaviour of Natural and Artificial N-Body Systems, 73–108.
© 1988 by Kluwer Academic Publishers.

neglected. However, in the XIX century it was also realised that the age of the Solar System is larger by six orders of magnitude than what was firmly believed at the times of Lagrange. Thus whenever the problem is the orbital evolution of the planets over their entire lifetime even the tiny third order effects cannot be neglected.

To assess the predictive value of the results concerning the *secular perturbation theories* over such a timescale, we need to take into account the fact that they actually assume that the solution of the equations of motion can be represented in the form of series. Let ε be a small parameter, expressing the size of perturbation; in the case of the mutual perturbations of the planets, ε contains the small factors m_i . Then it is assumed that the equations of motion and their solutions can be represented by a power series in ε with the order k portion – the coefficient of ε^k -- represented as a Fourier series in a number of fundamental arguments, which are in turn linear functions of the time. Such *formal series* can be handled more or less in the same way as the ordinary functions of time by the usual tools of calculus, and in particular it is always possible to integrate such a series term by term in an explicit way:

$$\int A \ sin(\alpha t + \sigma)dt = -\frac{A}{\alpha}cos(\alpha t + \sigma) + const \qquad (1)$$

In this setting, the statement due to Tisserand says that the formal series describing the time derivative of the major semiaxes a_i is of the form:

$$\frac{da_i}{dt} = \varepsilon g_1(t) + \varepsilon^2 g_2(t) + \varepsilon^3 g_3(t) + \ldots \qquad (2)$$

with g_1 and g_2 Fourier series containing only *short periodic terms*, that is the ones with periods of the order of the periods of the longitudes of the planets. However g_3 can contain terms with long periods of the order of the periods of the perihelia and the nodes (they are called *secular*), and the assumption that the value of g_3/a is of order unity is by no means justified. Indeed the series expansions of perturbation theory are often plagued by *small divisors* which are nothing else but the α, appearing under the fraction line in the right hand side of equation (1), whenever the period $2\pi/\alpha$ of the term is very long.

One such small divisor immediately appears in formula (2), although its relevance was pointed out quite recently (Message 1976): if g_3 contains a term with a secular period, that is with an argument νt ($\nu = O(\varepsilon)$, a slow frequency), then by integrating:

$$\int \varepsilon^3 g_3(t)dt = O(\varepsilon^2) \qquad (3)$$

and Tisserand theorem does not rule out changes in the major semiaxes of order ε^2, that is small but not negligibly small over timescales of millions of years.

To better understand these subtle effects one needs a tool which was not available to XIX century Celestial Mechanics, namely the possibility to compare the formal series with *observations*. Since by definition secular perturbations theories refer to the average behaviour over timescales longer than the duration of human civilisation, only computer simulations can provide the conceptual analogue of astronomical observation over timescales of millions of years. Numerical integrations of the orbits of the planets for many millions of years have become possible, at least for the outer planets (Jupiter to Pluto), in recent years (Kinoshita and Nakai, 1984; Milani et al., 1986; Applegate et al., 1986). The results of these integrations have been shaking many long held beliefs on the long term behaviour of the planetary orbits, the most striking case being that of the major semiaxes: the latter change their average value by a quantity which is far larger than the purely formal argument (2)-(3) would suggest.

In the next two Sections we shall try and explain how this is possible, and show that the most advanced versions of perturbation theory, namely the ones based upon the machinery of canonical transformation to normal form and Lie series, can explain this apparent anomaly by pointing to the relevant small divisor which makes the formal order estimate wrong as an order of magnitude estimate.

However, the success in explaining some subtle second order effects does not mean that analytical Celestial Mechanics is safe from further blows by the hands of supercomputers. When data for even longer integrations are analysed with sophisticated post-processing methods, as it was the case for the 100 million years integration of the outer planets *LONGSTOP 1B*, the very existence of analytical theories capable of predicting the orbital behaviour over such a long span of time is challenged in a more fundamental way.

To see the origin of the trouble, let us consider the other orbital elements,namely the eccentricities e_i and inclinations I_i and the corresponding angles ϖ_i and Ω_i (longitudes of the perihelia and the nodes). For these elements, even when the short periodic terms are averaged out, there are secular changes to every order in ε; the solution of first approximation for the secular perturbations is the *Lagrange solution*, and it is already a useful approximation to the behaviour of the planetary orbits over a few thousand years. However over a longer time span the Lagrange solution does not represent any more the orientation of the apsidal lines of the orbits; the frequency of precession of the perihelion of Saturn turns out to be wrong by about 20% and within a few tens of thousands years the analytical Lagrange solution and the output of a numerical integration are entirely out of phase for ϖ_6.

This discrepancy can be greatly reduced, if not entirely eliminated, by the use of an analytical theory of higher order which takes into account terms in the series expansion of the solution containing either higher powers of ε, or higher powers of the eccentricities and inclinations, or both. Again such a theory can be described in a very compact way within the formalism of Lie series transformations, and a somewhat simplified account of such a theory is presented in Section 4. There are significant problems of computational complexity (the number of terms in the series grows very fast with the order) but it is possible, as a matter of principle, to give an explicit series expansion of the solutions such that they formally solve the secular perturbation equations apart from remainders which are of very high degree in the small parameters, and therefore are supposed to be very small.

However these series solutions of the secular perturbation equations also contain small divisors. They arise in the same way described above, namely because of the term by term integration (1), but in this case the frequency α which appears as denominator is a combination of the secular frequencies; hence it is of order ε to start with, and possibly much smaller close to a *secular resonance*. In the process of solving recursively the equations of the secular perturbations with remainders of higher and higher order, higher and higher negative powers of the small denominators are accumulated in the remainder terms, until the smallness of the latter cannot be guaranteed any more.

This argument is related to the one used, in a more rigorous form, in the proofs of the famous theorems of Poincaré which state that there are not enough integrals to express in an explicit way the solutions of the equations of motion for the N-body problem for an arbitrarily long span of time. It is enough to point out that these integrals, being non-existent, cannot be represented by a series expansion unless the series are divergent, to see how deeply rooted is the trouble (Section 5).

The series expansions of perturbation theories, whatever the formalism, cannot have at the same time the three properties of being convergent, of describing a solution passing through a given initial condition, and of approaching the solution in a recursive way starting from a first approximation. The series expansions of the classical perturbation theories are always divergent; the more complicated series expansions of the Kolmogorov theory are sometimes convergent, but they cannot be used to find a solution with given initial condition. Whether it is possible to have a series expansion which is convergent and with fixed initial conditions is not known, but it is clear that it must be computed in an entirely new way; some possible new methods are discussed in Section 6.

These new methods are however applicable only provided considerable computer resources are availble. This is not only because sophisticated higher-order theories can be implemented only within a computer algebra system, but also because numerical integration is needed to assess the source and the seriousness of the danger, namely to monitor the smallness of the divisor and the occurrence of secular resonances, by computing the fundamental frequencies.

Being unable so far to compute explicitly the size of the remainder terms in the analytical theories (unless the frequencies are known, as in Giorgilli et al., 1987 and Simó, these proceedings) how do we know whether the problem of the divergence of the perturbative series is, for a specific dynamical problem and for a given span of time, a serious one or only a mathematical curiosity? To state the problem in another way, how do we know whether the remainder of a given theory is going to increase or to decrease when the same theory is pushed to a higher order?

The method we have developed within the *LONGSTOP* project to answer this question involves an unusual combination of numerical and analytical techniques to generate a *synthetic theory*. The latter summarises the output of a numerical integration in a form directly comparable to that of a hypothetical convergent analytical theory. Then a very good indication, if not a rigorous proof, of the character of the motion is given by the rate of decrease, or increase, of the terms of the synthetic theory with the order. When such a theory is constructed from the output of the *LONGSTOP 1B* 100 million years integration of the outer Solar System, the size of the terms does not appear to decrease at all. This can be interpreted as an indication that the theoretical divergence of the series is already relevant over such a timescale, although the effects are small (Section 6). Thus the *predictability horizon* (Lighthill, 1986) of the Solar System might have already been reached and the conceptual and computational tools of chaotic dynamics might be soon needed even for the computation of planetary orbits.

2 Averaging and Elimination of Short Periodic Terms

As a matter of principle, all the methods for constructing perturbation theories are more or less the same. However some are more convenient than others, and for such a complicated theory, as the one needed to describe the planets perturbing each other, a more convenient way to carry out the computation can make an essential difference. We shall therefore use the method of the *Lie series transformations* (Hori, 1966; Message, 1976, 1982) because of the following advantages: 1) it gives directly the series expansion of the solutions without going through the series expansion of the differential equations, that is the integration (1) is already done and this allows us to carry out the expansion to order ε^2 only, if the effect to be studied is of order ε^2, thus avoiding the small divisor problems of the kind (3); 2) it explicitly computes the solution in the form of a double series, a power series in ε and a Fourier series in the fundamental arguments.

To define a Lie series transformation three steps are required: given a canonical coordinate system $(q,p) = (q_1,\ldots,q_s,p_1,\ldots,p_s)$ and a Hamilton function $\chi(q,p)$, the derivative of any other function $H(q,p)$ along the solutions of the dynamical problem defined by χ and parametrised by τ can be computed by means of the well known Poisson bracket:

$$\frac{dH}{d\tau} = \sum_i \left(\frac{\partial H}{\partial q_i}\frac{\partial \chi}{\partial p_i} - \frac{\partial H}{\partial p_i}\frac{\partial \chi}{\partial q_i}\right) = \{H,\chi\}. \tag{4}$$

The second derivative of H can be computed as a double Poisson bracket:

$$\frac{d^2H}{d\tau^2} = \{\{H,\chi\},\chi\} \tag{5}$$

and so on. The second step is to follow the solutions of the Hamilton equations defined by χ from the initial conditions (q',p') at $\tau = 0$ to the state (q,p) at $\tau = 1$; this flow of χ is a canonical map:

$$F_\chi : (q',p') \longrightarrow (q,p). \tag{6}$$

Every function defined on the (q,p) space can be transformed by composition with F_χ into a function on the (q',p') space:

$$H(q,p) = H(F_\chi(q',p')) = T_\chi H(q',p') \tag{7}$$

with the operator T_χ acting on the function spaces in a contravariant way, that is it goes in the opposite direction with respect to F_χ . For instance the coordinate function q_i is transformed into $T_\chi q_i = q_i(F_\chi(q',p'))$, that is the expression of q_i as a function of (q',p'), which is not the coordinate function q'_i. Since the map F_χ is canonical, if there is a dynamics in the (q,p) space defined by the Hamiltonian H, then in the (q',p') space the same dynamics is defined by the transformed Hamiltonian $T_\chi H$.

The third step is the usual Taylor series formula, used to compute the transformed Hamiltonian $T_\chi H$ by expanding in a power series in the fictitious time τ, then setting $\tau = 1$; the first derivative of H computed on $F_\chi(q',p')$ is the Poisson bracket (4), the second derivative is the double Poisson bracket (5), etc.:

$$T_\chi H(q',p') = H(q',p') + \{H(q',p'),\chi(q',p')\} + \frac{1}{2}\{\{H,\chi\},\chi\} + \ldots = H'(q',p') \tag{8}$$

Setting aside (until Section 5) all the convergence problems, (8) is a series expansion of the transformed Hamiltonian. The purpose of the exercise is as follows: since we can choose χ as we wish, can we find a χ such that the transformed Hamiltonian $H'(q',p')$ is much simpler than the original $H(q,p)$? If we can do this in such a way that an explicit solution $q'(t),p'(t)$ can be computed in the (q',p') coordinates, then the solution of the original problem in the (q,p) space is explicitly given by the same transformation:

$$q_i(q',p') = T_\chi q_i = q'_i + \{q'_i,\chi\} + \ldots = q'_i + \frac{\partial\chi}{\partial p'_i} + \ldots$$

$$p_i(q',p') = T_\chi p_i = p'_i - \frac{\partial\chi}{\partial q'_i} - \frac{1}{2}\{\frac{\partial\chi}{\partial q_i},\chi\} - \ldots \tag{9}$$

Let us apply this method of the normal form to the problem of N planets orbiting around the Sun. The Hamilton function of this problem has two essential properties: firstly, it is of the form $H = H_0 + \varepsilon H_1$ with H_0 an integrable system and ε a small parameter; H_0 describes the motion of the planets around the Sun as N uncoupled 2-body systems and in a suitable action-angle coordinate system is a function of the actions only. Moreover H_0 does not really depend upon all the actions, but is highly *degenerate*: let $p = (\Lambda, \Theta)$ with Λ an N-dimensional vector containing the variables related to the major semiaxes, and Θ a 2N-dimensional vector with the variables related to the angular momenta, hence to the eccentricities and inclinations. Then $H_0 = H_0(\Lambda)$ while H_1 is a function of all the momenta as well as of all the conjugate angular coordinates $q = (\lambda, \theta)$; the λ_i are the mean longitudes, while the θ_i are the longitudes of the nodes and the perihelia:

$$H = H_0(\Lambda) + \varepsilon H_1(\lambda, \theta, \Lambda, \Theta). \tag{10}$$

We shall now expand the *generating function* χ in positive powers of ε:

$$\chi = \varepsilon\chi_1 + \varepsilon^2\chi_2 + \ldots \tag{11}$$

compute $H' = T_\chi H$ by (8) and reorder by the powers of ε:

$$H' = T_\chi H = H + \{H, \chi\} + \frac{1}{2}\{\{H, \chi\}, \chi\} + \ldots =$$

$$= H_0 + \varepsilon[\{H_0, \chi_1\} + H_1] + \varepsilon^2[\{H_0, \chi_2\} + \{H_1, \chi_1\} + \frac{1}{2}\{\{H_0, \chi_1\}, \chi_1\}] + \ldots \tag{12}$$

Our purpose is to simplify H' as much as possible, proceeding order by order: to order ε^1 we would like to choose χ_1 in such a way that $\{H_0, \chi_1\} + H_1$ is as simple as possible. To find the simplest normal form, we expand both the known function H_1 and the function to be determined χ_1 in a Fourier series with arguments λ and θ :

$$H_1 = \sum_{j,k} H_{1jk}(\Lambda, \Theta) cos(j \cdot \lambda + k \cdot \theta) \tag{13}$$

$$\chi_1 = \sum_{j,k} \chi_{1jk}(\Lambda', \Theta') sin(j \cdot \lambda' + k \cdot \theta') \tag{14}$$

where j and k are vectors with integer coefficients. Since the solution to order 0, i.e. the uncoupled 2-body solutions, are given by $\lambda = nt + const$ with the mean motions n_i :

$$n_i = \frac{\partial H_0}{\partial \Lambda_i}(\Lambda) \tag{15}$$

the Poisson bracket with H_0 is easy to compute:

$$\{H_0, \chi_1\} + H_1 = -n \cdot \frac{\partial \chi}{\partial \lambda} + H_1 \tag{16}$$

and formally term by term in the Fourier series (13) and (14) the equation to be solved is:

$$(n \cdot j)\chi_{1jk} = H_{1jk}. \tag{17}$$

From (17) we can conclude : 1) the terms of the form $H_{10k}\ cos(k \cdot \theta)$ cannot be eliminated by any choice of χ_1: they are the long periodic terms because only the slow angles θ appear; the coefficients χ_{10k} can be chosen $= 0$. 2) provided that no resonance $n \cdot j = 0$ for $j \neq 0$ occurs, it is possible to find a formal Fourier series for χ_1, such that $T_\chi H = H_0 + \varepsilon \overline{H_1} + \ldots$ with $\overline{H_1}$ the portion of H_1 not containing λ.

Having done this, we can look at the order ε^2 part of (12) and choose χ_2 in such a way that:

$$\{H_1, \chi_1\} + \frac{1}{2}\{\{H_0, \chi_1\}, \chi_1\} = H_{1,1} \tag{18}$$

$$\{H_0, \chi_2\} + H_{1,1} = \overline{H_{1,1}} \tag{19}$$

where again the overline denotes the long periodic part; equations like (17) will result, and the formal Fourier series for χ_2 will be such that:

$$H' = T_\chi H = H_0 + \varepsilon \overline{H_1} + \varepsilon^2 \overline{H_{1,1}} + \ldots \tag{20}$$

The procedure is recursive, in the sense that having chosen all the χ_j for $1 \leq j \leq k$ it is possible to solve for χ_{k+1} without recomputing all the previously calculated terms, and it provides a formal representation for χ as a double series in powers of ε and in trigonometric polynomials in λ, θ.

Again neglecting any convergence problem, we obtain:

$$T_\chi H = H'(\theta', \Lambda', \Theta') \tag{21}$$

that is H has been reduced to a normal form in which the longitudes do not appear.

3 Secular Perturbations on the Major Semiaxes

The dynamical problem defined by the Hamilton function H' of (21) is the problem of the *secular perturbations*; in this reduced problem in the space of the primed variables the λ' are cyclic and the Λ' are integrals, the significant dynamics reduces to the changes in (θ', Θ') which occur very slowly (the derivatives are $O(\varepsilon)$ or smaller).

From $\Lambda_i' = const$, can we conclude that Λ_i -hence the major semiaxes a_i- undergo only short periodic perturbations and are on the average constant? This would follow from the applicability of the principle of the averaging, which -loosely stated- says that the average of the solution is close to the solution of the averaged equations (for a precise statement, see Arnold, 1976, Chap. 10). Strictly speaking, the Hamiltonian of the secular perturbations $T_\chi H = H'$ is not the average $\overline{H} = H_0 + \varepsilon \overline{H_1}$ of the Hamiltonian, even if the two coincide up to order 1 (see (20)); however H' does represent the long term behaviour of the solutions. The explicit formula for the osculating elements related to the major semiaxes is:

$$\Lambda_i = \Lambda_i' - \varepsilon \frac{\partial \chi_1}{\partial \lambda_i'} - \varepsilon^2 \frac{\partial \chi_2}{\partial \lambda_i'} - \frac{1}{2}\varepsilon^2 \{ \frac{\partial \chi_1}{\partial \lambda_i'}, \chi_1 \} + \ldots \tag{22}$$

and the order 1 part of this formula has zero average because χ has been chosen with $\overline{\chi} = 0$; this means that the averaging principle is (formally) valid to order 1, and also proves Tisserand's theorem (because in taking the derivative of a secular term one more factor ε appears).

The same formula (22) also shows that the averaging principle fails at order 2 in ε; the essential reason is that the canonical tronsformation defined by χ is nonlinear, thus the short periodic terms introduced by T_χ can resonate with each other and produce long periodic beats. Two terms with arguments $j \cdot \lambda' + k \cdot \theta'$ and $j \cdot \lambda' + h \cdot \theta'$, that is with short periods very close to each other, give rise in the Poisson bracket $\{\partial \chi_1 / \partial \lambda_i', \chi\}$ to a long periodic term with argument $(k - h) \cdot \theta'$; this is Message's (1976) proof of Eginitis' theorem. However one would like to be able to give an explicit order of magnitude estimate of these second order effects, to be compared to the results of numerical integrations.

Since the short periodic first order effects are larger than the long periodic second order effects, to see the latter one would like to average out the former both in the theory expressed by formulas such as (22) and in the output of the numerical integration. As far as the theory is concerned, the overline operator is well defined for every formal Fourier series as the selection of the terms with zero coefficients for the fast variables λ', e.g.:

$$\overline{\Lambda_i} = -\frac{1}{2}\varepsilon^2 \overline{\{ \frac{\partial \chi_1}{\partial \lambda_i'}, \chi_1 \}} + O(\varepsilon^3) \tag{23}$$

As for the solution $\Lambda(t)$ of the full equations of motion there is no guarantee that it can be expanded in a convergent Fourier series; actually, to establish whether it is so is the problem of the stability of the Solar System in the formulation due to Weierstrass and Mittag - Lefter (Moser, 1973, Introduction), and to select the long periodic terms in a formal series which is actually divergent does not provide reliable information on the behaviour of the function $\Lambda(t)$.

The numerical analogue of the overline operator for a time series computed step by step is well known in the theory of digital signal processing: if a low-pass band filter is applied to the step by step output of the numerical integrator, a smoothed time series is obtained, in which the portion of the spectrum corresponding to the fast frequencies and their combinations is almost wiped out (actually all the amplitudes are reduced by a small *response factor*, e.g. 10^{-4}) (Carpino et al., 1987). Digital filters are essentially running averages with optimally chosen weights. They can be built to given specifications; however, in between the *dark band* (where the filter response is very small) and the *pass band* (where the filter response is very close to 1, to minimally distort the interersting long periodic signal) there is a *gap* where the filter response is allowed to smoothly change from close to 1 to close to 0. Since any line in the gap is distorted, the gap should correspond to a region of the spectrum actually void of lines.

Such a gap does actually occur in the frequency spectrum of the planetary elements, since the λ have frequencies $n = O(\varepsilon^0)$ (that is, periods up to a few hundred years) and the θ have frequencies $O(\varepsilon^1)$ (that is, periods longer than 40,000 years). There are some small divisors resulting from small integer combinations of two fast frequencies, such as $2n_5 - 5n_6$ (period about 900 years) and $n_7 - 2n_8$ (period about 4,300 years), but they do not fill the gap because they are few and not too small, i.e. they are not as small as $O(\sqrt{\varepsilon})$ (Pluto is not considered in this discussion, because it is indeed deeply resonant). Otherwise it would indeed be difficult to study the orbits of the planets, and may be difficult to live on them as well, since unstable orbits might result.

When a low-pass band digital filter was applied to the data points computed by numerical integration of the orbits of the outer planets for many millions of years (Nobili and Roxburgh, 1986; Milani et al., 1986), oscillations of the average value of the major semiaxes with periods typical of the secular arguments (e.g. 1,100,000 years) and surprisingly large amplitudes were found (see Figures 1 and 2). Since Celestial Mechanics is the most accurate of all sciences, we often need some explaining to make people working in other fields understand what we mean by a *very large discrepancy between theory and experiment*. These *secular perturbations in the major semiaxes* are indeed small, smaller by two orders of magnitude with respect to the short periodic perturbations (which result from the order 1 part of (22) and are well understood).

The right way to assess how small an effect should be is to use adimensional *small parameters*, and in the problem of the orbits of the planets these arise in three kinds. First there are small parameters, such as the ε employed in the previous formulas, describing how small the perturbation εH_1 is with respect to the integrable portion of the Hamiltonian H_0. They can be computed as combinations of adimensional ratios such as $m_i/m_\odot$ (mass ratio) and a_i/a_k (distance ratio) (Roy, 1979; Walker et al., 1980; Milani and Nobili, 1983); e.g. for the perturbation of Uranus on Neptune the relevant small parameter is:

$$\varepsilon_{78} \simeq (m_7/m_\odot)(a_7/a_8)^2 \simeq 1.8 \times 10^{-5} \tag{24}$$

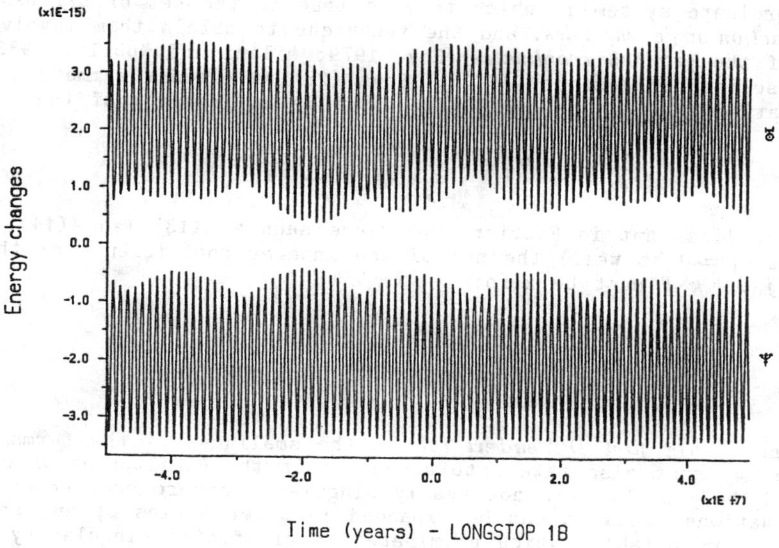

Figure 1 : Changes in the energies of Uranus (top) and Neptune (bottom); the two plots have been separated by subtracting two constants close to the average values of -3.36×10^{-10} and -2.54×10^{-10} $M_{\odot}\ AU^2/day^2$ respectively. The output of the *LONGSTOP 1B* numerical integration has been digitally filtered to eliminate all the effects with periods $\leq 100,000$ years.

while the effect of Neptune on Uranus is $\varepsilon_{87} \simeq 1.35 \times 10^{-5}$ and their square is indeed small! Moreover, the orbits of most planets are almost circular and almost coplanar, and as small parameters of the second kind we can use the eccentricities e_i and the inclinations I_i of the orbits (with respect to a plane perpendicular to the total angular momentum of the Solar system). These adimensional quantities are roughly of the order of $\sqrt{\varepsilon}$ (for some reason yet to be understood, not only the resonant Pluto but also Mercury does not obey this rule). In the expansion of the perturbing potential (13), as well as in the perturbative series (14), the powers of the eccentricities and the inclinations occur according to some special rules named after D'Alembert.

The D'Alembert rules are not too difficult to be understood, provided a clever choice is done for the coordinate system in such a way that a component of the total angular momentum c of the Solar system can be expressed as:

$$c = \sum_i \Theta_i + \sum_s \Lambda_s . \tag{25}$$

The coordinate system in which this is true is the set of *dimensional Delaunay action-angle variables*, and the technique to obtain them involves the use of the *jacobian coordinates* (Roy, 1979; Milani and Nobili, 1983) and is described in Message (1982). From (25) an elegant symmetry property arises; since the angular momentum is an integral of the motion :

$$\frac{dc}{dt} = \{c, H\} = -\sum_i \frac{\partial H}{\partial \theta_i} - \sum_s \frac{\partial H}{\partial \lambda_s} = 0 \tag{26}$$

and this implies that in Fourier expansions such as (13) (and (14), because $\{c, \chi\} = 0$ as well) the sum of the integer coefficients of the argument $j \cdot \lambda + k \cdot \theta$ must be zero:

$$\sum_s j_s + \sum_r k_r = 0. \tag{27}$$

To connect this *first D'Alembert rule* to the smallness of the terms in the series we must also take into account that the Hamiltonian H of the $(N+1)$-body problem is not really singular for zero eccentricity and inclinations, thus it can be expanded in power series of any set of nonsingular variables which eliminate the artificial singularity for $e_s = 0$ and $I_s = 0$. As an example the canonical analogue of polar coordinates, the *Poincaré variables*, can be used :

$$\eta_s = \sqrt{-2\Theta_s}\ cos\theta_s$$

$$\xi_s = \sqrt{-2\Theta_s}\ sin\theta_s \tag{28}$$

and they are approximately proportional to either some eccentricity or some inclination, that is the Poincaré variables are $O(\sqrt{\varepsilon})$. Every trigonometric term containing an argument θ can be expressed as a polynomial in the (η, ξ); therefore if the argument is $j \cdot \lambda + k \cdot \theta$ the order is at least $\varepsilon^{d/2}$ where d is the *order of the resonance* j :

$$d = |\sum_s j_s| = |\sum_r k_r|; \tag{29}$$

this is the *second D'Alembert rule*. The reason why a relative change in the major semiaxes of Uranus and Neptune of a few parts in a million looks too much is that a term in (23) follows the same D'Alembert rules, thus if it contains an argument of the form $\theta'_r - \theta'_s$ it also has the factor $e_r e_s$ besides ε^2; the relative size of the effect should then be $O({\varepsilon_{78}}^2 {e_7}^2) \simeq 10^{-12}$, and this formal estimate does not agree with the experimental data!

To solve the riddle of these *very large* secular changes in the semimajor axes we need to take into account the third class of small parameters occurring in the perturbation theories, namely the *small divisors*. The latter do not appear in the expansion of the perturbing potential (13), but only in the expansion of the generating function (14) because of the division to be performed in order to solve (17): that is, they

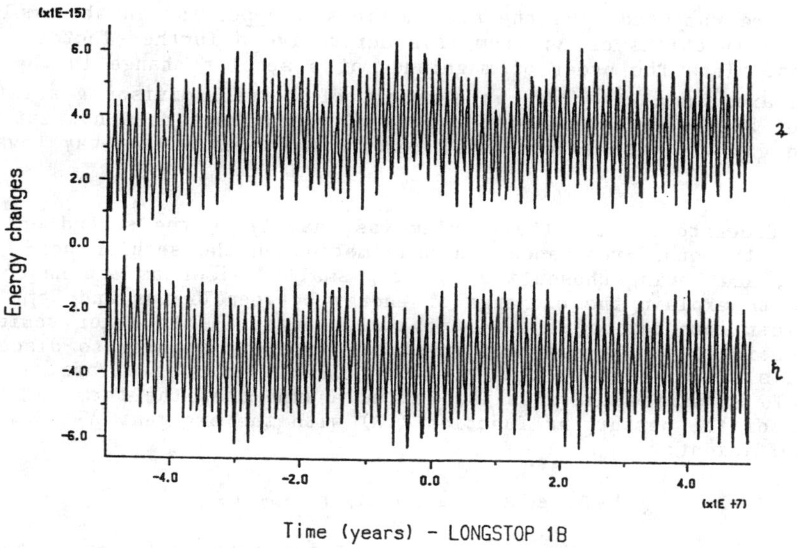

Figure 2 : Changes in the energies of Jupiter (top) and Saturn (bottom), filtered from the output of *LONGSTOP 1B*. The average values are -2.72×10^{-8} and -4.44×10^{-9} $M_{\odot}$ AU^2/day^2 respectively.

are the result of the integration term by term (1). When we form adimensional ratios $(j \cdot n)/n_i$ we find for the most prominent small divisors in the outer Solar System (Pluto excluded):

$$\delta_1 = \frac{2n_5 - 5n_6}{n_5} \simeq \frac{1}{70}$$

$$\delta_2 = \frac{n_7 - 2n_8}{n_7} \simeq -\frac{1}{50} \tag{30}$$

and the terms in χ with these combinations are correspondingly amplified by a factor $1/\delta_i$, forming the *great inequalities*. However a factor 50 or 70 is not enough to close the gap between theory and experiment, as far as the secular perturbations in the major semiaxes are concerned.

A close inspection of formula (23) allows us to find the main point of the solution of this problem, because it appears that the secular perturbations in the major semiaxes arise from the self-resonance of a term in the generating function χ, and if this term contains a small divisor δ then the factor $1/\delta$ will appear squared. Moreover, in computing the Poisson bracket the derivative of the terms with respect to the Λ are computed, and the mean motions n appearing in the small divisor are functions of Λ; from this derivative a further factor $1/\delta$ arises. Thus the order of magnitude of a secular change in the major semiaxes is $O(\varepsilon^2\varepsilon^{d/2}\delta^{-3})$ as a result of a small divisor δ arising because of a quasi-resonace of order d. This estimate does not entirely close the gap with the experimental results but goes a long way toward it.

Having understood where the problem was, namely in the second order effects of the quasi-resonances in mean motion on the secular perturbation, and having chosen the $n_7 - 2n_8$ small divisor as the best candidate to explain the exchange of energy between Uranus and Neptune, the explicit computation of the long term changes in the major semiaxes became possible, albeit somewhat involved; we are not going to discuss the details of such a computation, which can be found in Milani et al. (1987). To summarise the results, let us assume that the term in the expansion of the perturbing function (13) with the argument $\lambda_7 - 2\lambda_8 - \varpi_7$ has a coefficient:

$$\Gamma \ \Lambda_8 \ e_7 \ \varepsilon_{78} \ n_8 = \Gamma \ \Lambda_7 \ e_7 \ \varepsilon_{87} \ n_7 \tag{31}$$

where the number Γ, which is a function of the ratio a_7/a_8 only, needs either a little bit of computer algebra or a little bit of elliptic integrals to be computed (cumbersome, but classical). Then the relative changes in the 2-body energies h_7, h_8 of Uranus and Neptune with a secular argument of period about $1,100,000$ years are given by :

$$\frac{\Delta h_7}{h_7} = -\frac{3\Gamma^2 e_7^2}{\delta_2^3}[\varepsilon_{78}^2 + \frac{\Lambda_8}{2\Lambda_7}\varepsilon_{87}^2]$$

$$\frac{\Delta h_8}{h_8} = \frac{6\Gamma^2 e_7^2}{\delta_2^3}[\frac{\varepsilon_{78}^2}{2} + \frac{\Lambda_7}{\Lambda_8}\varepsilon_{87}^2] \tag{32}$$

The Lie series formalism, coupled with the very compact notation allowed by the Roy $\varepsilon_{i,}$, is very effective in simplifying this computation; but the real crucial test is the comparison of the *prediction* given by this formula with the *observation* provided by the digitally filtered output of the numerical integrations such as the 100 million years *LONGSTOP 1B* used for the Figures 1 and 2. Here the constant Γ of (31) and (32) acts as the last brick which allows the arch to stand: the D'Alembert rules prescribe the powers of the eccentricity to be found in each term, but they do not prescribe that the coefficient of the right power of e_i should be 1. (Actually there is a good argument by which one should expect the coefficients to grow with the degree in the eccentricity). In our case the coefficient Γ turns out to be $\simeq -3.05$, and its appearance squared in (32) provides the last missing order of magnitude: the comparison between the theory, in which a single term has been computed, and the experiments is as follows: for Uranus:

$$|\frac{\Delta h_7}{h_7}| = 3.7 \times 10^{-6}(exper.) = 3.4 \times 10^{-6}(theory) \tag{33}$$

and for Neptune:

$$|\frac{\Delta h_8}{h_8}| = 5.3 \times 10^{-6}(exper.) = 4.6 \times 10^{-6}(theory). \qquad (34)$$

which is of course not a perfect result, but seems to point at the possibility of not only understanding these second order effects, but also of computing them, maybe by means of computer algebra systems which would allow to compute not only the main term but many smaller ones as well.

Before concluding this section on this very optimistic note, we would like to add a few words of caution. Let us look again at Figures 1 and 2: can a second order theory, i.e. an expanded version of the computation mentioned above, really explain all the features of the long term behaviour of the major semiaxes? What about the even longer period terms, whose signature appears on the wavy behaviour of the maxima of the 1,100,000 years term? Maybe a second order success is not enough to be optimistic about the possibility of describing the orbits of the planets as a regular, quasiperiodic motion to be expanded in a Fourier series.

4 The Secular Perturbation Equations

To be able to find approximate solutions for the equation of motion described by the secular perturbation Hamiltonian H' we need to exploit again the D'Alembert properties. The symmetry property (26) applies to H' too; since it amounts to invariance with respect to a simultaneous rotation for all the orbits of the planets at once, this appears in the cartesian-type variables η', ξ' (defined as in (28) with the primed variables θ', Θ') as invariance with respect to a simultaneous rotation in all the planes (η'_s, ξ'_s). The only invariants of a system of 2-vectors with respect to rotation being the scalar and the vector products, the second D'Alembert rule can be stated for the secular Hamiltonian by saying that H' is a function of the combinations $\eta'_r\eta'_s + \xi'_r\xi'_s$ only; the odd combinations $\eta'_r\xi'_s - \eta'_s\xi'_r$ do not appear because H' is even, that is invariant with respect to a mirror reflection.

Thus the secular perturbation Hamiltonian can be expanded as a power series in η', ξ' with even degree terms only:

$$H' = H'_2 + H'_4 + H'_6 + \cdots \qquad (35)$$

moreover the degree 2 part H'_2 has the form:

$$H'_2 = \frac{1}{2}\sum_{s,r} A_{sr}[\eta'_s\eta'_r + \xi'_r\xi'_s] \qquad (36)$$

If all the terms in the Hamiltonian H' of degree 4 and higher are neglected, the solution of the linear differential equations defined by the Hamiltonian H'_2 can be easily computed by performing a rotation B which diagonalises the symmetric matrix A:

$$BAB^{-1} = diag[\nu_1, \nu_2, \cdots, \nu_s] \tag{37}$$

Then in a new set of variables $z = B\eta', w = B\xi'$ the dynamics can be described as a set of uncoupled linear oscillators with frequencies ν_r and the solutions are:

$$z_r = \sqrt{-2\Psi_r} cos(\nu_r t + \sigma_r)$$
$$w_r = \sqrt{-2\Psi_r} sin(\nu_r t + \sigma_r) \tag{38}$$

with integration constants Ψ, which describe the arbitrary amplitudes of the *proper modes* of oscillation, and arbitrary phases σ. When the proper modes are combined again with the matrix B^{-1} into the variables η', ξ' the classical epicyclic motion of the perihelia and nodes appears. The use of the rotation B and the meaning of the proper modes can be best understood in a simpler example, such as in the restricted 3-body problem (Henrard et al., 1986).

While there are as many proper modes as there are perihelia and nodes, the correspondence between one proper mode and either one perihelia or one node is not uniquely defined unless the matrix A, hence B, has a dominating principal diagonal. This is the case for the outer Solar System, with the only exception of the large off-diagonal terms linking the perihelia of Uranus and of Jupiter to one another. As a result, the perihelion of Uranus actually revolves around the Sun with the same proper frequency as the perihelion of Jupiter, and it is locked with the aphelion of Jupiter in a libration with a period of about 1,100,000 years and an amplitude of about 70 degrees (Milani and Nobili, 1985). In the inner Solar System the diagonal terms are not dominant at all, and the only way to understand the complicated interplay of coupled oscillations is to look at the proper modes themselves (Laskar, 1987a).

This coupled oscillator form of the solution for the secular advances of the perihelia and the nodes is often referred to as the *Lagrange solution*. However the solutions originally obtained by Lagrange and Laplace, as well as the more complete solution by Stockwell (1873), are obtained by neglecting not only the terms of degree four in the eccentricities and inclinations, but also the terms of order two in the small parameter ε, even if they belong to H'_2.

Actually the corrections to the frequencies ν due to the ε^2 terms are quite important; by inspecting equations (18), (19) and (20) we find that the ε^2 corrections result from Poisson brackets between two terms with the same combination of short periodic arguments λ, one in H_1 and one in χ_1, the latter with the corresponding divisor. As in Section 3, in performing the Poisson bracket one further negative power of the divisor appears, and for the terms with a small divisor δ such as the ones of (30) an $\varepsilon^2\delta^{-2}$ term is not much smaller than an ε term. By comparing Lagrange solutions for the Solar System computed with and without the second order corrections (Bretagnon, 1974; Laskar, 1987b), it is found that the fundamental frequencies ν can change by as much as 20% . For the same reason, asteroids with mean motions close, but not very close to the resonances (corresponding to the Kirkwood gaps) have perihelion and nodal frequencies completely different from the Lagrange theory values (Carpino et al., 1986). By comparing the second order analytical theories with numerical integrations (Laskar 1987b) we find that even the third order corrections (containing the $\varepsilon^3\delta^{-4}$ terms) are not negligible, but amount to as much as 1% of the fundamental frequencies.

Since the computational complexity of a complete third order solution is beyond the reach of the present generation of computers, a purely analytical theory capable of predicting the phases of the secular arguments over a time span of hundreds of million years does not appear feasible today. On the contrary, mixed techniques (involving the computation of an analytical theory by exploiting information provided by a numerical integration) seem more promising, as we shall discuss in Section 6.

Nevertheless, the solutions of the Lagrange type are very useful in describing the qualitative properties of the long term behaviour of the orbits of the outer planets; this would not be true of every N-body system, because three constraints have to be met. Firstly, the eccentricities and inclinations must be large with respect to the perturbation parameter ε: since the latter gives the order of magnitude of the short periodic perturbations in all the elements, the secular perturbation equations give significant informations only provided the amplitudes of the proper modes are larger. As an example, this is not the case for the satellite systems, where the proper modes, being free oscillations, are damped by dissipative effects (e.g. the tidal braking); there the proper modes have amplitudes of the order of ε and the quasi-resonant short periodic effects dominate, that is the frequency of circulation of the pericentres has nothing to do with the Lagrange solution frequencies (Laskar and Jacobson, 1987; Dermott et al., 1987; Fox et al., 1987).

The second constraint goes in the opposite way, that is the proper modes must have amplitudes small enough, so that the terms of higher degree in the eccentricities and inclinations are really negligible, or at least smaller. This is not the case for the asteroids, and the linear Lagrange-type theory does not describe well the behaviour of their orbits, not even for a timespan of the order of 100,000 years (Carpino et al., 1986; Knežević et al., 1987; Farinella et al., these proceedings). It turns out that the hypothesis used in the classical literature on secular perturbations of the orbits of the planets -namely that the eccentricities and inclinations are of the order of $\sqrt{\varepsilon}$ –is the most convenient one!

However, it is possible to compute approximate solutions which do not neglect the higher degree terms contained in H_4', in H_6', etc. This can be done within the same formalism of Lie series used in Section 2, following Message (1976, 1982). To understand why the same method applies, we need only to change coordinate system to one defined by the proper action variables Ψ_r and the fundamental secular arguments $\psi_r = \nu_r t + \sigma_r$. The latter are canonical variables, and the quadratic portion of the Hamiltonian is reduced to the Hamiltonian of the uncoupled proper modes of oscillations, that is:

$$H_2' = \sum_r \nu_r \Psi_r \tag{39}$$

Thus the Hamiltonian H' can be described as the sum of an integrable part H_2', function of the actions Ψ only, plus the perturbation function $H_4' + H_6' + \cdots$, and the machinery of the Lie series transformation with a suitable generating function χ' can be started again. In this case the perturbation is small if the action variables are small, that is the small parameter is of the same order : $\Psi = O(\varepsilon)$, and the generating function χ' to be solved for can be expanded in powers of $\eta', \xi' = O(\sqrt{\varepsilon})$:

$$\chi' = \chi_4' + \chi_6' + \cdots \tag{40}$$

$$T_{\chi'} H' = H' + \{H', \chi'\} + \frac{1}{2}\{\{H', \chi'\}, \chi'\} + \cdots \tag{41}$$

The next step in the computation looks the most innocuous of all, but as we shall see it is actually very dangerous: formula (41) is rewritten by ordering the terms in the transformed Hamiltonian by degree in η', ξ':

$$T_{\chi'} H' = H_2' + [\{H_2', \chi_4'\} + H_4'] + \textit{degree 6 and higher} \tag{42}$$

Then χ_4' can be chosen in such a way that all the terms of degree four which contain the angles ψ are eliminated; that is the equation:

$$\{H_2', \chi_4'\} + H_4' = \overline{H_4'} \tag{43}$$

is satisfied, where the overline operator indicates the average with respect to the angles ψ. Later on, another similar equation is solved for χ_6' to eliminate the angles ψ from the degree six portion of the transformed Hamiltonian, and a new recursive procedure is set up with the purpose of obtaining a transformed Hamiltonian which is integrable:

$$T_{\chi'} H' = H_2' + \overline{H_4'} + \cdots = H^*(\Psi^*) \tag{44}$$

This recursive procedure, which is not different from the Birkhoff normalisation apart from the use of a modern formalism, gives a formal solution to the problem of the secular perturbations because in the transformed variables ψ^*, Ψ^* the solutions are given by the proper actions Ψ^* constant and the fundamental secular arguments linear functions of time :

$$\psi_r^* = \nu_r^* t + \sigma_r^*$$
$$\nu_r^* = \frac{\partial H^*}{\partial \Psi_r^*} \tag{45}$$

For a complete planetary theory the number of terms becomes very large as the degree grows; nevertheless, at least in principle, both the integrable Hamiltonian H^* (hence the fundamental frequencies $\nu*$ by (45)) and the transformation :

$$F_{\chi'} : (\psi^*, \Psi^*) \longrightarrow (\psi', \Psi') \tag{46}$$

can be computed, not exactly (because the recursive procedure is actually infinite) but to such a high degree that the remainders are supposed to be very small. However, there is a third constraint to be met in order to to achieve a reliable prediction of the long term behaviour. We shall see in the next section that it is precisely the assumption that very high degree terms can be neglected that is the one which turns out to be the result of wishful thinking.

5 Divergence of the Perturbative Series

At a first glance, the recursive equations such as (43) used to compute a formal solution of the secular equations look simpler than the recursive equations such as (16) used to eliminate the short periodic arguments. An algorithm involving only polynomials, such as the one used to compute χ' as a power series in η', ξ', is simpler and moreover each step involves a finite amount of computation, because the polynomials of a given degree are defined by a finite number of coefficients. To see the difference, we can express the equations (43) by means of the action angle variables of the integrable quadratic Hamiltonian H_2', the already defined Ψ, ψ . If both the Hamiltonian H' and the unknown generating function χ' are expanded in Fourier series :

$$H'_s = \sum_{|k|\leq s} H'_{sk}(\Psi)\ cos\ (k\cdot\psi) \tag{47}$$

$$\chi'_s = \sum_{|k|\leq s} \chi'_{sk}(\Psi^*)\ sin\ (k\cdot\psi^*) \tag{48}$$

each of these expansions for the terms of degree s has a finite number of terms, namely only the ones with a combination of at most s secular arguments. This number grows very fast with the degree, but is anyway finite, while the expansions (13) and (14) are infinite. However the action-angle variables allow one very important similarity with the equations (16), (17) to be stressed; equation (43) becomes:

$$\{H'_2, \chi'_4\} = -\sum_r \nu_r \frac{\partial \chi'_4}{\partial \psi_r} = \overline{H'_4} - H'_4 \tag{49}$$

that is, term by term in the Fourier expansions (47) and (48) :

$$(\nu\cdot k)\ \chi'_{4k} = H'_{4k}\ ,\qquad 0 < |k| \leq s \tag{50}$$

with the *secular divisor* $\nu\cdot k = \sum \nu_r k_r$. Thus the series (48) for χ' can be formally computed to an arbitrarily high degree, but the convergence of the expansion $\chi' = \chi'_4 + \chi'_6 + \cdots$ is not guaranteed by the convergence of the expansion $H'_2 + H'_4 + H'_6 + \cdots$, because of the appearance of secular small divisors.

The easiest way to understand why the divergence must actually occur is to resort to a proof by contradiction. Let us suppose the power series expansion of χ' is convergent. Then the transformation $F_{\chi'}$ of (47) is a well defined analytical map, and so is the inverse transformation which can be easily computed in the same way:

$$(F_{\chi'})^{-1} = F_{-\chi'} : (\psi, \Psi) \longrightarrow (\psi^*, \Psi*) \tag{51}$$

and the dynamics in the ψ^*, Ψ^* space is defined by the Hamiltonian $H^*(\Psi^*)$, that is the Ψ^* portion of the map (51) gives as many integrals as there are degrees of freedom in the secular problem, and the latter is completely integrable. However, the N-body problem is not integrable (Poincaré 1892); the secular perturbation system is also not integrable (in this case the form of the non-integrability theorem due to Siegel (1941) has to be used). Thus the series for χ' cannot be convergent in any open neighbourhood of the origin in the η', ξ' space.

For over 60 years the theorems of Poincaré on the non-existence of integrals have been understood by the celestial mechanicians as implying that the series of the perturbation theories can only be used up to a limited order, hence cannot be arbitrarily accurate for an arbitrarily long span of time. After 1954 it became known that there is a large (in the sense of measure theory) portion of the phase space filled by invariant tori, on which the orbits are described by fundamental arguments linear functions of time (Arnold, 1963). That is, in this subset of the phase space the behaviour of the solution is identical to the behaviour of the solutions of an integrable system; moreover these tori can be explicitly constructed by means of a series expansion. In the original paper by Kolmogorov (1954) a sketch of the proof of the existence of the invariant tori is given by means of a formalism very similar to the Lie series expansions used above (for a Lie series version of the Kolmogorov theorem, see Benettin et al., 1984).

However it turns out that the existence of Kolmogorov type convergent perturbative series does not guarantee that whenever the initial conditions belong to an invariant torus the series expansions described above will be convergent at least on that point, or on that torus. On the contrary a stronger negative result can be easily proved : the power series expansion $\chi' = \chi'_4 + \chi'_6 + \cdots$ cannot be convergent on *any point* η', ξ' unless all, or at least some of the components are zero; in particular they cannot be convergent on a torus of the dimension prescribed by the Kolmogorov theorem (equal to the number of degrees of freedom).

The reason for this deceiving result is that a power series cannot converge on any set: by a time honoured result due to Abel, a power series in one complex variable can only converge on a disk, and the multidimensional analogue says that a multivariate power series can only converge on products of disks (with or without the boundary; Cartan, 1975, chap. 4, Proposition 2.2). In the real space, the convergence domains have the shape of a union of parallelepipeda; hence they always contain some open neighbourhood of the origin, unless they degenerate into lower dimensional parallelepipeda contained in some coordinate hyperplane. As an example, with only two planets the secular perturbation equations have four degrees of freedom and three independent integrals (and no more, by the non integrability theorems); thus the perturbation series are divergent in any neighbourhood of the origin. However the same problem in the plane has only two degrees of freedom and two integrals and is integrable: thus the series expansions of secular perturbation theory may well be convergent on the subspace obtained by setting all the inclinations to zero.

To understand this ubiquitous appearance of divergence phenomena we need to assess the effect of the secular small divisors on the higher and higher degree terms. When the equations analogue to (43) are computed for the higher degrees, the portion containing the highest degree parts of χ' is:

$$\{H'_2, \chi'_r\} = -\nu \cdot \frac{\partial \chi'_r}{\partial \psi} = -\{H'_4, \chi'_{r-2}\} + \textit{ many other terms} \tag{52}$$

Let us select one particular term in χ' containing the combination argument $k \cdot \psi^*$; let us suppose it appears in χ'_{r-2} , which is possible only provided $|k| \leq r-2$. Then the same combination will appear in χ'_r, with a coefficient satisfying the equation:

$$(\nu \cdot k)\chi'_{rk} = -(\frac{\partial \overline{H'_4}}{\partial \Psi} \cdot k)\chi'_{r-2k} + \cdots \tag{53}$$

Suppose we neglect the other terms, represented in (53) by $\cdots$; then the ratio between the term with the same frequency (defined by the multiindex k) in χ'_r and in χ'_{r-2} is given by:

$$d(k) = -(\nu \cdot k)^{-1}(\frac{\partial \overline{H'_4}}{\partial \Psi} \cdot k) + \cdots = \frac{\nu \cdot k - \nu^* \cdot k}{\nu \cdot k} + \cdots \tag{54}$$

where the last step is obtained from (45) :

$$\nu^* = \nu + \frac{\partial}{\partial \Psi^*}(\overline{H'_4} + \cdots) \tag{55}$$

If the ratio $|d(k)| \geq 1$ the term with combination argument $k \cdot \psi^*$ will have a coefficient given by a divergent series. On the other hand, if the small divisor $\nu \cdot k$ formed with the frequencies of the Lagrange solution and the small divisor $\nu^* \cdot k$ formed with the final frequencies have opposite sign, and if the $\cdots$ in (54) can be neglected, $|d(k)| \geq 1$. The occurrence of a combination frequency exactly equal to zero corresponds to a resonance, and (54) means that divergence phenomena will occur whenever the Lagrange frequencies ν and the final frequencies ν^* are on the opposite sides of a resonance. On the other hand, they always are, provided resonances of high enough order are considered; e.g. with two frequencies only, provided the ratio of the frequencies changes there is always a rational number in between the two ratios. Thus we can understand why the series always diverge, and also why they behave well until a critical order is reached, beyond which a $|d(k)| \geq 1$ appears and each successive approximation becames worse than the previous one.

What is then the meaning of the convergent series appearing in the theorems of Kolmogorov, Arnold and Moser ? To understand this we have to restart from the integrable Hamiltonian $H*(\Psi*)$ of (44) and to select a frequency vector ν^*, that is a single torus on which the solutions of the integrable system wind in the way prescribed by (45). Then it is conceivable to go back, that is to reconstruct the complete Hamiltonian H' by selecting a generating function χ^* such that :

$$T_{\chi^*} H^* = H' \tag{56}$$

By expanding again

$$\chi^* = \chi_4^* + \chi_6^* + \cdots \tag{57}$$

in powers of η^*, ξ^* (corresponding to ψ^*, Ψ^* by the usual polar to cartesian coordinate change similar to (28)), we can find a recursive set of equations such as:

$$\{H^*, \chi_4^*\} + \overline{H_4'} = H_4' \tag{58}$$

By solving (58) for a suitable χ_4^* we reintroduce into the Hamiltonian the original degree four terms, and by solving recursively similar equations we obtain a formal series for χ^* such that (56) is formally true. However, the series (57) is not any more a power series, because the equations such as (58) do not have polynomial solutions. By computing the Poisson bracket in the left hand side of (58) we find:

$$-\nu^* \cdot \frac{\partial \chi_4^*}{\partial \psi^*} = H_4' - \overline{H_4'} \tag{59}$$

which does not involve polynomials only, because ν^* is not a constant vector but a function of Ψ^* given by the derivative (45). Thus the series for $\chi*$ is not at all the opposite of the series for χ', because the coefficients for the Fourier expansion of χ^* will contain the divisors $\nu^* \cdot k$ while the series for χ' will contain the divisors $\nu \cdot k$. This statement appears somewhat paradoxical: the Lie series transformation being the composition with the flow for a time span 1, the inverse transformation is the composition with the flow for a time span -1, and to reverse the flow of time it should be enough to change the sign of the Hamiltonian as in (51); that is, we would expect the relationship $\chi^* = -\chi'$ to be true. Unfortunately, χ' does not exist (unless the system is integrable), thus χ^* is allowed to be very different. The difference can be understood by thinking of χ^* as obtained from $-\chi'$ by ordering the terms not only by degree but as well taking into account the smallness of the divisors; the reordering of the terms can indeed change the convergence behaviour of a series.

The statement of the Kolmogorov theorem can be given in this setting as follows: provided a ν^* is chosen in such a way that all the divisors $\nu^* \cdot k$ are not only non-zero, but also approaching zero slowly enough for $|k| \to \infty$, and provided the corresponding amplitudes (both in the Ψ^* and in the Ψ spaces) are small enough, χ^* is convergent on the single torus with $\Psi^* = const$ such that the frequencies have the chosen value ν^*.

Why does the Kolmogorov series χ^* not necessarily meet the same fate as the always divergent series χ' ? The proof of the theorem is not easy; however it is easy to understand the difference between the two series by repeating the computations such as (53), (54) for the Kolmogorov series. Then it is found that a small divisor cannot appear with an increasing negative power in the expression for a single coefficient of the Fourier series expansion of a Kolmogorov series.

The Kolmogorov type series are not power series, unless the frequencies ν^* are considered as fixed. They are sometimes called fixed frequencies expansions, but this terminology is somewhat confusing because also χ' is obtained by keeping the frequencies ν fixed. The point is, the frequencies ν are not the fundamental frequencies of the solution sought for, and by keeping them fixed the perturbation solution becomes more and more different from the true solution as the degree increases.

Having established that there are "good" perturbative series, such as the ones of the Kolmogorov type, which can be convergent (at least for the solutions belonging to invariant tori, that is in some subset of the phase space), and "bad" series , which always diverge, why don't we use the good ones? It turns out that life is not that easy, at least for celestial mechanicians. As we have stressed above, the point is not to keep the frequencies fixed, but rather to keep them fixed to the right value. The theorems of the Kolmogorov type do not provide a way to compute these frequencies; given a suitable set of frequencies, they compute in an effective (although complicated) way an invariant torus. Thus all the initial conditions belonging to that torus will share the same set of fundamental frequencies ν^*. However, *the map goes only one way*, from the frequencies to the initial conditions, and not the other way round; there is no known algorithm to compute the inverse map. This problem does not occur in other applications, such as in molecular dynamics, in which the observables are the fundamental frequencies. In our Solar System, because of the degeneracy of the integrable problem H_0, the secular fundamental frequencies correspond to periods longer than the duration of human civilisation. The only variables we can constrain by observation are the instantaneous values of the orbital elements, unless numerical integrations are seen as observations (Section 6).

This point needs to be stressed, also because most papers on the theorems of the Kolmogorov type give somewhat deceiving statements, by speaking of a given set of initial conditions with unperturbed frequencies fulfilling some non-resonance conditions, then proving the existence of an invariant torus *nearby* with the same frequencies. However *nearby* means that the difference is of the order of the size of the perturbation, which is also the same order of magnitude of the difference between ν and ν^*. The fact that the two constraints of keeping the initial conditions fixed and keeping the frequencies fixed are incompatible was well known to Poincaré; as Brouwer used to say, there is nothing new in Celestial Mechanics, only people who do not read the literature (however, Poincaré did not know that fixed frequencies expansions can converge; he wrote: *il est fort invraisemblable* (Poincaré 1893, section 149)).

Thus the problem of actually computing the fundamental frequencies corresponding to a given set of initial conditions has not been solved by perturbation theories; another possible approach is discussed in the next section.

6 Perturbation Theories and Numerical Integrations

In a non-integrable system -such as the N-body problem- KAM tori filled by conditionally periodic solutions and chaotic regions containing orbits with positive Lyapounov exponents occur in the same phase space. The question arises of how to establish whether a given initial condition belongs to the former or to the latter. However one has to remember that invariant tori do not fill any open set in the phase space; initial conditions being deduced with a finite accuracy from astronomical observations, the question as to whether our Solar System is bound to belong to an invariant torus for all time is not a physically meaningful question. On the other hand the total lifetime of our Solar System is only about 10^{10} years (from the formation of the planets to the red giant phase of the Sun) and we are really interested in computing the orbits of the planets over such a timespan only. Is there a constructive method, that is a finite algorithm, to perform this computation with a given accuracy?

As a matter of principle, the theorems of Kolmogorv, Arnold and Moser could be used: whether there is an invariant torus passing in the neighbourhood of the phase space containing the initial conditions compatible with the observations, and which are the frequencies ν^* on that torus, are meaningful questions. If these questions could be answered, Nekhoroshev type theorems could be used to give upper bounds for the difference between any orbit in the neigbourhood and the torus (see Giorgilli et al., 1987 and Simó, these proceedings). Unfortunately, the current state of the art does not provide any algorithm to compute analytically the frequencies ν^* in such a way that the error can be reliably estimated.

From the point of view of the rigorous mathematician the difficulties arising from the divergence of the perturbative series cannot be avoided, because they are in the very nature of the problem of non-integrable systems. On the contrary from the point of view of the astronomer it remains to be proven that all this discussion about non-integrability has any bearing on the problem of computing the planetary orbits for a physically meaningful time span. The best discussion on this point is still given by Poincaré (1893, Section 118): an asymptotic expansion of the solution can give a very accurate solution even being divergent, provided the computation is arrested before the terms start growing. The problem is how can we decide whether the computation can go further (e. g. to higher order and degree) or it has already gone too far?

While this specific argument has been developed later, it was on the basis of the same general idea that the *LONGSTOP* project was established in 1983, by initiative of A.E. Roy. There is no way to find out about the stability of the Solar System -over a timespan comparable to its real age- unless numerical integration is used.

Numerical integration has its own problems. Problem number one is the integration error, which also has to be estimated. However some progress has been done in this direction (see Nobili, these proceedings), and it turns out that the integration error can be kept to such a level that its effects, in particular on the secular variables such as η', ξ', are small and anyway comparable to the inaccuracies of the physical model of the Solar System (Nobili, 1987). A more subtle problem arises in the handling of the output of a numerical integration. The unprocessed output is a time series for the orbital elements of the planets, sampled at discrete intervals (spaced by the integration stepsize or a multiple thereof). This format is not at all suitable for the comparison with an analytical secular perturbation theory; actually, it is not suitable for any analysis, apart from very rough assessments of qualitative properties (e.g., no orbit crossing between the major planets over the integration timespan). Since a numerical integration for a timespan of 5×10^9 years is not yet possible because of the accumulation of the integration error, the output needs to be post-processed in order to extract the relevant information on the main periodic terms and on the values of the fundamental frequencies, so as to allow extrapolation to longer timespans.

Such a procedure has been developed within the *LONGSTOP* project and has been called *synthetic secular perturbation theory*. The first stage (which has to be performed in line with the numerical integration) is digital filtering to eliminate the short periodic oscillations, as discussed in Section 3. The smoothed time series are Fourier analysed, and the main long periodic oscillations are identified with the fundamental secular frequencies ν^* . Then the amplitudes of the larger terms are determined by a least square fit; the fit also improves the determination of the frequencies and the phases of the fundamental arguments ψ^*. The procedure is applied in an iterative way, by adding to the model to be fitted as many combination lines with arguments $k \cdot \psi^*$ as can be identified in the residuals of the previous step (Carpino et al., 1987).

The synthetic theory yields two results: an estimate of the fundamental frequencies ν^* (and phases σ^*) and the values of the amplitudes of the long periodic terms. The accuracy of the former is of course limited by the integration timespan T; however the iterative procedure we have devised has the property of yielding an accuracy better than π/T because the interference of combination lines is accounted for; checks are possible by means of proper modes analysis (Nobili and Milani, 1987; Nobili et al., 1988). The amplitudes give reliable informations only as long as the spectral lines found in the Fourier analysis of the residuals, at each stage in the procedure, can be identified with combination lines with frequencies $k \cdot \nu^*$; because of the limited frequency resolution, this is possible only up to some degree $|k|$, then the iterative procedure must stop. If the amplitude of the unidentified lines is small enough, and anyway smaller than the amplitude of the identified lower degree lines, the synthetic theory will be an efficient way to represent the solution. However, the latter requirement is just another way of presenting the small divisor problem.

Let us assume that the solution corresponding to the initial conditions of the numerical integration belongs to a KAM invariant torus. Then the frequencies estimated by the synthetic theory are the ν^* appearing in the statement of the Kolmogorov theorem, and the amplitudes estimated are the ones of the Kolmogorov series:

$$\eta' = T_\chi \cdot \eta^* \qquad ; \xi' = T_\chi \cdot \xi^* \tag{60}$$

which express the conditionally periodic solution as a convergent Fourier series in the fundamental arguments ψ^*. They can be compared with the corresponding amplitudes computed by a conventional power series expansion $\chi' = \chi_4' + \chi_6' + \cdots$, but they will not be equal: as soon as a small divisor $k \cdot \nu$ is encountered in χ', such that $k \cdot \nu^*$ has a significantly different value, that is as soon as for some k we have $d(k) \simeq 1$, the amplitudes will be very different.

On the other hand, the orbit might be chaotic. If the integration time span is long enough for the chaotic behaviour to be detected in the spectrum of the solution, the amplitude of the higher degree combinations will not decrease. This appearance of chaos can be detected in the synthetic theory in two ways. Firstly the theory itself will stop very soon: below a given order of magnitude, many of the prominent spectral lines will not be identified with theoretically permitted combinations of the fundamental frequencies ν^* up to a reasonable degree $|k|$. Moreover, one or more very small divisor $k \cdot \nu^*$ will appear with a relatively small order $|k|$. The corresponding critical argument $k \cdot \psi^*$ might either librate or circulate. In the former case it is of course possible that a stable conditionally periodic libration occurs, and a convergent Kolmogorov series might describe a torus imbedded in the phase space in a topologically non-equivalent way. If this is the case, a new fundamental frequency, which is the libration frequency of the critical argument, should appear in the spectrum.

Let us now look at the results of the most recent long numerical integrations of the orbits of the planets from this point of view.

LONGSTOP 1A is an integration of the outer planets for almost 10 million years; the synthetic theory has been published in Carpino et al. (1987). Apart from the results on the major semiaxes discussed in section 2 and in Milani et al. (1987), the main conclusion we could draw from the synthetic theory was that there are no chaotic phenomena strong enough to show up over such a timescale. The synthetic theory succeeded in fitting the data to an accuracy of 7×10^{-5} for the eccentricities and 1.5×10^{-5} for the inclinations of the four major outer planets, by identifying respectively 23 and 25 spectral lines with combinations of 1, 3 and 5 fundamental frequencies.

The *Digital Orrery* is a purpose built parallel computer, designed to achieve optimum efficiency in the numerical integration of N-body problems (Applegate et al., 1985). It was used to compute the orbits of the outer planets for over 200 million years. The output has

Table 1 – The Multiplet around the Fundamental Frequency g_6

Lines detected in the non-singular variables h,k in the LONGSTOP 1B output

Line Identification	Period Years	Jupiter —	Saturn Amplitudes	Uranus —
$g_6 - g_5 + g_7$	47,867.403	1.982×10^{-4}	-6.052×10^{-4}	5.80×10^{-5}
$g_5 - s_6 + s_7$	46,939.862	1.104×10^{-4}	-3.368×10^{-4}	1.61×10^{-5}
$2g_6 - g_7 + s_6 - s_8$	46,699.582	7.8×10^{-6}	-2.37×10^{-5}	—
$g_6 - x_3$	46,003.	-9.9×10^{-6}	3.05×10^{-5}	—
$g_6 - x_4$	45,977.	-9.8×10^{-6}	3.02×10^{-5}	—
$g_6 + x_2$	45,966.903	3.44×10^{-5}	-1.056×10^{-4}	—
$g_6 - x_5$	45,950.	-2.06×10^{-5}	6.33×10^{-5}	—
$g_6 + x_1$	45,928.094	4.21×10^{-5}	-1.292×10^{-4}	4.0×10^{-6}
$g_6 - x_6$	45,906.	-1.82×10^{-5}	5.58×10^{-5}	—
g_6	45,883.367	-1.57002×10^{-2}	4.82093×10^{-2}	-1.5471×10^{-3}
$g_6 + x_6$	45,861.	1.82×10^{-5}	-5.60×10^{-5}	—
$g_6 - x_1$	45,838.726	-4.22×10^{-5}	1.295×10^{-4}	-4.0×10^{-6}
$g_6 + x_5$	45,817.	2.05×10^{-5}	-6.28×10^{-5}	—
$g_6 - x_2$	45,800.133	-3.47×10^{-5}	1.065×10^{-4}	—
$g_6 - x_4$	45,790.	9.8×10^{-6}	-3.02×10^{-5}	—
$g_6 + x_3$	45,764.	1.01×10^{-5}	-3.09×10^{-5}	—
$g_7 - s_6 + s_8$	45,095.192	-8.5×10^{-6}	2.62×10^{-5}	—
$2g_6 - g_5 + s_6 - s_7$	44,873.382	-1.226×10^{-4}	3.792×10^{-4}	-1.18×10^{-5}
$g_6 + g_5 - g_7$	44,057.255	-1.936×10^{-4}	5.982×10^{-4}	7.2×10^{-6}

been Fourier analysed and spectral lines have been identified with combinations of 1 and 3 fundamental frequencies; some 5 frequencies combinations have also been detected (Applegate et al., 1986). The spectrum of the output contains many lines which have been detected but not identified by the authors; some could be 5 and even 7 frequencies combinations. The way the output is produced with the Digital Orrery does not allow to generate a synthetic theory as a fit to the data, thus it is not possible to assess how accurately the time series of the orbital elements is represented by the Fourier sum containing the identified lines; however the spectrum, graphically presented by means of windowing techniques, looks very much discrete. The Lyapounov exponent for Pluto has also been estimated by the technique of the two nearby orbits, and found to be smaller than 1 over 8 million years.

LONGSTOP 1B is an integration of the orbits of the outer planets for 100 million years; the equations of motion have been slightly complicated to model the main effects of the inner planets and of general relativity (Nobili, 1987). The output has been processed to yield a synthetic theory for the eccentricities and the inclinations with 58 and 76 identified lines respectively (with 1, 3, 5 and 7 frequencies) and r.m.s. residuals smaller than 1.5×10^{-4} and 7×10^{-6} respectively. The results have not been published at the time of this writing (Nobili et al., 1988); we have of course to apologize for this

Table 2 – Unidentified Long Period Frequencies Generating the g_6 Multiplet

Frequency (as in Table 1)	Period Years	Phase Degrees	Tentative Identification
x_1	−47,115,000.	89.928	—
x_2	−25,248,000.	129.574	—
x_3	17,620,000.	1.2	$3g_7 - 2g_5 - g_8$??
x_4	22,510,000.	158.6	—
x_5	31,660,000.	313.0	$2g_5 - 2g_7 + s_7 - s_8$
x_6	93,500,000.	271.5	—

delay, but we found something we did not understand, and some thinking (which led to the discussion presented here) was necessary.

The best example of what we found hard to understand is presented in the Tables 1 and 2; it can be described as a *multiplet* of spectral lines surrounding in a symmetric way the fundamental line g_6, which is one of the ν^*, and precisely the one corresponding to the proper mode which controls the perihelion of Saturn. In Table 1 the detected spectral lines are presented with their period and amplitude in the non-singular elements $h = e\ sin\ \varpi$, $k = e\ cos\ \varpi$ of Jupiter, Saturn and Uranus. The g_r and the s_r are the fundamental frequencies ν^* as determined by the synthetic theory (notation as in Bretagnon, 1974); on the contrary the x_r have not been identified in a reliable way with any combination of fundamental frequencies. Table 2 gives tentative identifications for some of the x_r; however most of the latter cannot be identified with combinations of up to 8 fundamental frequencies. The amplitudes of the unidentified lines are large enough to force the termination of the iterative procedure of identification and fitting; as a result the r.m.s. of the residuals for the *LONGSTOP 1B* theory is larger than that of the shorter *LONGSTOP 1A* integration.

The difference between the *LONGSTOP 1A* and the *LONGSTOP 1B* results does not necessarily mean that the small change in the physical model -introduced to represent in a more realistic way the outer Solar System for a longer time span- makes an essential difference. The multiplet of Table 1 is too close to the main g_6 line to be detected with the frequency resolution of a 10 million years integration: such a closely packed bunch of spectral lines can very well be fitted to a single line, until the timespan becomes long enough for the different lines to get out of phase by a large enough angle. On the other hand, a small change in the model and in the initial conditions could be important, since a narrow chaotic zone can be very close to an invariant torus.

The most likely interpretation of the *LONGSTOP 1B* findings is that the Kolmogorov series we are approximating by means of the synthetic theory is actually diverging, or at least its terms have stopped decreasing with the degree. Thus a 7 frequency line such as $g_6 + x_5$ can have a smaller amplitude than $g_6 + x_2$ which has at least 11 frequencies. If this interpretation is correct, two verifiable consequences follow.

Firstly, there must be at least one small divisor $k \cdot \nu^*$ which is responsible for the divergence (or at least, the temporary misbehaviour) of the Kolmogorov series. Moreover, if the series are really divergent, there must be a positive Lyapounov exponent.

The first test can be performed with the results of the synthetic theory itself. For each degree $r = |k|$ we can form all the divisors $k \cdot \nu^*$ and choose the one α_r closest to zero. If it is responsible for the misbehaviour of the Fourier expansion of the solution, it will appear in the unknown lines. It will also stick out of the crowd of the small divisors, e.g. in the sense that it will be smaller than α_{r+2}. One such divisor is $2g_5 - 2g_7 + s_7 - s_8$; it amounts to only 0.041 arcsec per year, that is, the corresponding long periodic effect has a period of over 31 million years. It appears in Table 2 as the only reasonably identified of the x_r unknowns, and it is not only α_6 but it is smaller than α_8.

Having identified a dangerously small divisor, the next step is to investigate the behaviour of the corresponding critical argument. However to form the latter we cannot simply use the combination $2\varpi_5 - 2\varpi_7 + \Omega_7 - \Omega_8$, because the longitude of the perihelion of Uranus ϖ_7 is not controlled by the fundamental frequency g_7, but by g_5, as discussed in Section 4. The angle $\varpi_5 - \varpi_7$ librates, as it is apparent from a plot in the plane $e_5 e_7\ cos(\varpi_5 - \varpi_7)$,$e_5 e_7\ sin(\varpi_5 - \varpi_7)$ (Figure 3); the argument corresponding to the combination $g_5 - g_7$ is well approximated by the angle τ of the figure. Figure 4 shows the time variation of the argument $2\tau + \Omega_7 - \Omega_8$; of course the plots of the two figures have been prepared by applying digital filtering in order to remove shorter period fluctuations, including the ones associated to the faster g_6 and s_6 frequencies. From Figure 4 we can conclude that the critical argument associated to the dangerous divisor x_5 is circulating, thus we can exclude that a resonant theory has to be applied to ease the convergence of the series. The slope of Figure 4 fits within 1% with the value of the combination frequency $2g_5 - 2g_7 + s_7 - s_8$, thus showing that the synthetic theory frequencies are very good.

A puzzle arises from the comparison of the results -relative to the divisor x_5 discussed above- between the *LONGSTOP 1B* integration and the *Digital Orrery* one. The result from the latter for the combination frequency $2g_5 - 2g_7 + s_7 - s_8$ corresponds to a period of 1,645 million years, i.e. it is numerically zero. Moreover, the multiplet of Table 1 does not appear. Should we conclude that in the *Digital Orrery* integration, because of the slightly different initial conditions and physical model, the corresponding critical argument is in libration? If the amplitude of this libration is small enough, the solution could be a conditionally periodic motion on a torus with a topologically different immersion in

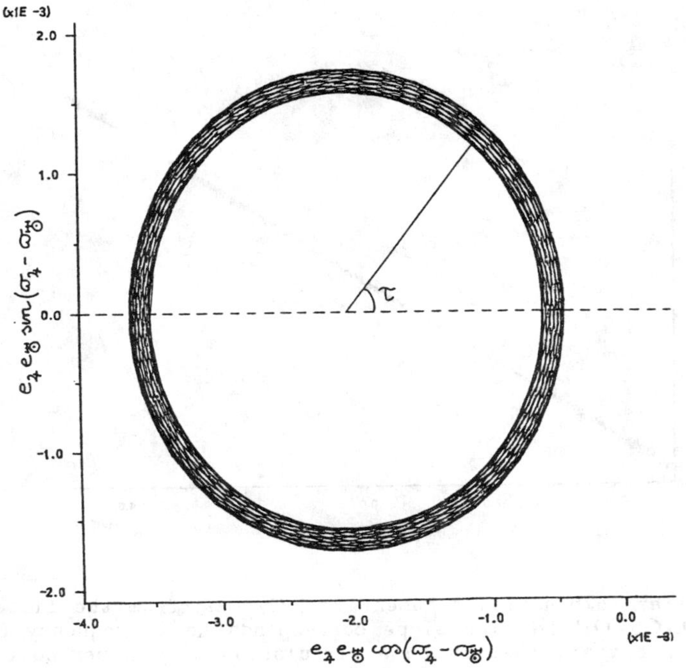

Figure 3 : The orbit of the outer Solar System is projected on the plane $e_5\ e_7\ cos(\varpi_5 - \varpi_7)$,. $e_5\ \ e_7\ sin(\varpi_5 - \varpi_7)$ to show the libration of the argument $\varpi_5 - \varpi_7$. The polar angle τ from the barycenter of the figure is a good approximation for the fundamental argument with frequency $g_5 - g_7$.

the phase space. Unfortunately the slow rate of output from the *Digital Orrery* makes it difficult to compute a very slowly changing critical argument.

As for the second consequence of a possible divergence of the Kolmogorov series, namely the existence of a positive Lyapounov exponent (for the definitions see Froeschlé, 1984), it must be pointed out that the expected size of such an exponent would be *of the order of* 1 over 31 million years, which is not only below the detectability threshold of the *Digital Orrery* computation, but indeed very hard to detect with the present generation of computers because of the accumulation of rounding off error.

The last integration I would like to mention has been completed only

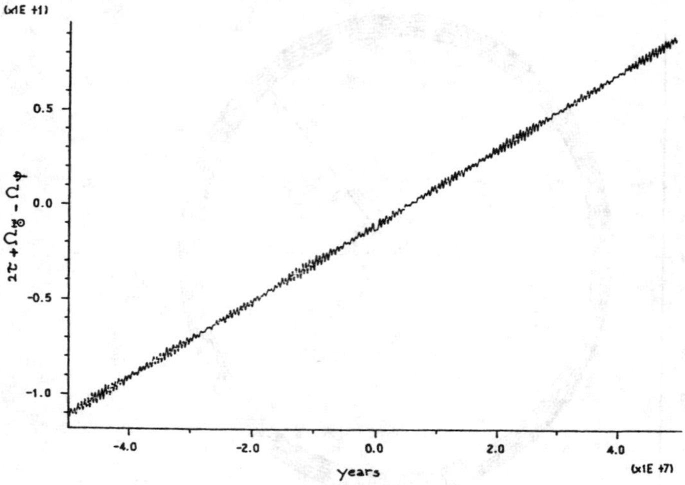

Figure 4 : The combination argument $2\tau + \Omega_7 - \Omega_8$ from the filtered output of *LONGSTOP 1B*. The slope corresponds to a frequency of 0.0409 arcsec per year, that is to a circulation with a period of about 31,690,000 years.

recently (Laskar, 1987a, 1987b), and is a numerical integration of the secular perturbation equations for all the major planets (Mercury to Neptune). The equations have been derived by performing something equivalent to the transformation T_χ of Section 2, up to order 2 in ε, by computer algebra. To avoid the secular small divisor problem, the secular equations have been integrated numerically with a step short with respect to the secular frequencies but very long with respect to the ones used in non-averaged integrations such as *LONGSTOP*. While there are still some problems related to the neglected third order terms, as discussed in Section 4, the results are by far the best available approximation to the orbits of the inner planets for a timespan of 30 million years. The output has been Fourier analysed, the fundamental frequencies determined, and a synthetic theory was formed by means of an iterative least square procedure. The results are shattering, from the point of view of the believers in the convergence of the series representation of the solutions. The solution for Mars shows the effects of the small divisor $2g_3 - 2g_4 + s_3 - s_4$ which is numerically zero; in the solution for Mercury *the second largest term* has a frequency which cannot be identified with any reasonable combination of fundamental frequencies.

It is not easy to draw conclusions on a problem on which such rapid progresses have been done and are being done. One tentative conclusion could be as follows. It appears -from the latest results- that in computing the orbits of the planets for a very long time span we are either very close or even beyond the limit of *naive computability*. By this I mean that both by numerical integration and by analytical solution by series expansions we encounter the effects of non-integrability of the dynamical system representing the planetary orbits. This effect appears as divergence of the series of the analytical theories, and as critical dependence upon the initial conditions in the numerical integrations. Poincaré wrote (1892, Introduction):

Cette époque, où l'on sera obligé de renoncer aux méthodes anciennes, est sans doute encore très éloignée; mais le théoricien est obligé de la devancer, puisque son oeuvre doit précéder, et souvent d'un grand nombre d'années, celle du calculateur numérique.

Since it appears that this *époque éloignée* is just now, to go on either by longer numerical integrations or by higher order theories does not look like the right thing to do. A *théoricien* like Poincaré not being available, we can only make wild guesses on what should be the next step. However, I believe essential progress can only come from the combined use of the two methods of numerical integration (with the most powerful supercomputers) and perturbation theory (to be computed by computer algebra).

One possible method to compute the orbits beyond the limit of naive computability could be as follows. First a numerical integration provides, by means of a synthetic theory, a good approximation to the values of the fundamental frequencies; it also provides information on the possible occurrence of secular resonances. Then a perturbation theory is constructed either with frequencies fixed to the values of the synthetic theory, or with variable frequencies to correct for effects acting over an even longer time span. For such a theory with *variable frequencies* there is no available convergence proof; but at least, *if the terms are all recomputed at each stage regardless for their degree*, there is no theorem guaranteeing its divergence. While this looks like a formidable task, it will never be done unless we begin.

Acknowledgements

This review paper is based upon research done in collaboration with many people within the *LONGSTOP* team, and especially A.M. Nobili and M. Carpino. Discussion with J. Henrard, Z. Knežević, J. Laskar, I. Percival and J. Wisdom have contributed important insights.

While this paper was being completed, and precisely on October 20, 1987, Andrei N. Kolmogorov, aged 84, died in Moscow. I wish to dedicate this paper to his memory.

References

Applegate, J.H., Douglas, M.R., Gursel, Y., Hunter, P., Seitz, C., and Sussman, G.J.: 1985, 'A Digital Orrery', *IEEE Transactions on Computers,* **c-34**, 822-831

Applegate, J.H.,Douglas, M.R., Gursel, Y., Sussman, G.J.and Wisdom, J.: 1986 'The outer solar system for 200 million years', *The Astronomical Journal* **92**, 176-194

Arnold, V.:1963, 'Small denominators and problems of stability of motion in classical and celestial mechanics', *Russian Mathematical Surveys* **18**, 85-191

Arnold, V.: 1976, *Méthodes Mathématiques de la Mécanique Classique*, MIR, Moscou

Benettin, G., Galgani, L. and Giorgilli, A.: 1984, 'A proof of Kolmogorov theorem on invariant tori using canonical transformations without inversion', *Nuovo Cimentobf 79b*, 201-223

Bretagnon, P.: 1974, 'Termes à longues périodes dans le système solaire', *Astronomy and Astrophysics* **30**, 141-154

Cartan,H.: 1975, '*Théorie élémentaire des fonctions analytiques d'une ou plusieurs variables complexes*, Hermann, Paris

Carpino, M., Gonczi, R., Farinella, P., Froeschlé, Ch., Froeschlé, C., Paolicchi, P. and Zappalá V.: 1986, 'The accuracy of proper orbital elements and the properties of the asteroid families: comparison with the linear theory', *Icarus* **68**, 55-76

Carpino, M., Milani, A. and Nobili, A.M.: 1987, 'Long-term numerical integrations and synthetic theories for the motion of the outer planets', *Astronomy and Astrophysics* **181**, 182-194

Dermott, S., Malhotra, R. and Murray, C.D.: 1987a, 'A chaotic route to melting Miranda', *Icarus*, submitted

Eginitis, M.D.: 1889, *Annales de l'Obs. Paris, Memoires* **19**

Fox, K., Malhotra, R., Murray, C.D. and Nicholson, P.D.: 1987, ' Secular Perturbations of the Uranian Satellites', preprint

Froeschlé, C.: 1984, 'The Lyapounov Characteristic Exponents and Applications to the Dimension of the Invariant Manifolds of a Chaotic Attractor', in *Stability of the Solar System and Its Minor Natural and Artificial Bodies*, Szebehely, V. editor, Reidel, Dordrecht, 265-282

Giorgilli, A., Delshams, A., Fontich, E., Galgani, L. and Simó, C.: 1987, 'Effective Stability for a Hamiltonian System Near an Elliptic

Equilibrium Point, with an Application to the Restricted Three Body Problem', preprint

Henrard, J., Lemaitre, A., Milani, A. and Murray, C.D.: 1986, 'The reducing transformation and apocentric librators', *Celestial Mechanics* **38**, 335-344

Hori, G.: 1966, 'Theory of General Perturbations with Unspecified Canonical Variables', *Publ. Astron. Soc. Japan* **18**, 287-296

Kinoshita, H. and Nakai, H.: 1984, 'Motions of the perihelions of Neptune and Pluto', *Celestial Mechanics* **34**, 203-217

Knežević, Z., Carpino, M., Farinella, P., Froeschlé, Cl., Froeschlé, Ch., Gonczi, R., Jovanović, B., Paolicchi, P. and Zappalá, V.: 1987, 'Asteroid short periodic perturbations and the accuracy of mean orbital elements', *Astronomy and Astrophysics*, in press

Kolmogorov, A.N.: 1954, 'The conservation of conditionally periodic motions with a small variation in the Hamiltonian' *Dokl. Akad. Nauk. SSSR* **98**, 527-530

Laskar, J.: 1987a, 'Secular evolution, proper modes and resonances in the inner Solar System', in *Resonances in the Solar System, IAU European Regional Astronomy Meeting, Prague 1987*, in press

Laskar, J.: 1987b, 'Secular evolution of the Solar System over 10 million years', *Astronomy and Astrophysics*, in press

Laskar, J. and Jacobson, R.A.: 1987, 'GUST86. An analytical ephemeris of the Uranian satellites' *Astronomy and Astrophysics*, in press

Lighthill, J.: 1986, 'The recently recognized failure of predictability in Newtonian dynamics', *Proceedings of the Royal Society of London A*, **407**, 35-50

Message, P.J.: 1976, 'Formal expressions for the motion of N planets in the plane, with the secular variations included, and an extension to Poisson's theorem', in *Long-Time Predictions in Dynamics*, Szebehely and Tapley eds., Reidel Pu. Co., Dordrecht, Holland, 279-293

Message, P.J.: 1982, 'Asymptotic series for planetary motion in periodic terms in three dimensions', *Celestial Mechanics* **26**, 25-39

Milani, A. and Nobili, A.M.: 1983, 'On the stability of hierarchical 4-body systems', *Celestial Mechanics* **31**, 241-291

Milani, A. and Nobili, A.M.: 1985, 'Resonant structure of the outer solar system', *Celestial Mechanics* **35**, 269-287

Milani, A.,Nobili, A.M., Fox K. and Carpino, M.: 1986, 'Long term changes in the semimajor axes of the outer planets', *Nature* **319**, 386-388

Milani, A., Nobili, A.M. and Carpino, M.: 1987, 'Secular variations of the semimajor axes: theory and experiments', *Astronomy and Astrophysics* **172**, 265-279

Moser, J.: 1973, *Stable and random motions in dynamical systems*, Princeton Univ. Press

Nobili, A.M.: 1987, 'Long term dynamics of the outer solar system: review of LONG-STOP project', in *The few-body problem*, Valtonen, M. ed., Reidel Pu. Co., Dordrecht, in press

Nobili, A.M. and Milani, A. : 1987, 'Long term energy variations and a 31 Myr period effect in the orbits of the outer planets', in *Resonances in the Solar System, IAU European Regional Astronomy Meeting, Prague 1987*, in press

Nobili, A.M., Milani, A. and Carpino, M.: 1988, 'Fundamental frequencies and small divisors in the orbits of the outer planets', in preparation

Nobili, A.M. and Roxburgh, I.W. : 1986, 'Simulation of general relativistic corrections in long term numerical integrations of planetary orbits', in *Relativity in Celestial Mechanics and Astrometry*, Kovalevsky, J. and Brumberg, V. eds., Reidel Pu. Co., Dordrecht, 105-111

Poincaré H.:1892, *Methodes Nouvelles de la Mechanique Celeste, Vol. I*, Gauthier-Villars, Paris

Poincaré H.:1893, *Methodes Nouvelles de la Mechanique Celeste, Vol. II*, Gauthier-Villars, Paris

Roy, A.E.: 1979, 'Empirical Stability Criteria in the Many-Body Problem', in *Instabilities in Dynamical Systems*, Szebehely, V. ed., Reidel Pu. Co., Dordrecht, 177-210

Siegel, C.L.: 1941, 'On the integrals of canonical systems' *Annals of Mathematics* **42**, 806-822

Stockwell, J.N.:1873, 'Memoir on the secular variation of the elements of the orbits of the eight principal planets', *Smithsonian Contributions to Knowledge* **18**

Tisserand, F.:1876, *Compt. Rend. Acad. Sci.* **82**, 442-

Walker, I.W., Emslie, G.A. and Roy, A.E.: 1980, 'Stability criteria in many body systems, I', *Celestial Mechanics* **22**, 371-402

THE ACCUMULATION OF INTEGRATION ERROR

Anna M Nobili
Gruppo di Meccanica Spaziale
Dipartimento di Matematica
Università di Pisa
Via Buonarroti 2
I-56100 Pisa, Italia

ABSTRACT The accumulation of integration error is discussed in relation to the LONGSTOP 1A and 1B numerical integrations of the orbits of the outer planets spanning 9.3 and 100 million years respectively. Various error sources are listed and our present understanding of their effects is discussed. It is shown that the error growth in the final integrations is due to the finite-precision arithmetic (the so called *rounding-off error*). A semi-empirical formula, containing an unknown numerical coefficient, is given to describe the accumulation of rounding-off error; by comparing it with the error found in the LONGSTOP integrations the value of the numerical coefficient is obtained.

1. VERY LONG NUMERICAL INTEGRATIONS

The availability to scientists of fast and powerful computer resources has recently made it possible to investigate, by means of direct computation of the solution, the behaviour of dynamical systems over very long intervals of time. However, numerical integration of the differential equations of motion is affected by several sources of error. If one expects the system to be mostly regular —hence predictable— over the timespan of the numerical integration, this is an experiment in deterministic dynamics and therefore all the error sources should be either got rid of, or anyway kept below the level of the relevant dynamical features. As a sort of paradox, whatever our understanding of the error sources and propagation, there will always be some kind of error left whose behaviour is not understood; were not so, the understanding of the mechanisms of its accumulation would have allowed us to eliminate it.

At present the error source which is not understood, and therefore sets the limit to the integration interval which can be explored by means of direct computation of the solution, is the so called *rounding-off error*, namely the error caused by the fact that the computer does not operate in exact arithmetic. The rounding-off error in longitude accumulates quadratically with the number N of the integration steps and is proportional to the relative machine accuracy $\varepsilon = 2^{-b}$ (b the number of bits in the mantissa of a floating point number as represented in the computer). By *very long numerical integration* we mean an integration spanning an interval of time close to the limit for which $N^2\varepsilon$ is larger than 1;

A. E. Roy (ed.), *Long-Term Dynamical Behaviour of Natural and Artificial N-Body Systems*, 109–116.
© 1988 by Kluwer Academic Publishers.

when reaching this limit ways should be found to reduce this source of error.

2. THE ROUNDING-OFF ERROR

The main sources of error in long term numerical integrations fall into 4 categories: the numerical algorithm used to discretize the equations of motion *(truncation error)*; the computer hardware and software *(rounding-off error)*; the physical model, initial conditions and reference system; the stability of the exact solution.

The physical model whose equations of motion are integrated numerically should contain all the essential features of the real system; the error in the initial conditions and the reference system can be reduced but is, to some extent, unavoidable in any real problem. For a discussion of the physical model, initial conditions and reference system of the outer solar system see Nobili et al., 1988.

The computed solution cannot be more stable than the real one; we confine ourselves to the study of error accumulation in the propagation of mostly regular solutions.

The computer does not represent real numbers with infinite precision; if the average value of the rounding-off error is exactly zero, its longitude effect should be: $\Delta\lambda \propto N^{3/2}$ (Brouwer, 1937). However, this does not seem to happen with real processors; in the case of the LONGSTOP 1A and 1B integrations, which have been carried out on the CRAY 1S of the University of London Computer Centre, the longitude error due to rounding-off turns out to be: $\Delta\lambda \propto N^2$. More precisely, we can describe the propagation of the rounding-off error in longitude with the formula:

$$\Delta\lambda \simeq c\varepsilon(nh)N^2 \tag{1}$$

with n the mean motion, h the stepsize and c is a numerical constant of order unity whose precise value is not known a priori. The value of h is chosen from considerations of numerical stability of the algorithm and so as to make the truncation error —given the orbital eccentricity of the nearest planet— smaller than rounding-off. In the LONGSTOP integrations h is 1/108 of Jupiter's period, hence $nh = 0.058$; in LONGSTOP 1B $N = 4.6 \times 10^8$. The CRAY-1S has a rather short mantissa ($b = 48$), thus $\varepsilon = 2^{-b} = 3.6 \times 10^{-15}$. With these numbers, formula (1) gives $(\Delta\lambda)_{tot} = c2500°$, which is too large an error to make such an integration meaningful. The rounding-off error as expressed in terms of the relative machine accuracy ε is proportional to the intensity of the force to be computed at each integration step; therefore it can be reduced by reducing the force. This can be done using Encke's method in which the force is computed with respect to some refernce keplerian orbit not too far from the actual one (Roy, 1982 §7.4); it is therefore only a small perturbation which must be computed at every step. The use of Encke's method in the case of a multistep integration algorithm is described in Milani and Nobili, 1988. In the LONGSTOP case it allows an error reduction by a factor $Q \simeq 6 \times 10^{-3}$; the formula for the accumulation of the rounding-off error becomes:

$$(\Delta\lambda)_{Encke} \simeq cQ\varepsilon(nh)N^2 \tag{2}$$

and in the LONGSTOP 1B case the errorr to be expected is $(\Delta\lambda)_{1B} \simeq c15°$. It is possible to make the truncation error smaller than rounding-off; therefore the comparison between the *observed error* (as deduced from the computed value of the total energy) and formula (2) will give the value of c. Formula (2), although empirical, will thus have a predictive value for the accumulation of rounding-off error.

3. THE TRUNCATION ERROR

The differential equation

$$\ddot{x} = f(x,t) \tag{3}$$

with x the position vector and f the force per unit mass, can be discretized using a multistep method in which the value of x at the $(k+1)^{th}$ step is given by:

$$x_{k+1} = x_k + (x_k - x_{k-1}) + h^2 \sum_{j=0}^{m} \gamma_j \nabla^j f_k + \cdots \tag{4}$$

where: $x_k = x(t_k)$, $f_k = f(x_k, t_k)$, $\nabla f(t) = f(t) - f(t-h)$ (the backward difference) and γ_j numerical coefficients which can be computed in a recursive way (Milani and Nobili, 1988). If formula (4) is truncated neglecting the dots (i.e. only up to the mth backward difference is used to compute the next step), the remainder at each step is:

$$R_{m+1} = h^2 \gamma_{m+1} \nabla^{m+1} f_k + \cdots \quad . \tag{5}$$

The *local truncation error* Δx_{k+1} —defined as the value computed with the truncated formula minus the *exact* value given by the infinite series— is *minus* the remainder (5). The generalized Lagrange formula:

$$\nabla^j f(t) = h^j D^j f(t^*) \quad , \quad t > t^* > t - jh \tag{6}$$

allows us to estimate the difference ∇^j with the derivative D^j at some preceding time. Thus,

$$\Delta x_{k+1} = -h^{m+3}\gamma_{m+1}D^{m+3}x(t^*) + \cdots \quad , \quad t > t^* > t-(m+1)h \tag{7}$$

which states that this method of computing the position is correct to order $(m+2)$, the remainder being $\propto h^{m+3}$. However, what really matters is the accumulation of the local truncation error (7) with the number of steps.

The accumulated effect of the truncation error can be computed by assuming that the computation of the next step by means of the truncated formula (4) is exact (i.e. there is no local error), but the system is perturbed by a force Δf which produces the displacement:

$$\Delta x_{k+1} = h^2\gamma_\circ\Delta f = h^2\Delta f \qquad (\gamma_\circ = 1) \tag{8}$$

For this displacement to equal the local truncation error (7) the perturbation Δf which simulates the effect of the truncation error must be:

$$\Delta f = -h^{m+1}\gamma_{m+1}D^{m+3}x(t^*) + \cdots \quad . \tag{9}$$

The problem is thus reduced to the dynamical problem of computing the accumulated along-track effect of Δf.

Let us consider the 2-body problem. The energy variation due to Δf is:

$$\dot{E} = <\Delta f, \dot{x}> \tag{10}$$

and the corresponding acceleration in longitude (to zero order in the eccentricity) is:

$$a\ddot{\lambda} = -\frac{3}{na}\dot{E} \tag{11}$$

(n the mean motion, a the semimajor axis). In our case:

$$a\ddot{\lambda} = \frac{3}{na}h^{m+1}\gamma_{m+1} < D^{m+3}x(t^*), \dot{x}(t) > + \cdots \quad . \tag{12}$$

If the orbit is circular ($e=0$, $\ddot{x}=-n^2x$) and m is even ($m=2s$):

$$a\ddot{\lambda} = -3(-1)^s a\gamma_{2s+1}(nh)^{2s+1}N^2\cos\delta + \cdots \tag{13}$$

with $\cos\delta \simeq 1$ because δ is the angle between velocity vectors at times differing by $(m+1)h$ at most. Therefore, the accumulated error in longitude ($m=2s$, $e=0$) is:

$$\Delta\lambda \simeq -\frac{3}{2}(-1)^s\gamma_{2s+1}(nh)^{2s+3}N^2 \tag{14}$$

If $e=0$ but m is odd ($m=2s+1$) we have:

$$a\ddot{\lambda} \simeq \frac{3}{na}h^{2s+2}\gamma_{2s+2} < D^{2s+4}x(t^*), \dot{x}(t) > \tag{15}$$

and therefore

$$a\ddot{\lambda} \simeq -3(-1)^s a\gamma_{2s+2}(nh)^{2s+2}n^2\sin\delta \tag{16}$$

with $\sin\delta \simeq \delta \simeq nhm/2$. For large enough values of m this term is more important than the next one and the use of an odd order method should be preferred; however, it might be numerically unstable. For instance, in LONGSTOP we used $m=12$ because $m=13$ turned out to be unstable (with the same stepsize). Thus, in the $e=0$ and m odd case the accumulated effect of truncation error in longitude is:

$$\Delta\lambda \simeq -\frac{3}{2}(-1)^s\gamma_{2s+2}(nh)^{2s+4}\frac{nhm}{2}N^2 \quad . \tag{17}$$

Let us now consider the case of an eccentric orbit, planar for semplicity, and m even. The coordinates of the position vector can be developed as follows (Kovalevsky, 1963):

$$x = a\left[\begin{pmatrix} -\frac{3}{2}e \\ 0 \end{pmatrix} + \sum_{p=1}^{\infty}\begin{pmatrix} C(p,e)\cos pl \\ S(p,e)\sqrt{1-e^2}\sin pl \end{pmatrix}\right] \tag{18}$$

where l is the mean anomaly and $C(p,e)$, $S(p,e)$ are combinations of Bessel functions as in Kovalevsky, 1963. Deriving (18) with respect to time, from (9) and (10), $\dot{E}$ is computed; taking the zero harmonic (because we are interested in the secular effect only) and averaging over the mean anomaly we get:

$$\overline{\dot{E}} \simeq (-1)^{s+1} a^2 n^3 \gamma_{m+1} (nh)^{m+1} Z(m,e) \tag{19}$$

with

$$Z(m,e) = \sum_{p=1}^{+\infty} \frac{1}{2} \left[C^2(p,e) + (1-e^2) S^2(p,e) \right] p^{m+4} \quad . \tag{20}$$

Therefore, the accumulated effect in longitude is the same as in the circular case apart for the factor $Z(m,e)$:

$$\Delta\lambda \simeq -\frac{3}{2}(-1)^s \gamma_{2s+1} (nh)^{2s+3} N^2 Z(m,e) \quad , \tag{21}$$

and similarly for the m odd case. Numerical experiments with a 2-body model confirm the validity of these predictions. The sign change every change of $m = 2s$ by 2 units and the dependence on the eccentricity can be used to check whether the error found is due to truncation. In the case of the outer planets, using $m = 12$ and $h = 1/108$ of Jupiter's period ($e = 0.05$, $Z(12, 0.05) \simeq 100$) it is found that the observed error is not due to truncation. The experiments are carried out for 10^5 steps and the error is computed with respect to a more accurate solution obtained using a smaller stepsize

4. THE OBSERVED ERROR IN THE LONGSTOP INTEGRATIONS

Were the numerical integration exact, the total energy of the outer planets should be constant; thus, the total energy as function of time gives the accumulation of the integration error whatever the source which generates it. Figure 1 shows such plots as obtained from the LONGSTOP 1A (Fig. 1a) and 1B (Fig. 1b) integrations. In the simplifying assumption that the error is mostly in Jupiter, the observed longitude error is related to the relative variation of the total energy $\Delta E_{tot}/E_{tot}$ through the equation:

$$\Delta\lambda_{obs} = \frac{3}{2}\pi \frac{\Delta E_{tot}}{E_{tot}} \frac{T/2}{P_{Jup}} \tag{22}$$

where T is the integration time span (the integration is carried out for $T/2$ forward and $T/2$ backward from present so as to have a smaller error over the interval T) and P_{Jup} is the orbital period of Jupiter. In LONGSTOP 1A ($T = 9.26$ Myr, $\Delta E_{tot}/E_{tot} \simeq 3 \times 10^{-10}$) we have:

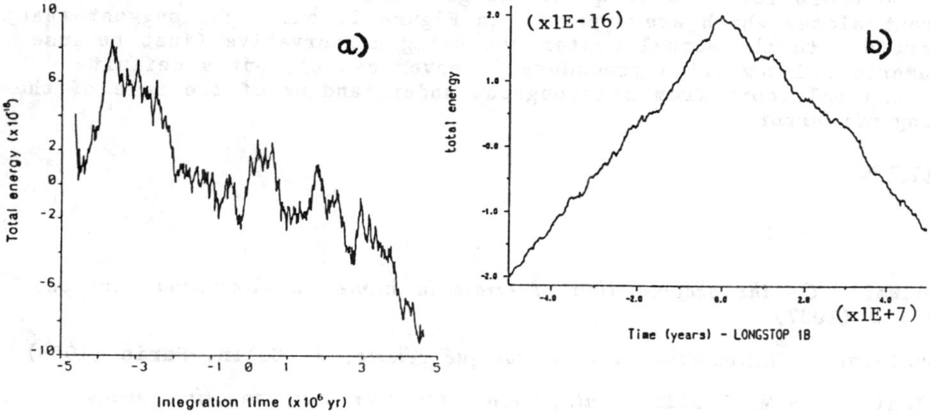

Figure 1. a) Total energy as function of time from the 9.26 Myr integration LONGSTOP 1A (Milani et al., 1986). The average value is -3.2×10^{-8} $M_{\odot}AU^2/day^2$, hence the relative energy variation is $(\Delta E_{tot}/E_{tot})_{1A} \simeq 3 \times 10^{-10}$. b) Total energy from the LONGSTOP 1B 100 Myr integration (Nobili et al., 1988); $(\Delta E_{tot}/E_{tot})_{1B} \simeq 1.3 \times 10^{-8}$

$$(\Delta\lambda_{obs})_{1A} \simeq 0^{\circ}.03 \quad . \tag{23}$$

As expected, the predicted truncation error (from (21)) is much smaller: $(\Delta\lambda_{trunc})_{1A} \simeq -0^{\circ}.0015$ whereas the predicted rounding-off (from (2)) is $(\Delta\lambda_{ro})_{1A} \simeq c0^{\circ}.13$, hence giving $c \simeq 0.2$. However, in the longer integration LONGSTOP 1B ($T = 100$ Myr, $\Delta E_{tot}/E_{tot} \simeq 1.3 \times 10^{-8}$) we have:

$$(\Delta\lambda_{obs})_{1B} \simeq 15^{\circ} \quad , \tag{24}$$

the predicted truncation error is still smaller ($(\Delta\lambda_{trunc})_{1B} \simeq 0^{\circ}\,17$) but the predicted rounding-off is just the same, giving $c \simeq 1$. We also notice from Figure 1 that the variation of the total energy in the LONGSTOP 1A integration does not show the V-shape, with the vertex at the present time, that one would expect. A V-shape does indeed appear in Figure 2, although the two slopes are not the same. We conclude that in very long numerical integrations the dominant source of error is rounding-off, that the error growth is quadratic (in longitude, linear in energy) and it can be expressed with the semi-empirical formula (2) with $c \simeq 1$; it is possible that the LONGSTOP 1A integration is still too short for a fully quadratic growth to dominate. As for the different slopes which are apparent in Figure 2, one might suggest that they are due to the actual system not being conservative (just because the numerical integration procedure is never exact), but a definite answer can only come from a throughout understanding of the role of the rounding-off error.

REFERENCES

D. Brouwer: *On the accumulation of errors in numerical integration*, Astron. J. **46** 199 (1937)

J. Kovalevsky: *Introduction a la mécanique céleste*, A. Colin, Paris (1963)

A. Milani and A.M. Nobili: *Integration error over very long time spans*, submitted (1988)

A. Milani, A.M. Nobili, K. Fox and M. Carpino: *Long-term changes in the semimajor axes of the outer planets*, Nature **319** 386 (1986)

A.M. Nobili, A. Milani and M. Carpino: *Fundamental frequencies and small divisors in the orbits of the outer planets*, submitted (1988)

A.E. Roy: *Orbital motion*, Adam Hilger Ltd., Bristol (1982)

NUMERICAL REGULARIZATION OF THE LAGRANGE PLANETARY EQUATIONS

Daniel E. Snow
Department of Aerospace Engineering and Engineering Mechanics
The University of Texas at Austin,
Austin, Texas, U.S.A.

ABSTRACT: The use of regularization in the numerical integration of Lagrange's Planetary Equations for a specific set of orbit elements is discussed. Comparisons between the integration of the position and velocity with the integration of the orbital elements for various dynamical systems are made on the basis of the speed and accuracy of the integrations. In general it is recommended that singularities appearing in the zeroth order terms be removed with a time transformation to improve the accuracy of numerical integrations.

INTRODUCTION

In the study of the stability of a dynamical system, the accuracy of the data generated is very important, since when evaluating the stability of a system, inaccurate data may lead to inaccurate results. Sometimes different dynamical systems require new methods of generating data to meet accuracy requirements. This paper discusses the initial attempts at developing a set of equations for a specific dynamic system which can be integrated accurately. Results which have general applications will also be drawn from these attempts.

The specific system studied is a spacecraft in a highly eccentric, highly inclined orbit about the Earth. The initial value of the semi-major axis of the spacecraft's orbit is approximately the mean Earth-Moon distance. The spacecraft experiences lunar and solar perturbations as well as perturbations due to the J_2 term in the Earth's geopotential field. Higher order geopotential terms have been found to have negligible effects upon the spacecraft (1). Due to the initial values of the orbital elements of the spacecraft the lunar perturbation can not be strictly classified, in the nomenclature of Stiefel and Scheifele, as that of the interior problem or the exterior problem (2). By this it is meant that the lunar orbit radius is neither an upper nor a lower bound on the spacecraft orbit radius. Therefore any long term study of a spacecraft with these initial conditions in this dynamical system must be numerical.

A. E. Roy (ed.), Long-Term Dynamical Behaviour of Natural and Artificial N-Body Systems, 117–122.
© 1988 by Kluwer Academic Publishers.

To determine the accuracy of a numerical integration some analytical relation between the variables must be found. Due to the complexity of the dynamic model mentioned, a useful constant of integration does not exist for the system. Therefore, two simpler models which possess constants of integration are chosen for preliminary study. These two models are the circular restricted problem of three bodies (3) and the oblate central body model (4). Only these two models are discussed in this paper.

SELECTION OF VARIABLES

The usual choice of variables for numerical integration is the Cartesian rectangular position and velocity components of the body of interest. However, equations also exist for the time rate of change of the Keplerian orbit elements. These equations are known as Lagrange's Planetary Equations (hereafter referred to as the Planetary Equations). Written in their standard formulation, these equations involve the partial derivatives of the disturbing function with respect to the orbit elements (4,5,6). The evaluation of these partial derivatives often makes the use of the Planetary Equations eather difficult. However, since the orbit elements generally change slowly, integrations of this system of equations are usually fast and accurate.

Previous experience suggests that the lunar perturbation is the most troublesome of the disturbances affecting the spacecraft. Bond and Janin state that there is some advantage to using eccentric anomaly when studying the effect of third-body perturbations upon a body (7). Therefore, the system of orbit elements used in this study are the semi-major axis (a), eccentricity (e), inclination (i), eccentric anomaly (E), argument of the periapse (ω), and the longitude of the ascending node (Ω). Even though the oblate central body model does not involve third-body perturbations, the above system of orbit elements is used for the study of this model since the real interest is in integrating the dynamic system initially discussed. This system does involve third-body perturbations.

REGULARIZATION

The presence of terms which become very large in the Planetary Equations presents problems when performing numerical integrations. These terms (called singularities hereafter) cause losses in both the speed and accuracy of the integration. Two types of singularities occur in the Planetary Equations. Geometric singularities are functions of the orbit elements which can become large (8). These singularities arise from the shape of orbit and the variables used to describe the state of motion. The second type of singularity which occurs is the dynamic singularity. Dynamic singularities are due to the forces in the dynamic system.

To eliminate the singularities in the equations of interest, regularization is used. In regularization a new independent variable s is introduced through a differential relation between s and the real time t. This differential relation is known as a time transformation. By a proper choice of the time transformation, the new system of equations will be free of singularities. Since the Planetary Equations are of first order, changing independent variables (as well as choosing the differential relation between s and t) is a relatively simple matter, as opposed to when the equations of interest are of second order (2,3). While the equations are often not being regularized in a mathematical sense (since all the singularities are not analytic in nature) the process described is similar to regularization and it shall be referred to as numerical regularization. It is noted that this definition of regularization differs from others which have appeared in the literature (9,10,11).

NUMERICAL RESULTS

The computer system used for this study is the University of Texas Dual Cyber 175/750 system. The numerical integration scheme is called ODE and is documented in the text by Shampine and Gordon (12).

In the circular restricted three-body model, the position and velocity of the spacecraft were integrated for three years. The CPU time used was called 1.0 for comparison purposes. The constant of integration for this model, known as the Jacobian integral, was compared throughout the integration and was found to agree with its initial value to six significant figures. Next, the Planetary Equations were integrated using the time transformation

$$dt = F(a,e,i)\, r\, r_2^2\, ds \quad ,$$

where

$F(a,e,i) = a^{-5/2}\, e\, (1-e^2)^{1/2} \sin(i)$
r = the satellite distance from the Earth
r_2 = the satellite distance from the Moon.

With this system of equations, the CPU time was 0.70 and the Jacobian agreed with its initial value throughout the integration to ten significant figures.

The oblate central body problem possesses two constants of integration which were used for the comparison; the constant of energy and the z component of angular momentum (4). When the position and velocity of the spacecraft were integrated there was agreement with the initial value of energy to eight significant figures throughout the integration. The z component of angular momentum stayed constant to twelve significant figures. Integrating the Planetary Equations using the time transformation

$$dt = F(a,e,i)\, r^5\, ds$$

produced a slight increase in accuracy (nine significant figures in the energy integral, twelve in the z component of angular momentum) but the CPU time used increased by forty percent. A study was then undertaken to determine the "best" integer value for the power of r in the time transformation. The range of values studied was from -2 to 6, inclusive. The best results were produced by using the time transformation

$$dt = F(a,e,i)\, r\, ds$$

While maintaining the accuracy achieved by using the previously discussed time transformation, the CPU time used was only one half of that in the integration of the position and velocity of the spacecraft.

ANALYSIS

To explain the results stated in the previous section, the Planetary Equations are examined with only the largest (or zeroth order) terms. After the time transformation under examination is applied to the Planetary Equations, these equations are expanded in Fourier Series with the new independent variable as the argument.

The time transformation for which the magnitude of the coefficients of the Fourier Series converges to zero the fastest will be the time transformation which optimizes the speed and accuracy of the numerical integration for the system of equations. The justification for this statement is twofold. First, a differencing process occurs in the integration subroutine to check if any error tolerances are met. This differencing process causes the terms with small coefficients to effectively vanish during the error tolerance check. The fact that the computer has a finite word length also means that terms with very small coefficients are ignorable. The maximum step size which maintains a high degree of accuracy is therefore directly related to the highest frequency term in the Fourier Series which has a non-negligible coefficient. To maximize the allowable step size we need to find the time transformation which minimizes the frequency of this term. This approach of only considering the zeroth order terms when choosing a time transformation has been used before (2,13).

From observation, the effect of the central body oblateness upon the orbital elements is always a first order effect. Therefore the elements a, e, i, ω, and Ω are constants during the analysis of the J_2 case. The time transformation which yields the desired result is

$$dt = F(a,e,i)\, r\, ds \quad .$$

With this time transformation the eccentric anomaly is a linear function of s and the relation between time and s is simply a form of Kepler's equation (noting that a, e, and i are constants).

When this same analysis is applied to the circular restricted three-body problem model, the result is naturally different. The third body effect can actually be the dominant term in the equations, depending upon the magnitude of the Moon-spacecraft distance. When this occurs, the third body "perturbation" is actually a zeroth order term and can not be ignored. An example of this is whenever a spacecraft orbits the Moon and the spacecraft's motion is described in an Earth centered coordinate system. While choosing the "best" time transformation is extremely difficult, one which involves the Moon-spacecraft distance should prove useful. The square of the Moon-spacecraft distance was chosen because this quantity occurs naturally in the Planetary Equations in this case.

CONCLUSIONS

Two types of conclusions can be drawn from the results stated. The first type pertains only to the specific systems studied. After these results have been summarized the second type, which are more general in nature, are stated. It is hoped that these results will be of use to the reader.

When using the system of orbit elements a, e, i, ω, Ω, and E to study the motion of a satellite with the initial conditions stated moving under the influence of an oblate central body, the time transformation

$$dt = F(a,e,i)\; r\; ds$$

with F(a,e,i) defined as previously, is the best time transformation involving an integer power of r. If the dynamic model describing the motion is instead the circular restricted three body problem model, the time transformation

$$dt = F(a,e,i)\; r\; r_2^2\; ds$$

is useful.

In general, the use of time transformations which remove singularities from the zeroth order terms will prove useful. However, caution must be exercised when trying to determine which terms are or are not zeroth order terms. It is noted that the above result also applies to regularization of the many-body problem(14).

ACKNOWLEDGEMENTS

This research was supported by Air Force Space Command contract #F05603-86-C-0014. The author would also like to thank Prof. V. Szebehely for his guidance during this project and Mr. D. Tuckness for generating some of the computer runs used in the analysis.

REFERENCES

1. Private Communication with Prof. V. Szebehely and Dr. J.J. F. Liu.
2. Stiefel, E. and Schiefele, G.: 1971, Linear and Regular Celestial Mechanics, Springer-Verlag, New York.
3. Szebehely, V.: 1967, Theory of Orbits, Academic Press, New York.
4. Brouwer, D. and Clemence, G.: 1961, Methods of Celestial Mechanics, Academic Press, New York.
5. Kovalevsky, J.: 1967, Introduction to Celestial Mechanics, Reidel, Dordrecht.
6. Roy, A.: 1978, Orbital Motion, Adam Hilger, Bristol.
7. Bond, V. and Janin, G.: 1981, Celest. Mech. 23, 159-172.
8. Born, G.: 1968, "The Derivation of a General Perturbation Solution and its Application to Orbit Determination", Ph.D. Dissertation, Univ. of Texas.
9. Saari, D.: 1974, Celest. Mech. 9, 55-72.
10. Tapley, B.: 1970, Celest. Mech. 2, 319-333.
11. Siegel, C. and Moser, J.: 1971, Lectures on Celestial Mechanics, Springer-Verlag, New York.
12. Shampine, L., and Gordon, M.: Computer Solution of Ordinary Differential Equations, Freeman, San Francisco.
13. Stiefel, E.: 1970, Celest. Mech. 2, 274-281.
14. Heggie, D.: In This Volume.

PART II

DYNAMICS OF NATURAL AND ARTIFICIAL SATELLITES

GENERAL FORMULAS FOR THREE-BODY RESONANCES

K. Aksnes
Norwegian Defence Research Establishment
P.O. Box N-2007 Kjeller and University
of Tromsø, N-9001 Tromsø

ABSTRACT

Analytic formulas are derived for the strengths and libration/circulation characteristics of resonances whose critical arguments take the form $\Theta = q\lambda_1 - (p+q)\lambda_2 + p\lambda_3$, where p and q are integers and λ_1, λ_2, and λ_3 are the mean longitudes of any three gravitationally interacting bodies revolving in coplanar orbits about the same primary. The formulas extend and generalize earlier results published by several authors on Laplacian resonances in the satellite systems of Jupiter and Uranus.

1 INTRODUCTION

This investigation grew out of an attempt by Dermott and Gold (1977) to explain the locations of the Uranian rings in terms of resonances between ring particles and pairs of satellites. It is assumed that the mean motions n_1, n_2, and n_3 of a massless ring particle and two satellites satisfy the relation

$$qn_1 - (p+q)n_2 + pn_3 = 0 \tag{1}$$

where p and q are non-zero integers.

Let now n_1, n_2, and n_3 be the mean motions of any three bodies of finite masses revolving about the same primary. It is easier to picture this three-body commensurability if we rewrite Equation (1) in the form

$$\frac{p}{n_1-n_2} = \frac{q}{n_2-n_3} = \frac{p+q}{n_1-n_3} = \frac{T}{2\pi}, \tag{2}$$

i.e.; during the interval T between consecutive alignments of the three bodies, body 1 makes p and p+q revolutions relative to body 2 and body 3, respectively, and body 2 makes q revolutions relative to body 3. We can therefore say that the mean motions are pairwise commensurate in a <u>rotating</u> frame - instead of in a fixed frame, as for two-body resonances.

A. E. Roy (ed.), Long-Term Dynamical Behaviour of Natural and Artificial N-Body Systems, 125–139.
© 1988 by Kluwer Academic Publishers.

It has been shown independently by Goldreich and Nicholson (1977) and Aksnes (1977, 1982) that these three-body resonances are probably too weak to explain the confinement of the Uranian rings at distances from Uranus most of which are strikingly close to the distances predicted by the Dermott-Gold ring theory. It is, nevertheless, of both practical and theoretical interest to derive general formulas for the three-body resonances, since there are other applications involving triplets of satellites or pairs of major planets plus one minor planet.

The present work provides a theoretical basis for the author's results just referred to and extends and generalizes earlier work by other authors who have studied the so-called Laplacian three-body resonances with q=1 and p=2 in Equation (1). These earlier results will be used to test the present theory.

As the name implies, the Laplacian three-body resonance was first studied by Laplace, and it involves the three inner Galilean satellites of Jupiter. Laplace found that the so-called critical argument,

$$\Theta = q\lambda_1 - (p+q)\lambda_2 + p\lambda_3 \quad , \tag{3}$$

where the λ's are the mean longitudes of the satellites, librates slowly and with a very small amplitude about the value 180°, thereby maintaining a stable configuration. The three inner Uranian satellites also nearly satisfy the Laplacian relation (q=1, p12), but in this case Θ appears to circulate (increases or decreases indefinitely with time). Harris (1949), Sinclair (1975), Greenberg (1976), and Veillet (1981) have investigated this problem both observationally and theoretically.

A dynamical problem equivalent to the Dermott-Gold ring problem but involving minor planets instead of ring particles, was investigated in considerable detail by Wilkens (1933), Okyay (1935), and Urban (1935). They found that of the about 1200 numbered minor planets at the time, many satisfy Equation (1) rather closely together with either Mars and Jupiter or Jupiter and Saturn, for values of $|p|$ and $|q|$ between 1 and 3. Unfortunately, their conclusions regarding libration/circulation are invalidated by computational errors. When repeating their calculations with correct formulas, no convincing case of a physically meaningful three-body resonance involving one minor and two major planets has been found.

It may be worth mentioning that the slow-moving Chiron is the only minor planet in a low-order (p=-2, q=3) near-commensurability with Saturn and Uranus.

2. THEORY

We begin by differentiating Equation (3) twice with respect to the time t:

$$\frac{d^2\Theta}{dt^2} = q\,\frac{d^2\lambda_1}{dt^2} - (p+q)\,\frac{d^2\lambda_2}{dt^2} + p\,\frac{d^2\lambda_3}{dt^2}\;. \tag{4}$$

For the special case p=2 and q=1 of a minor planet perturbed by Jupiter and Saturn, Wilkens (1933) has shown how to transform this differential equation into one describing the motion of a perturbed harmonic oscillator of "strength" c,

$$\frac{d^2\Theta}{dt^2} = c\,\sin\Theta + e\,\sin(\alpha t + \beta). \tag{5}$$

Here $e\,\sin(\alpha t + \beta)$ represents a dominant perturbation of long period - e.g. the "great inequality" due to the 2:5 near commensurability between the mean motions of Jupiter and Saturn. The influence of such a perturbation could be treated analytically, e.g. by a method devised by Brown and Shook (1933). We shall, however, ignore the term here. In that case, an integral of Equation (5) is

$$\frac{d\Theta}{dt} = (2C - 2c\,\cos\Theta)^{\frac{1}{2}}\;, \tag{6}$$

where the constant of integration C is to be obtained from the values of the orbital elements at t=0.

$$\frac{1}{2}\left(\frac{d\Theta}{dt}\right)_o^2 = \frac{1}{2}\,[qn_1 - (p+q)n_2 + pn_3]_o^2 = C - c\,\cos\Theta_o\;. \tag{7}$$

From Equation (6) it follows that Θ will librate if $|C/c| < 1$, the libration being about 0° if $c < 0$ and about 180° if $c > 0$, and circulate if $|C/c| > 1$. In both cases, Θ can be expressed in terms of elliptic integrals. Here we are only interested in the amplitude A and period P of the libration:

$$\begin{aligned} A &= \arccos(C/c) \\ p &= 4|c|^{-\frac{1}{2}} \int_o^{\pi/2} \frac{d\phi}{(1-\sin^2\frac{1}{2}A.\sin^2\phi)^{\frac{1}{2}}} \\ &= 4|c|^{-\frac{1}{2}}\,F(\sin\tfrac{1}{2}A,\,\tfrac{1}{2}\pi)\;, \end{aligned} \tag{8}$$

where F represents a hypergeometric series.

We note that as $A \to 0$, $P \to 2\pi\,|c|^{-\frac{1}{2}}$.

Wilkens (1933) derived a semi-analytic formula for c, for the special case mentioned, by making use of an existing numerical theory for the perturbations in the orbits of Jupiter and Saturn. An analytic self-contained formula for c will be derived here, valid for all possible three-body resonances between bodies of finite masses m_1, m_2, and m_3 (of which one is negligible in Wilkens' case).

We shall ignore the mutual inclinations of the orbits. Since the eccentricities are also assumed to be small, Wilkens believed that it was sufficient to consider only the terms in disturbing function of the zeroth order in the eccentricities. However, as we shall see, also the first-order terms actually contribute terms in c of order zero.

Since the mean longitudes λ_1, λ_2, and λ_3 will always occur in pairs in the arguments of the disturbing functions of the bodies, the critical argument Θ in Equation (5) must arise as the difference $\alpha_i - \beta_i$ from products of the kind $\sin\alpha_i \cos\beta_i$ or $\cos\alpha_i \sin\beta_i$ of second order in the masses. Here α_i and β_i can be paired in only three different ways:

$$
\begin{aligned}
\alpha_1^{(k)} &= (p+q-k)\lambda_1 - (p+q)\lambda_2 + k\tilde{\omega}_1 , \\
\beta_1^{(k)} &= (p-k)\lambda_1 - p\lambda_3 + k\tilde{\omega}_1 , \\
\alpha_2^{(k)} &= q\lambda_1 - (q+k)\lambda_2 + k\tilde{\omega}_2 , \\
\beta_2^{(k)} &= (p-k)\lambda_2 - p\lambda_3 + k\tilde{\omega}_2 , \\
\alpha_3^{(k)} &= q\lambda_1 - (q+k)\lambda_3 + k\tilde{\omega}_3 , \\
\beta_3^{(k)} &= (p+q)\lambda_2 - (p+q+k)\lambda_3 + k\tilde{\omega}_3 ,
\end{aligned}
\tag{9}
$$

where $\tilde{\omega}_1$, $\tilde{\omega}_2$, and $\tilde{\omega}_3$ denote the longitudes of the pericenters and $k=0, \pm1, \pm2, ---$. The terms in the disturbing function R_i of the i-th body of orders 0 and 1 in the eccentricities are (Brouwer and Clemence 1961, pp 490-494),

$$
\begin{aligned}
R_i &= G \sum_j m_j \left(\frac{1}{\Delta_{ij}} - \frac{\bar{r}_i \cdot \bar{r}_j}{r_j^3}\right) = R_i^{(o)} + R_i^{(1)} + --- , \\
R_i^{(o)} &= G\sum_j m_j \sum_{k=-\infty}^{\infty} A_{ij}^{(k)} \cos(k\lambda_i - k\lambda_j) , \\
R_i^{(1)} &= G\sum_j m_j \sum_{j=-\infty}^{\infty} \{e_i B_{ij}^{(k)} \cos[(k-1)\lambda_i - k\lambda_j + \tilde{\omega}_i] \\
&\quad + e_j C_{ij}^{(k)} \cos[k\lambda_i - (k+1)\lambda_j + \tilde{\omega}_j]\}.
\end{aligned}
\tag{10}
$$

Here

$$
G \approx G(1 + m_i) = n_i^2 a_i^3 ,
$$

$$
A_{ij}^{(k)} = f^{(k)} - \frac{1}{2} \frac{a_i}{a_j^2} (\delta_{1,-k} + \delta_{1,k}) ,
$$

$$B_{ij}^{(k)} = \left.\begin{matrix} g^{(k)} \\ h^{(-k)} \end{matrix}\right] - \frac{1}{2} \frac{a_i}{a_j^2} (\delta_{1,-k} - 3\delta_{1,k}) \quad , \tag{11}$$

$$C_{ij}^{(k)} = \left.\begin{matrix} h^{(k)} \\ g^{(-k)} \end{matrix}\right] - 2 \frac{a_i}{a_j^2} \delta_{1,k} \quad ,$$

where the upper values in front of the brackets are valid for $a_i < a_j$ (inner body perturbed by outer body) and the lower values for $a_i > a_j$ (outer body perturbed by inner body). The terms involving the Kronecker deltas ($\delta_{i,j}$ equals 1 for i=j and 0 for i≠j) come from the indirect part of the disturbing function whose direct part involves the functions

$$f^{(k)} = \frac{1}{2a} b_{1/2}^{(k)}(\alpha) \quad ,$$

$$g^{(k)} = - \frac{1}{2a} (2k + \alpha\frac{d}{d\alpha}) b_{1/2}^{(k)}(\alpha) \quad , \tag{12}$$

$$h^{(k)} = \frac{1}{2a} (2k + 1 + \alpha\frac{d}{d\alpha}) b_{1/2}^{(k)}(\alpha) \quad ,$$

where a denotes the biggest of the two semimajor axes a_i and a_j, and α equals the ratio of the smallest to the largest semimajor axis. The Laplace coefficient $b_{1/2}^{(k)}$ and its derivatives are defined in Appendix A.

We shall first derive an expression for the part $c^{(o)}$ of c in Equation (5) arising from $R_i^{(o)}$. When ignoring the eccentricities, we have from Lagrange's variational equations,

$$\frac{da_i}{dt} = \frac{2}{n_i a_i} \frac{\partial R_i^{(o)}}{\partial \lambda_i} \quad ,$$

$$\frac{d\lambda_i}{dt} = n_i + \frac{d\varepsilon_i}{dt} = n_i - \frac{2}{n_i a_i} \frac{\partial R_i^{(o)}}{\partial a_i} \tag{13}$$

where n_i is related to λ_i through the equation ,

$$\lambda_i = \int n_i dt + \varepsilon_i \ . \tag{14}$$

We find,

$$\frac{da_i}{dt} = \sum_j \sum_{k=-\infty}^{\infty} m_j a_{ij}^{(k)} \sin(k\lambda_i - k\lambda_j) \ ,$$

$$\frac{d\lambda_i}{dt} = n_i + \sum_j \sum_{k=-\infty}^{\infty} m_j \varepsilon_{ij}^{(k)} \cos(k\lambda_i - k\lambda_j) \tag{15}$$

$$= n_{io} + \sum_j \sum_{k=-\infty}^{\infty} m_j \lambda_{ij}^{(k)} \cos(k\lambda_i - k\lambda_j)$$

where n_{io} is the constant part of the osculating mean motion n_i, given by

$$n_i = n_{io} + \frac{\partial n_i}{\partial a_i} \Delta a_i = n_{io} - \frac{3n_i}{2a_i} \int \frac{da_i}{dt} dt \quad .$$

We readily deduce that

$$a_{ij}^{(k)} = -2n_i a_i^2 \; kA_{ij}^{(k)} \quad ,$$

$$\varepsilon_{ij}^{(k)} = -2n_i a_i^2 \; \frac{\partial A_{ij}^{(k)}}{\partial a_i} \quad , \tag{16}$$

$$\lambda_{ij}^{(k)} = \varepsilon_{ij}^{(k)} + \frac{3n_i \; a_{ij}^{(k)}}{2a_i k(n_{io} - n_{jo})} \quad ,$$

where n_{io} and n_{jo} are constant under differentiation.

We treat first the part

$$\frac{d^2\lambda_1}{dt^2} = \frac{dn_1}{dt} + \frac{d^2\varepsilon_1}{dt^2}$$

in Equation (4). Previous investigators have ignored the last term here, i.e., the acceleration of the mean longitude ε_1 at epoch, which may be significant. In view of Equations (9), (13), and (15), the only part of $d\varepsilon_1/dt$ that can contribute to $c^{(o)}$ is given by

$$\frac{d\varepsilon_1}{dt} = 2 \; [m_2 \varepsilon_{12}^{(p+q)} \cos\alpha_1^{(o)} + m_2 \varepsilon_{12}^{(q)} \cos\alpha_2^{(o)}$$

$$+ m_3 \varepsilon_{13}^{(q)} \cos\alpha_3^{(o)} + m_3 \varepsilon_{13}^{(p)} \cos\beta_1^{(o)} \;].$$

Differentiating this expression with respect to t and again ignoring terms which are not of interest, we find

$$\frac{d^2\varepsilon_1}{dt^2} = 2m_2 \; \{-(p+q)\varepsilon_{12}^{(p+q)} \frac{d\lambda_1}{dt} \sin\alpha_1^{(o)} + \frac{\partial \varepsilon_{12}^{(p+q)}}{\partial a_1} \frac{da_1}{dt} \cos\alpha_1^{(o)} \; \}$$

$$+ 2m_2 \{q\varepsilon_{12}^{(q)} \frac{d\lambda_2}{dt} \sin\alpha_2^{(o)} + \frac{\partial \varepsilon_{12}^{(q)}}{\partial a_2} \frac{da_2}{dt} \cos\alpha_2^{(o)} \; \}$$

$$+ 2m_3 \{q\varepsilon_{13}^{(q)} \frac{d\lambda_3}{dt} \sin\alpha_3^{(o)} + \frac{\partial \varepsilon_{13}^{(q)}}{\partial a_3} \frac{da_3}{dt} \cos\alpha_3^{(o)} \; \}$$

$$+ 2m_3 \{-p\varepsilon_{13}^{(p)} \frac{d\lambda_1}{dt} \sin\beta_1^{(o)} + \frac{\partial \varepsilon_{13}^{(p)}}{\partial a_1} \frac{da_1}{dt} \cos\beta_1^{(o)} \; \} \quad .$$

where we substitute for da_i/dt and $d\lambda_i/dt$ (in terms of $\lambda_{ij}^{(k)}$ only those terms in Equations (15) that will contribute terms in $d^2\varepsilon_1/dt^2$ of argument Θ .

Similarly, the relevant part of dn_1/dt is

$$\frac{dn_1}{dt} = -\frac{3}{2}\frac{n_1}{a_1}\frac{da_1}{dt}$$

$$= -3\frac{n_1}{a_1}\{ m_2 a_{12}^{(p+q)} \sin\alpha_1^{(o)} + m_2 a_{12}^{(q)} \sin\alpha_2^{(o)}$$

$$+ m_3 a_{13}^{(q)} \sin\alpha_3^{(o)} + m_3 a_{13}^{(p)} \sin\beta_1^{(o)} \} \quad .$$

We put $a_i = a_{io} + \Delta a_i$ and $\lambda_i = \lambda_{io} + \Delta\lambda_i$, where Δa_i and $\Delta\lambda_i$ are first-order perturbations, into this expression. We then expand it to the first order and obtain an expression similar to that for $d^2\varepsilon_1/dt^2$, but with sines and cosines interchanged and da_i/dt and $d\lambda_i/dt$ replaced by Δa_i and $\Delta\lambda_i$ given by

$$\Delta a_i = -\sum_j \sum_{k=-\infty}^{\infty} \frac{m_j a_{ij}^{(k)}}{k(n_i - n_j)} \cos(k\lambda_i - k\lambda_j) \quad , \tag{17}$$

$$\Delta\lambda_i = \sum_j \sum_{k=-\infty}^{\infty} \frac{m_j \lambda_{ij}^{(k)}}{k(n_i - n_j)} \sin(k\lambda_i - k\lambda_j) \ .$$

The final result is that the terms $qd^2\lambda_1/dt^2$ in Equation (4) contributes to $c^{(o)}$ an amount $c_1^{(o)}(p,q)$, given by

$$\begin{aligned} c^{(o)}(p,q) = 2qm_2m_3\{ & qb_{12}^{(q)} \lambda_{23}^{(p)} - \frac{\partial b_{12}^{(q)}}{\partial a_2} a_{23}^{(p)} \\ & + qb_{13}^{(q)} \lambda_{32}^{(p+q)} + \frac{\partial b_{13}^{(q)}}{\partial a_3} a_{32}^{(p+q)} \\ & - (p+q)b_{12}^{(p+q)} \lambda_{13}^{(p)} - \frac{\partial b_{12}^{(p+q)}}{\partial a_1} a_{13}^{(p)} \\ & + pb_{13}^{(p)} \lambda_{12}^{(p+q)} + \frac{\partial b_{13}^{(p)}}{\partial a_1} a_{12}^{(p+q)} \} \quad . \end{aligned} \tag{18}$$

where

$$b_{ij}^{(k)} = 2\varepsilon_{ij}^{(k)} - \lambda_{ij}^{(k)} \quad . \tag{19}$$

Note that Equations (15) imply that $a_{ij}^{(k)} = -a_{ij}^{(-k)}$, while $b_{ij}^{(k)}$, $\varepsilon_{ij}^{(k)}$, and $\lambda_{ij}^{(k)}$ do not change sign with k.

The two remaining parts of $c^{(o)}$ can be obtained from Equation (18) through cyclic permutation of the subscripts 1, 2, and 3 and appropriate changes of the arguments p and q (see Equation (4)):

$$c^{(o)} = c_1^{(o)}\ (p,q) + c_2^{(o)}(q,-p-q) + c_3^{(o)}(-p-q,p)\ , \tag{20}$$

where $c_1^{(o)}$, $c_2^{(o)}$, and $c_3^{(o)}$ are factored by m_2m_3, m_3m_1, and m_1m_2, respectively. The equations (18) - (20) are clearly valid irrespective of how the bodies have been ordered in Equation (1).

We now begin a parallel treatment to find the part $c^{(1)}$ of c due to $R_i^{(1)}$. Corresponding to the Equations (15), we have to the zeroth order in e_i:

$$\frac{de_i}{dt} = -\frac{\partial R_i^{(1)}/\partial\tilde{\omega}_i}{n_i a_i^2 e_i} = v_i \sum_j \sum_{k=-\infty}^{\infty} m_j B_{ij}^{(k)} \sin[(k-1)\lambda_i - k\lambda_j + \tilde{\omega}_i]\ .$$

$$e_i \frac{d\tilde{\omega}_i}{dt} = \frac{\partial R_i^{(1)}/\partial e_i}{n_i a_i^2} = v_i \sum_j \sum_{k=-\infty}^{\infty} m_j B_{ij}^{(k)} \cos[(k-1)\lambda_i - k\lambda_j + \tilde{\omega}_i]\ . \tag{21}$$

where $v_i = n_i a_i$. The only relevant part of the disturbing function for body 1 is

$$R^{(1)} = G \sum_{\pm(p,q)} [m_2\{e_1 B_{12}^{(p+q)} \cos\alpha_1^{(1)} + e_2\, C_{12}^{(q)} \cos\alpha_2^{(1)}\}$$
$$+ m_3\{e_3\, C_{13}^{(q)} \cos\alpha_3^{(1)} + e_1 B_{13}^{(p)} \cos\beta_1^{(1)}\}], \tag{22}$$

where the arguments are given by Equations (9) with $k=\pm1$. The summation here means that to the term in square brackets in Equation (22) is to be added the same term but with p replaced by -p and q by -q. The part of $d\varepsilon_1/dt$ of lowest order is then

$$\frac{d\varepsilon_1}{dt} = -\frac{2}{v_1}\frac{\partial R_1^{(1)}}{\partial a_1} + \frac{1}{2}\frac{e_1}{n_1 a_1^2}\frac{\partial R_1^{(1)}}{\partial e_1}$$
$$= -2v_1 a_1 \sum_{\pm(p,q)} \{m_2 e_1 \frac{\partial B_{12}^{(p+q)}}{\partial a_1} \cos\alpha_1^{(1)} + m_2 e_2 \frac{\partial C_{12}^{(q)}}{\partial a_1} \cos\alpha_2^{(1)}$$
$$+ m_3 e_3 \frac{\partial C_{13}^{(q)}}{\partial a_1} \cos\alpha_3^{(1)} + m_3 e_1 \frac{\partial B_{13}^{(p)}}{\partial a_1} \cos\beta_1^{(1)}\}\ .$$

In the last expression we have neglected the contributions arising from $\partial R_1^{(1)}/\partial e_1$ since they are found to cancel in the final Equation (24). We next differentiate this expression with respect to t and retain only the part that can contribute terms of the type c.sinΘ, where c is of order zero in the eccentricities:

$$\frac{d^2\varepsilon_1}{dt^2} = -2v_1 a_1 \Sigma_{\pm(p,q)} \{ m_2 \frac{\partial B_{12}^{(p+q)}}{\partial a_1} [\frac{de_1}{dt} \cos\alpha_1^{(1)} - e_1 \frac{d\tilde{\omega}_1}{dt} \sin\alpha_1^{(1)}]$$

$$+ m_2 \frac{\partial C_{12}^{(q)}}{\partial a_1} [\frac{de_2}{dt} \cos\alpha_2^{(1)} - e_2 \frac{d\tilde{\omega}_2}{dt} \sin\alpha_2^{(1)}]$$

$$+ m_3 \frac{\partial C_{13}^{(q)}}{\partial a} [\frac{de_3}{dt} \cos\alpha_3^{(1)} - e_3 \frac{d\tilde{\omega}_3}{dt} \sin\alpha_3^{(1)}]$$

$$+ m_3 \frac{\partial B_{13}^{(p)}}{\partial a_1} [\frac{de_1}{dt} \cos\beta_1^{(1)} - e_1 \frac{d\tilde{\omega}_1}{dt} \sin\beta_1^{(1)}] \} .$$

We substitute from Equations (21) only those terms that will make each bracket here contain the factor $\sin\Theta = \sin(\alpha_i - \beta_i)$.

The part of dn_1/dt of interest is

$$\frac{dn_1}{dt} = -\frac{3}{2}\frac{n_1}{a_1}\frac{da_1}{dt} = \frac{-3}{a_1^2}\frac{\partial R_1^{(1)}}{\partial \lambda_1}$$

$$= 3n_1 v_1 \Sigma_{\pm(p,q)} \{ m_2(p+q-1)e_1 B_{12}^{(p+q)} \sin\alpha_1^{(1)} + m_2 q e_2 C_{12}^{(q)} \sin\alpha_2^{(1)}$$

$$+ m_3 q e_3 C_{13}^{(q)} \sin\alpha_3^{(1)} + m_3(p-1)e_1 B_{13}^{(p)} \sin\beta_1^{(1)} \} .$$

We expand this expression to the first order in $e_i = e_{io} + \Delta e_i$ and $\tilde{\omega}_i = \tilde{\omega}_{io} + \Delta\tilde{\omega}_i$, keeping only the terms factored by Δe_i and $\Delta\tilde{\omega}_i$, as obtained from Equations (21):

$$\Delta e_i = -v_i \sum_j \sum_{k=-\infty}^{\infty} \frac{m_j B_{ij}^{(k)}}{(k-1)n_i - kn_j} \cos[(k-1)\lambda_i - k\lambda_j + \tilde{\omega}_i] ,$$

$$e_i \Delta\tilde{\omega}_i = v_i \sum_j \sum_{k=-\infty}^{\infty} \frac{m_j B_{ij}^{(k)}}{(k-1)n_i - kn_j} \sin[(k-1)\lambda_i - k\lambda_j + \tilde{\omega}_i] . \qquad (23)$$

The collected result is that the part $c^{(1)}(p,q)$ of $c^{(1)}$, arising from the term $q\,d^2\lambda_1/dt^2$, takes the form

$$c_1^{(1)}(p,q) = qv_1m_2m_3\Sigma_{\pm(p,q)}\{v_2B_{23}^{(p)}(2a_1\frac{\partial C_{12}^{(q)}}{\partial a_1} - 3n_1\tilde{C}_{12}^{(q)})$$

$$+ v_3B_{32}^{(-p-q)}(2a_1\frac{\partial C_{13}^{(q)}}{\partial a_1} - 3n_1\tilde{C}_{13}^{(q)})$$

$$+ v_1B_{13}^{(p)}(2a_1\frac{\partial B_{12}^{(p+q)}}{\partial a_1} - 3n_1\tilde{B}_{12}^{(p+q)}) \tag{24}$$

$$- v_1B_{12}^{(p+q)}(2a_1\frac{\partial B_{13}^{(p)}}{\partial a_1} - 3n_1\tilde{B}_{13}^{(p)})\} .$$

where

$$\tilde{B}_{ij}^{(k)} = \frac{(k-1)B_{ij}^{(k)}}{(k-1)n_i-kn_j} , \tag{25}$$

$$\tilde{C}_{ij}^{(k)} = \frac{kC_{ij}^{(k)}}{kn_i-(k+1)n_j}$$

Note the magnification caused by small denominators here for near-resonance cases.

By means of the expression (20) for $c^{(o)}$ and the analogous one for $c^{(1)}$,

$$c^{(1)} = c_1^{(1)}(p,q) + c_2^{(1)}(q,-p-q) + c_3^{(1)}(-p-q,p) , \tag{26}$$

we can finally evaluate c as

$$c = c^{(o)} + c^{(1)} . \tag{27}$$

3 APPLICATIONS TO THE LAPLACIAN RESONANCES

The theory has previously (Aksnes 1977, 1982) been used to investigate possible three-body resonances in the Uranian rings and amongst triplets of satellites of Jupiter, Saturn, and Uranus. We shall here look more closely at the Laplacian satellite resonances.

For the three inner Galilean satellites we have the pairwise near-commensurabilities,

$$n_1 - 2n_2 = n_2 - 2n_3 = 0.007 \text{ deg/day}, \tag{28}$$

and hence the exact triple commensurability,

$$n_1 - 3n_2 + 2n_3 = 0. \tag{29}$$

Then we obtain (with q=1, p=2)

$$\Theta = \lambda_1 - 3\lambda_2 + 2\lambda_3 ,$$

$$\ddot{\Theta} = c \sin\Theta , \qquad (30)$$

where, by means of Equations (18-20) and (24-27),

$$c = (3.31 m_2 m_3 + 11.93 m_3 m_1 + 2.05 m_1 m_2) \times 10^6 \text{ deg/day}^2 . \qquad (31)$$

If we insert the satellite mass values used by Tisserand (1896), we deduce a libration period,

$$p = \frac{360}{\sqrt{c}} = 2260 \text{ days.}$$

As mentioned earlier, all previous investigators have neglected the contribution to c from the accelerations $\ddot{\varepsilon}_i$ in the mean longitudes at epoch. When this contribution is excluded, p = 2274 days which agrees well with the value 2270 days quoted by Tisserand. In this special case, c is dominated by resonance magnification in the coefficients $\tilde{C}^{(1)}_{12}$, $\tilde{C}^{(1)}_{23}$, $\tilde{B}^{(2)}_{12}$, and $\tilde{B}^{(2)}_{23}$ which involve the small divisors in Equation (28) but no $\ddot{\varepsilon}_i$-terms.

For all initial conditions it is found that $\Theta_o \approx 180^o$ and $\dot{\Theta}_o \approx 0$ to within the limit of observation, so that Θ librates about 180° with an amplitude so small that it is scarcely detectable, in accordance with Equations (7) and (8).

For the Uranian satellites Miranda, Ariel, and Umbriel there are no pairwise commensurabilities since

$$n_1 - 2n_2 \approx n_2 - 2n_3 \quad -30.9 \text{ deg/day.}$$

while Equations (29) and (30) still hold. Greenberg (1976) has derived the following libration formula involving the first five multiples of Θ:

$$\ddot{\Theta} = m_2 m_3 n_1^2 [-11.53 \sin\Theta + 5.82 \sin 2\Theta$$
$$+ 1.06 \sin 3\Theta + 0.31 \sin 4\Theta + 0.11 \sin 5\Theta] \qquad (32)$$

This expression is here extended by means of the formulas in Section 2 which allow for additional contributions, neglected by Greenberg, due to Miranda's mass m_1 and $\ddot{\varepsilon}_i$:

$$\ddot{\Theta} = m_2 m_3 n_1^2 [(-11.16 - 36.07 \frac{m_1}{m_2} - 9.27 \frac{m_1}{m_3}) \sin\Theta$$
$$+ (6.74 + 31.66 \frac{m_1}{m_2} + 8.32 \frac{m_1}{m_3}) \sin 2\Theta$$

$$+ \left(1.56 + 8.39\frac{m_1}{m_2} + 2.49\frac{m_1}{m_3}\right)\sin 3\Theta$$

$$+ \left(0.52 + 3.00\frac{m_1}{m_2} + 0.95\frac{m_1}{m_3}\right)\sin 4\Theta \qquad (33)$$

$$+ \left(0.19 + 1.14\frac{m_1}{m_2} + 0.38\frac{m_1}{m_3}\right)\sin 5\Theta \;] \;.$$

The Voyager 2 flyby of Uranus (Tyler et al. 1986) has produced the mass ratios $m_1/m_2 = 0.055$ and $m_1/m_3 = 0.059$, and Harris (1949) has determined the initial condition $\Theta_o = 256.^o9 \pm 0.^o2$ and $\dot{\Theta}_o = (-0.08 \pm 0.0013)$ deg/day for the epoch JD = 2432760.0. By inserting these values in Equations (33) and (7), we find that the condition for libration about $\Theta=0$ is that $m_2m_3 > 2.3 \times 10^{-9}$, a limit ten times bigger than the mass product $m_2m_3 = (2.3 \pm 0.6) \times 10^{-10}$ determined from the Voyager results. It therefore seems certain now that the resonance argument Θ is, in fact, circulating, as has been concluded by several earlier investigators.

4 REFERENCES

Aksnes, K. (1977). Nature 269, 783.

Aksnes, K. (1982). In Planetary Rings (Proceedings of IAU Coll. No.75 in Toulouse, 30 Aug-2 Sept. 1982) Ed. by A. Brahic, CNES, Cepadues-Editions p.479, 1984.

Brouwer, D. and Clemence, G.M. (1961). Celestial Mechanics, Academic Press, New York/London.

Brown, E.W. and Shook, C.A. (1933). Planetary Theory, Cambridge Univ. Press (Dover Publ. 1964).

Dermott, S.F. and Gold, T. (1977). Nature 267, 590.

Goldreich, P. and Nicholson, P. (1977). Nature 269, 783.

Greenberg, R. (1976). Icarus 29, 427.

Harris, D.L. (1949). Dissertation, Univ. of Chicago.

Okyay, T. (1935). Astron. Nachrichten 255, 278.

Sinclair, A.T. (1975). Mon. Not. R. Astr. Soc. 171, 59.

Tisserand, F. (1896). Traité de Mecanique Celeste, Tome IV, Gauthier-Villars, Paris.

Tyler, G.L., Sweetnam, D.N., Anderson, J.D., Campbell, J.K., Eshleman, V.R., Hinson, D.P., Levy, G.S., Lindal, G.F., Marouf, E.A., and Simpson, R.A. (1986). Science 233, 79.

Urban, K. (1935). Astron. Nachrichten 256, 109.

Veillet, C. (1981). Astron. and Astrophys. 98, 218.

Wilkens, A. (1933). Sitzungsber. Math.-Naturwissens. Abteilung, Bayerischen Akad. der Wissens., München, Heft 1, p.71.

APPENDIX A. DERIVATIVES OF THE DISTURBING FUNCTION

In Equations (16), (18), and (24) the following partial derivatives, in terms of Laplace coefficients $b_{1/2}^{(k)}$ and their derivatives, are involved:

1) $\underline{a_i < a_j, \quad \alpha = a_i/a_j}$

$$\frac{\partial A_{ij}^{(k)}}{\partial a_i} = \frac{1}{2a_j^2} \frac{db_{1/2}^{(k)}}{d\alpha} - \frac{1}{2a_j^2} (\delta_{1,-k} + \delta_{1,k}) ,$$

$$\frac{\partial A_{ij}^{(k)}}{\partial a_j} = - \frac{1}{2a_j^2}(1 + \alpha \frac{d}{d\alpha})b_{1/2}^{(k)} + \frac{a_i}{a_j^3} (\delta_{1,-k} + \delta_{1.k}) ,$$

$$\frac{\partial^2 A_{ij}^{(k)}}{\partial a_i^2} = \frac{1}{2a_j^3} \frac{d^2 b_{1/2}^{(k)}}{d\alpha^2} ,$$

$$\frac{\partial^2 A_{ij}^{(k)}}{\partial a_i \partial a_j} = - \frac{1}{2a_j^3} (2 \frac{d}{d\alpha} + \alpha \frac{d^2}{d\alpha^2}) b_{1/2}^{(k)} + \frac{1}{a_j^3} (\delta_{1,-k} + \delta_{1,k}) ,$$

$$\frac{\partial B_{ij}^{(k)}}{\partial a_i} = - \frac{1}{2a_j^2} [(2k + 1) \frac{d}{d\alpha} + \alpha \frac{d^2}{d\alpha^2}]b_{1/2}^{(k)} - \frac{1}{2a_j^2} (\delta_{1,-k} - 3\delta_{1,k}) ,$$

$$\frac{\partial C_{ij}^{(k)}}{\partial a_i} = \frac{1}{2a_j^2} [(2k + 2) \frac{d}{d\alpha} + \alpha \frac{d^2}{d\alpha^2}] b_{1/2}^{(k)} - \frac{2}{a_j^2} \delta_{1,k} .$$

2) $\underline{a_i > a_j , \quad \alpha = a_j/a_i}$

$$\frac{\partial A_{ij}^{(k)}}{\partial a_i} = - \frac{1}{2a_i^2} (1 + \alpha \frac{d}{d\alpha}) b_{1/2}^{(k)} - \frac{1}{2a_j^2} (\delta_{1,-k} + \delta_{1,k}) ,$$

$$\frac{\partial A_{ij}^{(k)}}{\partial a_j} = \frac{1}{2a_i^2} \frac{db_{1/2}^{(k)}}{d\alpha} + \frac{a_i}{a_j^3} (\delta_{1,-k} + \delta_{1,k}) ,$$

$$\frac{\partial^2 A_{ij}^{(k)}}{\partial a_i^2} = \frac{1}{a_i^3} (1 + 2\alpha \frac{d}{d\alpha} + \frac{1}{2} \alpha^2 \frac{d^2}{d\alpha^2}) b_{1/2}^{(k)} ,$$

$$\frac{\partial^2 A_{ij}^{(k)}}{\partial a_i \partial a_j} = -\frac{1}{a_i^3}\left(\frac{d}{d\alpha} + \frac{1}{2}\alpha\frac{d^2}{d\alpha^2}\right)b_{1/2}^{(k)} + \frac{1}{a_j^3}(\delta_{1,-k} + \delta_{1,k}) \quad ,$$

$$\frac{\partial B_{ij}^{(k)}}{\partial a_i} = \frac{1}{2a_i^2}\left[2k - 1 + (2k-3)\alpha\frac{d}{d\alpha} - \alpha^2\frac{d^2}{d\alpha^2}\right] b_{1/2}^{(k)} - \frac{1}{2a_j^2}(\delta_{1,-k} - 3\delta_{1,k}),$$

$$\frac{\partial C_{ij}^{(k)}}{\partial a_i} = -\frac{1}{2a_i^2}\left[2k + (2k-2)\alpha\frac{d}{d\alpha} - \alpha^2\frac{d^2}{d\alpha^2}\right]b_{1/2}^{(k)} - \frac{2}{a_j^2}\delta_{1,k} \quad .$$

3) <u>$a_1 = a_i$ or $a_1 = a_j$</u>

$$\frac{\partial a_{ij}^{(k)}}{\partial a_1} = -kn_i a_i\left(\delta_{i,1}A_{ij}^{(k)} + 2a_i\frac{\partial A_{ij}^{(k)}}{\partial a_1}\right) \quad ,$$

$$\frac{\partial \varepsilon_{ij}^{(k)}}{\partial a_1} = -n_i a_i\left(\delta_{i,1}\frac{\partial A_{ij}^{(k)}}{\partial a_i} + 2a_i\frac{\partial A_{ij}^{(k)}}{\partial a_i \partial a_1}\right.\quad ,$$

$$\frac{\partial b_{ij}^{(k)}}{\partial a_1} = \frac{\partial \varepsilon_{ij}^{(k)}}{\partial a_1} + \frac{3n_i}{2a_i^2 k(n_{io} - n_{jo})}\left(\frac{5}{2}\delta_{i,1}a_{ij}^{(k)} - a_i\frac{\partial a_{ij}^{(k)}}{\partial a_1}\right) \quad .$$

4) <u>The Laplace coefficients</u> $b_{1/2}^{(k)}$ are given in integral form and as a hypergeometric series:

$$\frac{1}{2}b_{1/2}^{(k)} = \frac{2}{\pi}\alpha^k\int_0^{\pi/2}\frac{\sin^{2k}\theta\, dt}{\sqrt{1-\alpha^2\sin^2\theta}} = \frac{1\cdot 3\cdot 5\ ---(2k-1)\alpha^k}{2\cdot 4\cdot 6\ ---\ 2k}$$

$$\times\left[1 + \frac{1}{2}\frac{2k+1}{2k+2}\alpha^2 + \frac{1\cdot 3\cdot(2k+1)(2k+3)\alpha^4}{2\cdot 4\cdot(2k+2)(2k+4)} + ---\right] \quad .$$

This series can also be used to compute the first and second derivatives of $b_{1/2}^{(k)}$ by means of the relations,

$$(\alpha - \alpha^3)\frac{db_{1/2}^{(k-1)}}{d\alpha} = (k-1 + k\alpha^2)b_{1/2}^{(k-1)} - (2k-1)\alpha b_{1/2}^{(k)} \quad ,$$

$$(\alpha - \alpha^3) \frac{db_{1/2}^{(k)}}{d\alpha} = (2k - 1)\, \alpha b_{1/2}^{(k)} - [\, k + (k-1)\alpha^2] b_{1/2}^{(k)} \quad ,$$

$$(\alpha - \alpha^3) \frac{d^2 b_{1/2}^{(k)}}{d\alpha^2} = (2k - 1)\left(b_{1/2}^{(k-1)} + \alpha \frac{db_{1/2}^{(k-1)}}{d\alpha}\right)$$

$$- 2(k-1)\, \alpha b_{1/2}^{(k)} - [\, k+1 + (k-4)\alpha^2] \frac{db_{1/2}^{(k)}}{d\alpha} \; .$$

ANALYTICAL METHODS FOR THE RADIAL INTERMEDIARIES

Arribas, M., Sein-Echaluce, M.L.
Dpto. de Astronomia. Universidad de Zaragoza
50009 Zaragoza
Spain.

1. INTRODUCTION

In this paper the polar-nodal variables (r,θ,v,R,Θ,N) will be used. The variable r is the radial distance from the Earth's centre of mass to the satellite, θ is the argument of latitude, and n is the argument of the ascending node. The variables R,Θ, N are the conjugated momenta to the coordinates r,θ, v respectively.

It is assumed that the satellite moves in an axially symmetric gravitational field whose potential is of the form

$$V = -\frac{\mu}{r}\left(1 - \sum_{n=2}^{\infty} J_n \left(\frac{\alpha}{r}\right)^n P_n (\sin\phi)\right)$$

where α stands for the equatorial radius (we will take $\alpha = 1$), P_n $(\sin\phi)$ is the Legendre polynomial of degree n in sin f and f is the satellite's declination.

Denoting by

$$H = \frac{1}{2}\left(R^2 + \frac{\Theta^2}{r^2}\right) + V$$

the Hamiltonian function, after making two canonical transformations of Lie type, the result is Cid-Deprit's radial intermediary

$$J = \frac{1}{2}\left(R^2 + \frac{\Theta^2}{r^2}\right) - \frac{\mu}{r} - \varepsilon\mu\frac{(2-3\,s^2)}{4\,r^3}$$

to first order, where $\varepsilon = J_2$, $c = N/\Theta = \cos I$, $s^2 = 1-c^2$, I stands for the inclination of the satellite's orbital plane (Cid et al., 1986; Sein-Echaluce and Franco, 1987).

In order to integrate this Hamiltonian function, we apply two different methods: the regularization and linearization method and the stroboscopic method.

A. E. Roy (ed.), Long-Term Dynamical Behaviour of Natural and Artificial N-Body Systems, 141–147.
© 1988 by Kluwer Academic Publishers.

2. LINEARIZATION OF THE RADIAL INTERMEDIARY J

As it is well known in Celestial Mechanics, the Kepler problem is reduced to an harmonic oscillator making a change of variables $(r,t)\rightarrow(\rho,f)$ being $\rho = 1/r$ and f the true anomaly. Another possible transformation is $(r,t)\rightarrow(r,E)$ where E is the eccentric anomaly.

These transformations and analogous ones were used by Hansen, Gylden, Sundman and Levi-Civita in some problems (three body problem, theories of planetary orbits, etc.).

Some studies about regularizations and linearizations of central fields have been done by Szebehely (1976), Cid et al. (1986), Szebehely and Bond (1983), etc.

We have a potential $V(r,\Theta,N)$ with Θ = const. and N = const. and we look for the linearization of r by means of a change of variables $(r,t)\longrightarrow(\rho,\tau)$ given by

$$r = F(\rho) \quad , \quad dt = g(r)d\tau$$

where $F^* = dF/d\rho = 0$ and $g(r) > 0$.

The integral of the energy considering the potential V(r) can be expressed in the form:

$$\frac{1}{2}\left(\frac{dr}{dt}\right)^2 = h - V^*(r)$$

where

$$V^*(r) = \frac{\Theta^2}{2r^2} - \frac{\mu}{r} + V(r)$$

Making the change given by (1), the Szebehely-Bond equation results

$$\frac{d^2\rho}{d\tau^2} = \frac{1}{2}\,\frac{d}{d\rho}\left(\frac{g^2 V^*}{F^{*2}}\right) \tag{2}$$

We are looking for functions g and F such that the differential equation in the new variables (r,t) is reduced to a linear equation

$$d^2\rho/d\tau^2 = A\rho + B \tag{3}$$

where A and B are constant.

From the equations (2) and (3), we get

$$g(r) \equiv F^* \sqrt{\frac{Ar^2/2 + Br + C}{h - V^*(r)}} \tag{4}$$

being C a constant.

In our case, we have

$$V^*(r) = -b_1/r - b_2/r^2 - b_3/r^3$$

where $b_1 = \mu$, $b_2 = -\Theta/2$, $b_3 = -\varepsilon\mu(2-3s^2)/4$.

If we make a transformation given by $r = F(p) = p$, $dt = g(r)d\tau$, the expression

$$g(r) = \sqrt{\frac{Ar^2/2 + Br + C}{h - V^*(r)}}$$

is obtained.

If we consider "elliptic" motions, the equation $h-V^*(r) = 0$ has three real roots r_1, r_p, r_a, where $r_p > 0$, $r_a > 0$ and r_1 may be ≥ 0 or ≤ 0 depending on whether b_3 is ≥ 0 or ≤ 0 respectively (Sein-Echaluce, 1986). Then we obtain

$$g(r) = \frac{r^{3/2}}{\sqrt{(2h/A)(r-r_1)}} \tag{5}$$

(Ferrandiz et al., 1987)

Thus the differential equation that corresponds to our problem is

$$d^2r/d\tau^2 = Ar - A(r_p + r_a)/2 \tag{6}$$

which corresponds to an harmonic oscillator ($A < 0$) and its solution is given by

$$r = \bar{a}(1 - e\cos\bar{E}) \tag{7}$$

$$\bar{E} = \sqrt{-A}\tau, \quad \bar{a} = (r_a + r_p)/2 \quad , \quad \bar{e} = (r_a - r_p)/(r_a + r_p)$$

where we take $\bar{E} = 0$ when $\tau = 0$. The values a and e define an elliptic fictitious orbit in the perturbed motion and we can define, in analogous form such as in the unperturbed case, the "anomalies" $\bar{E}$ and $\bar{f}$ related by the same formula. We can note that if we choose $A = -1$, then $\tau = \bar{E}$ and the new independent variable is the eccentric anomaly.

2.1 Generalized Kepler equation. Calculation of t.

From the equation (5) we have

$$dt = \frac{r^{3/2}}{\sqrt{(2h/A)(r-r_1)}} d\tau$$

where r is given by the expression (7). Making the change $y = \cos E$

and integrating, we obtain the next results for the case $r_1 > 0$:

$$t-t_o = C_1F(\bar{f}/2,k)+C_2E(\bar{f}/2,k)+C_3\Pi(\bar{f}/2,\alpha^2,k)+C_4 \frac{\sin\bar{f}\sqrt{1-k^2\sin^2(\bar{f}/2)}}{1-\alpha^2\sin^2(\bar{f}/2)}$$

where $\bar{f}$ is the "true" anomaly, C_1, C_2, C_3, C_4, k, $\sin(\bar{f}/2)$ depending on r_p,r_a,r_1,h and F,E and Π are the normal elliptic integrals of first, second and third kind respectively. An analogous expression is obtained for the case $r_1 < 0$. (Sein-Echaluce, 1986).

2.2 Position of r in the orbital plane. Calculation of θ.

The equation $d\theta/dt = \partial H/\partial\Theta$ for the Hamiltonian J is

$$d\theta/dt = P_1/r^2 + P_2/r^3 \quad \text{with } P_1 = \Theta , \; P_2 = -\varepsilon\mu 3(1-s^2)/(2\Theta)$$

The solution for $r_1 < 0$ is given by

$$\theta-\theta_o = L_1F(\bar{f}/2,k) + L\,E(\bar{f}/2,k)$$

We get similar expressions for the case $r_1 > 0$. (Sein-Echaluce, 1986).

2.3 Motion of the node. Calculation of n.

The equation that corresponds to v in this problem is

$$dv/dt = Q/r^3 \quad \text{with} \quad Q = \varepsilon\mu 3c/(2\Theta)$$

whose integration gives the next results in the case $r_1 < 0$

$$v-v_o = K_1F(\bar{f}/2,k) + K_2E(\bar{f}/2,k)$$

the result is analogous for $r_1 > 0$.

So, for a given time t, we can know the value of the variables r,θ,v as well as those of the momenta R,Θ, N due to Θ and N being constant and R is obtained simply by the expression

$$R = \sqrt{2(h-V^*(r))}$$

3. APPLICATION OF THE STROBOSCOPIC METHOD

A different method to solve this problem consists of the application of the stroboscopic method (Roth, 1979). This is a semi-analytic method for the solution of the equations of a perturbed dynamical system which depends on several slow variables and one fast variable. The osculating elements are just known in a point of the orbit (perigee, apogee,...) and it is considered that the analytical theory is valid for one revolution. The process is repeated in the next revolution and so on.

The principal advantages of this method consist of the velocity of the calculations and a good precision reached. Besides, it is easy to consider all types of perturbations (third-body, airdrag, radiation pressure,...).

This method has been widely used in orbits propagations using classical elements (Janin, 1979) or equinoctial elements (Lecohier, 1985).

In this case, we present the first order stroboscopic method in Hill variables (r,θ,v,R,Θ,N). This way shows some differences with respect to the classical elements, since in the zero order some variables may not be constant and it is necessary to solve a differential system instead of quadratures, as we will see below. Recently, studies have been done in this sense by Sein-Echaluce et al. (1987).

3.1 First Order stroboscopic method.

The equations of motion of a particle subject to a perturbed Keplerian potential can be expressed in the general form

$$\frac{d\bar{x}}{dt} = \bar{F}_o + \varepsilon \bar{F}_1, \qquad \frac{dy}{dt} = G_o + \varepsilon G_1 \tag{8}$$

where $\bar{x} = (x_1,\ldots,x_n)$ is a set of variables and y is the fast angular variable.

Taking y as the independent variable, the Eq. (8) takes the form:

$$\frac{d\bar{x}}{dy} = \frac{\bar{F}_o + \varepsilon \bar{F}_1}{G_o + \varepsilon G_1}, \qquad \frac{dt}{dy} = \frac{1}{G_o + \varepsilon G_1} \tag{9}$$

whose solution is of the form:

$$\bar{x} = \bar{x}^0 + \varepsilon \bar{x}^1, \qquad t = t^0 + \varepsilon t^1 \tag{10}$$

We can develop the second members in the equations (8) in the neighbourhood of the non-perturbed solution $\bar{x}^0$, and then, taking into account the expressions (9), differentiating and identifying the coefficients of the powers of e, the next expressions are obtained

$$\frac{d\bar{x}^0}{dy} = \phi^0(\bar{x}^0, y), \qquad \frac{d\bar{x}^1}{dy} = \phi^1(\bar{x}^0, \bar{x}^1, y) \tag{11}$$

$$\frac{dt^0}{dy} = \psi^0(\bar{x}^0,y) \quad , \quad \frac{dt^1}{dy} = \psi^1(\bar{x}^0,\bar{x}^1,y) \tag{12}$$

(Sein-Echaluce et al., 1987).

When the equations of the perturbed motion (11) and (12) have been established, the stroboscopic method consists of obtaining the analytical solution when the angular variable increases its argument in 2π.

So with the conditions $\bar{x}^0(0) = \bar{x}_o(0)$, $\bar{x}^1(0) = 0$, we have:

$$\bar{x}\,(2\pi) = \bar{x}_1^1 = \int_0^{2\pi} \phi^1(\bar{x}^0,\bar{x}^1,y)\,dy$$

and the solution after the first revolution is obtained. If we repeat the process, at the end of n revolutions the relation is

$$\bar{x}_n = \bar{x}^0_{n-1} + \varepsilon\bar{x}^1_n \quad , \quad \text{where} \quad \bar{x}^0_{n-1} = \bar{x}_{n-1}$$

and in analogous form for the calculations of t by using the Eqs.(12).

If we apply the previous method to the Hamiltonian J and choose q as fast variable, we have for the zeroth order

$$\frac{dr^0}{d\theta} = \frac{R^0(r^0)^2}{\Theta} \quad , \quad \frac{dR^0}{d\theta} = \frac{\Theta}{r^0} - \frac{\mu}{\Theta} \quad , \quad \frac{dv^0}{d\theta} = 0 \tag{13}$$

and for the first order

$$\frac{dr^1}{d\theta} = \frac{2R^0r^0}{\Theta}\,r^1 + \frac{(r^0)^2}{\Theta}\,R^1 + \frac{3\mu N^2 r^0 R^0}{2\Theta^5}$$

$$\frac{dR^1}{d\theta} = \frac{\Theta}{(r^0)^2}\,r^1 + \frac{3\mu(2-3s^2)}{4\Theta(r^0)^2} - \frac{3\mu^2N^2}{2\Theta^5 r^0} + \frac{3\mu N^2}{2\Theta^5(r^0)^2} \tag{14}$$

$$\frac{dv^1}{d\theta} = \frac{3\mu N}{2r^0\Theta^3}$$

From the system (13) v^0 = cte. and we can obtain $r^0 = r^0(\theta)$ and $R^0 = R^0(\theta)$. Then, we can integrate (14) between 0 and 2π; the last equation is a quadrature and a linear differential system is obtained from the first and second equations. After solving it we will have v^1, r^1, R^1, and we will have the solution for the first revolution; the process is continued.

When the values of r^0, R^0, r^1, R^1 are known the equations corresponding to the time t are integrated for the first revolution. These equations are

$$\frac{dt^0}{d\theta} = \frac{(r^0)^2}{\Theta}, \quad \frac{dt^1}{d\theta} = \frac{3\mu N^2}{2\Theta^5} r^0 + \frac{2r^0}{\Theta} r^1$$

4. ACKNOWLEDGEMENTS

This work has been partially supported by Diputación General de Aragón, and the Comisión Asesora Interministerial Cientifica y Técnica of Spain.

5. REFERENCES

1. Cid, R., Ferrer, S. and Sein-Echaluce, M.L. (1986). Cel. Mech. 38, 191.
2. Ferrandiz, J.M., Ferrer, S. and Sein-Echaluce, M.L. (1987), to appear in Cel. Mech.
3. Janin, G. (1979). E.S.A. STM 208, 47 pp.
4. Lecohier, G. (1985). MAO Working Paper No.221
5. Roth, E.A. (1976). Journal of Applied Mat and Phys (ZAMP) 30
6. Sein-Echaluce, M.L.(1986). Dr. Thesis, Universidad de Zaragoza.
7. Sein-Echaluce, M.L., Elipe, A., Lanchares, V., and Arribas, M. (1987). XII Jornadas Luso-Españolas de Matematica. Braga. Portugal.
8. Sein-Echaluce, M.L., Abad, A. and Elipe, A. (1987), 87 AAS-Astrodynamics Conference Paper 87-442.
9. Szebehely, V. (1976). Cel. Mech. 14, 499-508.
10. Szebehely, V. and Bond, V. (1983). Cel. Mech. 30, 59-70.
11. Sein-Echaluce, M.L., Franco, J.M. (1987). NATO ASI. Cortina d'Ampezzo.

ANALYTICAL THEORY OF THE ROTATION OF AN ARTIFICIAL SATELLITE

E. Bois
CERGA
Avenue Copernic
06130 Grasse
France

ABSTRACT. This paper presents an analytical theory of the attitude evolution of an artificial satellite perturbed by given torques. The main constraint is that of a strongly perturbed slow rotation with the axis pointing far away from a position of equilibrium. Furthermore the three axes of inertia can be unequal. A formal solution is expanded in the second order according to the powers of a small parameter characteristic of the order of magnitude of the disturbing torques, and this theory is suitable whatever are the internal or external disturbing forces producing the torques. These torques are expanded in Fourier series and the theory applies whatever is the length of these series. The coefficients of the solution are given by an iterative formation law.

1. INTRODUCTION

The present work presents an analytical theory of the attitude evolution of an artificial satellite in perturbed rotation. The exploitation of many satellites involves indeed knowing with great accuracy the evolution of their attitude. In particular in the case of the HIPPARCOS satellite which is an astrometric satellite, that is a necessary condition for the success of the mission (see, for instance, Kovalevsky, 1984, or Kovalevsky and Bois, 1986). But here, first of all, we are faced with this problem of Celestial Mechanics: how to represent the rotational motion of a body with three unequal axes of inertia, submitted to disturbing torques, and with the axis pointing far away from a position of equilibrium.

The attitude evolution of a satellite is function of many parameters. In particular, it is connected to the ratio R of the magnitude of the total external torque over the magnitude of the rotational angular momentum. In 1971, Cochran propounded a formulation available whatever are the external torque and the satellite's dynamic shape but assuming that R is small. Crenshaw and Fitzpatrick, in 1968, have already adopted this formulation but just in order to study the gravity effects on the rotational motion of an uniaxial artificial satellite.

A. E. Roy (ed.), Long-Term Dynamical Behaviour of Natural and Artificial N-Body Systems, 149–154.
© 1988 by Kluwer Academic Publishers.

In 1972, the work of Stellmacher applies not only to the case where R is small but also in the case where R is large. Nevertheless this study is devoted to the effects of the torque of magnetic origin on the rotational motion of a satellite with inertial symmetry. In 1985, the theory of Van der Ha is suitable whatever are the satellite's inertia and the magnitude of R but for constant torques in the body-fixed system. In the present work the problem is generalized and the theory is suitable whatever are the satellite's dynamic shape and the orbital motion, whatever is the magnitude of R, small, large or very large, and whatever are the internal or external disturbing torques, constant, variable or strongly variable.

So the main constraint is that of a strongly perturbed slow rotation, the axis pointing far away from a position of equilibrium. The problem is indeed more general with the main hypothesis of a slow rotation. The solution will contain more terms than in the case of a fast rotation. In the end, this work aims at studying the rotational motion of a satellite, with any initial conditions, so long as the rotational axis is not too far away from its initial position.

The attitude will be described by three angles of position. The time derivative of the angle describing the satellite's rotation must be initially close to the nominal rotation velocity ω. At the origin of time, the other two angles are small or zero-value and their time derivatives are small quantities. This turns out to be the only hypotheses of this work.

2. THE BASIC FRAME

2.1 Systems of coordinates

Let OXYZ be the reference fixed frame with its origin at the satellite's center of mass. Let oxyz be the body-fixed system of coordinates defined by the satellite's principal axes of inertia.

2.2 The position angles

Known as the sequence of the Tait-Bryan angles, the adopted system of angles is composed of three direct rotations in the following order: α, β, γ. Because of its definition (Bois, 1986), this combination is called the 1.2.3-system. This system can answer to the conditions required by the problem:

- if the Oz-axis is the rotational axis, γ represents the proper rotation of the system,
- if the Oz-axis remains close to the OZ-axis, the angles α and β are small.

Thus composed of two small rotations and of a large one, the 1.2.3 -system proves to be adequate to express the required motion.

2.3 The equations of motion

In this 1.2.3-system and in the body axes, let us apply the fundamental law for the perturbed rotation of a solid body (Goldstein, 1964). One obtains then three differential equations of the second order in α,β and γ:

$$
\begin{aligned}
&(A\cos\beta\cos\gamma)\ddot{\alpha}+(A\sin\gamma)\ddot{\beta}+[(-A-B+C)\sin\beta\cos\gamma]\dot{\alpha}\dot{\beta}+[(-A+B-C)\cos\beta\sin\gamma]\dot{\alpha}\dot{\gamma}\\
&\qquad +[(A-B+C)\cos\gamma]\dot{\beta}\dot{\gamma}+[(B-C)\sin\beta\cos\beta\sin\gamma]\alpha^2=N_1\\
&(-B\cos\beta\sin\gamma)\ddot{\alpha}+(B\cos\gamma)\ddot{\beta}+[(A+B-C)\sin\beta\sin\gamma]\dot{\alpha}\dot{\beta}+[(A-B-C)\cos\beta\cos\gamma]\dot{\alpha}\dot{\gamma}\\
&\qquad +[(A-B-C)\sin\gamma]\dot{\beta}\dot{\gamma}+[(A-C)\sin\beta\cos\beta\cos\gamma]\dot{\alpha}^2=N_2\\
&(C\sin\beta)\ddot{\alpha}+C\ddot{\gamma}+\{[(-A+B)\cos 2\gamma+C]\cos\beta\}\dot{\alpha}\dot{\beta}+[(-A+B)\sin\gamma\cos\gamma](\dot{\beta}^2-\dot{\alpha}^2\cos^2\beta)=N_3
\end{aligned}
\tag{1}
$$

where A, B, C are the satellite's principal moments of inertia and so, the main matter of this work is about the resolution of these strongly non-linear equations. The right hand side (N_1, N_2, N_3) represents the disturbing torques.

3. PERTURBATIONS

The perturbations susceptible of producing torques are for example:
- the solar radiation pressure,
- the reactions of the gyroscopes (inside of the satellite),
- the gravity gradient,
- the magnetic field,

and other perturbations as the Earth's radiation pressure, etc... (Bois 1987, p.20-22). All the disturbing torques can be expanded in Fourier series with the general following form in α,β,γ:

$$
\begin{aligned}
N_t(\alpha,\beta,\gamma,\lambda) = \{ &\sum_{ijkl} [A_{ijkl}\cos(i\alpha+j\beta+k\gamma+l\lambda)\\
&+ {}^{-}B_{ijkl}\sin(i\alpha+j\beta+k\gamma+l\lambda)]\}.\varepsilon
\end{aligned}
\tag{2}
$$

where ε is a parameter characteristic of the order of magnitude. A and B are numerical coefficients, and λ refers to the orbital mean longitude position. If the dominating disturbing torques depend on a fixed direction in the space or if one seeks the attitude evolution by short intervals of time, then λ may be considered as a constant.

4. METHOD OF RESOLUTION

According to Poincaré's theorem, the disturbing function depending on the parameter ε , assumed to be small enough, the solution can be expanded in powers of this small parameter, in the following form:

$$\alpha = \alpha_o + \varepsilon\alpha_1 + \varepsilon^2\alpha_2 + \ldots$$
$$\beta = \beta_o + \varepsilon\beta_1 + \varepsilon^2\beta_2 + \ldots \qquad (3)$$
$$\gamma = \gamma_o + \varepsilon\gamma_1 + \varepsilon^2\gamma_2 + \ldots$$

-$(\alpha_o,\beta_o,\gamma_o)$ represents the zero-order solution, that is to say in the absence of any disturbing force. Let us take $\alpha_o = \beta_o = 0$ and $\gamma_o = \omega(t-t_o)+\bar{\gamma}_o$. That is a particular solution of the motion of Poinsot but this is the general solution of the zero-order of our problem, in accordance with our hypotheses (Bois, 1986, see p.318).
-$(\alpha_1,\beta_1,\gamma_1)$ represents the first-order solution and so on.

The first-order solution is then composed of linear functions and of Fourier series which only depend on γ. So, if we restrict ourselves to the first-order, after a substitution of the global form of the solution (3) in the initial differential equations (1), and after a similar approximation in the right hand side of these equations, one obtains a new differential system. After several transformations of these new equations, we have to solve a systematic process of identification, not very straight-forward for the first two equations but the analytical calculations have shown that the coefficients of the solution presented a kind of cross antisymmetry and that we could introduce two iterative formulae in function of the disturbing coefficients. So the method of resolution gives an analytic solution in which the angles α,β,γ are expressed literally according to the disturbing parameters. The coefficients of the solution are indeed given by combinations of the following iterative formation law:

$$S_j(x_i,y_i) = \frac{x_i(A+jB-C)+y_i(jA+B-C)}{2\omega[C^2-BC-AC-(j+1)(j-1)AB]}$$
$$D_j(x_i,y_i) = \frac{x_i(A+jB-C)-y_i(jA+B-C)}{2\omega[C^2-BC-AC-(j+1)(j-1)AB]} \qquad (4)$$

where x_i and y_i represent the different sets of disturbing parameters stemming from (2). The full solution is described in Bois (1986). Thus, the disturbing torques are expanded in Fourier series and the theory applies whatever is the length of these series.

Now, let us come back to the initial differential system (1). After many Taylor developments to the second-order in the periodical coefficients and in accordance with Poincaré's theorem, one gets the new differential equations for the second-order of the present theory. These new equations are a little more complicated with, for instance, the appearance of products of series. For the whole description of the second-order resolution method, one may consult Bois (1987) but let us say here that the main idea aims at trying to use again the iterative formation law (4) of the first-order resolution. Therefore the second-order solution is composed of linear functions, Fourier series, Poisson terms and in γ_2 of a parabolic function. Some terms are in the end given by using (4) as functions of new parameters which are in relation with the first-order disturbing parameters. But some other terms are even given directly as functions of the first-order terms without the use of (4).

5. APPLICATION

In order to appreciate the behaviour and the accuracy of the solution, we have applied the present theory to a realistic torque model accounting for the HIPPARCOS satellite structure. The dominating disturbing torques acting on the rotation of HIPPARCOS are the solar radiation pressure and gyroscope reaction torques. The model only depends on γ but it remains suitable the whole time of the mission. It is indeed a property of the HIPPARCOS nominal scanning law: the rotational axis of HIPPARCOS makes a constant angle with the Sun direction.

The values of the numerical solution stemming from a numerical integration based upon that HIPPARCOS model are compared with the values of the second-order analytical solution. The differences that remain are partly due to a non-rigorous identity in the choice of initial conditions and partly to the third-order terms. The curves of these differences are drawn on 600 seconds which is the probable time interval between resetting the initial conditions of the rotation of HIPPARCOS by gas jets. The differences are about one half second of arc for the three angles. And it is possible to improve the accuracy on γ in computing $\Delta\gamma$ around of a mean value of the secular terms of γ. Thus, one gets an accuracy better than 5 hundreds of second of arc. In the case of HIPPARCOS, it is very important to have a better accuracy on the fast angle which is closely connected with the determination of the x-coordinates of the stars on the scanning great circles. Beyond 600 seconds, the values of the differences increase but the regularity of these sinusoidal curves shows the good consistency of the theory (Bois, 1986, p.323). Furthermore the influence of the ratio "resultant intensity of the disturbing torques - rotational energy" on the attitude drifting is such that the smaller this ratio is, the more sufficient it is for a low order of the theory expansion to give an appropriate accuracy on the knowledge of the attitude evolution.

6. CONCLUSION

The theory described is not restricted to the particular conditions of HIPPARCOS but it extends to any perturbed rotation of a satellite. The theory is suitable whatever are the disturbing torques. Furthermore the three axes of inertia can be unequal.

Between the first-order and second-order solutions the accuracy is improved by a factor 20. That suggests the convergence of the theory.

The HIPPARCOS rotation velocity is extremely slow: 12 rev/day!

So for ω only 10 times less slow, with simply the first-order solution, the results are 100 times better.

The existence of an iterative formation law for the coefficients of the solution represents the advantage of this method. Beyond the saving of computation time (instead of an attitude reconstruction by numerical integration), the number of terms of the series of the right hand members may be as large as necessary. Hence, it is possible to take into account the effects of the usually neglected perturbations.

This analytical method also leads to confirm and generalize the hypotheses and the semi-numerical algorithm set up in the numerical processes of calculation of the HIPPARCOS attitude motion by the FAST Consortium in charge of the data reduction of HIPPARCOS (Donati, 1986).

BIBLIOGRAPHY

Bois, E., 1986, "First-order Theory of Satellite Attitude Motion, Application to HIPPARCOS", Celestial Mechanics, 39, p.309-327.
Bois, E., 1987, "Théorie Analytique de la Rotation d'un Satellite Artificiel". Thèse de Doctorat, Observatoire de Paris.
Cochran, J.E., 1971, "Rotational Equations of Motion for a Triaxial Rigid Body", AIAAJ. 9, p.1195-1197.
Crenshaw, J.W. and Fitzpatrick, P.M., 1968, "Gravity Effects on the Rotational Motion of a Uniaxial Artificial Satellite", AIAAJ. 6, p.2140-2145.
Donati, F., Canuto, E., Fassino, B. and Belforte, P., 1986, "High Accuracy On-ground Attitude Reconstitution for the ESA Astrometry HIPPARCOS Mission", Manuscripta Geodaetica, vol.11, No.2, p.119.
Goldstein, H., 1964, Méchanique Classique, Presses Universitaires de France, p.172.
Kovalevsky, J., 1984, "Prospects for Space Stellar Astrometry", Space Sci. Rev., 39, p.34.
Kovalevsky, J. and Bois, E., 1986, "Attitude Determination of the HIPPARCOS Satellite". Space Dynamics and Celestial Mechanics, p.345-354.
Stellmacher, I., 1972, "Influence du Champ Magnétique Terrestre sur le Mouvement d'un Satellite autour de son Centre de Gravité", Celestial Mechanics, 5, p.470.
Van der Ha, J., 1985, "Perturbation Solution of Attitude Motion under Body-Fixed Torques", Acta Astronautica, 12, p.861-869.

SOME MODELS FOR THE MOTION OF THE CO-ORBITAL SATELLITES OF SATURN

R. Broucke
University of Texas, Austin, TX, 78712
A. Konopliv
Jet Propulsion Laboratory, Pasadena, Cal., 91103

1. INTRODUCTION

The study of co-orbital motion of two small satellites in nearly the same orbit about a large central mass was sparked by the observations that the Voyager spacecraft took in 1980 and 1981 of Saturn's co-orbiting satellites Janus (1980S1) and Epimetheus (1980S3) (see Stone and Miner, 1981, and Smith et al. 1981, 1982), although there is evidence of previous optical observations of these satellites (Dollfus, 1981, Smith, 1980, Larson, 1981, Seidelman, 1981). The orbital elements of Janus and Epimetheus have been determined by Synnott et al. (1981) and the orbits are nearly circular with an eccentricity on the order of 0.01.

Because of the gravitational interaction between the two small satellites, it is possible that they may approach each other but not pass each other, and for this case, the relative motion between the two co-orbitals seems to be governed by repulsive gravity (see our Section 7). As the co-orbitals turn around, they experience a fairly complex exchange in the corresponding energy and angular momentum (Smith et al., 1982). The two principal types of motion that exist can be classified as horseshoe or tadpole motion (Garfinkel, 1977), and in fact we find that the problem of two co-orbitals is closely related to the restricted three-body problem with a small mass parameter (Dermott and Murray, 1981).

Yoder et al. (1983) extended Garfinkel's classification to the co-orbital problem by using a remarkably simple two-dimensional model (see Section 6). They look at the effect of eccentricities, inclinations, and perturbing satellites on the motion of co-orbitals, and especially the effect on the angular separation of the co-orbitals. Spirig and Waldvogel (1984) also looked at close encounters of co-orbitals and the transition between passing and no-passing orbits.

A. E. Roy (ed.), Long-Term Dynamical Behaviour of Natural and Artificial N-Body Systems, 155–169.
© 1988 by Kluwer Academic Publishers.

2. EQUATIONS OF MOTION

As a basic starting point, we take the General Problem of three bodies in the plane. We assume that the three masses are μ_o, μ_1, μ_2, with $\mu_o + \mu_1 + \mu_2 = 1$ and μ_o much larger than the two other masses. We may thus assume that μ_o represents the planet Saturn and $\mu_1, /\mu_2$ the two co-orbiting satellites. The two co-orbiting satellites are in nearly circular orbits with nearly the same radius, assumed to be unity. We basically use a Saturnocentric coordinate system rotating with unit angular velocity and in this system we define the position of the two co-orbital satellites μ_i by their polar coordinates (r_i, θ_i), i=1,2. The Lagrangian of this system is then:

$$\mathcal{L} = \frac{1}{2}\mu_1(1-\mu_1)(\dot{r}_1^2 + r_1^2\dot{\phi}_1^2) + \frac{1}{2}\mu_2(1-\mu_2)(\dot{r}_2^2 + r_2^2\dot{\phi}_2^2)$$
$$- \mu_1\mu_2[(\dot{r}_1\dot{r}_2 + r_1 r_2\dot{\phi}_1\dot{\phi}_2)\cos\theta_{21} + (r_1\dot{r}_2\dot{\phi}_1 - r_2\dot{r}_1\dot{\phi}_2)\sin\theta_{21}]$$
$$+ \frac{\mu_o\mu_1}{r_1} + \frac{\mu_o\mu_2}{r_2} + \frac{\mu_1\mu_2}{d} \quad . \tag{2.1}$$

We have defined here $\phi_2 = t + \theta_1$, $\phi_2 = t + \theta_2$, so that (ϕ_1, ϕ_2) are the polar angles with respect to the non-rotating frame of reference. The other two symbols are $\theta_{21} = \theta_2 - \theta_1$ and $d = r_{12}$. The equations of motion derived from this Lagrangian are

$$\ddot{r}_i - r_i(1 + \dot{\theta}_i)^2 + \frac{1-\mu_j}{r_i^2} = F_{r_i}$$
$$r_i\ddot{\theta}_i + 2r_i(1 + \dot{\theta}_i) = F_{\theta_i} \; , \tag{2.2}$$

where the subscript i = 1,2, and j = 2,1. Also, the two components of the perturbation on the right-hand side of the equations, are defined by:

$$F_{r_i} = -\mu_j\left[\frac{\cos\theta_{ji}}{r_j^2} + \frac{r_i - r_j\cos\theta_{ji}}{d^3}\right] \tag{2.3a}$$

$$F_{\theta_i} = -\mu_j r_j \sin\theta_{ji}\left(\frac{1}{r_j^3} - \frac{1}{d^3}\right) \tag{2.3b}$$

with

$$d^2 = r_1^2 + r_2^2 - 2r_1r_2\cos\theta_{21} \; . \tag{2.4}$$

However, we are mostly interested here in motion near the unit-circle, and we will thus introduce two small quantities ε_i, such that $r_i = 1 + \varepsilon_i$.
The equations of motion in $(\varepsilon_i, \theta_i)$ can then be written without any approximations:

$$\begin{cases} \ddot{\varepsilon}_i = 2\dot{\theta}_i + 3\varepsilon_i + \left[\dot{\theta}_i^2 + 2\varepsilon_i\dot{\theta}_i + \varepsilon_i\dot{\theta}_i^2 - \dfrac{3\varepsilon_i^2 + 2\varepsilon_i^3}{(1+\varepsilon_i)^2} \right] + \dfrac{\mu_j}{(1+\varepsilon_i)^2} + F_{r_i} , \\ \ddot{\theta}_i = -2\dot{\varepsilon}_i + \dfrac{2\dot{\varepsilon}_i(\varepsilon_i - \dot{\theta}_i)}{1 + \varepsilon_i} + F_{\theta_i} , \end{cases} \tag{2.5}$$

where, again, (i,j) are a permutation of $(1,2)$, and where the perturbing accelerations F_{r_i}, F_{θ_i} are as given above, with $\alpha = \theta_{21}/2$ and:

$$d^2 = 4(1+\varepsilon_1+\varepsilon_2+\varepsilon_1\varepsilon_2)\sin^2\alpha + (\varepsilon_2-\varepsilon_1)^2. \tag{2.6}$$

As was said above, no approximation has been made so far. We must remember that the $\varepsilon_i, \dot{\varepsilon}_i$ and $\dot{\theta}_i$ are all small, (near the unit circle). This will allow us to make different approximations and thus introduce several simplified models for the motion of the two co-orbiting satellites.

However, we will first make a few remarks here about the above equations of motion (2.5). These equations can truly be recommended for the numerical integration of the motion of the two small masses, because all the terms in the right hand sides are small, without any difference of large nearly equal terms. In fact the original equations of motion (2.2) have such a difficulty: they contain a difference of two quantities which are near one, the central gravitational force and the centrifugal force. In the equations (2.5) these terms have been cancelled out analytically and they no longer cause any loss of precision.

The difficulty is best visible in the radial equation (2.2a), where we see the term $(-\dot{r}_i)$ and another term r_i^{-2}, both nearly equal but opposite in sign. In the case of the real co-orbital satellites (masses 10^{-9}) the absolute values of these terms may differ only by about 10^{-9}. The key to avoiding this loss of precision is to use the identity

$$1 - \frac{1}{r_i^2} = \frac{\varepsilon_i(2 + \varepsilon_i)}{(1 + \varepsilon_i)^2} .$$

The leading factor on the right hand-side being the small quantity ε_i.

In fact, the basic simplifications in the course of our paper are based on the observations that a large number of our variables are of the order 10^{-9}. To be more precise, we essentially classify all of our numerical quantities in 4 groups. The first group contains θ_1, θ_2, θ_{21} as well as d, $(0 < d < 2)$.
These are large numbers and no approximations may be made to them. A second group of rather large quantities consists of r_0, r_1, r_2 all

very near to unity. Finally we consider two distinct groups of infinitesimal quantities; the first group contains $\dot{\theta}_1$ and $\dot{\theta}_2$ of the order of, say, 10^{-7}. The other group is the smallest of all: $\mu_1, \mu_2, \varepsilon_1, \varepsilon_2, \dot{\varepsilon}_1, \dot{\varepsilon}_2$. In the true Saturn co-orbital situation these would be of the order of 10^{-9}. The fact that we make a distinction between the $(\varepsilon_i, \dot{\varepsilon}_i)$-group and $(\dot{\theta}_1, \dot{\theta}_2)$ is a little arbitrary; it is simply based on the observed fact that near a circular orbit, the angles vary faster than the radial variables.

The next difficulty that occurs in the integration of the two small masses μ_i is related to the large difference between the short period and the long period near the triangular Lagrange points. The ratio between the two frequencies is here even larger than 1000; it is an absolutely major difficulty, as the presence of these fast high-frequency oscillations prevents us from performing accurate long-term integrations. We believe that the only solution here lies in the elimination of the short-period terms through averaging. Our own solution will be published later in another paper.

Figures 2.1 and 2.2 show some solutions of the general problem of three bodies integrated with the differential equations 2.5.

3. THE BASIC ROLE OF THE CLOHESSY-WILTSHIRE EQUATION

The second important remark which can be made here, is that (2.5) show that we are in the presence of Hill's equations, more precisely two coupled sets of Hill's equations. We will call them "Generalized Hill's equations". On the other hand, if in these Hill equations we neglect all non-linear terms, and all perturbations we are left with the well known Clohessy-Wiltshire (C.W.) equations which are familiar in the theory of rendez-vous, (see Kaplan, 1976):

$$\begin{cases} \ddot{\varepsilon}_i = 2\dot{\theta}_i + 3\varepsilon_i , & (3.1a) \\ \ddot{\theta}_i = -2\dot{\varepsilon}_i \quad ; \quad (i=1,2). & (3.1b) \end{cases}$$

These equations are with constant coefficients and their solution can be obtained at once. Omitting the subscripts, it can be written as:

$$\begin{cases} \varepsilon = \dot{\varepsilon}_o \sin t - (3\varepsilon_o + 2\dot{\theta}_o)\cos t - 2(2\varepsilon_o + \dot{\theta}_o), \\ \theta = 2\dot{\varepsilon}_o \cos t + 2(3\varepsilon_o + 2\dot{\theta}_o)\sin t - 3(2\varepsilon_o + \dot{\theta}_o)t + (\theta_o - 2\dot{\varepsilon}_o). \end{cases} \quad (3.2)$$

The subscript 0 refers here to the initial values. The presence of the characteristic secular term is to be noted in the θ-equation. This term can be avoided, however, by taking initial conditions such that $2\varepsilon_o + \dot{\theta}_o = 0$. A detailed study of the solution (2.8) of the C.W.-equations already shows quite a few of the features of the co-orbital motion. In figure (3.1) we display such a particular solution, corresponding to the initial values ($\varepsilon_o = 0.02$, $\theta_o = 0$, $\dot{\varepsilon}_o = -0.05$,

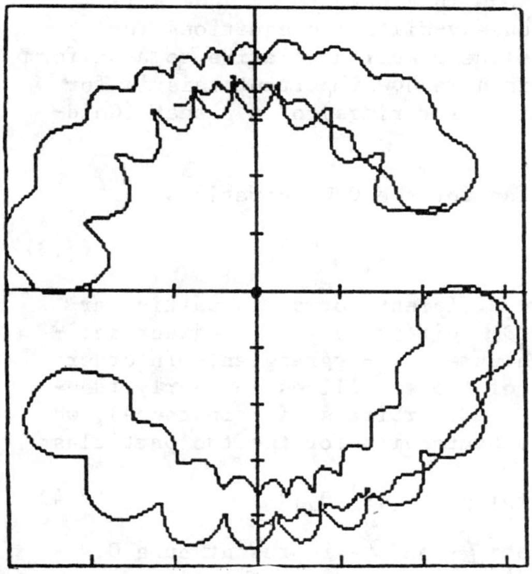

Figure 2.1

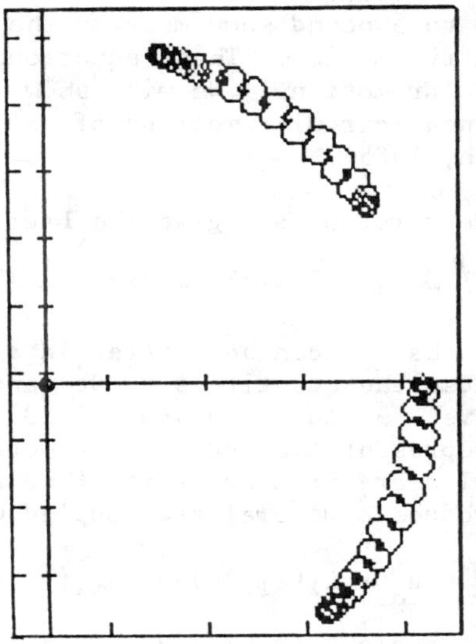

Figure 2.2

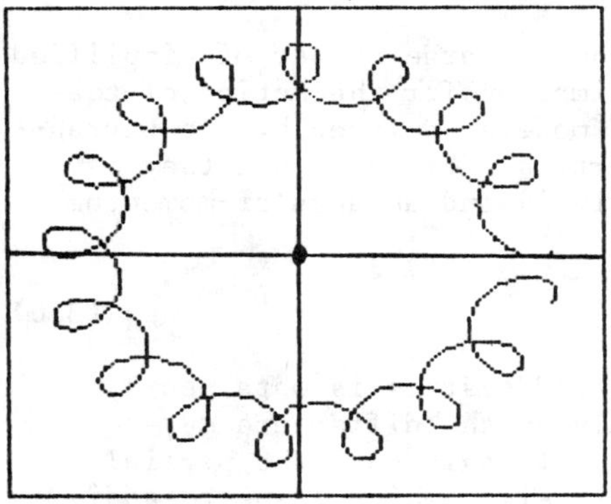

Figure 3.1

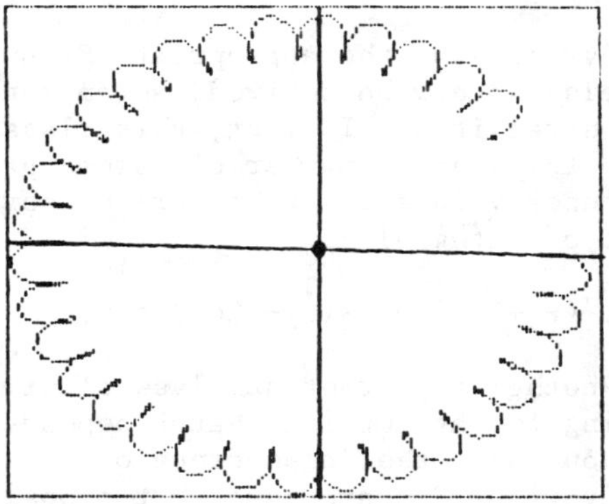

Figure 3.2

$\dot{\theta}_o = -0.06$). A similar orbit is shown in Figure (3.2). We still want to expound some more on the Clohessy-Wiltshire equations for relative motion. These equations define a motion relative to a uniform circular motion. We will show again here how important this is for the near-circular motions of satellites (or rings) of a planet (Goldreich, 1985).

We first of all give the Lagrangian for the C.W.-equations

$$\mathcal{L} = \frac{1}{2}(\dot{\varepsilon}^2 + \dot{\theta}^2) + 2\varepsilon\dot{\theta} + \frac{3}{2}\varepsilon^2 \tag{3.3}$$

Note that it can be written in a few different forms, in particular because the quantities $\varepsilon\dot{\theta}-\theta\dot{\varepsilon}$ and $2\varepsilon\dot{\theta}$ differ only by an exact derivative. Anyway we call (3.3) the pure C.W.-Lagrangian. In order to represent the interacting motion of two satellites in nearly identical near-circular orbits (Trojans or Co-orbitals, for instance), we introduce a generalized coupled C.W.-Lagrangian for the two particles:

$$\mathcal{L} = \mu_o\mu_1\mathcal{L}_1(\varepsilon_1,\theta_1) + \mu_o\mu_2\mathcal{L}_2(\varepsilon_2,\theta_2) + \mu_1\mu_2\, F(\varepsilon_1,\varepsilon_2,\theta_{21}) \tag{3.4}$$

where again $\theta_{21} = \theta_2 - \theta_1$. Here, both $\mathcal{L}_1$ and $\mathcal{L}_2$ represent pure C.W.-Lagrangians of the same form as (3.3). The equations of motion are:

$$\begin{cases} \ddot{\varepsilon}_1 = +2\dot{\theta}_1 + 3\varepsilon_1 + \mu_2 F_{\varepsilon_1} \\ \ddot{\theta}_1 = -2\dot{\varepsilon}_1 \qquad -\mu_2 F_\theta \end{cases} ; \qquad \begin{cases} \ddot{\varepsilon}_2 = +2\dot{\theta}_2 + 3\varepsilon_2 + \mu_1 F_{\varepsilon_2} , \\ \ddot{\theta}_2 = -2\dot{\varepsilon}_2 \qquad +\mu_1 F_\theta . \end{cases} \tag{3.5}$$

where F_{ε_i}, F_θ stand for $\partial F/\partial\varepsilon_i$ and $\partial F/\partial\theta_{21}$.

By choosing the appropriate F-function, a large number of simplified models can now be derived, as an approximation for the motion of the two satellites. In fact, this class of models, defined by the Lagrangian (3.4) has a number of common properties. First of all they have the energy integral (they are conservative!) and an angular-momentum type of integral:

$$\mu_1(\dot{\theta}_1 + 2\varepsilon_1) + \mu_2(\dot{\theta}_2 + 2\varepsilon_2) = C. \tag{3.6}$$

Another important subclass of simplified systems is obtained by making the assumption that F depends only on the difference $\varepsilon_{21} = \varepsilon_2 - \varepsilon_1$, without any other dependence on the ε_i. In this case the partial derivatives F_{ε_i} are equal but opposite in sign, which leads to additional simplifications. In fact, for this subclass of models, the equations of relative motion reduce to an independent system of order 4 (rather than 9), (with $\varepsilon_{21} = \varepsilon_2 - \varepsilon_1$):

$$\begin{cases} \ddot{\varepsilon}_{21} = 2\dot{\theta}_{21} + 3\varepsilon_{21} + (\mu_1 + \mu_2)\, F_\varepsilon(\varepsilon_{21},\theta_{21}) \\ \ddot{\theta}_{21} = -2\dot{\varepsilon}_{21} \qquad + (\mu_1 + \mu_2)\, F_\theta(\varepsilon_{21},\theta_{21}) \end{cases} \tag{3.7}$$

Finally, another property is that all these models are essentially integrable. Indeed, if we define the new variables

$$\begin{cases} \varepsilon = \mu_1\varepsilon_1 + \mu_2\varepsilon_2 , \\ \theta = \mu_1\theta_1 + \mu_2\theta_2 , \end{cases} \tag{3.8}$$

we easily deduce from the original equations of motion (3.7) that (ε,θ) satisfy the unperturbed C.W. equations (3.1) and have thus the solution (3.2).

In summary, we have thus a class of simple models which can give us much information on the mechanism of interaction of the co-orbital or other planetary satellites. Many of the simple models that will be described in the following sections can be related to the class of dynamical systems described by the Lagrangian (3.4). Let us also mention that in general, the function F will depend strongly on θ_{21} , but only weakly on ε_{21}. Therefore ε_{21} will often not be present in F.

4. THE SIMPLIFIED MODEL ONE

It is possible to simplify the previous equations of motion (2.5) and still maintain meaningful results, because, as we know, $\varepsilon_i, \dot{\varepsilon}_i, \dot{\theta}_i$ are small quantities. The equations contain the angles θ_i only in the difference $\theta_{21} = \theta_2 - \theta_2$ (and $\alpha = \theta_{21}/2$) and this is of course not a small quantity. In the same way, the distance between the two co-orbitals cannot be considered small, and we will thus not expand it, except that, if we neglect all our terms of order two and higher, we will have

$$d^2 = \varepsilon_{12}^2 + 2(1 - \cos\theta_{21}) \approx 4\sin^2\alpha . \tag{4.1}$$

If there is no close approach, the term $4\sin^2\alpha$ dominates the ε_{12}^2 which can therefore be neglected.

Equations (2.5) to first order in ε_i and $\dot{\theta}_i$ reduce to

$$\begin{cases} \ddot{\varepsilon}_i = 2\dot{\theta}_i + 3\varepsilon_i + \mu_j \left[2\sin^2\left(\frac{\theta_{21}}{2}\right) + \dfrac{(-1)^j\varepsilon_{21} - 2\sin^2\left(\frac{\theta_{21}}{2}\right)}{d^3}\right] & (4.2a) \\ \ddot{\theta}_i = -2\dot{\varepsilon}_i + (-1)^i \mu_j \sin\theta_{21}\left(1 - \frac{1}{d^3}\right) & (4.2b) \end{cases}$$

where (i=1 , j=2), (i=2, j=1), and $\theta_{21} = \theta_2 - \theta_1$, $\varepsilon_{21} = \varepsilon_2 - \varepsilon_1$.

With these equations, the coorbitals can pass each other since there is no singularity at $\theta_{12} = 0^\circ$. These equations are again Hill's equations for motion relative to a circular orbit with the coorbitals perturbing each other with their gravitational force. Equations (4.2) also give the collinear equilibrium points $(\bar{\varepsilon}_1, \bar{\varepsilon}_2)$ correct to the first order. The locations to first order are:

$$\bar{\varepsilon}_1 = -7\mu_2/12 \quad ; \quad \bar{\varepsilon}_2 = -7\mu_1/12 \quad ; \quad \theta_{21} = 0. \tag{4.3}$$

5. SOME OTHER SIMPLIFIED MODELS.

A simple but useful dynamical system is derived from the following Lagrangian which is a truncation of (3.4):

$$\mathcal{L} = \mu_o\mu_1\mathcal{L}_1 + \mu_o\mu_2\mathcal{L}_2 + \mu_1\mu_2\,[(\varepsilon_1+\varepsilon_2) - (1+\varepsilon_1+\varepsilon_2)\cos\theta_{21} + \frac{1}{d}\,], \tag{5.1}$$

where the distance d between the two co-orbital satellites is given by the full untruncated equation:

$$d^2 = r_1^2 + r_2^2 - 2r_1r_2\cos\theta_{21} \tag{5.2}$$

The corresponding equations of motions are:

$$\begin{cases} \ddot{\varepsilon}_1 = 2\dot{\theta}_1 + 3\varepsilon_1 + \dfrac{\mu_2}{\mu_o}\,[(1-\cos\theta_{21}) - (r_1 - r_2\cos\theta_{21})/d^3\,], \\ \ddot{\theta}_1 = -2\dot{\varepsilon}_1 + \dfrac{\mu_2}{\mu_o}\sin\theta_{21}[-(1+\varepsilon_1+\varepsilon_2) + r_1r_2/d^3\,], \\ \ddot{\varepsilon}_2 = 2\dot{\theta}_2 + 3\varepsilon_2 + \dfrac{\mu_1}{\mu_o}\,[(1-\cos\theta_{21}) - (r_2 - r_1\cos\theta_{21})/d^3\,], \\ \ddot{\theta}_2 = -2\dot{\varepsilon}_2 + \dfrac{\mu_1}{\mu_o}\sin\theta_{21}[+(1+\varepsilon_1+\varepsilon_2) - r_1r_2/d^3]. \end{cases} \tag{5.3}$$

Of course, we have again a constant angular momentum C and energy E, as well as the Jacobi integral (E-C).

Further simplifications of the present model consist for instance in adopting the following Lagrangian:

$$\mathcal{L} = \mu_o\mu_1\mathcal{L}_1 + \mu_o\mu_2\mathcal{L}_2 + \mu_1\mu_2\,(\frac{1}{d} - \cos\theta_{21}), \tag{5.4}$$

where d should be defined by

$$d^2 = 2(1+\varepsilon_1+\varepsilon_2)(1-\cos\theta_{21}). \tag{5.5}$$

The equations of motion are easily derived from this Lagrangian; they are just a tiny bit simpler than those in (5.3).

An even further simplification consists in assuming that (5.5) is reduced to

$$d^2 = 2(1-\cos\theta_{21}).$$

The equations of motion are then without any perturbation term in the radial equation:

$$\begin{cases} \ddot{\varepsilon}_i = +2\dot{\theta}_i + 3\varepsilon_i \\ \ddot{\theta}_1 = -2\dot{\varepsilon}_1 - \frac{\mu_2}{\mu_o}\sin\theta_{21}(1 - 1/d^3) , \\ \ddot{\theta}_2 = -2\dot{\varepsilon}_2 + \frac{\mu_1}{\mu_o}\sin\theta_{21}(1 - 1/d^3) . \end{cases} \tag{5.6}$$

For this model, there also exist periodicity conditions for both tadpole and horseshoe orbits. A sufficient periodicity condition is: if the initial velocities are normal to the unit circle ($\varepsilon_{i_o}=\dot{\varepsilon}_{i_o}=0$) and the particles return perpendicular to the unit circle at time

$$t_1 \ (\varepsilon_i(t=t_1) = 0, \quad \dot{\theta}_i(t=t_1) = 0),$$

then the orbit is periodic with period $2t_1$ and is symmetric about the unit circle.

For horseshoe orbits, another condition is: if $\theta_{21} = 180^o$ and the initial velocities are normal to the axis of symmetry ($\dot{\varepsilon}_{i_o} = 0$) and the trajectories cross normal to the unit circle at time t_1, then the orbit is periodic with period $4t_1$ and is symmetric about both the unit circle and the axis of symmetry. This theorem is similar to the mirror image theorem that holds for the general three-body problem (see Roy, 1982, and Roy and Ovenden, 1955). The numerical integrations that we performed with the above simple models show very little difference from the orbits obtained with the exact equations of motion of the general three-body problem. In the next section we show even further possible simplifications.

6. YODER'S SIMPLIFIED MODEL FOR COORBITAL MOTION

In this section we will describe in some detail a new simple model obtained by further simplifying the simplified model (5.6). Since the motion is very near the unit circle, we will set r_1 and r_2 exactly equal to unity. This is equivalent to let $\ddot{\varepsilon}_i = 0$, and

$$2\dot{\theta}_i + 3\varepsilon_i = 0 \quad ; \quad 2\ddot{\theta}_i + 3\dot{\varepsilon}_i = 0 \quad .$$

The two angular equations in (5.6) reduce therefore to

$$\ddot{\theta}_1 = +3\mu_2 \sin\theta_{21} . (1 - \frac{1}{d^3}) , \tag{6.1a}$$

$$\ddot{\theta}_2 = -3\mu_1 \sin\theta_{21} . (1 - \frac{1}{d^3}) , \tag{6.1b}$$

where $d = 2|\sin\alpha| = 2|\sin(\theta_{21}/2)| > 0$. In this model, the two coorbitals can of course not cross each other, since the denominator d would be zero. Therefore we will usually assume $0 \leq \theta_1(0) \leq \theta_2(0) \leq 2\pi$.

Yoder (1983) obtained essentially the same model, also by expanding about a circular reference orbit and assuming small eccentricity.

Many interesting properties can be derived for the model defined by the differential equations (6.1). The model is essentially very well understood thanks to the fact that it is integrable.

Three integrals of motion for equations (6.1) are easily derived and are (with $n_1 = \dot{\theta}_2 + 1$; $n_2 = \dot{\theta}_2 + 1$, $n_{21} = n_2 - n_1$)

$$\begin{cases} \mu_1(n_1 - 1) + \mu_2(n_2 - 1) = C_0 , & (6.2a) \\ \mu_1\theta_1 + \mu_2\theta_2 = C_0 t + C_1 , & (6.2b) \\ \frac{1}{3}\dfrac{n_{21}^2}{\mu_1+\mu_2} + 4\sin^2\left(\frac{\theta_{21}}{2}\right) + \dfrac{1}{\sin\left(\frac{\theta_{21}}{2}\right)} = C . & (6.2c) \end{cases}$$

Notice that the second integral contains the time explicitly. To obtain conservative integrals, another rotating coordinate system could be used.

Because of the singularity at $\theta_{21} = 0$, the coorbitals cannot pass and the orbits are periodic in θ_2 and θ_1 . So all initial conditions result in periodic solutions in a rotating coordinate system.

Thanks to the existence of the above integrals (6.2), the present simple model is completely integrable and many analytical results can be written down. We will limit ourselves to the following remarks.

Two sets of equilibrium points exist for equations (6.1). The first is the collinear equilibrium point where $\theta_{21} = 180^\circ$ and $n_{21} = 0$. The second is the Lagrangian triangular points where $\theta_{21} = 60^\circ$ and $n_{21} = 0$. The equilibrium points are tied in with the transition between different types of orbits, and the type of orbit can be determined by finding the bounds of θ_{21}.

To find the minimum and maximum values of θ_{21} describing the bounds of the periodic orbits, n_{21} is set equal to zero in equation (6.2c) and θ_{21} is solved for (see Yoder 1982). The extrema of θ_{21} are given by

$$x^3 - \frac{1}{4}Cx + \frac{1}{4} = 0 , \qquad (6.3)$$

where $C \gtrsim 3$ and

$$x = \sin\left(\frac{\theta_{21}}{2}\right) . \qquad (6.4)$$

All three solutions are real and are given by

$$\begin{cases} x_1 = \left(\frac{C}{3}\right)^{\frac{1}{2}} \cos\left(\frac{\beta}{3}\right) \\ x_2 = \left(\frac{C}{3}\right)^{\frac{1}{2}} \cos\left(\frac{\beta}{3} + \frac{2}{3}\pi\right) \\ x_3 = \left(\frac{C}{3}\right)^{\frac{1}{2}} \cos\left(\frac{\beta}{3} + \frac{4}{3}\pi\right) \end{cases} \quad (6.5)$$

where

$$\cos\beta = -\left(\frac{3}{C}\right)^{3/2} \quad (6.6)$$

From the last equation, β is constrained to be in the interval $\pi/2<\beta<\pi$. Then x_1 and x_3 are positive, x_2 is negative, and $x_1 > x_3$. For $C>5$, $x_1 > 1$ and x_3 is the only solution. From the solutions, the following four classes of orbits occur depending on C:

i. If $C>5$, the orbit is a horseshoe with $\theta_{21_{max}} = 2\pi-\theta_{21_{min}}$ and $\theta_{21_{min}} < \theta_a$. The orbits are symmetric about $\theta_{21}= 180^o$.

ii. If $C=5$, the coorbitals are at the collinear equilibrium points or are asymptotically approaching them.

iii. If $3<C<5$, the orbit is a tadpole with the closest approach θ_{min} bounded by $\theta_a < \theta_{min} < 60^o$ and the maximum separation θ_{max} bounded by $60^o < \theta_{max} < 180^o$.

iv. If $C=3$, the coorbitals are at the triangular equilibrium points ($\theta_{21}= 60^o$), where

$$\theta_a = 2 \sin^{-1} \left[\frac{1}{2} (2^{\frac{1}{2}} -1)\right] \approx 24^o \quad (6.7)$$

and is independent of the mass parameters. All orbits loop around the Lagrangian triangular points.

As shown by Yoder , the maximum and minimum values of r_1 and n_{21} occur at $\theta_{21} = 60^o$ for all orbits. At $\theta_{21}= 60^o$, the absolute value of the relative mean motion is

$$n_{21}= \left[\ 3(\mu_1+ \mu_2)\ (C-3)\ \right]^{\frac{1}{2}}. \quad (6.8)$$

For horseshoe orbits, $\theta_{21}= 180^0$ is a local minimum for the absolute value of n_{21} and is given by

$$n_{21}= \left[3(\mu_1 + \mu_2)\ (C-5)\ \right]^{\frac{1}{2}}. \quad (6.9)$$

The periods of the orbits are equal and since the orbits are symmetric about a circle near the unit circle, the period is given by

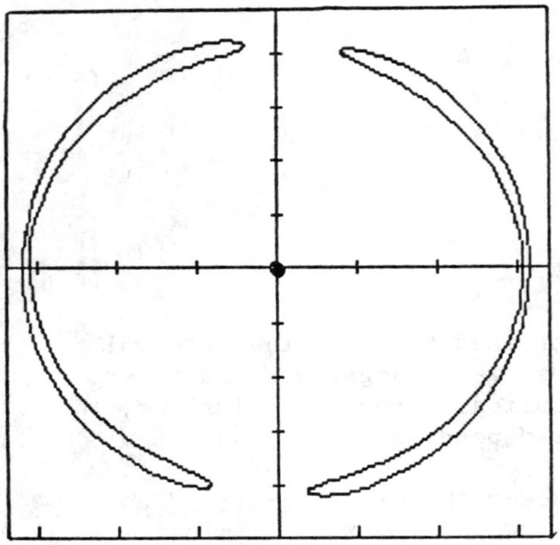

Figure 6.1

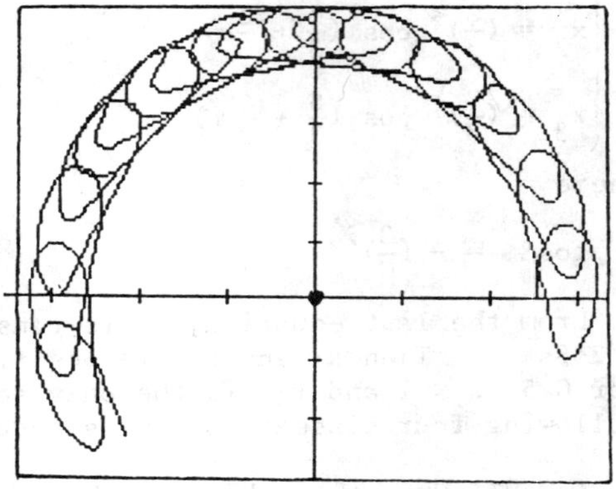

Figure 6.2

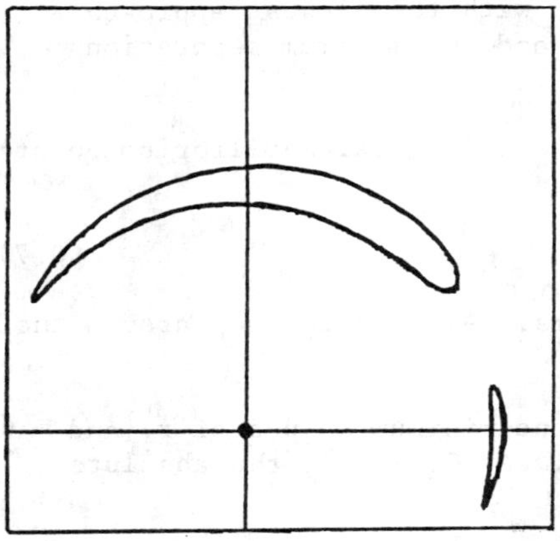

Figure 6.3

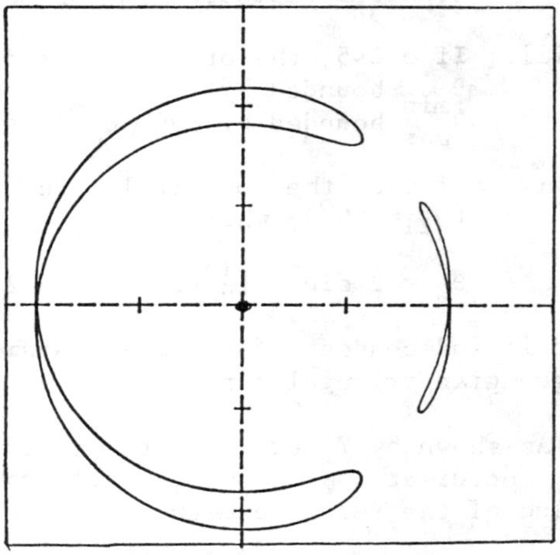

Figure 6.4

$$P = \frac{2}{[3(\mu_1+\mu_2)]^{\frac{1}{2}}} \int_{\theta_{21\,min}}^{\theta_{21\,max}} \frac{\pm d\theta}{[C-4\sin^2(\frac{\theta}{2}) - \frac{1}{\sin(\frac{\theta}{2})}]^{\frac{1}{2}}} \tag{6.10}$$

where the limits of integration are the maximum and minimum angles of separation. As C approaches 3, the period of the orbits approach

$$P = \frac{4\pi}{[27(\mu_1+\mu_2)]^{\frac{1}{2}}} \ . \tag{6.11}$$

The four figures 6.1 through 6.4 show a few orbits integrated with the Yoder model for the two co-orbital satellites.

7. REPULSIVE GRAVITY

A rather interesting and paradoxical phenomenon is found in the theory of co-orbital motion: the co-orbitals approach each other but then they repel each other. This repulsive gravity behaviour can be explained analytically with most of our simplified models. The result that is obtained is similar to the one given by Seidelman et al.(1984), (but not identical).

In Yoder's model, the distance between the co-orbitals is given by

$$d = 2 \sin \frac{\theta_{21}}{2} \tag{7.1}$$

The second derivative of d is found to be

$$\ddot{d} = -\frac{1}{2}\dot{\theta}_{21}^2 \sin\frac{\theta_{21}}{2} + \ddot{\theta}_{21}\cos\frac{\theta_{21}}{2} \ . \tag{7.2}$$

Using the above expression for d and the results of the previous section, the second derivative of d becomes

$$\ddot{d} = \frac{3(\mu_1+\mu_2)}{d^2} + \frac{3}{4}(\mu_1+\mu_2)\ [d - (4+C)d + 2d^3\ ]. \tag{7.3}$$

Near close encounter, the dominating term is $1/d^2$ and the motion is governed by a repulsive gravity with a magnitude three times greater than the actual attraction. An in depth examination will reveal that in fact we have here just another form of the Coriolis effects. Repulsive Gravity equations can also be derived from most of our other simplified models.

8. CONCLUSIONS

We have basically reviewed the equations of motion for two co-orbital satellites of a planet. The exact equations are first given in rotating polar coordinates, using the general three-body problem as a fundamental model. Next we describe several simplified models which can give us some insight in the mechanisms of interaction, while preserving the gross features of the motion. In particular, we showed the relationship with the Clohessy-Wiltshire equations of relative motion.

In a later publication, we intend to describe the connection with the Restricted Problem, and in particular with the Brown conjecture. We will also describe a new formulation which eliminates the short-period terms with an averaging technique, using Hill variables or Equinoctial Variables.

ACKNOWLEDGMENTS: The computations of this article were all performed with an IBM-PC donated to the University of Texas under the Quest Program.

REFERENCES

Aksnes, K., "The Tiny Satellites of Jupiter and Saturn", in V. Szebehely, "Stability of the Solar System and its minor bodies, Reidel, 1985, pp.3-16.

Dermott, S.F. and Murray, C.D., "The Dynamics of Tadpole and Horseshoe Orbits: I. Theory," Icarus 48, 1981, pp.1-11.

Dermott, S.F. and Murray, C.D., "The Dynamics of Tadpole and Horseshoe Orbits: II. The Coorbital Satellites of Saturn," Icarus 48, 1981, pp.12-22.

Dollfus, A. and Brunier, S., "Observations of Saturn's Inner Satellites and the Orbit of Janus in 1980," Icarus 48, 1981, pp.29-38.

Garfinkel, B., "Theory of the Trojan asteroids, Part I.," Astron. J., Vol.82, No.5, 1977, pp.368-379.

Goldreich, P. and Tremaine, S., "The Dynamics of Planetary Rings", in "Annual Reviews of Astronomy and Astrophysics", 1982, Vol.20, pp 249-283.

Harrington, R.S. and Seidelmann, P.K., "The Dynamics of the Saturnian Satellites 1980S1 and 1980S3," Icarus 47, 1981, pp.97-99.

Kaplan, M.H., Modern Spacecraft Dynamics and Control, John Wiley & Sons, New York, 1976.

Larson, S. M., Smith, B.A., Fountain, J.W., and Reitsema, H.J., "The 1966 Observations of the Coorbiting Satellites of Saturn, S10 and S11," Icarus 46, 1981, pp.175-180.

Roy, A.E. Orbital Motion, Adam Hilger, Bristol, 1982.

Roy, A.E. and Ovenden, M.W."On the Occurrence of Commensurable Mean Motions in the Solar System, Part II: The Mirror Theorem", M.N.R.A.S Vol.115, No.3, 1955, pp.296-309.

Seidelmann, P.K., Harrington, R.S., Pascu, D., Baum, W.A., Currie, D.G., Westphal, J.A., and Danielson, G.E. "Saturn Satellite Observations and Orbits from the 1980 Ring Plane Crossing", Icarus 47, 1981, pp 282-287.

Seidelmann, P.K., Harrington, R.S., and Szebehely, V., "Dynamics of Saturn's E. Ring", Icarus 58, 1984, pp.169-177.

Smith, B.A., Reitsema, H.J., Fountain, J.W., and Larson, S.M., "Saturn's Syzygistic Co-orbital Satellites", Bull. Amer. Astron. Soc. 12, 1980, pp.727-728.

Smith, B.A. and Voyager Imaging Team, "Encounter with Saturn: Voyager 1 Imaging Science Results", Science 212, 1981, pp.163-191.

Smith, B.A. and Voyager Imaging Team, "A New Look at the Saturn System: The Voyager 2 Images", Science 215, 1982, pp.504-537.

Spirig, F. and Waldvogel, J. "The Three-Body Problem with Two Small Masses: A Singular-Perturbation Approach to the Problem of Saturn's Coorbiting Satellites", Stability of the Solar System and Its Minor Natural and Artificial Bodies, V. Szebehely, ed., D. Reidel Publ. Co., 1984, pp.53-63.

Stone, E.C. and Miner, E.D., "Voyager 1 Encounter with the Saturnian System", Science 212, 1981, pp.159-163.

Synnott, S.P., Peters, C.F. Smith, B.A., and Morabito, L.A., "Orbits of small Satellites of Saturn", Science 212, 1981, pp.191-192.

Yoder, C.F., Colombo G., Synnott, S.P. and Yoder, K.A. "Theory of Motion of Saturn's Coorbiting Satellites", Icarus 53, 1983, pp.431-443.

LONG-PERIOD CHANGES IN THE SEMIMAJOR AXIS OF LAGEOS

M. Carpino
Osservatorio Astronomico di Brera-Milano,
Italy.

ABSTRACT.

LAGEOS is a passive satellite launched in 1976 with the purpose of determining accurately parameters of geophysical interest, such as plate tectonic movements, Earth's orientation and gravitational field. Its orbit (almost circular, at about 5900 km of height) and its shape (spherical, with low area-to-mass ratio) were chosen in order to minimize non-gravitational perturbations and to allow a very accurate orbit prediction. For the same purpose, the satellite is covered with corner-cube retroreflectors: it is tracked from a network of laser ranging stations with an accuracy of few centimetres. The whole mission was meant to be a very precise Celestial Mechanics experiment, a realization of the ideal test particle orbiting in the Earth's gravitational field. Nevertheless, the non-gravitational perturbations could not be eliminated completely. In particular, an unexplained drag-like force was soon discovered: it amounts to $3.4 \cdot 10^{-10}$ cm s^{-2} on the average (corresponding to a semimajor axis decrease of about 0.1 cm d^{-1}), but it exhibits also strong variations (with amplitude of the same order of magnitude as the average value) over timescales ranging from one month to 1-2 years.

The physical mechanisms responsible for these effects are various and not yet completely understood. As for the constant component of the drag force, the major candidates (with their estimated orders of magnitude) are:

- neutral ($0.2\text{-}0.5 \cdot 10^{-10}$ cm s^{-2}) and charge ($\sim 2.10^{-10}$ cm s^{-2}) particle drag (Afonso et al. 1980, 1985; Barlier et al. 1986);

- satellite thermal emission due to the infrared radiation from the Earth ($1\text{-}2\ 10^{-10}$ cm s^{-2}; Rubincam 1987).

As for the long-periodic variations, the effects which can contribute are:

A. E. Roy (ed.), Long-Term Dynamical Behaviour of Natural and Artificial N-Body Systems, 171–172.
© 1988 by Kluwer Academic Publishers.

Institut für Reine
und Angewandte Kernphysik
Kiel

- Earth reflected radiation pressure ($1.2 \cdot 10^{-10}$ cm s^{-2}; main periodicities 1050, 560 and 280 d; Anselmo et al. 1983, Barlier et al. 1986);

- satellite thermal emission due to solar radiation ($\sim 1.10^{-10}$ cm s^{-2}?)

- satellite thermal emission due to Earth infrared radiation ($\sim 0.4 \cdot 10^{-10}$ cm s^{-2}, main periodicity 1050 d; Rubincam 1987).

In general, it is not difficult to obtain estimates of the order of magnitude and of the periodicities of these effects; however, due to the complexity of the phenomena and the uncertainties in the physical parameters involved, it is usually impossible to model them with an accuracy sufficient for orbit prediction and/or determination. Therefore the existence of these effects constitutes a natural barrier to the accuracy reachable in the orbital models and, consequently, in the recovered geophysical information.

In the case of the Earth-reflected radiation pressure, for instance, it is possible to demonstrate, using simple symmetry considerations, that the diffused light component produces an along-track acceleration which averages out almost exactly over an entire orbital revolution of the satellite and therefore does not produce any significant long-periodic effect in semimajor axis. The only significant contribution is therefore from the specularly reflected component (though smaller by a factor ~ 4 than the diffused one). Its effects can be studied by using a semi-numerical approach: the along-track component of the force can be computed for every value of the relevant variables (the satellite orbital longitude λ and longitude of node Ω and the Sun longitude $\lambda_{\odot}$) under simplified assumptions about the reflectivity of the Earth's surface as a function of latitude and season; the mean effect is then obtained by averaging numerically over λ. The main periodic terms (computed with a two-dimensional Fourier transform) have frequencies $\dot{\Omega}$ (with a period of 1050 d), $\dot{\Omega} - \dot{\lambda}_{\odot}$ (560 d), $2\dot{\Omega} - 2\dot{\lambda}_0$ (280 d) and $\dot{\Omega} - 2\dot{\lambda}_0$ (221 d). All these components are actually detected in the semimajor axis residuals of LAGEOS, though the quantitative agreement with the amplitude and the phase predicted by the model can be considered satisfactory only for the $\dot{\Omega}$ and $\dot{\Omega} - \dot{\lambda}_0$ terms.

REFERENCES

Afonso G., Barlier F., Berger C. and Mignard F., C.R. Acad. Sc. Paris, Série B, 290, 445 (1980).
Afonso G., Barlier F., Berger C., Mignard F. and Walch J.J., Jour. Geophys. Res., 90, B11, 9381-98 (1985).
Anselmo L., Farinella P., Milani A. and Nobili A.M., Astron. Astrophys., 117, 3-8 (1983).
Barlier F., Carpino M., Farinella P., Mignard F., Milani A. and Nobili A.M., Annales Geophysicae, 4, A3, 193-210 (1986).
Rubincam D.P., Jour. Geophys. Res., 92, B2, 1287-94 (1987).

EFFECTS OF A NON-CIRCULAR SHEPHERD UPON A PLANETARY RING

F. Delhaise and M. Moons
Department de Mathematiques
Facultés Universitaires de Namur,
5000 Namur, Belgium.

1. INTRODUCTION

We consider the interactions between small satellites or ring particles which pass close to each other. Those perturbations are of the "impulse type" and not faint and continuous as in the case of planetary perturbations. This type of interaction is important because in many regions the morphology of Saturn's rings is primarily governed by close-by satellites' perturbations. The narrowness of Uranus' rings suggests strongly the presence of small satellites which confine them (Goldreich and Tremaine, 1979a). The modelisation of this type of interactions leads us to consider Hill's problem as Henon and Petit (1986) have done for the modelisation of the interaction between two ring particles, and as Waldvogel (1985) has done for the modelisation of the interaction between two coorbital satellites. Our contribution, in the case of a large impact parameter (e.g. in the case of Shepherd satellites) is the computation of the effects due to the non-circularity of the orbits.

2. NON-CIRCULAR HILL'S PROBLEM

Let m_o be the mass of the planet, m_1 and m_2 the masses of two ring particles or two shepherd satellites or respectively a satellite and a particle. We start from the equations of the general three-body problem expressed in terms of the Jacobian coordinates $\vec{X}$ and $\vec{\delta}$ where $\vec{X}$ describes the centre of mass of m_1 and m_2 with respect to m_o and $\vec{\delta}$ represents the position of m_2 with respect to m_1.

We then introduce Hill's problem which has the advantage of letting the ratio of masses m_1 and m_2 be arbitrary. However Hill's problem allows us to study only what happens in the zone of strongest perturbations defined in such a way that the distance between the two masses m_1 and m_2 is small enough to produce a non-negligible mutual interaction compared with the gravitational force coming from m_o.
If we assume that:

$$\| \vec{\delta} \| \leqslant m^{1/3} \| \vec{X} \|$$

A. E. Roy (ed.), Long-Term Dynamical Behaviour of Natural and Artificial N-Body Systems, 173–180.
© 1988 by Kluwer Academic Publishers.

where

$$m = \frac{m_1 + m_2}{m_o}$$

and if we take the total mass of the system and the gravitational constant equal to 1, we obtain the following equations of motion:

$$\ddot{\vec{X}} = - \frac{\vec{X}}{\|\vec{X}\|^3} + 0(m^{2/3}) \tag{2}$$

$$\ddot{\vec{\delta}} = -m \frac{\vec{\delta}}{\|\vec{\delta}\|^3} - \left(\frac{\vec{\delta}}{\|\vec{X}\|^3} - 3 \frac{(\vec{\delta} \mid \vec{X})}{\|\vec{X}\|^5} \vec{X}\right) + 0(m^{2/3}) \tag{3}$$

The equation in $\vec{\delta}$ is Hill's equation which has the particularity of being independent of the mass ratio m_1/m_2. From equation (2), we see that in the zone defined by the relation (1), the centre of mass of m_1 and m_2 follows approximately an orbit of the two-body problem. So its semi-major axis a' chosen equal to 1, its eccentricity e' and its argument of the pericentre g' are constant. Hence, if we know the variations of the orbital elements of one of the masses, we can calculate easily those of the other mass too, by means of the formulas:

$$a' = \frac{m_1 a_1 + m_2 a_2}{m_1 + m_2} \tag{4}$$

$$\vec{e}' = \frac{m_1 \vec{e}_1 + m_2 \vec{e}_2}{m_1 + m_2} \tag{5}$$

where a_i is the semi-major axis, g_i the argument of the pericentre, e_i the eccentricity and $\vec{e}_i = e_i(\cos g_i \vec{E}_1 + \sin g_i \vec{E}_2)$, the eccentric vector of the mass m_i. $(\vec{E}_1, \vec{E}_2)$ is an absolute reference frame. Hence we are going to concentrate only on the motion of m_2. We have assumed in equation (1) that the norm of $\vec{\delta}$ is of the order of $m^{1/3}$. Accordingly we change the scale of this vector:

$$\vec{\delta} = m^{1/3} \vec{x} . \tag{6}$$

We introduce also a mean position vector X_m for the centre of mass (see Figure 1) to avoid the use of pulsating coordinates. In relation to this vector, we consider a new reference frame $(\vec{f}_1 \vec{f}_2)$, centred in m_o and turning with $\vec{X}_m$ at an angular velocity equal to

$$\frac{d(\ell' + g')}{dt} .$$

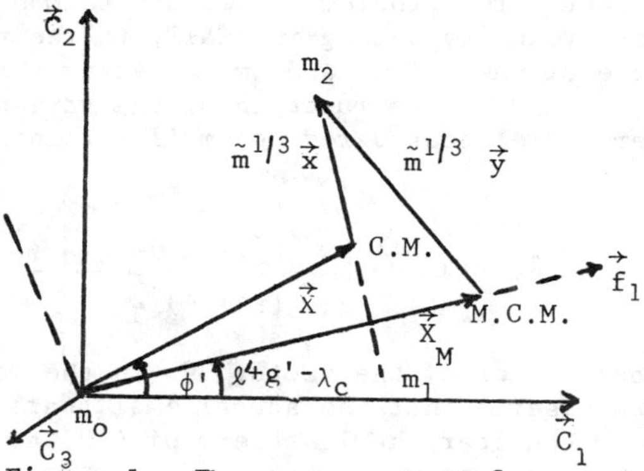

Figure 1: The mean centre of mass

$\vec{C}_1$: the direction of the conjunction.

ℓ' : the mean anomaly of the centre of mass.

g' : the argument of the pericentre of the centre of mass.

λ_c : the longitude of the conjunction.

$\phi' = f' + g'$, where f' is the true anomaly of the centre of mass.

$\tilde{m} = \dfrac{m_1^3}{(m_1+m_2)^2}$ the reduced mass.

The position of the mass m_2 with respect to that mean centre of mass is given by the vector $\tilde{m}^{1/3}\vec{y}$. If we write that last vector $\vec{y}$ as function of the vectors $\vec{x}$ and $\vec{X}$:

$$\tilde{m}^{1/3}\vec{y} = \tilde{m}^{1/3}\vec{x} - (\vec{X}_M - \vec{X}) \tag{7}$$

then the equation of motion satisfied by that vector can be deduced from the relations (2), (3) and (7):

$$\ddot{\vec{y}} = \frac{-\tilde{m}^{-1/3}(\vec{X}_M - \vec{X}) - \vec{y}}{\| \tilde{m}^{-1/3}(\vec{X}_M - \vec{X}) + \vec{y}\|^3} - \frac{\tilde{m}^{-1/3}\vec{X}_M + \vec{y}}{\|\vec{X}\|^3} + \tilde{m}^{-1/3}\vec{X}_M + 3\,\frac{(\tilde{m}^{-1/3}(\vec{X}_M - \vec{X}) + \vec{y}\,|\vec{X})}{\|\vec{X}\|^5}\vec{X} \tag{8}$$

From that last equation (8), it's easy to obtain the potential of the system, by means of the formula:

$$\ddot{\vec{y}} = -\,\vec{\mathrm{grad}}\,V \tag{9}$$

First of all, we reduce the problem to two dimensions by assuming that the mass m_2 and its velocity belong initially to the plane of the trajectory of the centre of mass of m_1 and m_2. After that, we express the vectors $\vec{X}_M$, $\vec{X}$ and $\vec{y}$ as function of the variables ℓ' and g' and expand in powers of e' considered as small retaining only first order term.

$$V = \frac{-1}{(\xi^2+\eta^2)^{\frac{1}{2}}} + \frac{\eta^2}{2} - \xi^2 + e'\,\tilde{m}^{-1/3}\left[\frac{\xi\cos\ell' - 2\eta\sin\ell'}{(\xi^2+\eta^2)^{3/2}}\right] \tag{10}$$

where (ξ,η) are the coordinates of the vector $\vec{y}$ in the moving reference frame $(\vec{f}_1,\vec{f}_2)$. We observe that the actual small parameter is $\dfrac{e'}{\|\vec{y}\|}$. If we want to consider the last term of (10) as a perturbation, we have to assume that:

$$e' \ll \tilde{m}^{-1/3}\,\|\vec{y}\|\,. \tag{11}$$

Afterwards we calculate the kinetic energy T, to obtain the expression of the Lagrangian L, from which we deduce the Hamiltonian of the system:

$$\begin{aligned} H = {}& \frac{1}{2}(p_\xi+\eta)^2 + \frac{1}{2}(p_\eta-\xi)^2 - \frac{3}{2}\xi^2 - \frac{1}{(\xi^2+\eta^2)^{\frac{1}{2}}} \\ & + e'\,\tilde{m}^{-1/3}\left[\frac{\xi\cos\ell' - 2\eta\sin\ell'}{(\xi^2+\eta^2)^{3/2}}\right] + L' \end{aligned} \tag{12}$$

where $p_\xi = \dfrac{\partial L}{\partial\dot{\xi}}$ the conjugated momentum to ξ.

$p_\eta = \dfrac{\partial L}{\partial\dot{\eta}}$ the conjugated momentum to η.

L' is the conjugated momentum to ℓ', which produces an autonomous Hamiltonian.

In order to simplify the terms of the kinetic energy, we do the following change of coordinates:

$$\begin{aligned} \xi &= x + 2\,p_y \qquad & p_\xi &= -y - p_x \\ \eta &= y + 2\,p_x \qquad & p_\eta &= -x - p_y \end{aligned} \tag{13}$$

Another advantage of these new variables is that we can see immediately what they represent in the first approximation. Indeed if we neglect the interaction between the masses m_1 and m_2, the Hamiltonian reads:

$$H = \frac{1}{2}(p_x^2 + x^2) - \frac{3}{2}p_y^2 \tag{14}$$

The variable y thus represents the linear motion of the mass m_2

with respect to the mean centre of mass, x and p_x represent its oscillatory motion and p_y is more or less the difference between the two semi-major axes a' and a_2, what we call by the impact parameter:

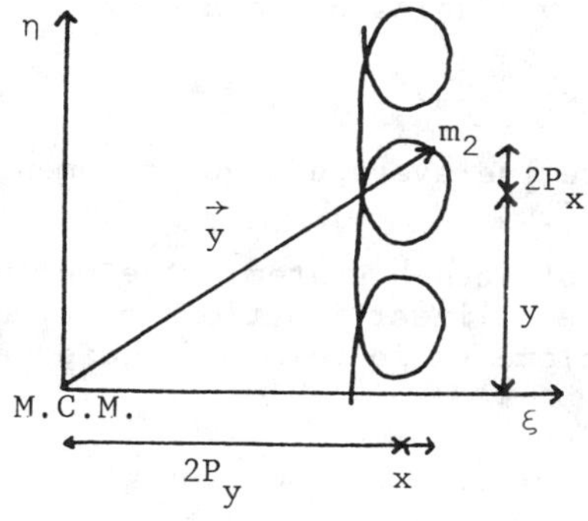

Figure 2: Geometrical interpretation of the variables.

3. THE CASE OF A LARGE IMPACT PARAMETER

The Hamiltonian related in (12) can be applied to several situations depending on the impact parameter's value. In the case of a small impact parameter we have for example, the horseshoe motion of two coorbital satellites (Waldvogel 1985). If the impact parameter is of the order of unity, the motion can be very irregular (Henon, Petit 1986). And in the case of a large impact parameter, the orbit of each body is slightly deflected by the other one. We have chosen to study that deviation. To concentrate on the large impact parameter case, we change again the scale. Assuming p_y large in absolute value (say of the order of $1/\varepsilon$), we propose:

$$x = \sqrt{2S_2}\,\sin(s_1+s_2) \qquad p_x = \sqrt{2S_2}\,\cos(s_1+s_2)$$

$$y = \mp\frac{3}{\varepsilon}\,s_1 \qquad p_y = \pm\frac{1}{\varepsilon} \mp \frac{\varepsilon}{3}\,(S_1-S_2-S_3)$$

$$\ell' = s_1 + s_3 \qquad L' = S_3 \tag{15}$$

with ε a small parameter, the superior sign being taken for the case where the mass m_2 is outside and the inferior sign for the opposite case.

In this way. y is also allowed to get large and we have introduced angle action coordinates for the oscillation in (x,p_x). This change of coordinates is canonical. In consequence, we obtain the Hamiltonian with a part which depends only on the momenta and the other part is the perturbation.

$$H = H_o(S_i) + \varepsilon H_1(s_i, S_i, \varepsilon) \qquad i = 1,2,3 \tag{16}$$

Therefore, we apply Lie's transformation method to resolve this problem. It's a method of perturbation which consists in making a change of coordinates close to the identity and canonical to obtain the Hamiltonian independent of the angular variables up to a certain order in ε.

$$H = \bar{S}_1 - \frac{1}{6}\varepsilon^2(\bar{S}_1 - \bar{S}_2 - \bar{S}_3)^2 \tag{17}$$

where $\bar{S}_i$ (i=1,2,3) are the average angular momenta deduced from the Lie's transformation.

We know the solution of such a system: the momenta are constant and the angular variables are linear functions of time. So it remains to calculate the orbital elements in terms of these variables whose behaviour is well-known.

$$\begin{aligned} a &= \bar{a} + \varepsilon\overline{(a\ ;\ W_1)} + \ldots \\ \vec{e} &= \bar{\vec{e}} + \varepsilon\overline{(\vec{e}\ ;\ W_1)} + \ldots \end{aligned} \tag{18}$$

where W_1 is the first order generator of the Lie's transformation and (-;-) is the operator Poisson bracket.

We obtain the variations of these elements by taking the difference between their value after the conjunction minus their value before the conjunction. Those expressions are written in the first order, by means of the operator "Poisson bracket" with the generatrice function W_1. And in the difference, defined previously, we have terms like:

$$h(\text{after}) \int_0^{\bar{s}_1(\text{after})} f\, d\bar{s}_1 - h(\text{before}) \int_0^{\bar{s}_1(\text{before})} f\, d\bar{s}_1 \tag{19}$$

where h and f are functions of the variables $\bar{s}_i$ and $\bar{S}_i$
"after" means : "after the conjunction"
and "before" means : "before the conjunction".

Let the integration's bounds be:

$$\begin{aligned} \bar{s}_1(\text{before}) &= -\infty \\ \bar{s}_1(\text{after}) &= +\infty \end{aligned} \tag{20}$$

Therefore if the function h, defined in (18), is independent of time, those terms become:

$$h \int_{-\infty}^{+\infty} f(\bar{s}_i, \bar{S}_i)\, d\bar{s}_1 \tag{21}$$

The functions f that we have to consider are of the form

$$\frac{\bar{s}_1 \sin(\bar{s}_1 + \bar{s}_2)}{(4 + 9\bar{s}_1^2)^{3/2}} \qquad (22)$$

from which it follows that the results will be expressed in terms of modified Bessel's functions.

4. ONE INTERACTION MAPPING

Two bodies revolve on keplerian orbits characterized by $(a_1, \vec{e}_1)$ and $(a_2, \vec{e}_2)$. They enter in interaction with a large impact parameter's value in such a way that their orbital elements undergo slight perturbations i.e.:

$$m^{1/3} \ll \frac{\Delta a}{a'} = \frac{|a_1 - a_2|}{a'} \ll 1$$
$$\|\Delta\vec{e}\| , \|\vec{e}\| \ll \frac{\Delta a}{a'} \qquad (23)$$

We note by $a_i + \partial a_i$ and $\vec{e}_i + \partial\vec{e}_i$ their values after the encounter. In the first order in eccentricity, these variations are:

$$\frac{\partial a_i}{a'} = \frac{2}{3}\alpha \frac{m_j}{m_o}\left(\frac{a'}{\Delta a_{ij}}\right)^3$$
$$\left[\mathrm{sign}(\Delta a_{ij}) 2(\Delta\vec{e}_{ij} \mid \vec{C}_2) + \alpha \frac{m_i + m_j}{m_o}\left(\frac{a'}{\Delta a_{ij}}\right)^2\right]$$
$$\partial\vec{e}_i = \mathrm{sign}(\Delta a_{ij}) \frac{m_j}{m_o}\left(\frac{a'}{\Delta a_{ij}}\right)^2$$
$$\left\{\alpha \vec{C}_2 + \left(\frac{a'}{\Delta a_{ij}}\right)\left[\beta(\Delta\vec{e}_{ij} \mid \vec{C}_2)\vec{C}_1 + \gamma(\Delta\vec{e}_{ij} \mid \vec{C}_1)\vec{C}_2\right]\right\} \qquad (24)$$

where $1 \leq i,j \leq 2$; $i \neq j$

$m_o \gg m_1, m_2$

$\Delta\vec{e}_{ij} = \vec{e}_i - \vec{e}_j$

$\Delta a_{ij} = a_i - a_j$

$\vec{C}_1$ the direction of the conjunction

$\vec{C}_2 = \vec{n} \times \vec{C}_1$ where $\vec{n}$ is a unit vector in the direction of the angular momentum

$$\alpha = 8\left[\frac{2}{9} K_o(\tfrac{2}{3}) + \frac{1}{9} K_1(\tfrac{2}{3})\right]$$

$$\beta = 8\left[\frac{20}{27} K_o(\tfrac{4}{3}) + \frac{19}{27} K_1(\tfrac{4}{3}) - \frac{1}{12}\right]$$

$$\gamma = 8\left[\frac{20}{27} K_o(\tfrac{4}{3}) + \frac{19}{27} K_1(\tfrac{4}{3}) + \frac{1}{12}\right]$$

K_o and K_1 are the modified Bessel's functions.

If we take in these formulae $\Delta\vec{e}_{ij} = \vec{0}$, we recover the results of Goldreich and Tremaine (see [2]). To this circular model, we have added the effects due to the non-circular orbits. Those effects are not negligible, especially for the variation in semi-major axis. Let us remark also that we arrive at the same conclusion as Showalter and Burns (1982) which is that the variations depend only on the difference between the eccentric vectors $\Delta\vec{e}_{ij}$, and not on the eccentric vectors $\vec{e}_i, \vec{e}_j$, independently. But this is not likely to be true at higher order in eccentricities.

ACKNOWLEDGEMENTS

We wish to thank Dr. Henrard for his comments and valuable suggestions.

REFERENCES

[1] N. Borderies, Dynamics of ring-satellite systems around Saturn and Uranus. Celest. Mech., 34: 297-327, 1984.

[2] P. Goldreich and S. Tremaine. The dynamics of planetary rings. Astron. Astrophys. 20:249-281, 1982.

[3] M. Henon and J.M. Petit. Series expansions for encounter - type solutions of Hill's problem. Celest. Mech., 38: 67-100, 1986.

[4] J. Henrard. Perturbation technique in the theory of non-linear oscillations and in celestial mechanics. Mathematical and Information Sciences Report No.44, BSRL, 1970.

[5] J.M. Petit and M. Henon. Satellite encounters. Icarus, **66**:536-555, 1986.

[6] M. Showalter and J. Burns. A numerical Study of Saturn's F-Ring, Icarus, 52: 526-544, 1982.

[7] F. Spirig and J. Waldvogel. The three-body problem with two small masses: A singular-perturbation approach to the problem of Saturn's coorbiting satellites. In: V. Szebehely, Stability of the Solar System and its Minor Natural and Artificial Bodies, Reidel, p.53-63, 1985.

GENERAL MOTION OF A TRIAXIAL RIGID BODY IN A NEWTONIAN FORCE FIELD

J.M. Ferrándiz and M.E. Sansaturio
Dpto. Matemática Aplicada a la Técnica
E.T.S. Ingenieros Industriales
Universidad de Valladolid (Spain)

1. INTRODUCTION

The study of the general motion of rigid bodies is of increasing importance since highly accurate solutions are required not only for the classical cases of precession and nutation of natural bodies but also for the attitude of artificial satellites. Intrinsically, this problem is of some difficulty. Although the general motion of a rigid body can be split up into the trnslation of its center of mass and the rotation about that point, both are usually coupled and therefore we are faced with a generally non-reducible 6-dimensional problem.

Nevertheless the reduction of it is possible in some interesting particular cases although it is never feasible in a simple way. Our ability of reducing problems strongly depends on their analytical formulation and specially on the sets of variables we use to formulate them. In the problem we are dealing with here, half of the variables we normally use are to determine the orbital motion of the center of mass whereas the other half define the rotation around that point. This is particularly useful when the translation is supposed not to be affected by the rotation, and we study the latter after having found a solution for the former. In many cases, this simplest approach provides accurate enough solutions to the relevant problem.

Even when we study both motions jointly, using such variables has the advantage that they are fairly known. However, there are also drawbacks since it is quite difficult to make good use of the symmetries of the problem to simplify it because the integrals usually have rather complicated expressions. Probably, the most desirable form for an integral appears when we have a conjugate momentum of a cyclic coordinate.

We may therefore ask the following question: does there exist a set of canonical variables to describe the general motion of a rigid body in which the integrals that are hoped for - or adiabatic invariants - have a simpler and handier form, but without complicating the analytical expressions arising in the formulation of the problem?

A. E. Roy (ed.), Long-Term Dynamical Behaviour of Natural and Artificial N-Body Systems, 181–187.
© 1988 by Kluwer Academic Publishers.

In the simplest particular case of this problem in which we have a triaxial rigid body under the gravitational attraction of a material point the answer is: yes.

We will obtain a new set of canonical variables to describe the roto-translatory motion of a rigid body which is similar to the usual ones coming from considering separately variables to describe translation and rotation. In this new set of variables, two integrals correspond to the conjugate momenta of two cyclic coordinates.

2. DERIVATION OF THE VARIABLES

We will start from the Hill variables (r,u,h,p_r,p_u,p_h) to describe the translation of the center of mass of the rigid body and the Andoyer variables $(\lambda,\mu,\nu, p_\lambda,p_\mu,p_\nu)$ to determine its rotation around that point.

Let us introduce new coordinates $\lambda_1,\lambda_2,\lambda_3$, ξ and the angles $\varepsilon_2,\varepsilon_3$ whose definition is shown in the following figure:

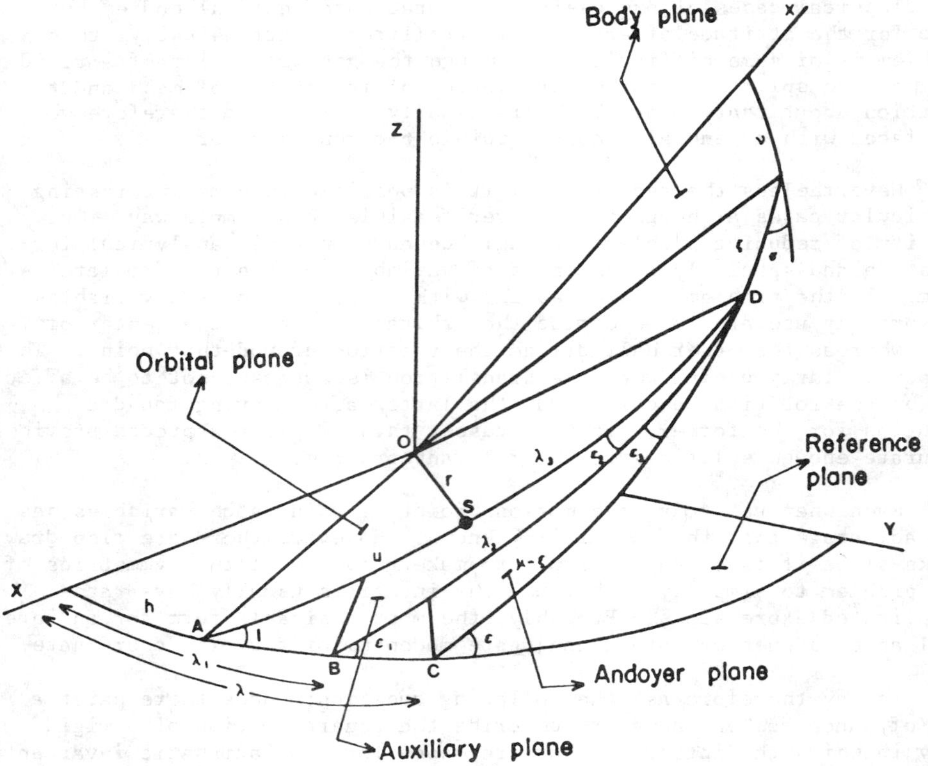

The differential formulae of spherical trigonometry applied to the spherical triangles ADB and BDC allow us to obtain the following identity

$$p_\mu d\mu + p_\lambda d\lambda + p_u du + p_h dh = p_\mu d\xi - p_u d\lambda_3 + (p_u \cos\varepsilon_3 + p_\mu \cos\varepsilon_2) d\lambda_2$$
$$+ (p_\mu \cos\varepsilon + p_u \cos I) d\lambda_1 + \sin\lambda_2 (p_\mu \sin\varepsilon_2 - p_u \sin\varepsilon_3) d\varepsilon_1 . \qquad (1)$$

Let us introduce the relations among the new variables

$$p_\mu \sin\varepsilon_2 - p_u \sin\varepsilon_3 = 0$$
$$p_\mu \cos\varepsilon_2 + p_u \cos\varepsilon_3 = M_T \quad . \qquad (2)$$

In that way, ε_2 and ε_3 become parameters which depend only on p_μ, p_u and M_T that can be considered as new momenta.

Therefore, we deduce that the transformation

$$(r,u,h,\lambda,\mu,\nu,p_r,p_u,p_h,p_\lambda,p_\mu,p_\nu) \sim (r,\lambda_1,\lambda_2,\lambda_3,\xi,\nu,p_r,p_1,p_2,p_3,p_\xi,p_\nu)$$

is canonical, where the momenta are given by the equation

$$p_1 = p_\lambda + p_h \qquad p_2 = M_T \qquad p_3 = -p_u \qquad p_\xi = p_\mu \quad . \qquad (3)$$

Let us note that the total angular momentum vector is expressed in the auxiliary plane by means of the equality

$$M = \begin{pmatrix} 0 \\ p_u \sin\varepsilon_3 - p_\mu \sin\varepsilon_2 \\ p_u \cos\varepsilon_3 + p_\mu \cos\varepsilon_2 \end{pmatrix} \qquad (4)$$

Then, the relationship $p_\mu \sin\varepsilon_2 = p_u \sin\varepsilon_3$ allows us to regard M_T as the magnitude of such a vector M and the auxiliary plane as the invariable one.

3. APPLICATION TO THE MOTION OF A TRIAXIAL RIGID BODY ATTRACTED BY A MASS-POINT.

Let us consider now the problem of the roto-translatory motion of a triaxial rigid body attracted by a mass-point. We will denote m_a and m_s the respective masses of both bodies while $m = m_s m_a/(m_s+m_a)$ will stand for the reduced mass of the system and A, B, C for the principal moments of inertia.

From (3), it is evident that the expression of the total kinetic energy in the new system of variables is immediately obtained from the corresponding expression in terms of the Hill-Andoyer variables by simply changing the notation. In this way, if V denotes the non-Keplerian potential the Hamiltonian is written as

$$H = \frac{1}{2m}\left(p_r^2 + \frac{p_3^2}{r^2}\right) + \frac{m}{2}\left\{(p_\xi^2 - p_\nu^2)\left[\frac{\sin^2\nu}{A} + \frac{\cos^2\nu}{B}\right] + \frac{p_\nu^2}{C}\right\} - \frac{km}{r} + V , \qquad (5)$$

where $k = G^*(m_a+m_s)$ and G^* is the gravitational constant.

From now onwards, we will assume the distribution of mass of the triaxial rigid body to be quasi-spherical so that V becomes a disturbing term.

If we restrict ourselves to the first order of perturbation its expansion turns out to be

$$V = -\frac{k}{m_a r^3}[(A-B)(1 - 3\cos^2\alpha)+(C-A)(1-3\cos^2\gamma)]. \qquad (6)$$

Here, $\cos\alpha$, $\cos\beta$ and $\cos\gamma$ are the direction cosines of m_s expressed in the system of principal axes of m_a. Their expression in the new set of variables can be obtained, for instance, from the equality

$$\begin{pmatrix}\cos\alpha\\ \cos\beta\\ \cos\gamma\end{pmatrix} = R_3^t(\nu)R_1^t(\sigma)R_3^t(\xi)R_1^t(\varepsilon_2+\varepsilon_3)R_3^t(\lambda_3) \begin{pmatrix}1\\0\\0\end{pmatrix} \qquad (7)$$

$R_i(x)$ standing for a positive rotation of angle x around the (i)-axis.

Note that the parameters $\sigma,\varepsilon_2,\varepsilon_3$ are related to the canonical variables by means of the relations

$$\cos\sigma = \frac{p_\nu}{p_\xi} , \quad \cos\varepsilon_2 = \frac{p_\xi + p_3\cos(\varepsilon_2+\varepsilon_3)}{p_2} , \quad \cos\varepsilon_3 = \frac{p_\xi\cos(\varepsilon_2+\varepsilon_3)+p_3}{p_2}$$

and

$$\cos(\varepsilon_2 + \varepsilon_3) = \frac{p_3^2 + p_\xi^2 - p_2^2}{2p_\xi p_3} . \qquad (8)$$

It is important to notice that to obtain the previous expressions in terms of the variables of Hill and Andoyer, the number of required rotations would have been seven, i.e. two more, so that the developments would be considerably longer.

Let us remark that the Hamiltonian H does not depend on λ_1,λ_2 and therefore they are two cyclic coordinates, the corresponding momenta p_1 (Z-component of the total angular momentum vector M) and p_2 (modulus of M) being constant.

In what follows, we will carry out an approximate integration of the problem. For that, it is convenient to substitute the (r,u,p_r,p_3)

variables for others (l,g,L,G) which are analogous to those of Delaunay. The transformation is formally the same and we will not detail it here. (for instance, see Deprit, 1981 pp 113-117).

Taking into account the equality

$$\frac{-k^2m^3}{2L^2} = \frac{1}{2m}\left(p_r^2 + \frac{p_3^2}{r^2}\right) - \frac{km}{r}, \tag{9}$$

the Hamiltonian can be cast in the sum of its non-perturbed part and the perturbed one. Grouping together the orbital kinetic energy and that of rotation in the first part, we have

$$H = H_o + H_1$$

with

$$H_o = \frac{-k^2m^3}{2L^2} + \frac{m}{4}\left[\frac{1}{A} + \frac{1}{B}\right] p_\xi^2 \tag{10}$$

$$H_1 = \frac{-m}{4}(p_\xi^2 - p_\nu^2)\left[\frac{1}{A} - \frac{1}{B}\right]\cos 2\nu + \frac{m}{4} p_\nu^2\left[\frac{2}{C} - \frac{1}{A} - \frac{1}{B}\right] - \frac{k^4}{m_a L^6}\left[\frac{2C-A-B}{2}P + (A-B)Q\right],$$

where

$$P = \sum_{k\in\mathbb{Z}^+\cup\{0\}} C_k^* \cos kl + \sum_{(i,j,k)\in J} D_{ijk} \cos(i\xi + jg + kl)$$

$$Q = \sum_{(i,j,j,q)\in J'} E_{ijkq} \cos(i\xi + jg + kl + q\nu) \tag{11}$$

$$J = \{-2,-1,0,1,2\} \times \{0,2\} \times \mathbb{Z} \qquad J' = J \times \{-2,2\},$$

the coefficients D_{ijk}, E_{ijkq} being functions of the momenta and of Hansen's coefficients.

Now, we will carry out an approximate analytical integration by perturbation methods. First of all, we use an infinitesimal contact transformation to eliminate the angular variables, except ν, from the Hamiltonian. The generating function and the equations of the transformation can be easily calculated up to the first order, in absence of resonances (for instance, by means of Deprit's method) and will not be reproduced here.

After such a transformation the Hamiltonian in the new variables turns out to be

$$K = \frac{-k^2m^3}{2L^2} + \frac{m}{2}(p_\xi^2 - p_\nu^2)\left[\frac{\sin^2\nu}{A} + \frac{\cos^2\nu}{B}\right] + \frac{p_\nu^2}{C} + \tag{12}$$

$$\frac{-k^4}{8m_a L^3 G^3}(3\cos^2(\varepsilon_2+\varepsilon_3)-1)[(2C-A-B)(3\cos^2\sigma-1)+3(B-A)\sin^2\sigma\cos 2\nu] .$$

As K only depends on the coordinate ν, all the momenta, except p_ν, are constant as well as the auxiliary angle $(\varepsilon_2+\varepsilon_3)$. The associate Hamiltonian equations can be integrated in closed form (Cid and Ferrándiz, 1980). For it, we carry out a non-canonical transformation substituting the (ν,p_ν) variables for the couple (x,X) defined as follows

$$x = \sqrt{p_\xi^2 - p_\nu^2}\ \cos\nu , \qquad X = \sqrt{p_\xi^2 - p_\nu^2}\ \sin\nu. \tag{13}$$

In this way, all the equations keep their Hamiltonian form except those corresponding to x, X and ξ . Likewise, if we introduce a new time variable s by means of the differential equation below

$$\frac{ds}{dt} = p_\nu\ \omega , \tag{14}$$

we get the solution

$$x = \beta\ C_o \sin(s-s_o) , \qquad X = C_o \cos(s-s_o) . \tag{15}$$

In the above formulae, constants ω,β depend on the distribution of mass and the current values of the integrals, and C_o also depends on the value of p at the time s_o.

The relationship between the real and fictitious time is obtained by means of an elliptic integral and then, the final solution for x and X is given by the equations

$$x = \beta C_o \operatorname{sn} \rho(t-t_o) , \quad X = C_o \operatorname{cn} \rho(t-t_o) , \quad \rho = \frac{\sqrt{p^2 - C_o^2}}{\omega} . \tag{16}$$

The rest of the variables $(l,g,\lambda_1,\lambda_2,\xi)$ can also be obtained as functions of t, in closed form, by means of elliptic functions.

Let us finally point out that the canonical set of variables we have just introduced here can be used to simplify the study of other more complicated problems. For instance, we may assume that the triaxial body is attracted by a spheroid with J_2, or consider other perturbations. The advantages specially come from the reduction in the expansions of the potential or from the slow variation of the momenta providing the modulus of the angular momentum or its Z-component. Work on such problems is being done currently by the authors.

4. ACKNOWLEDGEMENTS

The authors are deeply grateful to Prof. Cid for his invaluable comments and suggestions. In fact, the ideas in the paper are based, to a large extent, on some talks he had with the authors some years ago. This work has been partially supported by D.G.A. Proy. No.CB4/85 and CAICYT Proy. No. 779-84.

REFERENCES

Cid, R. and Ferrándiz, J.M.: 1980, Publ. Sección de Matemáticas de la Universidad Autónoma de Barcelona, 22, pp 231-235 (Spanish Language).

Deprit, A.: 1981. Celest. Mech. 24, pp 111-153.

THE MOTION OF THE ORBIT PLANE OF IAPETUS

David Harper
Computer Laboratory Chadwick Building
The University
P.O. Box 147
Liverpool L69 3BX
England

1. INTRODUCTION

The secular motion of the orbit plane of Iapetus is characterised by perturbations due to the actions of the Sun, Titan and (to a lesser degree) the oblateness of Saturn. If we restrict the disturbing function to those terms containing only the inclination and node of Iapetus we may eliminate short-period terms and medium-period Solar perturbations (see for example Sinclair (1974), Harper et al (in submission)) and consider only the secular part of the motion of the orbit plane. Referring the position of the orbit to an arbitrary fixed reference plane, we may show that the equation for the perturbation in the inclination is as follows, to the second power of the inclination.

$$dI/dt = 2n\{\chi_s \; I_s \; \sin(\Omega-\Omega_s) + \chi_t \; I_t \; \sin(\Omega-\Omega_t) + \\ + \chi_e \; I_e \; \sin(\Omega-\Omega_e)\} \qquad (1)$$

where χ_s, χ_t and χ_e are dimensionless coefficients giving the relative sizes of the disturbing effects. They are given below with their numerical values.

$$\chi_s = (3/8)\; \mu_s \; (a/a_s)^3 = 2.037 \; 10^{-5} \qquad (2)$$
$$\chi_e = (3/4)\; J_2 \; (a_e/a)^2 = 0.345 \; 10^{-5} \qquad (3)$$
$$\chi_t = (1/8)\; \mu_t \; \alpha b^{(1)}_{3/2} = 1.330 \; 10^{-5} \qquad (4)$$

We make the assumption that Titan and the oblateness act in the same plane so that $I_t = I_e$ and $\Omega_t = \Omega_e$. This is a good approximation since the orbit plane of Titan lies within 1° of the equator plane of Saturn. It simplifies (1) to the form

$$dI/dt = 2n\{\chi_s \; I_s \; \sin(\Omega-\Omega_s) + \chi_t^* \; I_t \; \sin(\Omega-\Omega_t)\} \qquad (5)$$

where we have written $\chi_t^* = \chi_t + \chi_e$. If we choose the reference plane so that it passes through the intersection of the orbit planes of Titan

A. E. Roy (ed.), Long-Term Dynamical Behaviour of Natural and Artificial N-Body Systems, 189–192.
© 1988 by Kluwer Academic Publishers.

and the Sun then we may write

$$\Omega_s = \Omega_t + 180^o \tag{6}$$

and

$$dI/dt = 2n(\chi_s I_s - \chi_t^* I_t) \sin(\Omega - \Omega_s). \tag{7}$$

Clearly we may make dI/dt vanish if

$$\chi_s I_s - \chi_t^* I_t \equiv 0. \tag{8}$$

This determines the position of the reference plane between the orbit planes of Titan and the Sun. The orbit plane of Iapetus maintains a constant inclination to this fixed plane, which is known as the Laplacian plane of the orbit.

The position of the Laplacian plane is determined by the positions of the orbit planes of the disturbing bodies and by the relative sizes of their perturbations, as given by the χ coefficients. In the case of Iapetus the position of the Laplacian plane is

$$\Omega_L = 163^o.738 \quad ; \quad I_L = 13^o.3614$$

referred to the mean ecliptic and equinox of 1950.0.

The rate of precession of the orbit of Iapetus is constant and it is determined by the mean motion of Iapetus and the χ coefficients. Denoting it by λ we have

$$\lambda = -2\, n\, (\chi_s + \chi_t^*) \tag{9}$$

The numerical value of this quantity is $-11^o.3478$ per Julian century.

2. COMPARISON WITH DATA

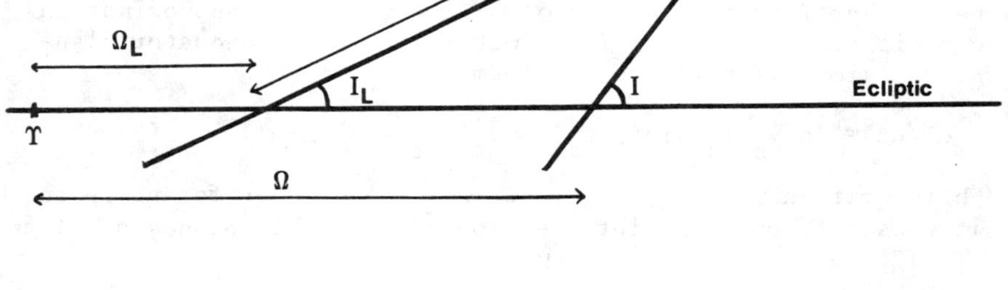

We may determine the inclination of the orbit of Iapetus to its Laplacian plane by comparing the theory with observed values of the node and inclination of the orbit upon the ecliptic. If we denote the inclination of the orbit to the Laplacian plane by I' (a constant) and its position with respect to the ecliptic and equinox of 1950.0 at any instant by Ω and I then we may write

$$\sin(\Omega-\Omega_L)\sin I = \sin I' \sin \Psi \tag{10}$$

$$\cos(\Omega-\Omega_L)\sin I = \cos I' \sin I_L + \sin I' \cos I_L \cos \Psi \tag{11}$$

$$\cos I = \cos I' \cos I_L - \sin I' \sin I_L \cos \Psi \tag{12}$$

where Ψ is defined as in figure 1 and is given by

$$\Psi = \Psi_o + \lambda(t - t_o) \tag{13}$$

and ψ_o is the value of ψ at the initial epoch t_o. Comparison with the data allows us to find Ψ_o and I' and hence Ω and I at the initial epoch, which we denote by Ω_o and I_o.

We use the following observed values of node and inclination. They are opposition mean values which have been corrected by subtracting long-period Solar perturbations so that only the secular part remains. The reference frame is the ecliptic and equinox of 1950.0.

Date	Node	Inclination
1787.70	$146^\circ.50627$	$19^\circ.26139$
1832.50	$145^\circ.05501$	$18^\circ.85991$
1857.50	$144^\circ.32894$	$18^\circ.70378$
1876.70	$143^\circ.39184$	$18^\circ.54219$
1880.20	$143^\circ.35909$	$18^\circ.54550$
1885.60	$143^\circ.10195$	$18^\circ.46443$
1917.20	$141^\circ.96303$	$18^\circ.15884$
1918.20	$141^\circ.91871$	$18^\circ.15050$
1926.40	$141^\circ.55608$	$18^\circ.06129$
1927.40	$141^\circ.57161$	$18^\circ.05516$
1973.00	$139^\circ.89829$	$17^\circ.56719$

We employ a least-squares fitting process in which we solve to find Ω_o, I_o and μ_t. The initial epoch is 1885.25. The node and inclination are dependent upon μ_t via the position of the Laplacian plane (equations (8) and (4)) and via the rate of precession (equation (9)). We obtain the following values:

$$\Omega_o = 143^\circ.084 \pm 0^\circ.040$$
$$I_o = 18^\circ.449 \pm 0^\circ.013$$
$$\mu_t = (2.33 \pm 0.063)\ 10^{-4}$$

and the root-mean-square residuals are:

in sin I $\Delta\Omega$ $0^{\circ}.0368$
in ΔI $0^{\circ}.0240$

The value of μ_t agrees well with previous similar determinations. For example, Jeffreys (1953) obtained $(2.357 \pm 0.052)\ 10^{-4}$ from a study of the motion of the orbit plane of Iapetus.

The theory can be extended to three or more disturbing forces by the use of the transformation $p = \sin I \sin \Omega$, $q = \sin I \cos \Omega$ (Harper 1987). Under the action of several disturbing forces, the orbit plane can again be shown to maintain a constant inclination to a certain fixed plane whose position is determined by the planes of action and relative sizes of the disturbing forces. Thus the Laplacian plane is still a useful concept in cases where a satellite is acted upon by three or more forces of similar size.

REFERENCES

Harper, D., Sinclair, A.T., Taylor, D.B. and Shen K.X. (in submission): Astron. Astrophys. (to be published)
Jeffreys, H. (1953): Mon. Not. Roy. Astron. Soc. 113, 81.
Sinclair, A.T. (1974): Mon. Not. Roy. Astron. Soc. 169, 591.
Harper, D. (1987): Ph.D. thesis, University of Liverpool (unpublished), Chapter 3.

ASTRODYNAMICS IN THE REAL-WORLD ENVIRONMENT
(The whole truth and nothing but the truth)

Denise A. Kaya, Lori A. McCarter
and Joseph J. F. Liu
HQ Air Force Space Command,
Co., USA.

When a researcher develops a new orbit theory for estimating the motion of an Earth artificial satellite, it is necessary that he verifies the validity and the accuracy of that theory. To do this, one needs to know the true orbit so that the differences between the predicted orbit and the true orbit can be measured. But how does one go about determining this "true orbit"? A commonly accepted procedure is to generate a reference for comparison by numerically integrating the exact equations of motion adopted by the theory. Using a set of given initial conditions, one can build a reference orbit for the theory composed of numerically generated positions and velocities, or an equivalent set of orbital elements, at a sequence of time points. It is well known that the accuracy of such a reference orbit is not only highly dependent on the exactness of the mathematical formulation of the dynamical system, but also on computational roundoff and truncation errors. With proper implementation, however, these errors can be minimized and the numerically generated ephemerides may be treated as a reliable reference which simulates a "true" orbit over a limited time period. This hypothetical reference orbit will be used for the verification and validation study that is necessary for the initial examination of any analytical solution.

The ultimate test for the utility of the new theory in a real-world environment is its ability to predict where the satellite is observed to be. How well it can do this is the true test of its ability to model the real-world. But, if the new theory does well compared to a reference orbit shouldn't it also do well against observational data? Aren't both data bases a representation of the truth? In principle, this is what one would like and is often led to believe. In the real-world environment, it is necessary that the truth be that which is observed.

Why is this so? The initial conditions of a satellite are not precisely known and a first guess is estimated by using a small collection of observation and an initial orbit determination algorithm.

A. E. Roy (ed.), Long-Term Dynamical Behaviour of Natural and Artificial N-Body Systems, 193–196.
© 1988 by Kluwer Academic Publishers.

The accuracy of the initial conditions can be improved by using a precision orbit determination algorithm, commonly referred to as a differential correction, as more observations become available. Once the initial conditions are obtained, a prediction can then be made against subsequent observations. Frequent comparisons have been made in the past of various theories for predicting satellite motion. These theories, in general, can be placed in one of three categories: numerical, analytical and semi-analytical. The major forces included in the equations of motion of all three theories are those of the Earth's gravity, lunar and solar perturbations and atmospheric drag. A numerical theory generally incorporates very complete forcing functions (e.g., a higher order geopotential function, an empirical atmospheric density model such as a Jacchia model and accurate third-body perturbing forces, etc.) and integrates the equations of motion using a small step size. An analytical theory employs the same force model but in a truncated form (e.g., a geopotential model which may include only the effects of J2, J3, J4 and a few of the more significant half-day and full-day resonance terms, an atmospheric density model which is usually represented by a simple analytical function, and is derived with a perturbation method. A semi-analytical theory is formulated using elements of both theories and the transformed variational equations obtained by a perturbation method are integrated numerically using a much larger step size. Generally speaking, a theory with a more complete force model and a more precise integration should produce more accurate solutions. Therefore, the numerical theory should provide more accurate solutions than the analytical or semi-analytical theory. It is well known that computational errors are introduced in the numerical and analytical integrations due to truncation and roundoff, therefore the accuracy of all three theories for satellite prediction degrades with time. It may be necessary that the satellite's initial conditions be frequently updated using the information provided by more recent observational data. Then why do observations represent the truth in the real-world of astrodynamics? Observations are the only known entities and the only measurements used to deduce initial conditions. In addition, the comparison of a theory's prediction to observations is the only measurement available to quantify the quality of that prediction. Therefore, satellite observations describe the orbit that one is constrained to use as the "truth" in modelling the real-world.

Let us act out a scenario. We are the researcher who has developed this wonderful new theory for orbit prediction. To test our theory we try to predict against a reference orbit that we have generated. The results are excellent and we are very excited about the theory's potential. For operational implementation, the theory must be tested against an observational data base. Due to our previous success using a reference orbit, we are very confident in our theory and feel that this test will be easy to accomplish. The results come in and we are surprised that for some cases our theory doesn't perform as well as we expected. At first we think that there must be some mistake. But, after determining that there was no error, we are most humbled.

What is the cause of this "true" anomaly? The obvious answer may be that our theory was not as complete as we first envisoned it to be. There may be some element of truth to this but it is not the entire explanation. We had mentioned previously that, when comparing numerical, analytical and semi-analytical theories, the numerical one is more accurate than the semi-analytical and analytical ones. When using these theories to predict against reference orbits, this is normally the case. However, when we use observational data, occasionally the semi-analytical and analytical theories do almost as well as and sometimes better than the numerical theory. For many of these cases we have determined that the satellite has manoeuvered or that a solar flare has occurred, leaving all theories at a disadvantage. For other cases we found our data to be poor. Our data is collected by various types of sensors distributed throughout the world. Sensor inaccuracies, poor weather conditions which prohibit data collection, human/operator errors, insufficient sensor coverages or unpredictable environmental changes will all degrade our data base. As a consequence, the data may have gaps or the observations may have limited accuracy. Again all theories are at a disadvantage. However, there are a few cases where no obvious explanation exists and we are left puzzled.

Satellite observations have various degrees of inaccuracies and uncertainties. In addition, not all physical forces are modelled or modelled accurately. Many of these inadequacies and uncertainties have been well addressed by researchers and practitioners in the field of celestial mechanics and astrodynamics: others are taken for granted. The repeated process of updating the initial conditions can help in correcting inherent model errors and filtering out past observational errors. However, uncertainty in the model and the data still exists and, as a consequence, we often obtain results much different than what we believe should be true. A few real-world examples are to be demonstrated to illustrate this concept. Although normalized values are presented, these examples should provide some substance to show the "true" difficulties in a real-world environment.

In conclusion, it is necessary to use a reference orbit to verify and validate a new orbit theory. However, if that theory is to be used to predict a real-world satellite, it must also be tested using observational data. If this new theory is computationally more efficient or improves prediction accuracy, it is eligible as a replacement of existing models used for operational support. Independent of the theory one uses, it is not possible to truly predict a satellite's motion since we do not have precise initial conditions, we do not know the forces exactly, we are not capable of predicting the environmental parameters and our dynamical system is always highly nonlinear and nonintegrable. Our observations include errors that make it even more difficult to know the satellite's "true" position. Because of these uncertainties, it is difficult to determine what the truth really is. Perhaps in the future it will be possible to minimize some of these uncertainties. For the present, we are constrained to define the real truth to be that which we observe.

REFERENCES

Szebehely, V.: 1986. 'Uncertainties and Instabilities in Celestial Mechanics', Particle Accelerators, Vol.19, pp.43-47.
Szebehely, V.: 1986, 'New Nondeterministic Celestial Mechanics', Space Dynamics and Celestial Mechanics, pp. 3-14.

A FINITE-TIME STABILITY CRITERION FOR SUN-PERTURBED PLANETARY SATELLITES

A. E. Roy and B. A. Steves,
Department of Physics and Astronomy,
Glasgow University, Glasgow, U.K.

ABSTRACT

The inability of the c^2H stability criterion in the general three-body problem to guarantee Hill (or hierarchical) stability in the case of a planetary satellite perturbed by the Sun when the planet's orbital eccentricity is non-zero leads to the search for a finite-time stability criterion applicable to such cases. Its application to the main satellites of Jupiter, Saturn and Uranus shows that stability lifetimes of at least 10^6 years may be demonstrated for them.

1. INTRODUCTION

In recent years a number of authors (see for example Bozis 1976, Easton 1971 and 1975, Golubev 1968, Marchal 1971, Marchal and Saari 1975, Roy 1979, Roy et al. 1984, Smale 1970, Szebehely and Zare 1977, Valsecchi et al. 1984, Zare 1976, 1977) have made use of the general three-body problem c^2H term, a combination of the total angular momentum $\underset{\sim}{c}$ and the total energy H integrals of the system, in order to examine the hierarchical stability (i.e. Hill stability) of a given three-body system. In such a system, two of the bodies, of masses m_1 and m_2 , move in disturbed Keplerian elliptic orbits about their common centre of mass, while the third body, of mass m_3 , moves in a much larger disturbed Keplerian orbit about the centre of mass of the first two bodies.

If the orbits do not cross, it is possible, by computing c^2H_{crit} from the initial conditions, to show that the binary formed by m_1 and m_2 will never be disrupted by m_3 if $c^2H \leqslant c^2H_{crit}$. The third body m_3 cannot, in its evolving orbit, cross the changing orbit of m_1 and m_2 , though it may escape to infinity. The system is then said to have hierarchical stability. In the restricted three-body problem, moreover, m_2 is a particle of infinitesimally small mass in orbit about m_1 and is disturbed gravitationally by m_3 moving in a circular orbit about m_1 . The well-known work of G. W. Hill (1878), utilising the concept of the

A. E. Roy (ed.), Long-Term Dynamical Behaviour of Natural and Artificial N-Body Systems, 197–215.
© 1988 by Kluwer Academic Publishers.

surfaces of zero velocity or Hill surfaces derived from the Jacobi integral C, showed that if the surface about m_1 is closed, the particle is prevented from escaping from m_1 and entering into a trajectory about m_2. The particle in this case is customarily said to have Hill stability. In the general three-body problem, the c^2H_{crit} value is equivalent to the value of the Jacobi constant for which the two lobes of the Hill surfaces surrounding each of the massive particles just barely touch. In the general problem, c^2H_{crit} is calculated from the relevant collinear configuration of the three massive particles which is one of three collinear equilibrium configurations dictated by the masses and their linear arrangements (see for example Walker and Roy 1981).

In the restricted problem, it is important to note that, even if there is no guarantee that the massless particle will be kept in orbit about m_1, the particle does not necessarily escape. Its chance of doing so is dictated by its orbital evolution and the width of the 'neck' connecting the now coalesced surfaces about m_1 and m_2. The escape may be immediate or may take any length of time extending out to an infinitely long period. Similarly, in the general three-body problem, if $c^2H > c^2H_{crit}$, there is no guarantee of hierarchical stability, but the break-up of the original hierarchy (i.e. m_2 is in orbit about m_1, while m_3 is in a larger orbit about the centre of mass of m_1 and m_2) may occur in a time interval that can vary enormously, depending upon the initial conditions of the system. In a recent series of papers (Roy 1979, 1982, 1983, Roy et al. 1984, Roy et al, 1985, Walker 1983, Walker et al. 1980, Walker and Roy 1983a, 1983b), the existence of a considerable region of empirical stability is demonstrated in cases where $c^2H > c^2H_{crit}$.

Empirical stability, in other words the maintenance of the hierarchy in a general three-body system for a considerable time even when no c^2H guarantee can be given, becomes important in studies of the solar system's sub-systems of bodies. Among such sub-systems are: (1) the Sun and two planets; (2) a planet and two satellites; (3) a planet, its satellite and the Sun.

No example of these three types of sub-systems exists by itself. We are always dealing with an n-body problem where $n>3$ and where there is no analytical guarantee of hierarchical stability. Nevertheless, in these triple systems analytical hierarchical stability can be said to exist until the value of c^2H for such a triple system is brought, by perturbations of other bodies, above the value of c^2H_{crit} for the triple system (see Milani and Nobili 1983a,b). It should be noted that the critical value of c^2H, viz c^2H_{crit}, is a function only of the three masses and therefore is constant. It may be said, then, that a necessary but not sufficient condition for the hierarchical stability of an n-body system, where $n>3$, is that the system's three-body subsystems have c^2H stability.

Now in the solar system, when the values of c^2H_{crit} are computed for examples of types (1) and (2) from the orbital elements and masses, it is found that in all cases c^2H is comfortably less than c^2H_{crit}.

For examples of type (3), namely planet-satellite-Sun systems, $c^2H < c^2H_{crit}$ for all the major satellites of the planets as long as the eccentricities of the planets' orbits about the Sun are neglected. The exceptions to the rule, the retrograde satellites of Jupiter and Saturn, are probably captured bodies. In the case of the Earth's moon, however, the c^2H stability criterion is barely satisfied.

When we take into account the planets' orbital eccentricities, the results change drastically. All planet-satellite-Sun triples (see Valsecchi et al. 1984) fail to satisfy the c^2H hierarchical stability criterion. This result probably shows how restrictive the criterion is rather than indicates the instability of planetary satellites against solar perturbations. A different approach is needed in order to re-examine the problem and the present paper is largely a report on work in progress to find and evaluate such an approach.

2. A FINITE TIME STABILITY CRITERION

Instead of seeking a stability criterion valid for all time, we try to provide one which not only holds in any particular case for a finite time but also provides a means of estimating that time. At first, the approach is illustrated by applying it to the case where the Sun is assumed to move in a circular orbit about the planet, with the satellite having negligible mass. The satellite moves in a disturbed elliptical orbit about the planet in the plane of the Sun's planetocentric orbit. In this case we have the circular coplanar restricted three-body problem. The method is then applied to the case where the Sun is assumed to move in a fixed elliptic orbit about the planet, in other words the elliptic restricted coplanar three-body problem.

2.1 The circular coplanar restricted three-body problem

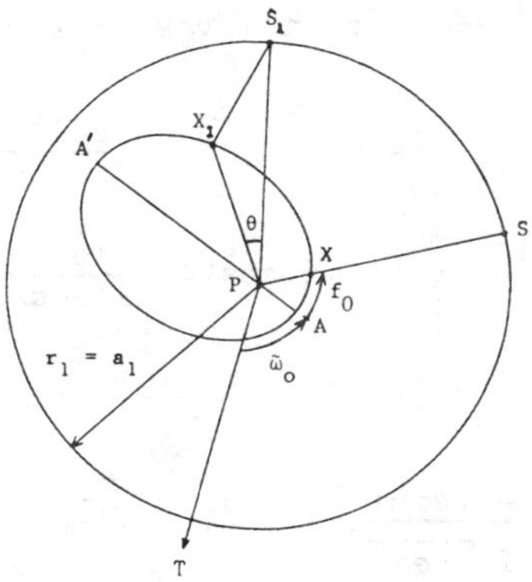

Figure 1: The coplanar circular restricted three-body problem where the planet (P) - satellite (X) system is disturbed by the Sun(S), assumed to be moving in a fixed circular orbit. The initial longitude of the satellite's pericentre $\tilde{\omega}_o$ is arbitrarily chosen to be zero, so that PA lies along PT.

In Figure 1 P, S and X denote the positions of the planet, Sun and satellite with PX = r, PS = r_1, XS = Δ and $r \ll r_1$. Let the osculating orbital elements of the satellite orbit be a (the semi-major axis), e (the eccentricity), $\tilde{\omega}$ (the longitude of pericentre) and ε (the mean longitude of the epoch). Let the masses of the planet, Sun and satellite be m, m_1 and m' with m' negligible compred to m and m_1 and with $m \ll m_1$.

If n and n_1 are the mean motions of the satellite and the Sun respectively then, taking units such that the gravitational constant has the value unity, we have the following relations:

$$n^2a^3 = m+m' \; ; \quad n_1^2\, a_1^3 = m_1 + m + m' \; ; \quad r_1 = a_1 = \text{constant}. \tag{1}$$

The Roy-Walker empirical stability parameters ε_{32} and ε^{23}(Roy 1979) are given by:

$$\varepsilon_{32} = \mu_3\, \alpha^3 \quad , \quad \varepsilon^{23} = \mu(1-\mu)\alpha^2$$

where $\mu_3 = m_1/(m+m')$, $\mu = m'/(m+m')$, $1-\mu = m/(m+m')$ and $\alpha = a/a_1$. Using Equations 1 and the relation $m' \ll m \ll m_1$, the empirical stability ε_{32} can be approximated as:

$$\varepsilon_{32} \sim (n_1/n)^2 = \nu^2 \; ;$$

the other empirical stability parameter ε^{23} is effectively zero.

Let the sidereal periods of the Sun and the satellite be T_1 and T respectively and let their synodic period be T_s. The changes in the satellite's orbital elements due to solar perturbation may be obtained from the Lagrange planetary equations, viz.

$$\frac{da}{dt} = 2\varepsilon_{32} \frac{n}{\sqrt{1-e^2}} \{r_1(\frac{1}{\Delta^3}-\frac{1}{r_1^3}) \, [e \sin f \cos\theta - (1+e\cos f)\sin\theta] - \frac{er}{\Delta^3}\sin f\}$$

$$\frac{de}{dt} = \varepsilon_{32}\frac{n}{a}\sqrt{1-e^2}\{r_1(\frac{1}{\Delta^3}-\frac{1}{r_1^3}) \, [\sin f \cos\theta - (\frac{e(1+\cos^2 f)+2\cos f}{1+e\cos f})\sin\theta] - \frac{r}{\Delta^3}\sin f\}$$

$$\frac{d\tilde{\omega}}{dt} = -\varepsilon_{32}\frac{n\sqrt{1-e^2}}{ae}\{r_1(\frac{1}{\Delta^3}-\frac{1}{r_1^3})[\cos f\cos\theta + (\frac{2+e\cos f}{1+e\cos f})\sin f \sin\theta] - \frac{r}{\Delta^3}\cos f\} \tag{2}$$

$$\frac{d\varepsilon}{dt} = -2\varepsilon_{32}\frac{n}{a}\{r_1(\frac{1}{\Delta^3}-\frac{1}{r_1^3}) \, [(\frac{r}{a} - \frac{3\,ent\sin f}{2\sqrt{1-e^2}})\cos\theta + \frac{3a\sqrt{1-e^2}nt}{2r}\sin\theta]$$

$$- \frac{r}{\Delta^3}(\frac{r}{a} - \frac{3ent\sin f}{2\sqrt{1-e^2}})\} + (1-\sqrt{1-e^2})\frac{d\tilde{\omega}}{dt}$$

Equations (2) may be integrated numerically or solved analytically to a desired degree of approximation. For the present purpose, however, it is advantageous to change the independent variable in Equations 2 from time t to the true anomaly f, and subsequently to θ , the angle between the planetocentric radius vectors of the satellite and the Sun, (see Figure 1) so that

$$\theta = \ell - \ell_1 = f + \tilde{\omega} - \ell_1$$

where ℓ and ℓ_1 are the longitudes of the satellite and the Sun respectively.

The first step, using the relation

$$r^2 \frac{df}{dt} = h = na^2\sqrt{1-e^2} \text{ and setting } a_1 = 1 \text{ so that } \alpha = a,$$

gives

$$\frac{da}{df} = \frac{2\varepsilon_{32}}{1-e^2} (\frac{r}{a})^2\{r_1(\frac{1}{\Delta^3}-\frac{1}{r_1^3}) [e \sin f \cos\theta - (1+e\cos f)\sin\theta] - \frac{er}{\Delta^3} \sin f\}$$

$$\frac{de}{df} = \frac{\varepsilon_{32}}{a} (\frac{r}{a})^2\{r_1(\frac{1}{\Delta^3}-\frac{1}{r_1^3})[\sin f \cos\theta - (\frac{e(1+\cos^2 f)+2\cos f}{1+e\cos f})\sin\theta] - \frac{r}{\Delta^3}\sin f\}$$

$$\frac{d\tilde{\omega}}{df} = \frac{-\varepsilon_{32}}{ea}(\frac{r}{a})^2 \{r_1(\frac{1}{\Delta^3}-\frac{1}{r_1^3})[\cos f \cos\theta + (\frac{2+e\cos f}{1+e\cos f})\sin f \sin\theta] - \frac{r}{\Delta^3}\cos f\} \qquad (3)$$

$$\frac{d\varepsilon}{df} = \frac{-2\varepsilon_{32}}{a\sqrt{1-e^2}} (\frac{r}{a})^2\{r_1(\frac{1}{\Delta^3}-\frac{1}{r_1^3})[(\frac{r}{a} - \frac{3ent\sin f}{2\sqrt{1-e^2}}) \cos\theta + \frac{3nt\sqrt{1-e^2}}{2r} \sin\theta]$$

$$-\frac{r}{\Delta^3}(\frac{r}{a} - \frac{3ent\sin f}{2\sqrt{1-e^2}}) \} + (1 - \sqrt{1-e^2}) \frac{d\tilde{\omega}}{df}$$

Equations (3) are exact and can easily be integrated numerically. The second step is then carried out by using the following transformation,

$$\frac{d\sigma}{d\theta} = (\frac{d\theta}{df})^{-1} \frac{d\sigma}{df}$$

where $\sigma = a, e, \tilde{\omega}$ or ε ; $\quad \frac{d\theta}{df} = 1 - \frac{d\ell_1}{df}$.

ℓ_1 as a function of f can be found through a series expansion. The differential equations $d\sigma/d\theta$ are then integrated analytically after expanding them to the desired degree of approximation about the small values of $\alpha = a/a_1$, e, and $\nu = n_1/n$. Both numerical and analytical integration procedures were applied to the problem as a check on the accuracy of the solutions. In this investigation, we are not interested so much in the exact positions of the bodies at any given time, but rather in the minimum time the Sun requires to render the satellite's orbit parabolic. Up until this point, all the equations and the method of analysis described have been applicable to both the case where the Sun is assumed to move in a fixed circular orbit and the case where the Sun is assumed to move in a fixed elliptical orbit. We now apply the method to the circular case.

Step 1. Suppose that at time $t=0$ a conjunction PXS occurs at a true anomaly of value f_0. The initial longitude of pericentre $\tilde{\omega}_o$ as measured from the direction PT, where T is the vernal equinox, can be arbitrarily chosen to be zero (i.e. PA lies along PT), since all initial orientations of the satellite's elliptical orbit with respect to the Sun's circular orbit produce identical configurations. Thus at a conjunction the initial true anomaly of the Sun f_{10} is equal to f_0. The problem now contains four independent initial parameters: the satellite's orbital displacement f_0, the satellite's semimajor axis a, the satellite's eccentricity e, and the mass ratio of the three bodies through

$$\nu = n_1/n \approx \sqrt{\mu_3 \alpha^3} = \sqrt{\varepsilon_{32}} \quad .$$

We wish to obtain the changes Δe, Δa, $\Delta\tilde{\omega}$, $\Delta\varepsilon$ that occur in the elements a, e, $\tilde{\omega}$, ε over one synodic period taken from the opposition of the Sun and the satellite prior to this conjunction, to the opposition immediately after the conjunction. The equations (4) have therefore to be integrated from $\theta = -\pi$ to $\theta = +\pi$. We assume that over one synodic period the satellite's orbital elements remain constant on the right hand sides of equations (4) throughout the integration. r_1 is set equal to a_1 for the circular case. Then, integrating and expanding equation (4) to second order about the small quantities $\alpha = a/a_1$, $\nu = n\ /n$, and e (i.e. $\alpha^i e^j \nu^k$ where $i+j+k \leqslant 2$) the following results are obtained:

TABLE 1

$\Delta a = 0$

$\Delta e = \pi\varepsilon_{32} \times$ (sin)

Coefficient	L_o
$\frac{15}{8} a - \frac{39}{32} a\nu + 4\nu + \frac{40}{3}\nu^2$	1
$\frac{-15}{2} e - \frac{33}{2} e\nu$	2

$e\Delta\tilde{\omega} = \pi\varepsilon_{32} \times$ (cos)

Coefficient	L_o
$\frac{3}{2} e\ (1+\nu)$	0
$\frac{-15}{8} a + \frac{39}{32} a\nu - 4\nu - \frac{40}{3}\nu^2$	1
$\frac{15}{2} e + \frac{33}{2} e\nu$	2

$\Delta\varepsilon = \pi\varepsilon_{32} \times$ (cos)

Coefficient	L_o
$\frac{5}{2} + 7\nu + \frac{23}{2}\nu^2 - \frac{15}{2} e^2 - 6a(1+2\nu)$	0
$\frac{-33}{16} ae + 46\ e\nu + 12e$	1
$\frac{-27}{2} e^2$	2

where $L_0 = f_0 + 2e\nu \sin f_0 + \ldots\ldots$

The column headed 'Coefficient" contains the coefficient of the sine or cosine term, while the other column contains the coefficients of the arguments of the cosine or sine terms given at the top of the column. For example, the second term in Δe is

$$\pi\varepsilon_{32}\left(- \frac{15}{2} e - \frac{33}{2} e\nu \right) \sin 2 L_0$$

The loss of accuracy incurred when the two nearly equal quantities $1/\Delta^3$ and $1/r_1^3$ are subtracted (see Equations (2) and (3)) is avoided by the binomial expansion:

$$\frac{1}{\Delta^3} - \frac{1}{r_1^3} = \frac{\rho}{r_1^3}\Big[3 \cos\theta + \frac{3}{2}\rho(5 \cos^2\theta - 1) + \frac{5}{2}\rho^2(7 \cos^3\theta - 3 \cos\theta)$$
$$+ \frac{15}{8} \rho^3(1 - 14 \cos^2\theta + 21 \cos^4\theta)\Big] \text{ , where } \rho = \frac{r}{r_1}$$

For a well-behaved satellite, Δa is very small. The danger, therefore, to the satellite's stability will arise, not from changes, in a, but from changes in the orbital eccentricity e. If e reaches unity, the satellite will escape or collide with its planet, in the practical case of a finite-sized planet.

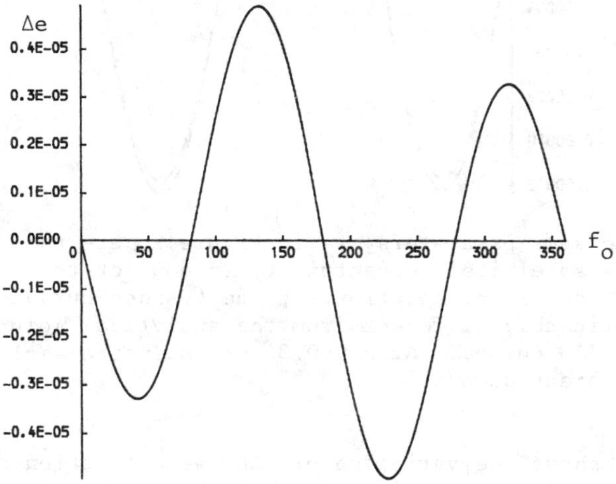

Figure 2: The variation in the change in the satellite's eccentricity Δe over one synodic period for different starting values of the initial conjunction true anomaly f_o, in the case where the Sun is assumed to be moving in a fixed circular orbit. The data used here is that of the Jupiter, Galilean satellite system where $\alpha = 0.0025$, $e = 0.01$ and $\mu_3 = 1{,}100$. Both the numerical and analytical solutions for Δe are shown here to lie virtually on the same curve.

We now sample the size of Δe by repeating the evaluation of Δe for different initial starting values of the conjunction true anomaly f_0. Figure 2 shows the variation in the value of Δe for f_0 values ranging from 0° to 360° for both our numerical and analytical solutions. The analytical solution was found to be virtually identical to the numerical solution for typical solar system values of the initial independent parameters α, e and μ_3, the relative differences between the two solutions averaging about 10^{-5}. In particular, Figure 2 shows the results for the Galilean satellites of Jupiter, where α, e and μ_3 are taken to be 0.0025, 0.01, and 1,100 respectively.

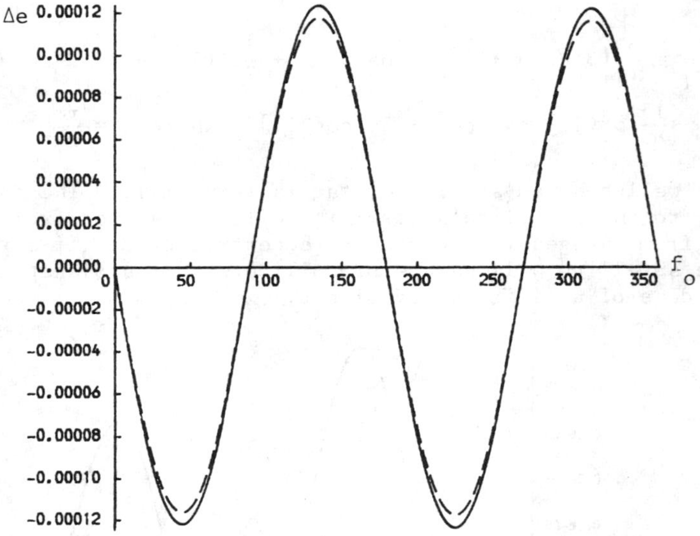

Figure 3: The same parameters, as in Figure 2, are used except that the satellite's eccentricity is now set to e = 0.3. The numerical solution for Δe (dashed curve) is now noticeably different from the analytical solution for Δe (solid curve). At e = 0.3 our analytical solution starts to break down.

Figure 3 shows the variation of Δe as a function of f_0 for the same Jupiter-Galilean satellite system with the exception that now the eccentricity of the satellite's orbit has been increased to e = 0.3. At e = 0.3, we see that the numerical and analytical solutions start to deviate noticeably from each other, having relative differences of about 10^{-2}. Beyond e = 0.3, for typical solar system values for α and μ_3, our analytical solution tends to break down.

Let the maximum size of Δe be Δe_{max}. Through a simple optimization of Δe, we find that the maximum Δe is given by:

Table 2: The Minimum Time needed for solar perturbations to increase a satellite's orbital eccentricity from e_o to e_T.

Planet-Satellite System	α	e	μ_3	e_1	T_s (days)	T_1 (years)	CIRCULAR CASE		ELLIPTICAL CASE	
							T_D (years)	T_c (years)	T_D (years)	T_c (years)
Jupiter-Callisto	0.0025	0.01	1,100	0.05	20	12	4.32×10^2	1.45×10^7	3.72×10^2	Currently being computed
Saturn-Iapetus	0.0025	0.03	3,500	0.06	20	30	9.25×10^1	2.25×10^6	7.73×10^1	
Uranus-Umbriel	0.0003	0.004	23,000	0.05	5	84	3.95×10^3	3.35×10^{10}	3.41×10^3	

$$\Delta e_{max} = \pi \varepsilon_{32} (A_1 - 2A_2 \cos L_o) \sin L_o$$

where

$$\cos L_o = \frac{A_1 - \sqrt{A_1^2 + 32A_2^2}}{8A_2}, \quad \sin L_o = \sqrt{\frac{1}{2} - \frac{A_1}{4A_2} \cos L_o}$$

and where

$$A_1 = \frac{15}{8} a - \frac{39}{32} a\nu + 4\nu + \frac{40}{3} \nu^2, \quad A_2 = \frac{15}{2} e + \frac{33}{2} e\nu$$

Depending on the values of α, e and μ_3 used, $L_0 \approx f_0$ is typically about 135° for $\Delta e \approx \Delta e_{max}$.

Suppose that each successive synodic period adds the Δe_{max} for the current value of e to that current value of e. This supposition is of course ridiculously pessimistic regarding the satellite's projected lifetime in orbit,but gives a first estimate from which to proceed. If our method of calculating Δe is valid for values of e up to $e_T=0.3$, then the time interval $T_D = NT_S$ can be easily computed, where N is the number of synodic periods needed and T_D is the time taken in order for e to grow from e_0 to e_T under the above pessimistic assumption. Table 2 shows the values of T_D calculated for three separate trial cases in the solar system. For preliminary testing of our method of analysis, values for α, e and μ_3 for the Jupiter, Saturn,Uranus satellite systems have been used which will give the largest Δe_{max} (i.e. we use the largest values of α, e and μ_3) and hence the shortest minimum duration of the planet-satellite system. Thus the initial three trial runs give the calculated minimum durations for all the Galilean satellites, all the satellites of Saturn excluding Phoebe and Hyperion which have highly eccentric orbits and all the satellites of Uranus. Calculations show that for these satellites of Jupiter, Saturn and Uranus, $T_D = NT_S > T_1$, where T_1 is not only the Sun's sidereal period of revolution about the planet, but is also close to the time taken for a conjunction line to make one revolution with respect to the satellite orbit's major axis.

Step 2. Figure 2 shows that as f_0 varies from 0° to 360°, the change in Δe is essentially cyclic. Therefore if the satellite remains in orbit for a length of time T_1 equivalent to f_0 moving through 360 degrees, the net change in e, Δe, will be very much smaller than that computed by the pessimistic summing of the Δe_{max} every synodic period. We define the time T_1 taken for the conjunction line to move through a full cycle of 360 degrees to be the period of a conjunction line cycle. If $T_D > T_1$, we may jump to this next, less pessimistic and more realistic stage of assessment of the minimum time it will take for the eccentricity to increase from e_0 to e_T.

Even if $T_D \sim T_1$ or is less than T_1, there is no guarantee that instability will eventually result. Nevertheless, in this paper we confine our attention to the case $T_D > T_1$.

The analytical theory is now used to map synodic period to synodic period to obtain, for an initial value f_0 of the true anomaly at the first conjunction,

(i) the change of the elements $\Delta\sigma$, where $\sigma = \alpha, e, \tilde{\omega}, \varepsilon$ over a complete conjunction line cycle where the revolution is taken with respect to the satellite orbit's major axis,

(ii) the maximum value of e, viz e_{max}, reached within the conjunction line cycle.

Note: If e_{max} exceeds the range of e within which the analytical theory holds, numerical integration of Equations (4) is required during that part of the cycle where e lies outside that range.

It might be thought that we should now choose the f_0 of the first conjunction which gives the largest value of Δe in a conjunction cycle and suppose, in our pessimistic way, that this value of Δe is added onto e every conjunction cycle. This is an unnecessarily pessimistic procedure.

The angle between two consecutive conjunctions is $\phi = n_1 T_s$, so that a conjunction must occur each cycle at an f_0 no more than $\phi/2 = \frac{1}{2} n_1 T_s$ from the pericentre of the satellite's orbit. It is the Δe occurring in a cycle based on a conjunction of true anomaly $f_{0(max)} = \frac{1}{2} \cdot n_1 T_s$ that should be used in the pessimistic summing procedure.

Applying the previous procedure when we summed the Δe_{max} obtained in a single synodic period, but now summing the Δe_{max} change per conjunction cycle, the minimum time $T_c = N'T_1$ it will take for the satellite orbit's eccentricity to grow from e_0 to e_T is computed. It is found that for all major satellites of Jupiter, Saturn and Uranus, $N' \gg 1$. Table 2 shows the calculated values of T_c.

By this stage we have effectively applied procedures that have removed the short-period behaviour (i.e. terms with periods in the neighbourhood of the satellite's sidereal period) and also much longer periodic behaviour (terms with periods in the neighbourhood of the Sun's sidereal period). Can we go still further in lengthening the minimum time it will take for the Sun to tear the satellite away from the planet?

Step 3. The Roy-Ovenden mirror theorem states that if a system of point masses moving under their mutual gravitational attractions comes at a particular epoch into a dynamical configuration such that all the relative velocity vectors of the masses are perpendicular to all their corresponding relative radius vectors, then the system's behaviour after that epoch is a mirror image of its behaviour prior to that epoch. Roy and Ovenden (1954) pointed out that in practice only two different kinds of mirror configurations exist. In one configuration, the masses are arranged collinearly, with their velocity vectors perpendicular to the straight line through the masses. In the other configuration, the masses lie in a plane, with their velocity vestors perpendicular to that plane.[1]

---

1. C. Marchal, at a lecture during the proceedings of which this is a record, pointed out that these mirror conditions were known to Poincaré. Though Marchal was unable to give an exact reference, it would seem this is yet another example of the old saying that there are no new discoveries in celestial mechanics, only a failure to read the literature!

In an elliptic orbit, of eccentricity e, the angle γ between the velocity vector $\underset{\sim}{v}$ and the radius vector $\underset{\sim}{r}$ at a true anomaly f is given by

$$\left.\begin{aligned} \sin\gamma &= \frac{1 + e\cos f}{(1 + 2e\cos f + e^2)^{\frac{1}{2}}} \\ \cos\gamma &= \frac{-e\sin f}{(1 + 2e\cos f + e^2)^{\frac{1}{2}}} \end{aligned}\right\} \quad (5)$$

Obviously γ_{max} occurs when $\cos f = -e$ (i.e. $r = a$).

Also

$$\sin\gamma_{max} = \sqrt{1 - e^2}$$

$$\cos\gamma_{max} = -e$$

Now the angle between two consecutive conjunctions is $\phi = n_1 T_s$ where T_s is the synodic period of the satelitte. The maximum angle between the apse line of the satellite's orbit and the first conjunction after it or the last conjunction before it is therefore $\phi/2 = T_s/T_1 \sim \pi n_1/n$ since, for a satellite, $T \sim T_s$.

If we replace γ by $90^o + q$ in (5) we have

$$\cos q = \frac{1 + e\cos f}{(1 + 2e\cos f + e^2)^{\frac{1}{2}}}$$

$$\sin q = \frac{e\sin f}{(1 + 2e\cos f + e^2)^{\frac{1}{2}}}$$

where q is the deviation from 90° of the angle between the radius vector and the velocity vector at the true anomaly f. For the conjunctions immediately following or proceeding the pericentre, f is then necessarily small, being given by f_c, where

$$f_c \lesssim \pi \frac{n_1}{n} = \pi\nu .$$

Then for small f_c, $q \sim e\sin f_c \sim e f_c$.

Each time the conjunction line is near the apse line, the system passes through a close approximation to a Roy-Ovenden mirror configuration. Hence whatever has happened to the eccentricity during the first half of a conjunction cycle of duration T_1 must be almost completely reversed in the second half. Now we have seen (Table 2) that for all the major satellites of Jupiter, Saturn and Uranus $T_D = NT_1$ with N being a large number. This means that within the time interval T_D there will occur N conjunction cycles, each cycle of which contains a conjunction occurring near the apse line. Unless the periods of the satellite and the Sun are exactly commensurable, the angle f_c within the period T_D will wander between $-\pi\nu$ and $+\pi\nu$. If N is large, there will be a statistical certainty that a value $f_c \sim \pi\nu/N$ will occur, so that the angle q, the deviation from 90° of the angle between the satellite's velocity vector and radius vector, will be, on at least one occasion when there is a nearest configuration to the apse line, as small as $q \sim e\pi\nu/N$. In fact, an opposition of the Sun and the satellite is just as good as a conjunction as far as a mirror condition is concerned. Likewise a conjunction or opposition near pericentre or apocentre is equally 'useful' in almost reversing completely the perturbations that have been built up by the Sun's gravitational field. Hence we should modify f_c to $\pi\nu/2N$ and thus get

$$q \sim \frac{e\pi\nu}{2N}.$$

If we take typical values of the parameters, say $e = .03$ $\nu = n_1/n = 5\times10^{-3}$, $N \sim 40$, then q is no more than 6×10^{-6} rad., or 1.2 arcsec., a deviation from 90° so small that the solar perturbations must be almost completely reversed, thus pushing to even longer durations the survival of the satellite in orbit about the planet. Work is in progress to extend the satellite's survival time by this method.

2.2 The elliptic coplanar restricted three-body problem

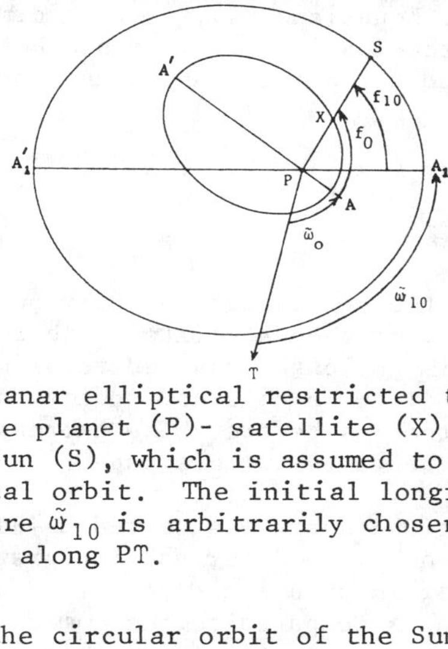

Figure 4: The coplanar elliptical restricted three-body problem where the planet (P)- satellite (X) system is disturbed by the Sun (S), which is assumed to be moving in a fixed elliptical orbit. The initial longitude of the Sun's pericentre $\tilde{\omega}_{10}$ is arbitrarily chosen to be zero, so that PA_1 lies along PT.

In this case the circular orbit of the Sun is replaced by a fixed elliptic orbit centred on the planet (Figure 4). Let the Sun's orbital elements be, in the usual notation, a_1, e_1, $\tilde{\omega}_1$, ε_1. Then we have the usual two-body relations giving the Sun's coordinates and velocity components as functions of these elements and the time t. For example,

$$r_1 = \frac{a_1(1-e_1^2)}{1+e_1\cos f_1} \; ; \; V_1^2 = \mu_1\left(\frac{2}{r_1} - \frac{1}{a_1}\right) \; ; \; r_1^2\frac{df_1}{dt} = h_1 = n_1a_1^2\sqrt{1-e_1^2}$$

and so on. Also, $T_1 = 2\pi/n_1$.

It may be noted in passing that no integral exists for the elliptic coplanar restricted three-body problem, unlike the previous case which possesses the Jacobi integral. We now proceed to follow the steps taken in the circular case.

Step 1. At a conjunction, the true anomalies of the satellite and the Sun are said to be f_0 and f_{10} respectively, while their longitudes of pericentre are $\tilde{\omega}_0$ and $\tilde{\omega}_{10}$ as measured from the direction PT, where T is the vernal equinox. Without loss of generality, since the Sun's orbital major axis if fixed, we can immediately replace PT by PA_1 as the reference direction. This means that $\tilde{\omega}_1 = \tilde{\omega}_{10} = 0$.

Also, at the initial conjunction,

$$f_0 + \tilde{\omega}_0 = f_{10} \quad , \qquad \text{or}$$

$$\tilde{\omega}_0 = f_{10} - f_0 .$$

We now have five independent starting parameters, f_0, f_{10}, α,e and ν as opposed to the four parameters f_0,α,e and ν we had in the circular case.

The Lagrange planetary Equations (4), in the form where the independent variable θ is the longitude difference between the radius vectors of the satellite and the Sun are unchanged, but θ is now given by

$$\theta = \ell - \ell_1 = f + \tilde{\omega} - (f_1 + \tilde{\omega}_1) = f - f_1 + \tilde{\omega} .$$

since $\tilde{\omega}_1 = \tilde{\omega}_{10} = 0$.

The expansions giving the net changes $\Delta\sigma$ over one synodic period in the elements σ = a,e,$\tilde{\omega}$, ε of the satellite's orbit are now more complicated since the Sun is no longer assumed to be moving in a circular orbit. Nevertheless the $\Delta\sigma$ can be obtained to the desired order of accuracy using the same method as before. The expressions obtained are very long and will not be given in this paper.

This general perturbation theory was tested as before by comparing its results with those given by the numerical integration of the Lagrange's planetary Equations (3). Again we are not concerned with high accuracy. The theoretical integration as compared to the numerical integration remains accurate enough for our purposes, viz. they agree to within an order of five per cent, for all values of e and e_1 up to 0.3.

Then, as before, the maximum value of the net change in e,Δe_{max}, over one synodic period T_s is found from the values of Δe obtained through sample calculations where f_0 and f_{10} are made to range from 0° to 360°. (Figure 5). Figure 5 shows the same variation of Δe with f_0 as in Figures 2 and 3, but for a range of starting initial conjunction true anomalies f_{10} for the Sun where the Sun is assumed to be moving in a fixed elliptical orbit. It is interesting to note that the case where the Sun is assumed to be moving in a fixed circular orbit (as denoted in Figure 5 by the dashed line) is no different from the equivalent elliptical case but merely falls approximately at the centre of a range of similar curves for the elliptical case.

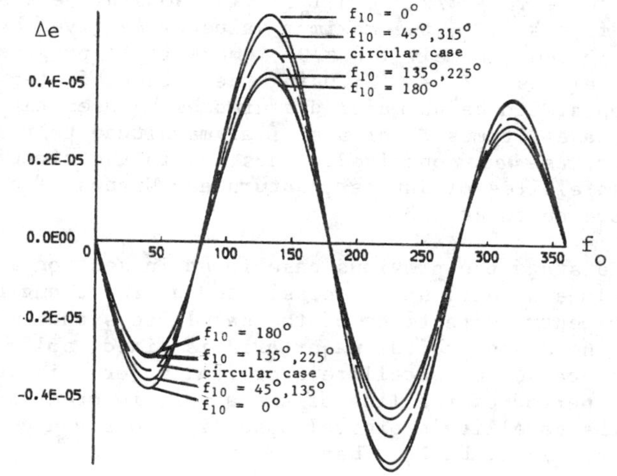

Figure 5: The variation in the change in the satellite's eccentricity Δe over one synodic period with the satellite's initial conjunction true anomaly f_0 for given values of the Sun's initial conjunction true anomaly f_{10} , in the case where the Sun is assumed to be moving in a fixed elliptical orbit. The data used here is that of the Jupiter-Galilean satellite system where $\alpha = 0.0025$, $e = 0.01$, $\mu_3 = 1{,}100$ and $e_1 = 0.05$. The dashed curve represents the equivalent case where the Sun is assumed to be moving in a fixed circular orbit. This curve also lies approximately in the same position as the curve for $f_{10} = 90^\circ$, 270° in the elliptical case.

Again, supposing pessimistically that Δe_{max} is added on to the value of e every synodic period and supposing that the change in e because of this addition is taken into account when computing the new value of Δe_{max}, we obtain the minimum time T_D in which the Sun can secularly perturb the satellite's eccentricity from its initial value e_0 to a value e_T taken to be 0.3. It should be noted that T_s is no longer constant but varies about a mean value $\bar{T}_s$.

Again it is found, for all the major satellites of Jupiter, Saturn and Uranus, that $T_D = N\bar{T}_s$, with $N \gg 1$. In particular, $T_D > T_1$ where the Sun's sidereal period T_1 is also approximately the period of revolution of the conjunction line with respect to the satellite's major axis. See Table 2.

<u>Step 2.</u> Since $T_D \gg T_1$, we can advance a step further to compute $T_c = N'T_1$. T_c is the minimum time taken by the Sun and N' is the number of the conjunction line cycles needed, to raise the satellite's orbital eccentricity from e_0 to $e_T = 0.3$, if Δe_{max} is now taken to be the net change in e over one cycle of the conjunction line with respect to the

satellite's pericentre. Here, as before, the value of the satellite's true anomaly at the initial conjunction f_0 of the connunction line cycle is taken to be $f_0 = \phi/2 = \frac{1}{2} n_1 T_s$. In choosing Δe's size, we also take, for $f_0 = \frac{1}{2} n_1 T_s$, the maximum value of Δe given by f_{10} in the range $0 < f_{10} < 360^\circ$, namely $f_{10} = 0^\circ$. Work is in progress to obtain for the elliptical case, $T_c = N'T_1$ using the method of step 1. As before there appears to be no major differences between the circular and elliptical cases, times T_c of a similar magnitude to those found for the circular case being obtained. Thus, as in the circular case, for the major satellites of Jupiter, Saturn and Uranus, $T_c >> T_1$ which allows us to move on to step 3.

Step 3. At this stage the previous case found in section 2.1 must be modified to include a third step necessitated by the changing angle between the pericentre directions of the satellite's and the Sun's orbits. While the Sun's orbital major axis is fixed, solar perturbations on the satellite cause the satellite's orbital major axis to advance. Let the synodic period of the line of the apses, in other words, the time taken by the satellite's orbital apse line to advance 360° on the solar orbit's apse line, be $T_{\tilde{\omega}}$. Then

$$T_{\tilde{\omega}} = \frac{2\pi}{\dot{\tilde{\omega}} - \dot{\tilde{\omega}}_1} = \frac{2\pi}{\dot{\tilde{\omega}}} \qquad \text{since } \dot{\tilde{\omega}}_1 = 0$$

We can then write as a rough approximation

$$T_{\tilde{\omega}} = \frac{2\pi \; T_s}{\overline{\Delta\tilde{\omega}}}$$

where $\overline{\Delta\tilde{\omega}}$ is the mean change in $\tilde{\omega}$ over one synodic period of the satellite and is preferably calculated over many conjunction cycles. We expect that, in general,

$$T_s \text{ and } T_1 << T_{\tilde{\omega}} << T_c .$$

Let $T_{\tilde{\omega}} = MT_1$, where $1 << M << N'$. Then $2N'/M$ is the number of times the two lines of the apses are collinear in the time T_c taken by the Sun to increase the satellite's orbital eccentricity from e_0 to e_T. (Note: The apses are collinear twice in a synodic period of the apses).

Step 4. We now apply the Roy-Ovenden mirror theorem. A perfect mirror configuration would occur if a conjunction or an opposition of the Sun and the satellite occurred with both bodies at pericentre and/or apocentre and the lines of the apses were collinear. This will rarely, if ever, happen.

There will, nevertheless, be occasions when close approaches to mirror configurations take place. Consider an occasion at epoch τ when the apses are collinear. The worst conjunction or opposition for a mirror configuration at this moment would be one that occurred at 90° from the common apse line. It is obvious that in this case the conjunction or opposition 'spoke' nearest the apses would be occupied by the Sun and the satellite at a time $t = T/4$ after time τ. By this time the apses would have separated by the small angle $\phi_{\tilde{\omega}}$, given by

$$\phi_{\tilde{\omega}} = \frac{2\pi\, T_1}{4T_{\tilde{\omega}}} = \frac{\pi}{2}\,\frac{T_1}{T_{\tilde{\omega}}} .$$

This means there will always be in each apse line cycle a conjunction or opposition within an angle $\phi_{\tilde{\omega}}$ of the apse lines. At that moment, using (7)

$$\sin q \sim e \sin f \sim ef$$

$$q \sim e\,\phi_{\tilde{\omega}} = \frac{\pi e_1\, T_1}{2T_{\tilde{\omega}}}$$

But $T_{\tilde{\omega}} = MT_1$, therefore

$$q = \frac{\pi e}{2M} .$$

Now before the Sun can increase the satellite's eccentricity from e_0 to e_T, a time T_c will have elapsed where $T_c = N'T_1$. Unless a hard commensurability exists, therefore, there is a statistical certainty that on at least one occasion in that time, the angle q will be as small as

$$q \sim \frac{\pi e}{2M}\,\frac{N'}{M} = \frac{\pi e N'}{2M^2}$$

This angle, for the main satellites of Jupiter, Saturn and Uranus, is so small that a very close approach to a mirror configuration is almost certain to occur within a time T_c, resulting in an almost perfect reversal of all solar perturbations. This time of survival of the satellite can then, with confidence, be taken to be very much in excess of the value of T_c computed for that satellite.

A method is now being sought for by the present authors that will use the synodic period of the line of the apses $T_{\tilde{\omega}}$ and the Roy-Ovenden mirror theorem to gain even longer minimum durations for planet-satellite systems under solar perturbations.

3. DISCUSSION AND CONCLUSIONS

In this preliminary account of the application of the finite-time stability criterion method, it is seen that as far as solar perturbations of the major satellites of Jupiter, Saturn and Uranus are concerned the minimum time needed to detach the satellites from their planet is much the same whether the eccentricity of the planet's orbit is taken into account or not.

Preliminary results suggest that the Sun cannot, within the lifetime of the solar system, detach the five major satellites of Uranus from Uranus, even when the eccentricity of Uranus' orbit is included in the problem.

Where the major satellites of Jupiter are concerned, namely Amalthea, Io, Europa, Ganymede and Callisto, the minimum time taken by the Sun to enhance appreciably the eccentricities of their orbits seems to be much longer than any given by previous studies of stability. The minimum duration of Jupiter's satellites is still several orders of magnitude smaller than the age of the solar system, but the present authors hope to lengthen this minimum time appreciably in future work.

In the present paper, the existence of more than one satellite in the satellite system and their effects upon each other has not been taken into account and it would be of some interest, for example, to consider the mediating effects of the Laplace relation on the evolution of the relevant Galilean satellites under solar perturbations.

Saturn's main satellites, Mimas, Enceladus, Tethys, Dione, Rhea, Titan and Iapetus, likewise, seem stable against solar perturbations for millions of years. Again future work may well increase this minimum time.

The Earth's Moon is a complex case and has not been included in the present work, but will hopefully be the subject of a future paper.

It is well-known that the introduction of the eccentricity of the Sun's orbit, by turning the circular restricted coplanar three-body problem into the elliptic restricted coplanar three-body problem, eliminates the possibility of using the Jacobi integral and the resulting solution giving the stability guarantee of a closed Hill surface about the planet. Likewise, in the general three-body problem, the guarantee of $c^2H \leq c^2H_{crit}$ fails for those satellites we have treated, when the solar orbit's eccentricity is taken into account. It would thus appear that the c^2H criterion is far too stringent a test for most of the real cases of interest in the satellite systems and that a more modest search for a simple finite-time stability criterion can produce estimates closer to our beliefs in the long lifetimes in orbit of these objects as supported by planetological studies of their surface histories.

REFERENCES

Bozis, G. (1976). Astrophys. Sp. Sci., 43, 355.
Easton, R. (1971). J. Diff. Eq., 10, 371.
Easton, R. (1975). J. Diff. Eq., 19, 258.
Golubev, V.G. (1968). Doklady Akad. Nauk. SSSR, 180, 308.
Hill, G.W. (1878). Am. J. Math., 1, 129, 245.
Marchal, C. (1971). Astron. Astrophys. 10, 278.
Marchal, C. and Saari, D. (1975). Celest. Mech., 12, 115.
Milani, A. and Nobili, A. (1983a). Celest. Mech., 31, 213.
Milani, A. and Nobili, A. (1983b). Celest. Mech., 31, 241.
Roy, A.E. (1979). Empirical stability criteria in the many-body problem. In V. Szebehely (ed.) Instabilities in Dynamical Systems. Reidel.
Roy, A.E. (1982), The stability of N-body hierarchical dynamical systems In V. Szebehely (ed.), Applications of Modern Dynamics to Celestial Mechanics and Astrodynamics, Reidel.
Roy, A.E. (1983). Asymptotic approach to mirror conditions as a trapping mechanism in N-body hierarchical dynamical systems. In V.V. Markellos and Y. Kozai (eds.).Dynamical Trapping and Evolution in the Solar System. Reidel.
Roy, A.E., Carusi, A., Valsecchi, G. and Walker, I.W. (1984). Astron. Astrophys., 141, 25.
Roy, A.E. amd Ovenden, M.W. (1954). Mon. Not. Roy. Astron. Soc., 114, 232.
Roy, A.E., Walker, I. W. and McDonald, A.J.C. (1985). Studies in the Stability of hierarchical dynamical systems. In V. Szebehely, (ed.) Stability of the Solar System and its Minor Natural and Artificial Bodies. Reidel.
Smale, S. (1970). Invent. Math., 11, 45.
Szebehely, V.G. and Zare, K. (1977). Astron. Astrophys. 58, 145.
Valsecchi, G.B., Carusi, A. and Roy, A.E. (1984). Celest. Mech. 32, 217-230.
Walker, I.W. (1983). Celest. Mech., 29, 149.
Walker, I.W., Emslie, A.G. and Roy, A.E. (1980). Celest. Mech., 22, 371.
Walker, I.W. and Roy, A.E. (1981). Celest. Mech., 24, 195-225.
Walker, I.W. and Roy, A.E. (1983a). Celest. Mech., 29, 117.
Walker, I.W. and Roy, A.E. (1983b). Celest. Mech., 29, 267.
Zare, K. (1976). Celest. Mech., 14, 73.
Zare, K. (1977). Celest. Mech., 16, 35.

REFERENCES

[illegible]
[illegible] (1989) [illegible]
[illegible]
[illegible]
[illegible]
[illegible]
[illegible] and [illegible]
[illegible]
[illegible]
Rov[illegible] (1979) [illegible] these groups [illegible]
[illegible]
Roy, [illegible] (198[illegible]) [illegible] ical esc[illegible]
[illegible]
[illegible]
Roy, [illegible] (19[illegible]) [illegible] Symbolic [illegible]
[illegible] (ed.), [illegible] Science, [illegible]
[illegible]
[illegible]
[illegible]
Roy, [illegible] and [illegible] A.[illegible] [illegible]
[illegible]
[illegible]
[illegible]
Spearhead, [illegible]
Valdés [illegible] Roy, [illegible]
Wal[illegible] (198[illegible]) [illegible]
Walker, [illegible] (198[illegible]) [illegible]
[illegible]
Williams, [illegible] and Roy, [illegible]
Wal[illegible] and Ro[illegible]
[illegible]
Zar[illegible]

A NEW RADIAL INTERMEDIARY AND ITS NUMERICAL INTEGRATION

Sein-Echaluce M.L.* and Franco J.M.**
*Departamento de Astronomia. Universidad de Zaragoza, Spain
**Departamento de Matematica Aplicada. Universidad de Zaragoza, Spain.

ABSTRACT: The zonal artificial satellite is presented in polar-nodal variables considering all the harmonics and, taking into account the importance that the radial intermediaries have in the satellite theory, we present another alternative to the second order analytical methods of Deprit (1981), Alfriend and Coffey (1984). This process consists of the application of two Lie-transformations in order to simplify the Hamiltonian and obtaining finally a new radial intermediary which only contains the even harmonics. A twelfth-order Cowell method with a fixed step size of 100 seconds is applied to compare the solution of the Cid's and Deprit's intermediaries (until first order) with respect to the main problem solution.

INTRODUCTION

The classical approach to analytic satellite theory consists of the elimination of the short and long period terms, in closed form (Brouwer, 1959; Kozai, 1962; Lahulla, 1970) or by using development of the eccentricity (Deprit and Rom, 1970; Kinoshita, 1977).

A different scheme has been developed by Deprit (1981), Coffey and Deprit (1982) and Alfriend and Coffey (1984) which is based in the elimination of the parallax and the elimination of the perigee, (in polar nodal variables), in order to obtain a new Hamiltonian which only depends on the radial variables and which is called radial intermediary (Deprit, 1981).

In this paper, by using this scheme in a similar way, we develop an analytic second-order solution of the zonal earth artificial satellite, considering all the zonal harmonics, by simplifying the Hamiltonian and obtaining a new radial intermediary (which is integrated by applying other different methods as an alternative to the Delaunay normalization Deprit, 1981).

The simplification of the Hamiltonian is made by means of two Lie-transformations. The first one, called Cid-Deprit transformation,

A. E. Roy (ed.), Long-Term Dynamical Behaviour of Natural and Artificial N-Body Systems, 217–222.
© 1988 by Kluwer Academic Publishers.

is based on the elimination of the argument of latitude in the first order (Cid and Lahulla, 1969) and the elimination of the parallax in the second order (Deprit, 1981). The second one consists of the elimination of the argument of perigee (Alfriend and Coffey, 1984).

The new radial intermediary presents some advantages with respect to the intermediaries obtained by other authors. It only contains the even harmonics and the theory is not singular for small inclination and small eccentricity.

The integration of this given radial intermediary has been made using different methods: Krylov-Bogoliubov method (Caballero et al. 1986), time transformations (Ferrandiz et al., 1987; Ferrer and Sein-Echaluce, 1986), the stroboscopic method (Sein-Echaluce et al., 1987) and numerical methods.

In this paper we integrate this intermediary (until first order) and the one given by Deprit (1981), and study the behaviour of these solutions with respect to the main problem solution. We use a twelfth-order Cowell method with a fixed step size of 100 seconds.

NEW RADIAL INTERMEDIARY

We consider a satellite moving in an axially symmetric gravitational field, and we use the polar nodal variables $(r,\theta, v,R,\Theta, N)$ where r is the radial distance from the Earth's centre of mass to the satellite $\theta = \omega+f$ is the argument of latitude, v is the argument of the ascending node and R, Θ = modulus of the angular momentum and $N = \Theta \cos I$ are the momenta for r, θ and v respectively. The variable I stands for the inclination of the satellite's orbital plane, f is the true anomaly and ω is the argument of perigee.

The Hamiltonian function in those variables (see Cid et al. 1986) is given by

$$H = H_{0,0} + \varepsilon H_{1,0} + \frac{\varepsilon^2}{2} H_{2,0} ,$$

being

$$H_{0,0} = \frac{1}{2}\left(R^2 + \frac{\Theta^2}{r^2}\right) - \frac{\mu}{r} , \quad H_{1,0} = - \frac{\mu}{r^3}(B_2^0 + B_2^2 \cos 2\theta),$$

$$H_{2,0} = \frac{2\mu}{r^2} \sum_{n=2}^{\infty} \{J'_{2n-1} \left(\frac{1}{r}\right)^{2n-2} \sum_{k=0}^{n-1} B_{2n-1}^{2k+1} \sin (2k+1)\theta +$$

$$+ J'_{2n}\left(\frac{1}{r}\right)^{2n-1} \sum_{k=0}^{n} B_{2n}^{2k} \cos 2k\theta \},$$

with $\varepsilon = - J_2$, $J'_n = J_n/J_2^2$ and $B_i^j = B_i^j (I)$ (see Cid et al. 1986).

We will make a Lie-transformation in the sense of Deprit (1969) and will choose the new Hamiltonian in the following manner:

(a) we require that the first order be equal to the radial intermediary given by Cid and Lahulla (1969),
(b) at the second order and beyond, we proceed, as Deprit did for the elimination of the parallax, by requiring that the Hamiltonian be of the form $F(L,G,H,g)/r^2$ (Deprit, 1981).

The generating function of this transformation is simpler than that one corresponding to the elimination of the parallax and does not contain any mixed terms.

The next step consists of applying a new Lie-transformation following the process of Alfriend and Coffey (1984), in which they remove the troublesome argument of perigee without affecting the short period terms. But whereas they only dealt with the main problem, we study the whole problem. The resulting Hamiltonian is our new radial intermediary given by

$$H = \frac{1}{2}(R^2 + \frac{\Theta^2}{r^2}) - \frac{\mu}{r} + \varepsilon\frac{\mu}{r^3}(\frac{1}{2} - \frac{3}{4}s^2)$$

$$+ \frac{\varepsilon^2}{2}\frac{\Theta^2}{p^4 r^2}\{\frac{3s^2(4s^2-3)}{8} + 2\sum_{n=2}^{\infty}\frac{J'_{2n}}{p^{2(n-2)}}B^0_{2n}\} + O(\varepsilon^2 e^2)$$

being

$$e^2 = (\frac{\Theta}{\mu})^2\{R^2 + \Theta^2(\frac{1}{r} - \frac{\mu}{\Theta^2})^2\}, \quad p = \frac{\Theta^2}{\mu}, \quad c = \frac{N}{\Theta}, \quad s^2 = 1 - c^2$$

This is a radial intermediary with one degree of freedom in the radial variables (r,R); it depends only on the J_{2n} $(n > 2)$ zonal harmonics and it is not singular for small inclination and eccentricity.

COMPARISON OF INTERMEDIARIES

We are going to present a qualitative study of this new radial intermediary and the one given by other authors.

Firstly, the radial intermediary proposed by Cid and Lahulla (1969) is

$$J_C = \frac{1}{2}(R^2 + \frac{\Theta^2}{r^2}) - \frac{\mu}{r} + \varepsilon\frac{\mu}{r^3}(\frac{1}{2} - \frac{3}{4}s^2),$$

which corresponds to the first order intermediary obtained in this paper (for the main problem), namely

$$J = \frac{1}{2}(R^2 + \frac{\Theta^2}{r^2}) - \frac{\mu}{r} + \varepsilon\frac{\mu}{r^3}(\frac{1}{2} - \frac{3}{4}s^2)$$

$$+ \frac{\varepsilon^2}{2}\frac{\Theta^2}{p^4 r^2}\{\frac{3s^2(4s^2-3)}{8} + \frac{3}{8}e^2(\frac{23}{8}s^2 - 2)\} \qquad (1)$$

Deprit (1981) has taken as the basic radial intermediary

$$J_D = \frac{1}{2}\left(R^2 + \frac{\Theta^2}{r^2}\right) - \frac{\mu}{r} + \varepsilon \frac{\Theta^2}{p^2 r^2}\left(\frac{1}{2} - \frac{3}{4}s^2\right),$$

which corresponds to the first order radial intermediary of Alfriend and Coffey (1984)

$$H = J_D + \varepsilon^2 \frac{3\Theta^2}{16\, r^2 p^4}\left\{\frac{9}{4}s^4 + 7s^2 - \frac{10}{3} + e^2\left(\frac{5}{8}s^4 + s^2 - 1\right)\right\} \qquad (2)$$

Comparing these two intermediaries we see that in the Eq.(1) a second order term, independent of the I, appears and in the first order, for small inclinations, the radial (1) tends to the whole problem. Thus, Eq.(1) is better suited than Eq.(2) to study the motion at small inclinations.

After that we study the numerical behaviour of these two intermediaries up to first order, with respect to the main problem which has the following expression

$$H_m = \frac{1}{2}\left(R^2 + \frac{\Theta^2}{r^2}\right) - \frac{\mu}{r} + \varepsilon\frac{\mu}{r^3}\left[\frac{1}{2} - \frac{3}{4}s^2(1 - \cos 2\theta)\right]$$

Starting the process, we consider a Keplerian osculating orbit with these values

$$a = 9000 \text{ km.},\ \omega = 0^\circ,\quad v = 0^\circ,\ T = 0 \text{ (perigee epoch)}$$

Before making the numerical integration we have to obtain the averaged initial values for the Cid and Deprit models, by means of the generating function of the Lie-transformation which are given (up to first order) by these two expressions respectively

$$W_c = \frac{3\,\Theta^2}{8\,p^2} s^2\left\{\sin 2\theta + \frac{Cp}{\Theta}\left(\sin\theta + \frac{1}{3}\sin 3\theta\right) + \frac{Sp}{\Theta}\left(\cos\theta - \frac{1}{3}\cos 3\theta\right)\right\}$$

$$W_D = \frac{3}{8}s^2 \sin 2\theta + \frac{Cp}{\Theta}\left[\left(\frac{1}{2} - \frac{3}{8}s^2\right)\sin\theta + \frac{1}{8}s^2\sin 3\theta\right]$$

$$- \frac{Sp}{\Theta}\left[\left(\frac{1}{2} - \frac{9}{8}s^2\right)\cos\theta + \frac{1}{8}s^2\cos 3\theta\right]$$

where $C(r,\theta,R,\Theta)$ and $S(r,\theta,R,\Theta)$ are the state functions (see Cid et al. 1986).

A twelfth-order Cowell method with fixed step size of 100 seconds has been used to integrate (in the predictor-corrector form (PECE)) these three intermediaries (a generalization of this method for the problems $y'' = f(t,y,y')$ has been made by Franco and Palacios, 1986). The initialization values of this numerical method have been obtained by using a tenth-order Nystrom method with a fixed step size of 50 seconds.

After integrating the intermediaries and having the numerical solutions we must invert the transformations in order to obtain the real solutions of these two intermediaries and compare with the numerical solution of the main problem.

The errors have been calculated in euclidean form for the difference vector, namely

$$e_c = \|\vec{x}_c - \vec{x}\|_2 , \quad e_D = \|\vec{x}_D - \vec{x}\|_2$$

where x_c, x_D and x are the solutions in cartesian form of the Cid's and Deprit's intermediaries and the main problem respectively.

We can obtain the following conclusions of the results: for small inclinations the Cid's intermediary is the nearest to the main problem and when the eccentricity increases, the errors scale increases too.

More exhaustive studies are being made at this moment.

ACKNOWLEDGEMENTS

This work has been supported by DGA Proy. CB4/85 and CAICYT n° 779-84.

REFERENCES

1. Alfriend, K.T. and Coffey, S.L.: 1984. "Elimination of the perigee in the satellite problem". Celes. Mech. 32, 163-172.
2. Brouwer, D.: 1959, "Solution of the Problem of Artificial Satellite Theory without Drag". Astron. J. 64, 378-397.
3. Caballero, J.A., Ferrer, S. and Sein-Echaluce, M.L.: 1986, "Second Order Solution of a Radial Intermediary in Satellite Theory", Bhatnagar ed., Space Dynamics and Celestial Mechanics, 243-253.
4. Cid, R. and Lahulla, J.F.: 1969. "Perturbaciones de corto periodo en el movimiento de un satélite artificial en functión de las variables de Hill". Rev. Acad. Ciencias de Zaragoza, 24, 159-165.
5. Cid, R. y. Lahulla, J.F.: 1969. "Perturbaciones de corto periodio en el movimiento de un satélite artificial, en función de las variables de Hill". Rev. Acad. Cienc. de Zaragoza 24, 159-165.
6. Coffey, S.L., and Deprit, A.: 1982. "Third-order solution to the main problem in satellite theory". J. Guidance, Control and Dynamics 5, 366-371.
7. Deprit, A.: 1969. "Canonical transformations depending on a small parameter". Celes. Mech. 1, 12-30.
8. Deprit, A.: 1981. "The elimination of the parallax in satellite theory". Celes. Mech. 24, 111-153.
9. Deprit, A. and Rom, A.: 1970. The Main Problem of Artificial Satellite Theory for Small and Moderate Eccentricities". Celes. Mech. 2, 166-206.

10. Ferrándiz, J.M., Ferrer, S. and Sein-Echaluce, M.L.: 1987. "The Generalized Elliptic Anomalies", to appear in Celes. Mech.
11. Ferrer, S. and Sein-Echaluce, M.L.: 1986. "Times Transformation and Linearization on Radial Intermediaries in the Zonal Earth Artificial Satellite Theory". ESA SP-255, Dec.1986.
12. Franco, J.M. and Palacios, M.: 1986, "Une classe de methodes numeriques pour l'integration directe d'equations differentielles du type y" = f(t,y,y')". Proceedings of the XIX Congres d'Analyse Numerique, Toulouse.
13. Kinoshita, H.: 1977. "Third Order Solution of an Artificial Satellite Theory". NASA-CR-154509, Report No.379, Nov.1979.
14. Kozai, V.: 1962, "Second-Order Solution of Artificial Satellite Theory Without Air Drag", Astron. J., 67, 446-461.
15. Lahulla, J.F.: 1970. "Movimiento de satélites artificials. Eliminación de pequeños divisores en una teoría de segundo orden". Doctoral Thesis. Zaragoza.
16. Sein-Echaluce, M.L., Abad, A. and Elipe, A.: 1987. "Application of the Stroboscopy method to the Radial Intermediaries in Satellite Theory". IAF-87-334 Brighton.

Co-orbital Satellites and Hill's Lunar Problem

by Jörg Waldvogel and Franz Spirig
Swiss Federal Institute of Technology (ETH), Zürich, Switzerland

Abstract

Hill's lunar problem has been revived by the fact that it describes the close encounters of coorbital satellites in a rotating frame of reference (besides approximating the motion of the moon). The complete description of coorbital motion requires matching of asymptotic expansions in the framework of singular perturbation theory. In this paper we derive 3 (partially new) series solutions of Hill's lunar problem that pass through infinity. Two of them approximate the motion globally for large or small energies, respectively, whereas the third one is an expansion valid for large values of time.

1. Introduction

In the framework of singular perturbation theory the motion of coorbital satellites (e.g. Janus and Epimetheus or the F Ring Shepherds of Saturn) may be approximated by two independent two-body problems when they are far apart (outer solution). This approximation breaks down whenever the distance of the two satellites becomes small, and an inner solution must be used instead, see [4]. It turns out that the inner approximation is a solution of Hill's lunar problem [2]. In classical celestial mechanics this equation was used for approximating the motion of the moon; therefore Hill was mainly interested in periodic solutions. In our application, however, the main interest lies in solutions beginning and ending at infinity. The present paper is concerned with asymptotics of Hill's lunar equation.

2. Hill's Lunar Problem

In this section we give a brief account of the coordinate systems used, and of the limiting process leading from the planar problem of three bodies to Hill's lunar problem. For more details the reader is referred to [4]. We consider the motion of 3 bodies m_j (j=0,1,2) located at positions $\underline{x}_j \in \mathbf{R}^2$, with the center of mass at rest at the origin, $\Sigma m_j \underline{x}_j = 0$. For describing the relative motion near the close encounters relative positions

(1) $$\underline{r}_1 = \underline{x}_1 - \underline{x}_0 , \quad \underline{r}_2 = \underline{x}_2 - \underline{x}_0$$

and Jacobian coordinates

A. E. Roy (ed.), Long-Term Dynamical Behaviour of Natural and Artificial N-Body Systems, 223–234.
© 1988 by Kluwer Academic Publishers.

(2) $$\underline{R} = \frac{m_1 \underline{r}_1 + m_2 \underline{r}_2}{m_1 + m_2}, \qquad \underline{d} = \underline{x}_2 - \underline{x}_1$$

(3) $$\underline{r}_1 = \underline{R} - \frac{m_2 \underline{d}}{m_1 + m_2}, \qquad \underline{r}_2 = \underline{R} + \frac{m_1 \underline{d}}{m_1 + m_2}$$

are introduced. The equations of motion are then obtained from the Lagrangean $L(\underline{R},\underline{d},\dot{\underline{R}},\dot{\underline{d}})=T+V$ with

(4) $$2T = \frac{m_0(m_1+m_2)}{m_0+m_1+m_2}\dot{\underline{R}}^2 + \frac{m_1 m_2}{m_1+m_2}\dot{\underline{d}}^2$$
$$V = \frac{m_0 m_1}{|\underline{r}_1|} + \frac{m_0 m_2}{|\underline{r}_2|} + \frac{m_1 m_2}{|\underline{d}|}$$

as

(5) $$\left(\frac{\partial T}{\partial \dot{\underline{R}}}\right)^{\cdot} = \frac{\partial V}{\partial \underline{R}}, \qquad \left(\frac{\partial T}{\partial \dot{\underline{d}}}\right)^{\cdot} = \frac{\partial V}{\partial \underline{d}},$$

where dots denote differentiation with respect to time t. At this point we assume m_1, m_2 to be small compared to m_0, such that

(6) $$\varepsilon = \frac{m_1+m_2}{m_0} << 1 .$$

By introducing a properly scaled relative position vector $\underline{r}$ by

(7) $$\underline{d} = \varepsilon^{1/3}\underline{r}$$

and by going to the limit $\varepsilon \longrightarrow 0$ we obtain the inner system

(8) $$\ddot{\underline{R}} = -\underline{R}\,|\underline{R}|^{-3}$$

(9) $$\ddot{\underline{r}} = -\underline{r}\left(|\underline{r}|^{-3}+|\underline{R}|^{-3}\right) + 3\underline{R}(\underline{r}\,\underline{R})\,|\underline{R}|^{-5} .$$

The first equation, (8), states that while the two particles m_1, m_2 are sufficiently close their center of mass $\underline{R}$ describes a Keplerian orbit about the central body m_0.

The second equation (9) which describes the relative motion of m_2 with respect to m_1 is best discussed by using the complex coordinate $z=x+iy\in \mathbf{C}$ according to

(10) $$r = R\,z, \qquad R\in \mathbf{C},\ r\in \mathbf{C}$$

with the obvious meaning of R, r (z describes the relative motion in a "rotating and pulsating" coordinate system). According to [4] the resulting differential equation is

$$\ddot{z} + 2i\dot{z} = f(s)\left(3x - z|z|^{-3}\right)$$

(11)

$$f(s) = \frac{1}{1+e\cos s}, \qquad \dot{} \equiv \frac{d}{ds},$$

where e, s are the eccentricity and true anomaly of the Keplerian orbit of R.

By again using real notation and by restricting ourselves to the circular case e=0, s=const·t we obtain from (11)

$$\ddot{x} - 2\dot{y} - 3x = -x\rho^{-3}$$

(12)

$$\ddot{y} + 2\dot{x} = -y\rho^{-3}, \qquad \rho = \sqrt{x^2+y^2},$$

Hill's classical lunar equation [2]. Equ. (12) admits the Jacobian integral

(13) $$\frac{1}{2}\left(\dot{x}^2+\dot{y}^2\right) - \frac{3}{2}x^2 - \rho^{-1} = h$$

where h is the Jacobi constant.

For convenience we introduce the 4-vectors

(14) $$\underline{x} = (x, y, \dot{x}, \dot{y})^T, \qquad \underline{f} = \left(0, 0, -x\rho^{-3}, -y\rho^{-3}\right)^T$$

and the matrix

$$A = \begin{pmatrix} 0 & 0 & 1 & 0 \\ 0 & 0 & 0 & 1 \\ 3 & 0 & 0 & 2 \\ 0 & 0 & -2 & 0 \end{pmatrix}.$$

Hill's lunar equation (12) then becomes

(16) $$\dot{\underline{x}} = A\underline{x} + \underline{f}.$$

In view of later applications the matrix A will be transformed to normal form by a similarity transformation. From the characteristic polynomial

$$p(\lambda) = \det(A - \lambda I) = \lambda^4 + \lambda^2$$

the eigenvalues are found to be $\lambda_1=i$, $\lambda_2=-i$, $\lambda_3=\lambda_4=0$. The matrix A turns out to be defective, and for convenience (see Section 4) we choose its real canonical form as

(17) $$D = \begin{pmatrix} 0 & -1 & 0 & 0 \\ 1 & 0 & 0 & 0 \\ 0 & 0 & 0 & 0 \\ 0 & 0 & -1 & 0 \end{pmatrix}.$$

With the transformation matrices

$$T = \begin{pmatrix} -1 & 0 & 2/3 & 0 \\ 0 & 2 & 0 & 1 \\ 0 & 1 & 0 & 0 \\ 2 & 0 & -1 & 0 \end{pmatrix}, \qquad T^{-1} = \begin{pmatrix} 3 & 0 & 0 & 2 \\ 0 & 0 & 1 & 0 \\ 6 & 0 & 0 & 3 \\ 0 & 1 & -2 & 0 \end{pmatrix} \tag{18}$$

we have

$$A = TDT^{-1}. \tag{19}$$

With the new variable $\underline{u} \in \mathbf{R}^4$,

$$\underline{u} = T^{-1}\underline{x}, \tag{20}$$

Hill's lunar problem finally becomes

$$\dot{\underline{u}} = D\,\underline{u} + \underline{g}, \tag{21}$$

where all the nonlinearities are concentrated in

$$\underline{g} = T^{-1}\underline{f} = \left(-2y\rho^{-3}, -x\rho^{-3}, -3y\rho^{-3}, 2x\rho^{-3}\right)^T. \tag{22}$$

In terms of $\underline{u}$ the old variables are written as

$$\begin{aligned} x &= -u_1 + \frac{2}{3}u_3, & \dot{x} &= u_2 \\ y &= 2u_2 + u_4, & \dot{y} &= 2u_1 - u_3, & \rho &= \sqrt{x^2+y^2}, \end{aligned} \tag{23}$$

and the Jacobi integral (13) becomes

$$\frac{1}{2}\left(u_1^2+u_2^2\right) - \frac{1}{6}u_3^2 - \rho^{-1} = h. \tag{24}$$

3. Regular perturbation expansion

Particular types of solutions are obtained by introducing scaling transformations. Here we are interested in solutions where both coordinates x, y are large. These solutions globally describe the relative motion of the F Ring Shephers at close encounters.

Introducing the paramter $\gamma >> 1$ we assume the expansion

$$\underline{u}(s) = \gamma\,\underline{u}_0(s) + \gamma^{-2}\underline{u}_1(s) + \gamma^{-5}\underline{u}_2(s) + \ldots, \tag{25}$$

spacing the powers of γ in steps of 3 for convenience. Substitution into (21) yields the perturbation equations

$$\dot{\underline{u}}_k = D\underline{u}_k + \underline{g}_k, \tag{26}$$

where the coefficients $\underline{g}_j$ of the expansion

(27) $$\underline{g} = \gamma \underline{g}_0 + \gamma^{-2}\underline{g}_1 + \gamma^{-5}\underline{g}_2 + \ldots$$

are obtained by substitution of (25) into (23), (22); in particular $\underline{g}_0=0$. Therefore

(28) $$\underline{u}_0(s) = e^{Ds}\underline{c}_0$$

where $\underline{c}_0=\underline{u}_0(0)$ is the vector of the 4 orbital elements, and, by the power series method

(29) $$e^{Ds} = \begin{pmatrix} \cos s & -\sin s & 0 & 0 \\ \sin s & \cos s & 0 & 0 \\ 0 & 0 & 1 & 0 \\ 0 & 0 & -s & 1 \end{pmatrix}.$$

The higher order coefficients are obtained by varying the constant:

(30) $$\underline{u}_k(s) = e^{Ds}\underline{c}_k(s), \qquad k = 1, 2, \ldots$$
$$\dot{\underline{c}}_k = e^{-Ds}\underline{g}_k, \quad \underline{c}_k(0) = 0.$$

4. Asymptotics for s $\longrightarrow \pm\infty$

In this section the local behaviour of an arbitrary solution of the normal form (21) at infinity is investigated. For convenience we use the complex variable $u_0 = u_1 + iu_2 \in \mathbb{C}$ instead of u_1, u_2. Explicitly, Equ. (21) then reads as

(31) $$\begin{aligned} \dot{u}_0 - iu_0 &= F, & F &= -(2y+ix)\rho^{-3} \\ \dot{u}_3 &= G, & G &= -3y\rho^{-3} \\ \dot{u}_4 + u_3 &= H, & H &= 2x\rho^{-3} \end{aligned}$$

where x, y, ρ are given by Equ. (23).

It turns out that Equs. (31) do not admit formal solutions in terms of power series; logarithmic terms will be present. In the following we describe an elegant method that avoids logarithmic terms in the series. We seek expansions with respect to a new independent variable τ defined by the differential relation

$$\dot{\tau} = q(\tau) = \sum_{k=0}^{\infty} q_k \tau^{-k}.$$

In contrast to [1] where a certain infinite sum is used, it will be shown that the finite transformation

(32) $$\dot{\tau} = q_0 + q_1\tau^{-1}$$

with properly chosen q_0, q_1 will allow the formal series

(33) $$u_0 = \sum_{k=0}^{\infty} a_k(s)\tau^{-k}, \qquad u_3 = \sum_{k=0}^{\infty} b_k(s)\tau^{-k}, \qquad u_4 = \sum_{k=0}^{\infty} c_k(s)\tau^{1-k}$$

with certain trigonometric polynomials $a_j(s)$, $b_j(s)$, $c_j(s)$ to satisfy (31). Once q_0, q_1 are determined time s is given by

(34) $$s - s_1 = \frac{1}{q_0}\left(\tau - \frac{q_1}{q_0}\log\left|\tau + \frac{q_1}{q_0}\right|\right),$$

as follows from (32) (s_1 is the first integration constant).

The next step is to substitute (33) into (31) (using (32)) and equate the coefficients of like powers of τ. For this purpose it is necessary to express the coefficients of the expansions

(35) $$F = \sum_{k=0}^{\infty} F_k\tau^{-k}, \qquad G = \sum_{k=0}^{\infty} G_k\tau^{-k}, \qquad H = \sum_{k=0}^{\infty} H_k\tau^{1-k}$$

in terms of the a_k, b_k, c_k. For the moment we restrict ourselves to the branch $\tau<0$, $c_0<0$ such that, e.g.

$$\rho^{-3} = (c_0\tau)^{-3} - 3(c_0\tau)^{-4}\left[c_1 + i(\bar{a}_0 - a_0)\right] + O(\tau^{-5}).$$

In this way the following infinte set of differential equations for the coefficient functions $a_k(s)$, $b_k(s)$, $c_k(s)$ and q_0, q_1 is obtained:

(36) $$\dot{a}_k + (1-k)q_0a_{k-1} + (2-k)q_1a_{k-2} - ia_k = F_k$$

(37) $$\dot{b}_k + (1-k)q_0b_{k-1} + (2-k)q_1b_{k-2} = G_k \qquad (k = 0, 1, \ldots)$$

(38) $$\dot{c}_k + (2-k)q_0c_{k-1} + (3-k)q_1c_{k-2} + b_{k-1} = H_k$$

with $a_k=b_k=c_k=0$ for $k<0$. A formal solution is constructed by first solving (36), (37), (38) for $k=0,1,2,\ldots$ introducing integration constants α_k, β_k, γ_k. These integration constants as well as q_0, q_1 are determined by the requirement that $a_k(s)$, $b_k(s)$, $c_k(s)$ contain *no secular terms*. The 3 remaining parameters a, s_0, b of the family of formal solutions will come up naturally as the quantities which still remain undetermined during this process.

We carry out the first 3 steps explicitly; $k=0$ yields $\dot{a}_0 - ia_0=0$, $\dot{b}_0=0$, $\dot{c}_0=0$. With the abbreviation

(39) $$E = E(s) = a\exp\left[i(s-s_0)\right]$$

we obtain the solution

(40) $$a_0 = \alpha_0E, \qquad b_0 = \beta_0, \qquad c_0 = \gamma_0.$$

The value k=1 yields $\dot{a}_1 - ia_1 = 0$, $\dot{b}_1 = 0$, $\dot{c}_1 + q_0 c_0 + b_0 = 0$ with the solution

$$(41)\quad a_1 = \alpha_1 E, \quad b_1 = \beta_1, \quad c_1 = \gamma_1 .$$

Secular terms in c_1 are avoided by the condition

$$(42)\qquad q_0 c_0 + b_0 = 0 .$$

In the same way k=2 yields $\dot{a}_2 - ia_2 - q_0 a_1 = -2c_0^{-2}$, $\dot{b}_2 - q_0 b_1 = -3c_0^{-2}$, $\dot{c}_2 + q_1 c_0 + b_1 = 0$ with the solution

$$(43)\qquad \begin{array}{ll} a_2 = \alpha_2 E - 2ic_0^{-2}, & \\ b_2 = \beta_2, & \beta_1 = 3q_0^{-1} c_0^{-2}, \\ c_2 = \gamma_2, & q_1 c_0 + b_1 = 0 . \end{array}$$

The conditions in (42) and (43) are satisfied, e.g., by $c_0 = -1$, $q_0 = b_0$, $q_1 = b_1$. In general, the integration constants α_k, β_k, γ_k are chosen such that $a_{k+1}(s)$, $b_{k+1}(s)$, $c_{k+1}(s)$ contain no secular terms (are periodic in s). In particular, we may choose $c_1 = \gamma_1 = 0$ without loss of generality since the integration constant s_1 was introduced in (34). The results up to order k=4 are collected in Table 1; by means of an algebraic processor (SYMBAL) the order k=7 was easily reached. Large coefficients in the high order terms seem to indicate divergence of the series.

Theorem 1. *With the time transformation given by (32) and* $q_0 = b_0$, $q_1 = b_1$ *the formal expansion (33) exists to arbitrarily large order k.*

It remains to be shown that the infinite system of differential equations (36), (37), (38) has a solution with periodic functions $a_n(s)$, $b_n(s)$, $c_n(s)$. Using complex notation, it will be shown inductively that a_n, b_n, c_n (n=0,1,...) may be written as polynomials in E, $\bar{E}$, where

$$(44)\qquad E \cdot \bar{E} = a^2$$

according to (39). Assume that $a_k = \mathrm{pol}(E,\bar{E}) + \alpha_k E$, $b_k = \mathrm{pol}(E,\bar{E}) + \beta_k$, $c_k = \mathrm{pol}(E,\bar{E}) + \gamma_k$ for k=0,1,...,n-1, where α_{n-1}, β_{n-1}, γ_{n-1} are still undetermined. Then F_n, G_n, H_n are polynomials in E, $\bar{E}$, and Equ. (36), e.g., is of the form

$$(45)\qquad \dot{a}_n - ia_n = \sum_{k,l \geq 0} p_{kl} E^k \bar{E}^l .$$

Under conditions to be discussed below this equation is solved by

$$(46)\qquad a_n = \sum_{k,l \geq 0} \frac{p_{kl}}{i(k-l-1)} E^k \bar{E}^l + \alpha_n E .$$

This series is valid if the right-hand side of (45) contains no terms that would give rise to a vanishing denominator in (46),

(47) $$E\sum_{l\geq 0} p_{l+1,l}\, a^{2l} = 0 .$$

This, however, is a non-degenerate linear condition for the still unknown integration constant α_{n-1}. Analogous considerations for b_n, c_n conclude the proof.

To conclude this section we mention that the above series may be easily extended to a branch with $c_0>0$ by a time reversal. The equations of this section remain valid if u_0, E are replaced by their complex conjugates and y, u_4, s, τ, q_1, s_0, s_1 undergo a sign change (x, u_3, a, b remain unchanged). Finally, the Jacobi constant h of Equ. (13) or Equ. (24), written in terms of the parameters a, b, s_0, s_1 becomes

(48) $$h = \frac{a^2}{2} - \frac{b^2}{6} .$$

5. Singular-perturbation expansion for h ⟶ 0

In the following the scaling tranformation of Section 3 will be modified by assuming the parameter γ to be small and by transforming the original Hill equations (12) by

(49) $$x = \gamma\xi , \qquad y = \gamma^{-2}\eta , \qquad s = \gamma^{-3}\sigma ,$$

thus introducing scaled variables ξ, η, σ. In this way Equs. (12), (13) become

(50) $$\begin{aligned} 2\eta' + 3\xi &= \mu(\xi'' + \xi r^{-3}) , & r^2 &= \eta^2 + \mu\xi^2 \\ \eta'' + 2\xi' + \eta r^{-3} &= 0 , & \mu &= \gamma^6 \\ \frac{1}{2}\eta'^2 - \frac{3}{2}\xi^2 - r^{-1} - \gamma^{-2}h &= -\frac{\mu}{2}\xi'^2 , \end{aligned}$$

where primes denote differentiation with respect to σ. Our goal is to expand smooth solutions of (50) with respect to σ in the limit $\mu \longrightarrow 0$, $h \longrightarrow 0$, $\gamma^{-2}h$ fixed. For convenience we put $h=-0.375\gamma^2$.

These solutions correspond to the adiabatic invariant mentioned in [1], [3]; they describe the relative motion of two small coorbital satellites in 1:1 resonance (e.g. Janus and Epimetheus of Saturn) during their close encounter.

By assuming expansions

(51) $$\xi(\sigma) = \sum_{k=0}^{\infty} \xi_k(\sigma)\mu^k , \qquad \eta(\sigma) = \sum_{k=0}^{\infty} \eta_k(\sigma)\mu^k$$

the zeroth-order equations

(52) $$2\eta_0' + 3\xi_0 = 0, \qquad \eta_0'' + 2\xi_0' + (\eta_0|\eta_0|)^{-1} = 0,$$

(53) $$\frac{1}{2}\eta_0'^2 - \frac{3}{2}\xi_0^2 - |\eta_0|^{-1} + \frac{3}{8} = 0$$

are obtained from Equ. (50). With no loss of generality we will restrict ourselves to the case $\eta_0>0$ (the other case, $\eta_0<0$, may be obtained by means of a sign change in η and σ). An explicit solution of (52), (53) is given by the orbit

(54) $$\eta_0 = \frac{8/3}{1-\xi_0^2}$$

and the differential relation

(55) $$\xi_0' = -2\eta_0^{-2} = -\frac{9}{32}\left(1-\xi_0^2\right)^2,$$

or equivalently,

(56) $$\sigma = -\frac{16}{9}\left(\operatorname{Artanh}\xi_0 + \frac{\xi_0}{1-\xi_0^2}\right).$$

This suggests to write the singular perturbation problem (50) in terms of the parameter ξ_0 instead of σ. With the notation

(57) $$\zeta := \xi_0, \qquad Q := \eta_0^{-1} = \frac{3}{8}(1-\zeta^2)$$

we obtain from (55), (52)

(58) $$\xi_0' = -2Q^2, \qquad \eta_0' = -\frac{2}{3}\zeta,$$

and the derivatives transform as

(59) $$\frac{d}{d\sigma} = -2Q^2\frac{d}{d\zeta}.$$

Now, the expansions (51) are substituted into the first two equations of (50). With the series coefficients f_k, g_k defined by

(60) $$\mu(\xi r^{-3} + \xi'') = \sum_{k=1}^{\infty} f_k\mu^k$$
$$\eta r^{-3} = \eta_0^{-2} + \sum_{k=1}^{\infty}\left(g_k - 2\eta_k\eta_0^{-3}\right)\mu^k$$

we obtain the differential equations

$$(61)\qquad \begin{aligned} &2\eta_k' + 3\xi_k - f_k = 0 \\ &\eta_k'' + 2\xi_k' - 2\eta_0^{-3}\eta_k + g_k = 0\,, \qquad\qquad k = 1, 2, \ldots\,. \end{aligned}$$

Elimination of ξ_k and transformation of the independent variable by means of (59) yields

$$(62)\qquad \frac{1-\zeta^2}{2}\,\ddot{\eta}_k - 2\zeta\dot{\eta}_k + 2\eta_k = Q^{-3}g_k - \frac{4}{3}Q^{-1}\dot{f}_k\,, \qquad \dot{} \equiv \frac{d}{d\zeta}\,.$$

It may be shown inductively [5] that the right-hand side of (62) is an even polynomial in ζ for every $k\geq 1$,

$$(63)\qquad Q^{-3}g_k - \frac{4}{3}Q^{-1}\dot{f}_k = \sum_{j=0}^{m} c_{2j}\zeta^{2j}\,.$$

Then, Equ. (62) has a particular solution in the form of an even polynomial,

$$(64)\qquad \eta_k = \sum_{j=0}^{m} \gamma_{2j}\zeta^{2j}\,,$$

where the coefficients γ_j are given by the recurrence relation

$$(65)\qquad \gamma_l = \frac{(l+2)(l+1)\gamma_{l+2} - 2c_l}{(l-1)(l+4)}\,, \qquad \gamma_{2m+2} = 0\,, \qquad l = 2m, 2m-2, \ldots, 2, 0\,.$$

Hence the following theorem has been established in a constructive way:

Theorem 2. *Hill's lunar equation (12) in the scaling* $x=\gamma\xi$, $y=\gamma^{-2}\eta$, $s=\gamma^{-3}\sigma$ *possesses a formal solution*

$$\xi = \sum_{k=0}^{\infty} \xi_k(\zeta)\mu^k, \qquad \eta = \sum_{k=0}^{\infty} \eta_k(\zeta)\mu^k\,, \qquad \mu = \gamma^6$$

with

$$\xi_0(\zeta) = \zeta\,, \qquad \eta_0(\zeta) = \frac{8/3}{1-\zeta^2}\,,$$

$$\sigma = -\frac{16}{9}\left(\operatorname{Artanh}\zeta + \frac{\zeta}{1-\zeta^2}\right)$$

which exists to arbitrarily large order k.

Practically the expansion has been carried up to order $k=5$ by means of SYMBAL. For $k>0$ $\eta_k(\zeta)$ is an even polynomial of degree $6k-2$.

In particular

Table 1: Asymptotics of Hill's problem as $s \to \infty$.

$$\ddot{x} - 2\dot{y} - 3x + x\rho^{-3} = 0 , \quad \ddot{y} + 2\dot{x} + y\rho^{-3} = 0, \quad \rho = \sqrt{x^2 + y^2} , \quad \cdot \equiv \frac{d}{ds}$$

Parameter: τ , Integration constants: a, b, s_o, s_1

Jacobi integral $h = \frac{1}{2}(\dot{x}^2 + \dot{y}^2) - \frac{3}{2} x^2 - \rho^{-1} = \frac{a^2}{2} - \frac{b^2}{6}$

Time $s - s_1 = \frac{1}{b} (\tau + \frac{3\gamma_o}{b^2} \log|\tau - \frac{3\gamma_o}{b^2}|)$

Auxiliary quantity $E = a. \exp [i (s - s_o)]$

Series coefficients

k	0	1	2	3	4
a_k	$\alpha_o E$	$\alpha_1 E$	$\alpha_2 E - 2i$	$\alpha_3 E + \frac{9}{4}\bar{E} + \frac{10}{3} b$	$\alpha_4 E + 18ib^{-4} + 10b^{-1} - \frac{21i}{2} a^2 - \frac{21i}{4} E^2 + \frac{11i}{4} \bar{E} + \frac{11i}{8} b \bar{E} + \frac{34i}{3} b^2$
b_k	β_o	β_1	β_2	$\beta_3 + 6(E+\bar{E})$	$\beta_4 - 15ib(E-\bar{E}) - \frac{81i}{16} (E^2 - \bar{E}^2)$
c_k	γ_o	γ_1	γ_2	γ_3	$\gamma_4 + 5i (E - \bar{E})$
α_k	1	0	$-\frac{7i}{4} b^{-1}$	$\frac{7i}{2} b^{-3}$	$\frac{63i}{16} b^{-5} - \frac{49}{32} b^{-2} - \frac{297i}{64} a^2 b^{-1} + \frac{13i}{12} b$
β_k	b	$3b^{-1}$	$-\frac{9}{2} b^{-3}$	$\frac{21}{4} a^2 b^{-1} - \frac{2}{3} b$	$\frac{81}{8} b^{-7} - \frac{63}{8} a^2 b^{-3} - 7b^{-1}$
γ_k	-1	0	$-\frac{9}{2} b^{-4}$	$\frac{27}{4} b^{-6} + \frac{21}{8} a^2 b^{-2} + \frac{1}{3}$	$-\frac{81}{8} b^{-8} - \frac{63}{8} a^2 b^{-4} - \frac{5}{3} b^{-2}$

Series $u_o = \sum_{k=0}^{\infty} a_k(s) \tau^{-k}$, $u_3 = \sum_{k=0}^{\infty} b_k(s) \tau^{-k}$, $u_4 = \sum_{k=0}^{\infty} c_k(s) \tau^{1-k}$

Coordinates $x = \frac{2}{3} u_3 - \text{Re } u_o$, $y = u_4 + 2 \text{ Im } u_o$, $\dot{x} = \text{Im } u_o$, $\dot{y} = - u_3 + 2 \text{ Re } u_o$

$$\xi_1(\zeta) = \frac{Q^2}{3}\left(\frac{81}{8}\zeta - \frac{61}{8}\zeta^3\right), \qquad Q = \frac{3}{8}(1 - \zeta^2),$$

$$\eta_1(\zeta) = -\frac{9}{32} + \frac{3}{2}\zeta^2 - \frac{19}{32}\zeta^4,$$

$$\xi_2(\zeta) = \frac{Q^2}{3}\left(\frac{41067}{2048}\zeta - \frac{108081}{1024}\zeta^3 + \frac{503721}{2560}\zeta^5 - \frac{1102929}{7168}\zeta^7 + \frac{88977}{2048}\zeta^9\right)$$

$$\eta_2(\zeta) = -\frac{243}{1024} + \frac{25839}{8192}\zeta^2 - \frac{34821}{4096}\zeta^4 + \frac{109131}{10240}\zeta^6 - \frac{180273}{28672}\zeta^8 + \frac{11685}{8192}\zeta^{10}.$$

The rapid growth of the coefficients seems to indicate divergence of the series.

References

[1] Hénon, M. and Petit, J.M.: Series Expansions for Encounter-Type Solutions of Hill's Problem. Celest. Mech. 38 (1986), 67-100.

[2] Hill, G.W.: On the Part of the Motion of the Lunar Perigee which is a Fraction of the Mean Motions of the Sun and Moon. Acta Math. 8 (1886), 1; also: Collected Mathematical Works of G.W. Hill. Vol. 1, p. 243. Carnegie Inst. of Wash., Washington, D.C., 1905.

[3] Petit, J.M. and Hénon, M.: Satellite Encounters. Icarus 66 (1986), 536-555.

[4] Spirig, F. and Waldvogel, J.: The Three-Body Problem with Two Small Masses: A Singular-Perturbation Approach to the Problem of Saturn's Coorbiting Satellites. In: V. Szebehely (ed), Stability of the Solar System and its Minor Natural and Artificial Bodies. Reidel, 1985, 53-63.

[5] Spirig, F. and Waldvogel, J.: A Theory of Coorbital Satellites Using Hill's Lunar Equation. In preparation, ZAMP 1988/89.

PART III

DYNAMICS OF ASTEROIDS

THE PUZZLE OF ASTEROID FAMILIES

P. Farinella*, Cl. Froeschle** and Z. Knežević***
* GMS, Dipartimento di Matematica, Universita di Pisa, Italy
** Observatoire de Nice, Nice, France
*** Astronomska Opservatorija, Beograd, Yugoslavia

Seventy years after their discovery by K. Hirayama, asteroid families still provide a number of puzzling problems to solar system dynamicists, being at the same time a potential source of valuable information on the internal structure and the collisional history of bodies representing an intermediate stage in the growth process from planetesimals to major planets. Hirayama (1918) first pointed out the existence of several groups of asteroids whose orbital parameters were clustered at a statistically significant level, provided they had been properly corrected to account for planetary gravitational perturbations (the so-called 'proper elements' being the outcome of this delicate correction procedure). He also suggested that the origin of the asteroids belonging to the same family could be traced back to an unique event, namely a primordial explosive breakup of a parent body, ejecting fragments into independent heliocentric orbits with relative velocities much lower than their orbital speed. In the last few decades, physical studies have shown that the members of the most populous families bear some similarity in their surface composition (Zellner et al., 1985) and their rotational behaviour (Binzel, 1987), supporting the hypothesis of a common origin and opening up for the first time the opportunity of directly probing via astronomical observations into the inner structure of the parent bodies. Moreover, collisional fragmentation has been proven to represent a plausible formation mechanism (Fujiwara, 1982; Farinella et al., 1982), both from the point of view of the probability of occurrence during the solar system's lifetime, and with respect to the expected outcomes of such catastrophic events, which have been simulated in the laboratory by small-scale experiments on high velocity impacts against rocky bodies (Fujiwara, 1986).

However, a number of important questions remain to be answered. The most basic one is: how many families can we indeed recognize today? Different investigators have obtained contrasting results, both in the assignment of individual asteroids to various families, and in the overall number of families 'extracted' from the background (or 'field') asteroids (Carusi and Valsecchi, 1982). This number

A. E. Roy (ed.), Long-Term Dynamical Behaviour of Natural and Artificial N-Body Systems, 237–244.
© 1988 by Kluwer Academic Publishers.

ranges from 5 according to the most prudent estimates, to more than 100 according to the most optimistic ones; any objective assessment is hampered by differences in the procedures adopted to derive proper elements and to identify non-random clusterings. Recently, many doubts have arisen on the reliability of many families identified by Williams and having just a few member asteroids, since from the point of view of their spectrophotometric properties (and of the inferred mineralogical compositions) it seems unlikely that they can be interpreted as fragments of a cosmochemically plausible parent body (Chapman 1986). On the other hand, such 'small' families should exist, since going to small target sizes, collisional fragmentation becomes more and more likely, and even for sizeable targets, nothing prevents the escape of only a few fragments following a barely catastrophic impact. A related open question regards the age of the families. Dynamical arguments set a lower limit of the order of 10^6 yr, but the solar system is 10^3 times older, and both cratering records on planetary and satellite surfaces and collisional evolution models (see, e.g., Davis et al., 1985) imply that collisions were much more frequent in the early phases than now. The problem is complicated by the fact that we do not know whether a 'primordial' family would still be recognizable now, or it would have been spread throughout the field asteroids by some slowly-acting dynamical diffusion process. Thus, it is unknown whether families represent the outcomes of relatively frequent events, or instead of a few exceptionally violent collisions. Other problems show up when one tries to predict in a quantitative way the expected collisional outcomes for asteroids by scaling up from the sizes of laboratory targets ($\sim$10 cm) to those of the precursors of families ($\sim$100 km). In the latter case self-gravity provides a significant contribution to the overall binding energy, and a disruptive event dispersing fragments 'to infinity' requires a larger amount of specific energy than for a small-scale target; therefore, unless for some unknown reason the solid-state impact strength of asteroids is much higher than for small rocky bodies, the family-forming fragmentation events should shatter the target asteroids to an extent which is clearly inconsistent with the observed size distributions in the families (Davis et al., 1985). Finally, it has been found (Zappala et al., 1984) that even for the few populous and 'reliable' families, the relative velocities needed to explain the observed distributions of proper elements are systematically higher than the velocities observed in laboratory experiments leading to similar outcomes; furthermore, as already noticed by Brouwer (1951), these velocities are very anisotropically distributed in a reference system related to the orbital motion of the parent body, a result at odds with the collisional origin theory, which would predict no significant systematic anisotropy of this kind.

In order to solve most of these problems, the starting step is an assessment of the accuracy and reliability of the available proper elements. In fact, it appears very likely that families are older than the time span over which we can use with confidence the existing secular perturbation theories to quantitatively predict the evolution of the orbits. This issue is further complicated by the frequent occurence among asteroid orbits of relatively high eccentricities and/or inclinations and of near-commensurabilities, which severely limit the accuracy of any theory based on series expansions of the perturbing function (see Milani, 1987, this volume). It is therefore understandable (though disappointing) that most investigators of families derived and used proper elements without carrying out any quantitive estimate of their accuracy and long-term stability. At a qualitative level, it is well known that the proper element related to the orbital energy (or the semimajor axis) is far less sensitive to planetary perturbations and more readily derived than the proper elements related to eccentricity and inclination, and this difference could well explain the above-mentioned anisotropy in the relative velocity distribution within each family (Brouwer, 1951; Carpino et al., 1986). However, this explanation does not solve the problem of why the average relative velocities (even the along-track components, related just to semimajor axis differences) are higher by a factor ~ 5 than those found in the experiments. This problem, like the inconsistencies found when scaling with target size the critical specific impact energy required for catastrophic disruption, is probably related to our insufficient understanding of the complex physical processes associated with impact fragmentation.

We have recently tackled the dynamical part of the problem, namely the issues concerning derivation and accuracy of asteroid proper elements (for a recent detailed review of this entire subject, see Froeschle et al., 1987), by a step-by-step approach. Firstly (see Carpino et al., 1986), we have carried out numerical integrations of the orbits of many real and fictitious family asteroids, taking into account perturbations by Jupiter (moving along a non-circular orbit) and in some cases by other planets. The formation of families has been simulated by starting a 'swarm' of test bodies at the same position but with small velocity differences; then the orbital evolution has been monitored and proper elements have been derived at different times, using in this first stage a linear secular perturbation theory (retaining in the perturbing potential only terms of first order with respect to the planet-to-Sun mass ratios and of second degree with respect to orbital eccentricities and inclinations). Significant long-period ($\sim 10^4$ yr) variations of these proper elements (which were used for instance by Brouwer, 1951, for family classification purposes) have been evidenced, and their origin clearly lies in the neglected higher-degree perturbative terms. Strong disturbances due to the proximity of mean motion commensurabilities are also commonplace. For example, the orbital evolution of the members of the populous Themis family is strongly affected by the fact that their semimajor axes are only by a few percents smaller than the value corresponding to the 2:1 mean motion resonance with Jupiter and to the associated Kirkwood gap. In this case, we have shown that fragments generated in the family-forming collisional event might have been injected into nearly-resonant orbits characterised by chaotic-type evolution, thus providing a plausible dynamical path from the main asteroid belt to planet-crossing orbits like those of meteorites and of Aten-Apollo-Amor objects. The comparison of numerical integrations with the linear secular perturbation theory can also be used for another purpose, i.e., to infer constraints on the ages of families from the long-term behaviour of the sum between the proper longitude of perihelion and the proper nodal longitude, which is an integral of motion in the frame of the linear theory (and a slowly-varying argument for the actual orbits).

In a subsequent stage, we derived the asteroid proper elements by means of the theory of secular perturbations by Yuasa (1973). This theory takes into account terms in the disturbing function of up to the fourth degree of eccentricity and inclination, in the first order with respect to perturbing mass, and of up to the second degree in the second order. The elimination of perturbations and the derivation of proper elements are made in two steps: first, the short-periodic terms are eliminated by means of a canonical transformation in terms of Delaunay's variables; then, by switching to Poincaré's variables, another canonical transformation is performed to eliminate the long-periodic perturbations. The solution is obtained in an explicit form, convenient for a straightforward determination of proper elements.

The results of the first step, namely of the elimination of short-periodic perturbations, yielding the so-called "mean" elements, are analysed in Knežević et al. (1987), where the derivation procedure is explained in more detail and its accuracy is tested in a number of different cases. The estimated residual variations of the mean semimajor axes (which are assumed to coincide with the proper values, since the long-term effects for them are negligible, at least at order 1 in the perturbing mass - see Milani, 1987, this volume) generally do not exceed a few parts in 10^{-4} AU, a result which can be considered as satisfying. Similar accuracies apply in most cases to the eccentricities and inclinations as well. Small-divisors related to near-resonances, however, can sometimes seriously affect the determination of mean elements. As for the high-degree terms neglected in the theory, a preliminary estimate of the critical eccentricities and inclinations, beyond which the theory is no more capable of providing mean elements of acceptable accuracy, is described in Knežević and Jovanović (1987).

Following the second step (the elimination of long-periodic terms to get proper elements), a new accuracy estimate was carried out by using as a comparison the numerical integrated orbits by Carpino et al. (1986), referring to the Sun-Jupiter-asteroid restricted, planar, elliptical 3-body case, and covering time spans of the same order as the periods of circulation of the asteroid perihelia and nodes. By applying to these data the analytical method for removing the perturbations, we obtained, for a number of epochs within the covered span of time, the corresponding proper elements. Being by definition constants of integration, proper elements would be expected to remain the same, regardless the epoch of their determination. Due to the limitations of the theory used for their derivation, however, this is not the case, and thus the deviation of proper elements from constancy can be employed as an indicator of their (in) accuracy and long-term (un)reliability (at least, for the time span covered by integrations). As reference data for such a comparison, one can also use the results of a Fourier analysis of the same integrations, as already made by Carpino et al. (1986) in studying the accuracy of proper elements derived by means of the linear theory.

In Fig.1, a typical example of the outcome of our procedure is shown for the asteroid Eos. The figure shows a plot of Eos' osculating eccentricity variation in a time span of 2 x 10^4 yr (top), the mean eccentricity (freed of the short periodic perturbations; middle), and the proper eccentricity (bottom). At the right bottom corner of the figure, the differences Δ between the maximum and minimum values of the proper eccentricity are given, thus providing an estimate of the deviation of proper eccentricity from constancy. The epochs of successive points differ by about 500 yr. Similar figures can be produced for other orbital elements as well.

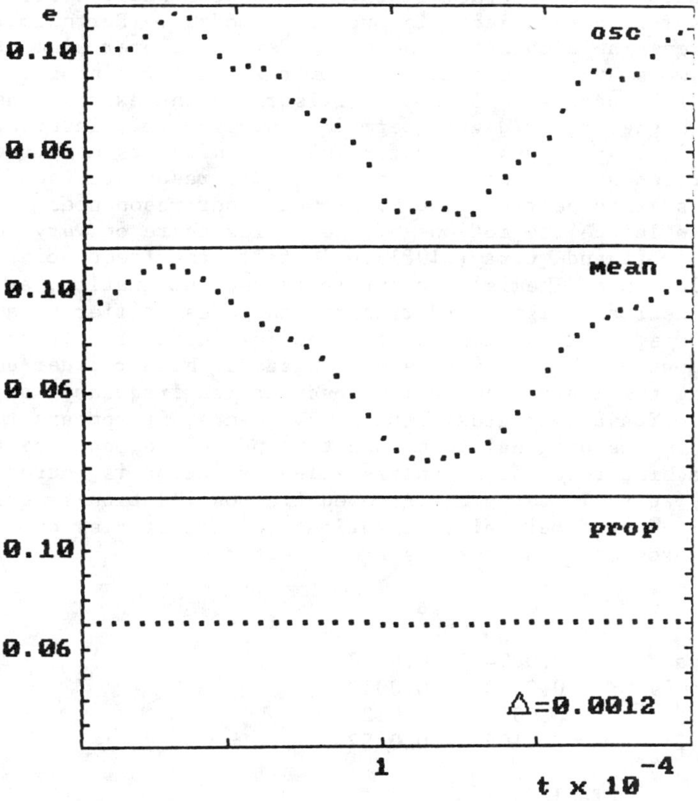

FIGURE 1

As can be seen in Fig.1, although expected to be constant, the proper eccentricity of Eos still exhibits a slight variation of amplitude 6.1×10^{-4}. This is, however, a very good result. If one takes into account that a 10^{-3} accuracy is required for proper elements in order to obtain reliable family classifications, and that the corresponding variation of the proper eccentricity derived from the linear theory is an order of magnitude larger (its amplitude being 6.2×10^{-3}), the improvement achieved by applying the higher order/degree theory appears to be very significant. Notice, moreover, that the amplitude of the first harmonic inferred from Fourier analysis of the numerically integrated orbits (Carpino et al., 1986) and amounting to 0.070725 for Eos' eccentricity, is also in very close agreement with the corresponding analytical result.

Unfortunately, not all the results of the rederivation of proper elements by means of the higher order/degree theory are as satisfactory as that described above. Due to the fact that already the linear theory supplies fairly accurate results for asteroids of low eccentricity/inclination, the accuracy improvement of the newly derived proper elements for these objects is much more modest. Furthermore, for nearly commensurable objects, the newly derived proper elements are sometimes even worse than those coming from the simpler linear theory, owing again to the appearance of small divisors in the expressions for computation of higher order/degree terms. A particularly interesting situation in this respect was found for Themis, which, as mentioned earlier, is located quite close to the strong 2:1 resonance, and in a region of the asteroid belt crowded by higher order resonances, that are likely to influence the motions of the bodies there on very long time scales (Dermott and Murray, 1981). In fact, the frequencies of the free oscillation of Themis's eccentricity and inclination, as derived from linear and higher order/degree theories, differ by as much as 50%. This, in turn, implies that no meaningful result can be obtained by computing the amplitudes and phases of higher order/degree terms containing the linear-theory free-oscillation frequency, as it is apparent from Yuasa's original equations. Hence, it appears that Yuasa's theory in its original form cannot be pushed beyond a certain limit in approaching resonances, unless some correction is introduced, for example, by suitably using an improved free-oscillation frequency (e.g., one derived by numerical integrations). Some studies on this subject are in progress, and results are expected soon.

Asteroid	Δe	Δe_{lin}
Flora	0.0004	0.0027
Koronis	0.0010	0.0017
Eos	0.0012	0.0124
Themis	0.0105	0.0057

Table 1.

A summary of the results derived by means of the higher order/degree theory of Yuasa for the proper eccentricity of the four major members of the largest Hirayama families, is shown in Table I in terms of the accuracy parameter Δ. The corresponding values, coming from the linear theory, are also given for a comparison. The results for the proper inclination are not included, since they are less interesting and informative due to the simple 3-body model employed in the numerical integrations. In conclusion, it can be stated that, for dynamically normal asteroids, the higher order/degree theory provides proper elements of significantly better accuracy than the classical linear theory; if this accuracy is not degraded on time scales longer than those covered by the available numerical integrations by 2 or 3 orders of magnitude, a much more reliable family classification will be possible on the basis of these elements. For asteroids close to resonances, however, the problem common to all analytical theories arises - deterioration of accuracy, due to the appearance of small divisors; both theoretical and numerical studies are necessary, and are to be properly combined, in order to handle this problem and to further improve the results.

ACKNOWLEDGEMENTS

We are grateful to M. Carpino, Ch. Froeschlé, R. Gonczi, B. Jovanović, A. Milani, P. Paolicchi, and V. Zappalá, with whom we have been collaborating on the issues described in this paper.

REFERENCES

Binzel, R.P. 1987, Icarus, in press.
Brouwer, D. 1951, Astron.,J. 56, 9.
Carpino, M., Farinella, P., Froeschlé, Ch., Froeschlé, Cl., Gonczi, R., Paolicchi, P., and Zappalá, V. 1986, Icarus, 68, 55.
Carusi, A., and Valsecchi, G.B. 1982, Astron. Astrophys., 115, 327.
Chapman, C.R. 1986, J. Ital. Astron. Soc., 57, 103.
Davis, D.R., Chapman, C.R., Weidenschilling, S.J., and Greenberg, R. 1985, Icarus, 62, 30.
Dermott, S.F., and Murray, C.D. 1981, Nature, 290, 664.
Farinella, P., Paolicchi, P., and Zappalá, V. 1982, Icarus, 52, 409.
Froeschlé, Cl., Farinella, P., Carpino, M., Froeschlé, Ch., Gonczi, R., Paolicchi, P., Zappalá, V., and Knežević, Z. 1987, in The Few-Body Problem, Proc. IAU Coll. No.96 (Turku, Finland, June 1987), M. Valtonen ed., Reidel, Dordrecht, in press.
Fujiwara, A. 1982, Icarus, 52, 434.
Fujiwara, A. 1986, J. Ital. Astron. Soc., 57, 47.
Hirayama, K. 1918, Astron. J., 31, 185.
Knežević, Z., Carpino, M., Farinella, P., Froeschlé, Ch., Froeschlé, Cl., Gonczi, R., Jovanović, B., Paolicchi, P., and Zappala, V. 1987, Astron. Astrophys., in press.

Knežević, Z., and Jovanović, B. 1987, Proc. 10th IAU ERMA (Prague, Czechoslovakia, August 1987), in press.
Yuasa, M. 1973, Publ. Astron. Soc. Japan, 25, 399.
Williams, J.G. 1979, in Asteroids (T. Gehrels, Ed.) p. 1040, Univ. of Arizona Press, Tucson.
Zappala, V., Farinella, P., Knežević, Z., and Paolicchi, P. 1984, Icarus, 59, 261.
Zellner, B., Tholen, D.J., and Tedesco, E.F. 1985, Icarus, 61, 355.

HIGH ECCENTRICITY LIBRATION

S. Ferraz-Mello
Universidade de Sao Paulo
Instituto Astronomico e Geofisico
Caixa Postal 30627
01051 Sao Paulo, SP, Brazil.

ABSTRACT: An analytical averaging theory founded on the elliptic restricted three-body problem and valid for high-eccentricity (HE) librations shows the existence of forbidden intervals for the values of the <u>mean</u> semi-major axis at resonances. The librators may cross these 'gaps' but cannot remain there. The results are compared to data of some of the actual Hildas.

We consider in this paper the averaged motion of a first-order resonant asteroid in the plane of the complex variable

$$z = e.\exp i\theta$$

where

$$\theta = (p+1)L' - pL - \tilde{\omega},$$

$e, L, \tilde{\omega}$ are the asteroidal eccentricity, mean longitude and longitude of the perihelion, L' is the mean longitude of Jupiter, p an integer and i the imaginary unit.

In a first approximation, the equation of motion of a resonant asteroid in this plane is

$$z = k_0 + A.\exp i\alpha + \beta e' \exp i\theta'$$

where

$$\theta' = (p+1)L' - pL - \tilde{\omega}',$$

k_0 and β are constants, A and α are functions of t. e' and $\tilde{\omega}'$ are the eccentricity and the longitude of the perihelion of Jupiter. The morphogenesis of the motions in the plane

$$z_0 = k_0 + a \exp i\alpha$$

has been studied in 1903 by Andoyer and has been reconsidered in recent times by many authors: Sessin (1981), Henrard and Lemaître (1983), Sessin and Ferraz-Mello (1984) and Ferraz-Mello (1985 and 1988). When an asteroid moves in the inner side of a (p+1):p resonance, that is,

A. E. Roy (ed.), Long-Term Dynamical Behaviour of Natural and Artificial N-Body Systems, 245–250.
© 1988 by Kluwer Academic Publishers.

faster than the corresponding resonant asteroid, we have the situation shown in Fig.1(a). The motions are orbits around the center S_1. This orbit may be either a circulation, if it includes the origin of the plane, or, otherwise, a low-eccentricity libration. It is not superfluous to insist on the fact that these orbits form a structurally stable family and that the orbit passing through the origin is not a true separatrix. As the asteroid approaches the resonance, k_0 increases following the approximate law

$$k_0 = -\frac{n\alpha m'}{2} \frac{[2(p+1)b_{1/2}^{p+1}(\alpha) + \alpha \frac{d}{d\alpha} b_{1/2}^{p+1}]}{(p+1)n' - pn - g}$$

where n and a are the mean motion and the semi-major axis of the librator of zero amplitude, m', n' and a' are the Jovian mass, mean motion and semi-major axis, $\alpha = a/a'$, b the coefficients of Laplace, and

$$g = \frac{n\alpha^2 m'}{4} b_{3/2}^{1}(\alpha).$$

(Ferraz-Mello, 1987a). Moreover, the innermost ovals start bending, becoming bean-shaped. This situation leads to the catastrophe shown in the middle of Fig.1, when a cusp, which will give rise to a saddle point and a new center, appears. From this point on the family of orbits around S_1 is enclosed by a true separatrix (homoclinic orbits emanating from the saddle point), as shown in Fig.1(b).

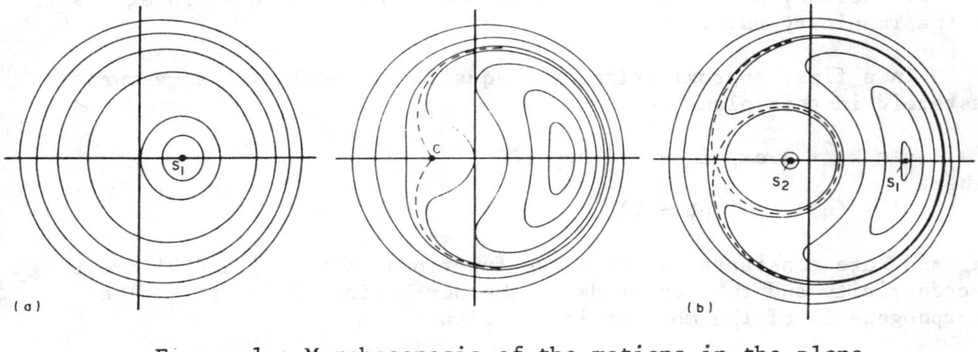

Figure 1 : Morphogenesis of the motions in the plane $z_0 = k_0 + A \exp i\alpha$.

As shown by Ferraz-Mello (1988), the center S_1 goes to infinity as the exact resonance is reached. As a consequence the librational motions in deep resonance are always associated with high eccentricities, as shown in Table I.

Table I: High eccentricity (HE) Librators

Librators	Resonance	Eccentricity
Alindas	3:1 with Jupiter	0.5 - 0.6
Hildas	3:2 " "	0.1 - 0.3
Griquas	2:1 " "	0.2 - 0.5
Pluto	2:3 " Neptune	0.25
Hyperion	3:4 " Titan	0.1

The techniques used to study HE librations are a new expansion of the disturbing function (Ferraz-Mello, 1987b) and the averaging method of Hori. The new expansion of the disturbing function is founded on a simple idea first used by Woltjer (1928): if we intend to study motions about S_1 the natural procedure is to expand the disturbing function R about S_1 and not about the origin (z=0) as done in the classical Laplacian expansion. The disturbing function is then expanded as

$$R = R_0 + R_1(k-k_0) + R_2 h + R'e' \cos\theta' + \text{etc}$$

where

$$k + ih = z = e.\exp i\theta.$$

This development is a Taylor series expansion about $z = k_0$ and $e' = 0$. The coefficients R_i and R' depend on the semimajor axes a and a'. In a first step they depend also on the mean synodic longitude L-L', but, for the sake of constructing formal long-period solutions they are averaged over it by numerical quadratures. After this averaging a Hamiltonian formed by secular, long period and resonant terms is obtained. The formal solution of the new dynamical system depends on the successful accomplishment of one basic step: the identification of an integrable Hamiltonian (Hori's averaging kernel), having the same topological structure as the region of the phase space under study. A kernel taking into account the eccentricity of Jupiter was first obtained by Sessin (1981) and allowed us to average the elliptic asteroidal problem near a first-order resonance for both low and high eccentricities (Ferraz-Mello 1987a and 1988).

The results of the theory were summed up into three laws: one gives the period of the proper oscillation about S_1 and the others give k_0 and β as functions of the semi-major axis of the orbit of zero-amplitude libration (that is, of the <u>mean</u> semi-major axis of the low amplitude orbits about S_1). The position of S_1 plays a determining role in shaping the asteroidal belt near resonances and the law giving its abcissa k_0 has been called <u>law of structure</u>. It is

$$[(p+1)n' - p(n_0+\delta n)]\, n_0 a_0^2 = R_1(a_0)/k_0$$

where n_0 is the mean motion corresponding to a_0, and

$$\delta n = - (2/n_0 a_0)(dR_0/da_0).$$

The results obtained with this formula can be compared to those obtained by numerical averaging and integration of the equations of motion by Schubart (1968 and personal communication). The continuous line in Fig.2 shows the law of structure $k_0(a_0)$. The dots give the arithmetic means of the minimum and maximum values of the semi-major axis of the actual Hildas as determined by Schubart as a function of his parameter e_p.

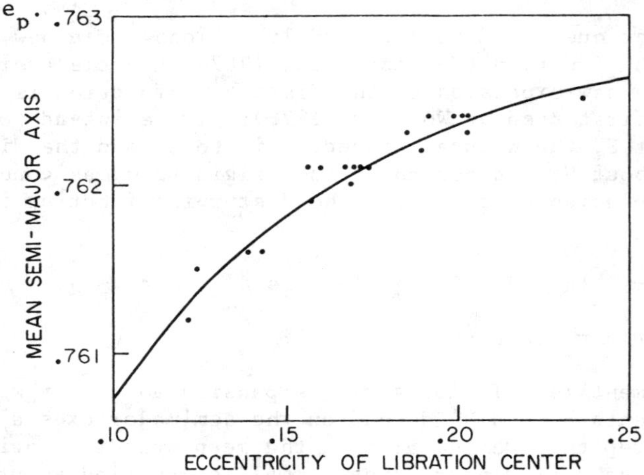

Figure 2: Solid line: law of structure $k_0(a_0)$. Dots: Results obtained by numerical averaging integrations by Schubart (see text).

The continuous line in Figure 3 shows the law of structure in a wide region about the resonance 3:2. The pericentric branch, in the inner side of the resonance. corresponds to libration solutions where $\theta = 0$ (the conjunctions happen when the asteroid is at the perihelion). The apocentric branch, in the outer side of the resonance, corresponds to $\theta = \pi$ (the conjunctions happen when the asteroid is at the aphelion; centre S_2 in Figure 1(b)).

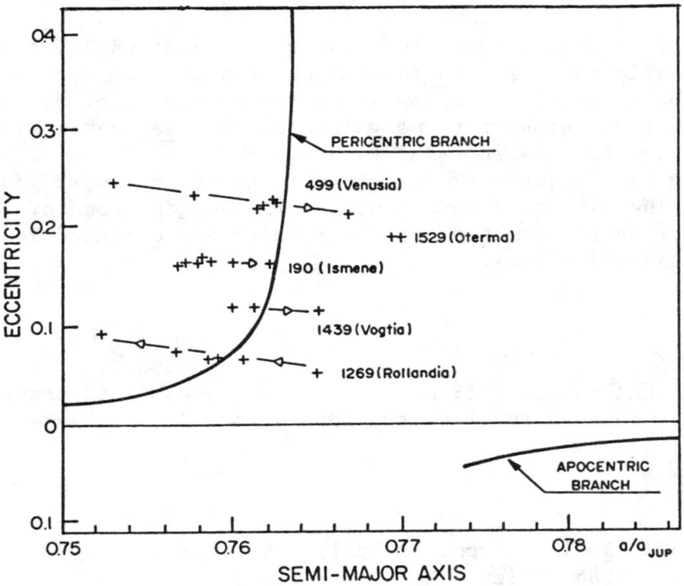

Figure 3: Thick line: law of structure $k_0(a_0)$. Crosses: eccentricities and semi-major axes of some Hildas, since their discovery, as determined from the observations.

Figure 3 shows the existence of an interval about the exact resonance $3n' - 2n = 0$ where no stationary solution exist. However the presence of asteroids in that interval is not excluded and, in fact, at any time almost half of the Hildas will be incursioning in it. The motion of each librating Hilda is an oscillation about one center in the pericentric branch (there are no apocentric Hildas). This fact is clearly shown in Fig.3 where we plotted the actual motion of some Hildas: (190) Ismene, (499) Venusia, (1269) Rollandia, (1439) Vogtia and (1529) Oterma, taken from the lists published by the Berliner Astronomisches Jahrbuch (1881 to 1918), Bahnelemente und Oppositionsephemeriden der Kleinen Planeten (1912 to 1945) and Efemeridy Malikh Planet (1947 to 1987). All crosses represent orbits actually determined from the observations (orbits just propagated by analytical or numerical methods were discarded). Figure 3 confirms the theoretical results of Ferraz-Mello (1988) after which HE librators may have large variations in the semi-major axis. Indeed, some of the Hildas may cross completely the structural gap as is the case of (1529) Oterma whose semi-major axis is presently around 0.771 a' and is still increasing (the maximum is expected to be reached in the next few years).

CONCLUSION

The analytical modelling of the planar motion of HE librators leads to a law relating the mean semi-major axis and the eccentricity at the libration centers. This law shows the existence of forbidden intervals (structural gaps) for the values of the mean semi-major axes. For the resonance 3:2 this interval is 0.763 - 0.773 a' (3.970 - 4.019 A.U.). However the presence of librators in this structural gap is not excluded. The librators may incursion in the gap, some of them being able to cross it completely. The results are compared to data of some of the actual Hildas.

ACKNOWLEDGEMENT

The documental investigations leading to the asteroid tracks in Figure 3 were done at the Institut för Astronomie, Universität Wien.

REFERENCES

Andoyer, H.: 1903, Bull. Astron. 20, 321.
Ferraz-Mello, S.: 1985, Celest. Mech. 35, 209.
Ferraz-Mello, S.: 1987a, Astron. J. 94, 208.
Ferraz-Mello, S.: 1987b, Astron. Astrophys. 183, 397.
Ferraz-Mello, S.: 1988, Astron. J. (to be published)
Henrard, J. and Lemaitre A.: 1983, Celest. Mech. 30, 197.
Schubart, J.: 1968, Astron. J. 73, 99.
Sessin, W.: 1981, Dr. Thesis, Univ. Sao Paulo.
Sessin, W. and Ferraz-Mello, S.: 1984, Celest. Mech. 32, 307.
Woltjer Jr., J.: 1928, Ann. Sterrew. Leiden 16 (3).

EVOLUTION OF ASTEROIDAL ORBITS LOCATED IN THE MAIN SECULAR RESONANCES ν_5, ν_6 and ν_{16}

Ch. Froeschlé, H. Scholl
Observatoire de Nice, B.P. 139,
F-06003 Nice Cedex France

ABSTRACT. We investigated the orbital evolution of fictitious asteroids located in and near the three main secular resonances ν_5, ν_6 and ν_{16}. Our results confirm a suggestion by Wetherill (1979) concerning the delivery of Apollo asteroids: asteroids located in the secular resonances ν_5 and ν_6 may increase strongly their eccentricities and therefore may become Mars-crossers, Earth-crossers and even Venus-crossers. For the secular resonance ν_{16}, large increases in inclination are obtained. In case of the ν_{16} resonance, three different regions for secular resonant motion are found similar to the case of mean motion resonances: transitions between these three regions occur. According to our numerical results, secular resonances should be considered as sources of meteorites as conjectured by Wetherill and Williams (1979).

We know that the major planets excite secular perturbations in the path of small bodies orbiting the Sun. Strong secular perturbations occur when the precession rate $\dot{\tilde{\omega}}$ (perihelion longitude) or $\dot{\Omega}$ (longitude of node) of an asteroidal orbit is nearly equal to the precession rate $\dot{\tilde{\omega}}$ or $\dot{\Omega}$ of a planetary orbit. This is called a secular resonance case. In the asteroidal belt, there are three dominant secular resonances ν_5 , ν_6 and ν_{16}. The rates of these secular resonances are:

$$\begin{aligned} \dot{\nu}_5 &\simeq \dot{\tilde{\omega}}_J \\ \dot{\nu}_6 &\simeq \dot{\tilde{\omega}}_S \\ \dot{\nu}_{16} &\simeq \dot{\Omega}_J \simeq \dot{\Omega}_S \end{aligned} \qquad (1)$$

where the subscripts J and S refer respectively to Jupiter and Saturn.

The resonance conditions are given by:

$$\begin{aligned} \dot{\tilde{\omega}} - \dot{\tilde{\omega}}_J &\simeq 0 \quad \text{for } \nu_5 \\ \dot{\tilde{\omega}} - \dot{\tilde{\omega}}_S &\simeq 0 \quad \text{for } \nu_6 \\ \dot{\Omega} - \dot{\Omega}^S_J &\simeq 0 \quad \text{for } \nu_{16} \end{aligned} \qquad (2)$$

A. E. Roy (ed.), Long-Term Dynamical Behaviour of Natural and Artificial N-Body Systems, 251–256.
© 1988 by Kluwer Academic Publishers.

It is sufficient to show that the arguments $\tilde{\omega} - \tilde{\omega}_J$, $\tilde{\omega} - \tilde{\omega}_S$ and $\Omega - \Omega_J$ librate around a constant value in order to demonstrate a ν_5, ν_6 or respectively a ν_{16} resonance case. According to Williams (1969), such a libration can occur only either about 0° or about 180°.

We integrated numerically over 1 Myr in the frame of the four body problem Sun - Jupiter - Saturn - asteroid the orbits of model asteroids. Our numerical integrations were based on the n-body program of Schubart and Stumpff (1966).

- Results for the ν_5 resonance.

We exhibit one example of a model asteroid whose starting values are given in Table I. As shown in Figure 1, the secular argument $\tilde{\omega} - \tilde{\omega}_J$ librates around 0°. A maximum of eccentricity of nearly 0.4 is obtained.

Res	Body	a	e	i^o	$(\Omega-\Omega_J)^o$	$(\tilde{\omega}-\tilde{\omega}_J)^o$	$(\tilde{\omega}-\tilde{\omega}_S)^o$
ν_5	19	2.6	0.14	31.87	246	359	281
ν_6	6	2.05	0.17	4.63	335	180	102

Table I - Starting values of the model asteroids.

In analogy with mean motion resonances we plotted the orbit in a polar diagram with axes defined by (Charlier, 1902):

$$\psi 1^J = [2(1 - \sqrt{1 - e^2})]^{\frac{1}{2}} \cos(\tilde{\omega} - \tilde{\omega}_J)$$
$$\psi 2^J = [2(1 - \sqrt{1 - e^2})]^{\frac{1}{2}} \sin(\tilde{\omega} - \tilde{\omega}_J) \qquad (3)$$

Again the libration is demonstrated. There are loops superimposed, which have the same periods as the oscillations in Fig.1. The same periods can be seen in figures presenting the evolution with time of $\tilde{\omega}_S$ and $\tilde{\omega}_J$ (see for instance Cohen et al., 1973).

- Results for the ν_6 resonance.

The time evolution of the ν_6 resonance argument $\tilde{\omega} - \tilde{\omega}_S$ is plotted in Fig.3 for a model asteroid. We see clearly the libration of $\tilde{\omega} - \tilde{\omega}_S$ about 180°, with an amplitude exceeding 90° and with a period of the order of 1 Myr. More information of orbital evolution is obtained by plotting the eccentricity against the argument $\tilde{\omega} - \tilde{\omega}_S$ (Fig.4). The orbit rotates clockwise about a mean central point located on an axis defined by $\tilde{\omega} - \tilde{\omega}_S = 180^o$. We want to underline that we obtain a maximal eccentricity of 0.62, which yields a perihelion distance of 0.78 AU. The corresponding orbit crosses the Earth-orbit, and approaches closely the orbit of Venus. This example demonstrates the importance of the resonance ν_6 for the direct delivery of meteorites from the asteroidal belt to the Earth.

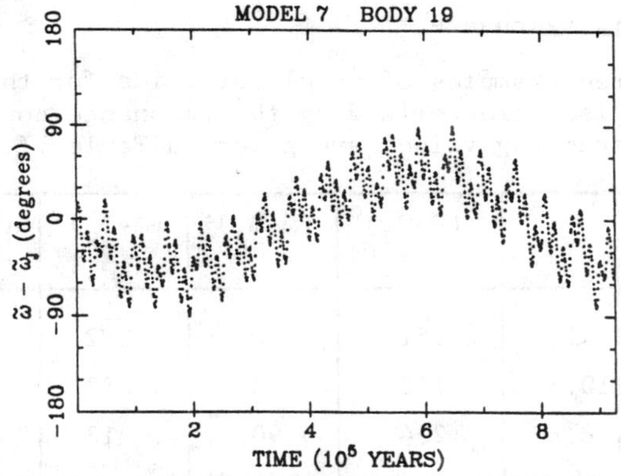

Figure 1.- The time-evolution of longitude of perihelion $\tilde{\varpi}$ with respect to Jupiter's longitude of perihelion $\tilde{\varpi}_J$ over 1 Myr for body 19 of Table I.

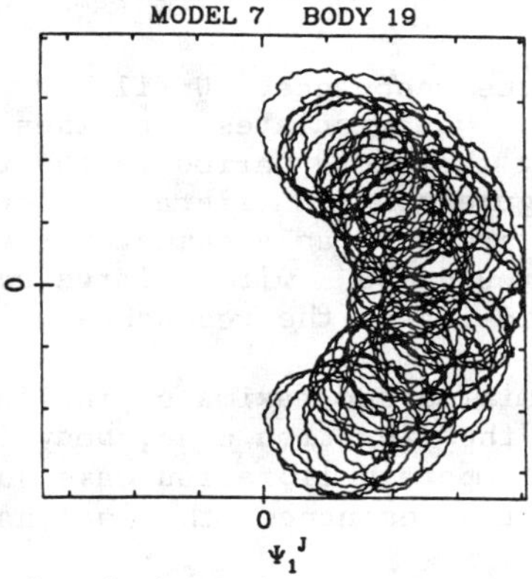

Figure 2.- Orbital evolution in a polar diagram for the librator body 19.

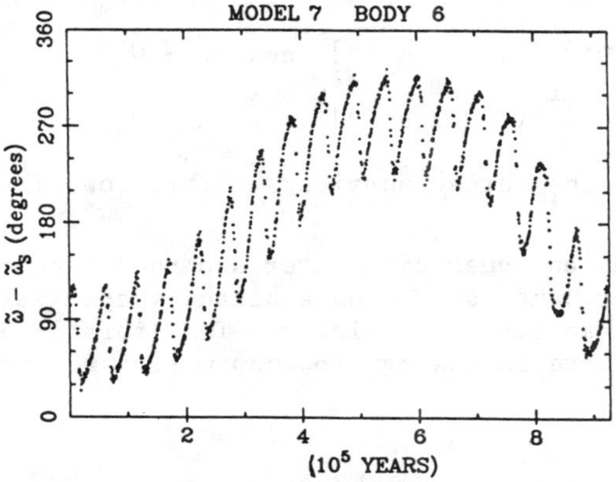

Figure 3.- The time-evolution of longitude of perihelion $\tilde{\varpi}$ with respect to Saturn's longitude of perihelion $\tilde{\varpi}_S$ over 1 Myr for body 6 of Table I.

- Results for the ν_{16} resonance

We present three examples of model asteroids for the secular resonance ν_{16}, which is characterized by the resonance condition $\Omega - \Omega_J \simeq$ constant. The starting values are given in Table II.

Body	a	e	i^o	$(\Omega-\Omega_J)^o$	$(\tilde{\omega}-\tilde{\omega}_J)^o$	$(\tilde{\omega}-\tilde{\omega}_S)^o$	i^o_{min}	i^o_{max}
8	2.05	0.17	4.37	151	0	282	4	23
13	2.05	0.17	10.37	152	0	282	10	30
16	2.05	0.17	8.89	254	90	12	8.9	22

Table II- starting values of the model asteroids located in the ν_{16} secular resonance.

Body 8 is entering the resonance. Until $\simeq 2 \times 10^5$ years the corresponding argument $\Omega - \Omega_J$ circulates, and then librates around 180°. Body 13 is located during the whole period in the ν_{16} resonance. According to Figure 5, the argument $\Omega - \Omega_J$ librates about 180°. The third model asteroid (body 16) presents an argument $\Omega - \Omega_J$ which librates until $\sim 8 \times 10^5$ years around 180° with a large amplitude, and then circulates. This asteroid leaves the resonance.

Table II shows the minima and maxima of inclinations. It is interesting to note that the libration case, body 13, reveals a smaller variation of i than the temporary libration case body 16. Furthermore, when the asteroid leaves the resonance, the inclination remains high. ($\sim 25^o$).

When the orbits are plotted in the polar diagram (Fig.6)

$$\begin{aligned} \psi 3^J &= [2(1 - e^2)^{\frac{1}{2}} (1 - \cos i)]^{\frac{1}{2}} \cos (\Omega - \Omega_J) \\ \psi 4^J &= [2(1 - e^2)^{\frac{1}{2}} (1 - \cos i)]^{\frac{1}{2}} \sin (\Omega - \Omega_J) \end{aligned} \qquad (4)$$

We see clearly again the three behaviours, libration, inner circulation, outer circulation.

Body 8 starts as an inner circulator and then becomes a librator. Body 16, on the other hand, starts on a banana-shaped curve as a librator and becomes an outer circulator. The librator body 13 moves on a banana-shaped curve in analogy to mean motion resonant orbits.

CONCLUSIONS

We showed that the secular resonances ν_5 and ν_6 may cause large increases in eccentricity up to 0.40 in the ν_5 resonance, and up to

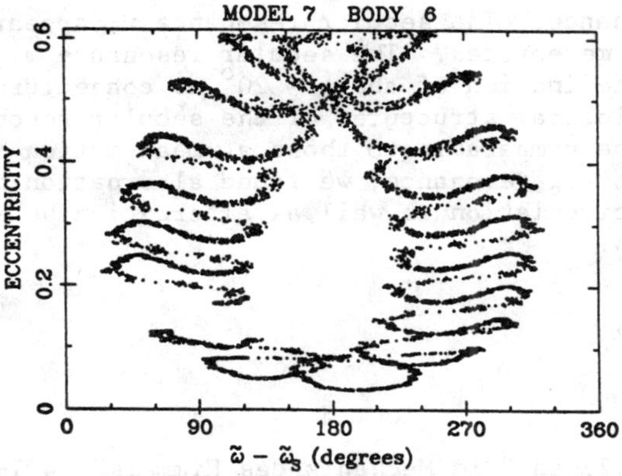

Figure 4.- Variation of osculating eccentricity vs. the argument $\tilde{\omega} - \tilde{\omega}_S$ for body 6 of Table I.

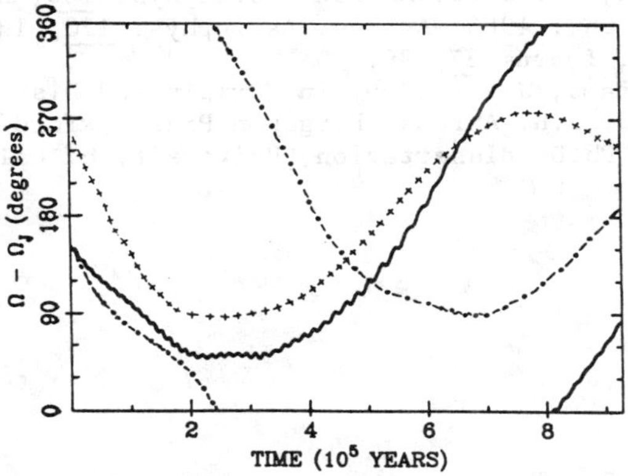

Figure 5.- The time evolution of longitude of ascending node Ω with respect to Jupiter's longitude of ascending node Ω_J over 1 Myr: -.-. body 8; +++ body 13; ___ body 16 of Table II.

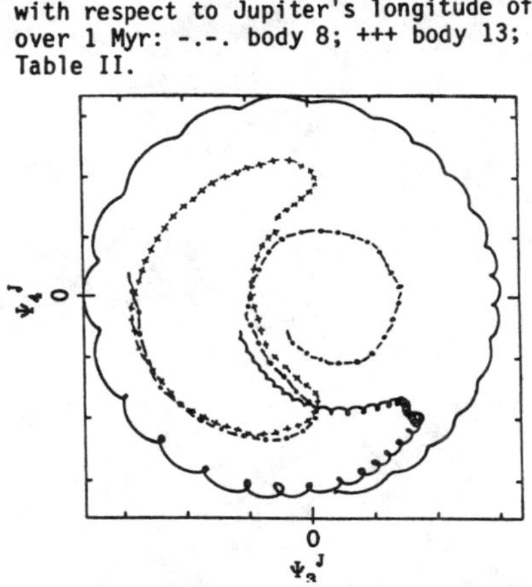

Figure 6.- Orbital evolution in a polar diagram for: -.-. body 8; +++ body 13; ___ body 16 of Table II.

0.62 for the ν_6 resonance. The secular resonance ν_{16} appears to be an important source for meteorites. The secular resonance ν_{16} causes large increases in inclination of about $\sim 20^o$ as conjectured by Wetherill. The topological structures of the secular resonances ν_5, ν_6 and ν_{16} seem to be comparable to those at mean motion resonances. In particular for the ν_{16} resonance, we found alternation between libration and outer circulation as well as alternation between libration and inner circulation.

REFERENCES

Charlier, C.V.L.: 1902, in "Die Mechanik des Himmels", 421-436, Verlag von Veit Leipzig.
Froeschlé, Ch., Scholl, H.: 1986, Astron. Astrophys. 166, 326-332.
Scholl, H. Froeschlé, Ch.: 1986, Astron. Astrophys. 170, 138-144.
Wetherill, G.W.: 1979, Icarus 37, 96.
Wetherill, G.W., Williams, J.G.: 1979, in "Origin and Distribution of the Elements", ed. L.H. Ahreus, Pergamon Press Oxford/New York.
Williams, J.G.: 1969, Ph.D. dissertation, University of California.

ALGEBRAIC MAPPINGS NEAR A RESONANCE WITH AN APPLICATION TO ASTEROID MOTION

John D. Hadjidemetriou,
University of Thessaloniki,
Thessaloniki, Greece.

ABSTRACT: A method is presented to obtain an algebraic mapping for a Hamiltonian system with two degrees of freedom near a resonance. The theory of adiabatic invariants in Hamiltonian systems with a slowly varying parameter is extended to mappings of the above type. This method is used to study the evolution of asteroid orbits on passage through a resonance as the energy decreases slowly. Sudden changes of the orbital elements are observed in certain cases and these can be related to the generation of gaps. A phenomenon of magnification of the effects of a small chaotic region is also observed.

1. INTRODUCTION

Many subsystems in the Solar System are described by a Hamiltonian which is close to an integrable one,

$$H = H_0 + \varepsilon H_1, \qquad (1)$$

where H_0 is the Hamiltonian of the integrable part and εH_1 is the perturbation. Several methods exist for the study of this type of motion. A method widely used is to transform the flow in phase space determined by (1) to an equivalent mapping on a suitable surface of section. In this way the dimensions are reduced by two and this makes the study easier. Also the mapping is faster and more accurate than the solution of the differential equations.

We shall present here a method to construct an algebraic mapping which has all the main properties of the actual system. This will be illustrated by considering a Hamiltonian system with two degrees of freedom near resonance, but the method can be applied to any number of degrees of freedom. The construction of the mapping is not based on an approximate solution of the differential equations of (1). It is applicable in the cases where the main qualitative effect of the perturbation εH_1 can be obtained analytically by methods other than approximation techniques. The numerical values of the parameters involved are then found from numerical integrations, so that this method could

A. E. Roy (ed.), Long-Term Dynamical Behaviour of Natural and Artificial N-Body Systems, 257–276.
© 1988 by Kluwer Academic Publishers.

be considered as semianalytic. This method will be applied to study asteroid orbits near the 3:1 resonance with Jupiter. In this case the main qualitative effect of the perturbation is obtained by the theory of Krein.

The mapping constructed in the above way can be used to find the behaviour of all similar systems, because the inclusion of several types of perturbations is easy. In this way we can find what properties are basic, unaffected by the various approximation techniques used for the study of the Hamiltonian (1). In this sense the obtained mapping can be considered as a model to study the generic properties of all similar Hamiltonian systems near a resonance. In addition, the study of such a mapping has an interest of its own.

2. TRANSFORMATION TO A MAPPING

2.1 Action-angle variables in the unperturbed case.

We assume that the Hamiltonian is expressed as

$$H = H_0(x_1, x_2, p_1, p_2) + \varepsilon H_1(x_1, x_2, p_1, p_2), \tag{2}$$

where H_0 is integrable. We shall introduce action-angle variables $J_1, J_2, \theta_1, \theta_2$ to the unperturbed Hamiltonian H_0 (Landau and Lifshitz. 1973, also Hadjidemetriou 1986). The new Hamiltonian is a function of the actions only, $H_0(J_1, J_2)$ and the solution is

$$\theta_1 = n_1 t + \theta_{10}, \quad \theta_2 = n_2 t + \theta_{20}, \quad J_1 = J_{10}, \quad J_2 = J_{20} \tag{3}$$

where θ_{i0}, J_{i0} are constants and

$$n_1 = \partial H_0/\partial J_1, \qquad n_2 = \partial H_0/\partial J_2, \tag{4}$$

are the frequencies of the system. The solution (3) represents motion on a 2-torus with radii J_1, J_2 and angles θ_1, θ_2, as shown in Figure 1. The canonical system of equations from which the solution (3) is obtained is

$$\dot\theta_1 = n_1, \quad \dot\theta_2 = n_2, \quad \dot J_1 = 0, \quad \dot J_2 = 0. \tag{5}$$

Let us consider now the surface of section

$$H_o = h, \quad \theta_2 = 0 \tag{6}$$

In this way we define a mapping in the variables J_1, θ_1, of the form

$$J_1 \to J_1, \quad \theta_1 \to \theta_1 + f(J_1). \tag{7}$$

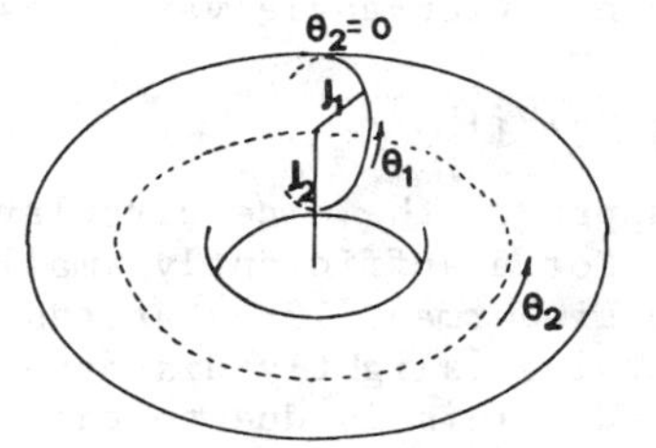

Figure 1: The 2-torus on which the motion (3) takes place

This is a twist mapping whose invariant curves are the circles J_1 = constant and is area preserving since it is derived from a Hamiltonian. The rotation angle $f(J_1)$ is constant for a fixed J_1 and the time interval between two consecutive intersections is given by

$$T_2 = 2\pi/n_2. \tag{8}$$

Consequently, the angle θ_1 advances by

$$\Delta\theta = 2\pi\ n_1/n_2 \tag{9}$$

from one intersection to the other.

2.2 Resonant and nonresonant periodic orbits

In the set of the unperturbed motions given by (3), we distinguish two types of periodic orbits: resonant and nonresonant. The nonresonant periodic orbits correspond to $J_{10} = 0$. They are represented by the dotted line in Figure 1, i.e. the 2-torus in this case has been reduced to a circle. We shall call it the "central periodic orbit", and its period is equal to $2\pi/n_2$. This periodic orbit is represented by the central fixed point $J_1 = 0$ in the corresponding unperturbed mapping.

When $J_1 \neq 0$ we have, for each fixed set of values J_1, J_2, a ratio n_1/n_2 of the frequencies. We distinguish now between two cases: a rational and an irrational value. When the ratio is irrational, the orbit covers densely the 2-torus. If however it is rational,

$$n_1/n_2 = r/s, \tag{10}$$

the orbit is periodic and is represented on the surface of section by a circular, resonant, invariant curve J_1 = constant. All points on this invariant curve are fixed points of the mapping (multiple, in general). This type of periodic orbit will be called a "resonant periodic orbit".

2.3 The perturbed mapping

We apply now to H_o the perturbation εH_1, i.e. the Hamiltonian is

$H = H_0 + \varepsilon H_1$, and expressed in action-angle variables takes the form

$$H = H_0(J_1, J_2) + \varepsilon H_1(J_1, J_2, \theta_1, \theta_2). \tag{11}$$

The question is what will happen to the nice circular invariant curves of the unperturbed problem. For a sufficiently small perturbation the KAM theorem (Lichtenberg and Lieberman, 1983) states that all the irrational invariant curves survive (slightly distorted), but the rational invariant curves are destroyed. This is due to the fact that only a finite number of the fixed points of the rational invariant curve survive, half of them being stable and half unstable, according to the Poincaré-Birkhoff fixed point theorem (Arnold and Avez, 1968; Berry, 1978). This changes completely the topology of phase space near a resonance. We should mention however that this change of the topology has practical significance for resonances (10) with small integers r,s. The above will become clear with the examples of the next sections.

As we mentioned before, the central fixed point $J_1 = 0$ is, in general, non-resonant. For particular values of the energy however, it is resonant and at particular resonances predicted by the theory of Krein it comes unstable, thus changing also the topology of phase space near the origin.

3. A NEW APPROACH TO FIND THE PERTURBED MAPPING

The standard method to obtain the solution of the perturbed problem is to apply a perturbation technique to the equations of motion. The advantage of this method is that the solution obtained is correct, within a certain approximation, for a finite interval of time (Wisdom 1982, 1983, 1985; Henrard 1987; Murray, 1986; Ferraz-Mello, 1987). However, in general, the effect of the neglected terms and the approximations made, is not known. It may happen that different approximations may produce qualitatively different results, especially in chaotic regions (Henrard and Lemaitre, 1987). Despite these deficiencies, this is the only way to approximate closely the solution.

A method widely used, especially to study motion near a resonance, is the method of averaging. This is especially useful in systems with two degrees of freedom, because the averaged problem is of one degree of freedom and consequently it is integrable (Schubart, 1964; Henrard and Lemaitre, 1983; Lemaitre, 1984). The level curves in this case, as obtained from the energy integral, are the trajectories of the averaged problem. In this way we can obtain the topology of the phase space of the perturbed problem.

A different approach to study the perturbed problem is to construct an algebraic mapping, based on the following idea: Instead of solving approximately the perturbed system H, we find the exact solution of the integrable part H_0, and its corresponding mapping. We then perturb this mapping in such a way that it has all the main properties of the actual system. The problem which arises here is: what are these main properties? How do we know them since we do not want to solve, even approximately, the system? This method can be applied therefore to the cases where the basic properties of the system can be obtained analytically by a method other than the usual perturbation techniques. The quantitative adjustments of the parameters can then be made by numerical integrations of particular orbits.

The structure of phase space on a surface of section depends critically on the stability properties of the fixed points. Consequently, it is important to know where the fixed points are and whether or not instability is generated by the perturbation. The prediction where instability will be generated will be made by the theory of Krein (see Yakabovich and Starzhinski, 1975). Thus, the problem is to find a suitable perturbation to the unperturbed mapping, so that its topology coincides with that of the actual problem. Once we succeed, we have some obvious advantages: the mapping is simple and consequently it is fast and accurate. It also contains the main features of the actual system and these were included without the use of any approximation method. There are, however, disadvantages, which should be realized in order to draw the correct consclusions: The mapping we derive in the above way is not found directly from the differential equations, even by an approximation method. In fact, this mapping represents, in a sense, a different problem. But an approximate solution of the actual system is not a different problem? This mapping can be used to study the basic properties of the actual system in a clear way because it is simple. The actual system is much more complicated because it contains all the details which are essential for an accurate description of the motion. These details however are not essential for the generic properties of the system.

4. FAMILIES OF PERIODIC ORBITS - GENERATION OF INSTABILITY

In the previous sections we used action-angle variables to represent resonant and nonresonant motion. However, at the origin $J_1 = 0$, which is the central periodic orbit, these variables present a singularity because θ_1 is not defined. For this reason we shall use a new set of canonical variables, defined by

$$x_1 = \sqrt{2J_1}\sin\theta_1 \ , \ p_1 = \sqrt{2J_1}\cos\theta_1 \ , \ x_2 = \sqrt{2J_2}\sin\theta_2, \ p_2 = \sqrt{2J_2}\cos\theta_2. \tag{12}$$

The canonical equations (5) are then transformed to

$$\begin{aligned} \dot{x}_1 &= n_1 p_1 \ , \quad & \dot{x}_2 &= n_2 p_2 \ , \\ \dot{p}_1 &= -n_1 x_1 \ , \quad & \dot{p}_2 &= -n_2 x_2 \ , \end{aligned} \tag{13}$$

where the frequencies n_1, n_2 are functions of $J_i = (x_i^2 + p_i^2)/2$, i=1,2. Although these four equations are coupled through n_1, n_2, the motion in the $x_1 x_2$ plane is two harmonic oscillations in the x_1 axis and the x_2 axis, respectively, with frequencies n_1, n_2 which depend on the initial conditions (because $x_i^2 + p_i^2$ = constant in (13), i=1,2,).

The central periodic orbit $J_1 = 0$ is, in the variables (12),

$$x_1 = 0, \quad p_1 = 0, \quad x_2 = A \sin n_2 t \;, \quad p_2 = A \cos n_2 t \tag{14}$$

i.e. an oscillation along the x_2 axis with frequency n_2, which depends on the initial conditions. This solution belongs to a monoparametric family of periodic orbits along which the energy varies. For arbitrary initial conditions we have oscillations both along the x_1 and x_2 axis, with frequencies n_1, n_2, respectively. For $n_1/n_2 = r/s$ the motion is a resonant periodic orbit.

Concerning the stability of the central periodic orbit (14), we find, by linear analysis, that the eigenvalues are

$$\lambda_1 = \lambda_2 = 1 \quad , \quad \lambda_{3,4} = e^{\pm i\phi} \quad , \quad \phi = 2\pi n_1/n_2 \tag{15}$$

The surface of section (6) in action-angle variables corresponds, in the variables (12), to the surface of section $H_0 = h$, $x_2 = 0$. The geometric meaning is that we take the consecutive intersections of the orbit with the x_1 axis, so that we have a mapping in the plane $x_1 p_1$. The invariant curves in this plane are the circles $x_1^2 + p_1^2$ = constant, as we find from (13), and the periodic orbit (14) is the central point $x_1 = p_1 = 0$ in the $x_1 p_1$ plane.

All the central fixed points, for any value of the energy, are evidently stable in the unperturbed case and we want to know how the stability evolves when the perturbation is applied. We note that for a fixed value of the energy, the angle ϕ has a certain value and as h varies, the eigenvalues λ_3, λ_4 move on the unit circle, in the unperturbed case. If $\phi \neq 0, \pi$ the perturbation cannot drive λ_3, λ_4 outside the unit circle (because of the properties: $\lambda_3 \lambda_9 = 1$, $\lambda_3^* = \lambda_4$). Also, the pair $\lambda_1 = \lambda_2 = 1$ survives the perturbation due to the existence of the energy integral. Consequently, instability cannot be generated when $\phi = 0, \pi$. If however $\phi \to 0, \pi$, the eigenvalues λ_3, λ_4 meet at the point ± 1 and the perturbation generates instability, driving them outside the unit circle, as shown in Fig.2

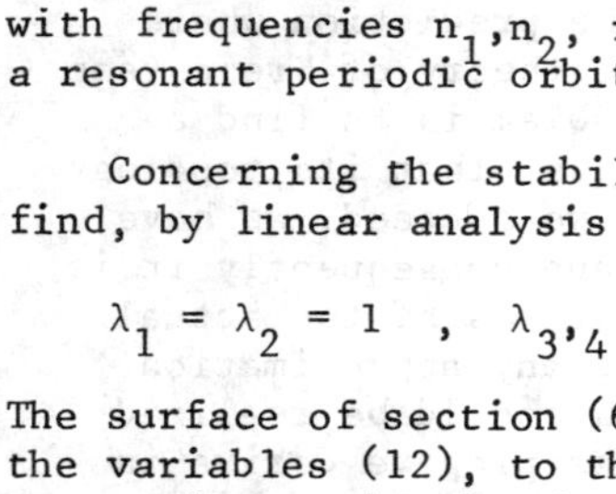

Fig.2: Generation of instability at -1

(Hadjidemetriou, 1985). This destabilization of the central fixed point takes place only when

$$n_1/n_2 = k/2 \quad , \quad k = 1,2,\ldots\ldots, \tag{16}$$

because only at these particular resonances the eigenvalues λ_3, λ_4 meet on the unit circle at +1 or -1. Examples of generation of instability by this mechanism will be given in the following, for asteroid orbits.

5. ADIABATIC INVARIANTS IN A MAPPING

The mapping (7) and the corresponding perturbed mapping depend on the energy constant h. We wish now to study the evolution of the system as the energy varies slowly. There are several physical models in the Solar System which imply a slow variation of a parameter (Torbett and Smoluchovski, 1982; Gonzi et al., 1982).

Let us consider the area preserving mapping

$$x_{n+1} = F_1(x_n,y_n;h) \quad , \quad y_{n+1} = F_2(x_n,y_n;h) \; , \tag{17}$$

depending on a parameter h. We assume that for an interval of the parameter h and in a certain region in the xy plane the invariant curves are closed and the consecutive points of the mapping are close to each other. In this case the mapping (17) can be expressed in the form

$$x_{n+1} = x_n + \delta f(x_n,y_n) \quad , \quad y_{n+1} = y_n + \delta g(x_n,y_n) \; , \tag{18}$$

where δ is sufficiently small. This mapping can be considered to be generated by a canonical transformation derived from the generating function

$$F_2 = x_{n+1}\, y_n + \delta G(x_{n+1},y_n) \; , \tag{19}$$

by the relations $x_{n+1} = \partial F_2/\partial y_n$, $y_{n+1} = \partial F_2/\partial x_{n+1}$, i.e.

$$x_{n+1} = x_n - \delta(\partial G/\partial y_n),$$
$$y_{n+1} = y_n + \delta(\partial G/\partial x_{n+1}).$$

Linearizing in δ we find the mapping

$$\begin{aligned} x_{n+1} &= x_n - \delta(\partial G(x_n,y_n)/\partial y_n), \\ y_{n+1} &= y_n - \delta(\partial G(x_n,y_n)/\partial x_n), \end{aligned} \tag{20}$$

which is area preserving to a linear approximation in δ.

Consider now the Hamiltonian $H = G(x,y)$, where G is the same function as that in (19), corresponding to a fictitious system with one degree of freedom. The corresponding canonical equations are $dy/dt = \partial G/\partial x$, $dx/dt = -\partial G/\partial y$, from which we find, for $\tau << 1$,

$$x' = x-\tau(\partial G/\partial y) \quad , \quad y' = y+\tau(\partial G/\partial x). \tag{21}$$

The relations (21) define a mapping in the xy plane, which maps the point x(t), y(t) to the point $x(t+\tau)$, $y(t+\tau)$. The time interval τ is considered small, of the order of δ in (20). Comparing (20) and (21) we see that there is a correspondence between the mapping (20) and the Hamiltonian system (21).

Let us assume now that the function G depends on a parameter h which varies slowly and that the trajectories of the Hamiltonian system $H = G(x,y:h)$ are closed in a certain region. Then the theory of adiabatic invariants applies for the evolution of the Hamiltonian system (Landau and Lifshitz, 1973; Henrard 1982, 1985a,b, 1987). The trajectories in the xy plane are no longer closed but for a short time interval the system follows closely a closed trajectory, the guiding trajectory, corresponding to a fixed value of h. After a long time interval the same is true, but now the guiding trajectory is different from the old one. The theory of adiabatic invariants states that the area enclosed by a closed trajectory is invariant. Since now the mapping (20) is similar to the mapping (21), the same behaviour is expected for the mapping (20) as well. Thus, finally, we come to the conclusion that the theory of adiabatic invariants can be extended to area-preserving mappings with a slowly varying parameter, provided the assumptions made for the mapping are satisfied. We shall see in the following that this is indeed the case for the mapping describing asteroid motion.

6. ASTEROID ORBITS IN THE PLANE - UNPERTURBED MOTION

6.1 The Hamiltonian in rotating coordinates

We shall apply the theory developed in the previous sections to the study of asteroid orbits. The unperturbed orbit will be in this case a Keplerian orbit around the Sun and the perturbation will be the gravitational attraction from Jupiter. The orbit of Jupiter will be considered as circular and the asteroids will revolve around the Sun in the same direction as Jupiter. The motion will be studied in a rotating frame xOy whose x-axis is the line Sun-Jupiter. This is the model of the well-known restricted circular three body problem.

The Hamiltonian, in polar coordinates, is

$$H = H_0 + \mu H_1 \,{}_{2} , \tag{22}$$

where

$$H_0 = \frac{1}{2}\left(p_r^2 + \frac{p_\phi}{r}\right) - \frac{GM}{r} - n' \, p_\phi \tag{23}$$

is the unperturbed part, describing Keplerian motion in the rotating frame xOy, M is the mass of the Sun, $\mu = m/(M+m)$ where m is the mass of Jupiter and n' is the angular velocity of xOy.

6.2 Action-angle variables

We shall study elliptic orbits only. The action-angle variables in H_o can be found in the standard way (Landau and Lifshitz, 1973; Berry, 1978):

$$J_1 = - p_\phi + GM/\sqrt{-2(h-n' \, p_\phi)} \quad , \quad J_2 = p_\phi , \tag{24}$$

where $h = H_0$. The Hamiltonian H_0 takes now the form

$$H_0 = - n' J_2 - (GM)^2/[2(J_1+J_2)^2] \tag{25}$$

and the two frequencies of the system are

$$n_1 = \partial H_0/\partial J_1=(GM)^2/(J_1+J_2)^3, \quad n_2 = \partial H_0/\partial J_2=(GM)^2/(J_1+J_2)^3 - n'. \tag{26}$$

A better geometrical meaning can be obtained if we express them in terms of the elements of the orbit. Using the formulas of Keplerian orbit (Roy, 1982) we find

$$\begin{aligned} J_1 &= \sqrt{GMa}\,(1 - \sqrt{1-e^2}) \\ J_2 &= \sqrt{GMa}\,(1-e^2) \\ n_1 &= n \\ n_2 &= n-n' \end{aligned} \tag{27}$$

where a is the smeimajor axis, e the eccentricity and n the mean motion of the asteroid. Thus, we have for the angles

$$\theta_1 = n_1 t + \theta_{10} \quad , \quad \theta_2 = (n-n')t + \theta_{20} \ . \tag{28}$$

The Hamiltonian (25) is also expressed in the form

$$H_0 = - \frac{GM}{2a} - n' \sqrt{GMa} + n' J_1. \tag{29}$$

6.3 The unperturbed mapping

Working as explained in 2.2 and using (27), we find the unperturbed mapping

$$J_1 \to J_1 , \quad \theta_1 \to \theta_1 + 2\pi/(1-n'/n) \tag{30}$$

on the surface of section h = constant, $\theta_2 = 0$. The mapping (30) can be also expressed in the form

$$J_1 \to J_1 , \quad \theta_1 \to \theta_1 + 2\pi/(1-(a/a')^{3/2}), \tag{31}$$

where a' is the semimajor axis of Jupiter. If now we use the normalized GM=1, n'=1, a'=1, we find for the energy integral

$$- \frac{1}{2a} - \sqrt{a} + J_1 = h \tag{32}$$

and for the mapping (30)

$$J_1 \to J_1 \quad , \quad \theta_1 \to \theta_1 + f(J_1; h), \tag{33}$$

where the rotation angle f is given by

$$f(J_1;h) = 2\pi/(1-a^{3/2}) \tag{34}$$

and is a function of J_1 and h defined implicitly from (32).

The invariant curves of (45) are the circles J_1 = constant. We obtain now from the first of (27), for the eccentricity,

$$e = \sqrt{1-(1 - j_1/\sqrt{a})^2} . \qquad (35)$$

From this formula it is clear that the central fixed point $J_1 = 0$ corresponds to a circular orbit (e=0) and all the invariant curves J_1 = const. correspond to elliptic orbits with the same semimajor axis and different orientation. We can also find that on a given surface of section corresponding to a fixed energy h, the value of the eccentricity increases as J_1 increases. Also the ratio n/n' is minimum at the origin (depending on h) and increases as J_1 increases.

7. ASTEROID ORBITS NEAR THE 3:1 RESONANCE FOR $\mu>0$

From the last two equations (27) we find that for n/n' = 3/1, we have $n_1/n_2 = 3/2$. Consequently, from (15), we obtain $\lambda_3 = \lambda_4 = -1$. Thus, instability is generated when $\mu>0$ by the mechanism of Fig.2. This means that on the family of circular orbits there exists a small unstable region AB (Fig.3). At the end points of this region we have a bifurcation of two families, respectively, of elliptic resonant periodic orbits at the 3:1 resonance. One of them is stable and the other unstable (Hadjidemetriou and Ichtiaroglou, 1984). In the standard normalization of the restricted problem the energies at the points A,B are $h_A = -1.7343$, $h_B = -1.7332$ and the maximum unstable eigenvalue is $\lambda = -1.0054$, for $\mu = .001$. From Fig.3 we can easily find that the structure of phase space at the energy levels h_1, h_2, h_3 are those shown in Fig.4. The algebraic mapping that we want to construct must have the topology of the figures of Fig.4 at the corresponding energy levels.

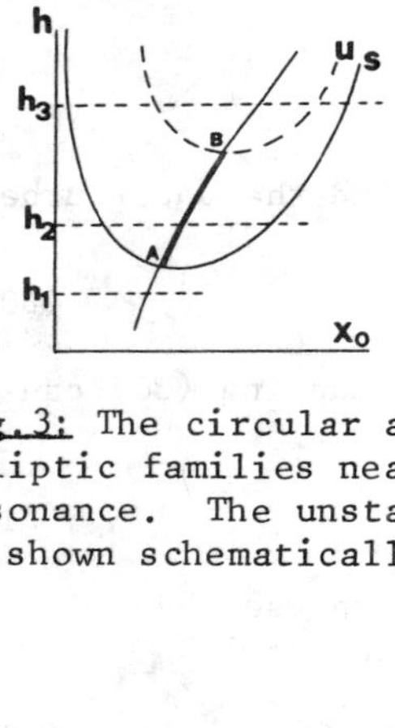

Fig.3: The circular and the elliptic families near the 3:1 resonance. The unstable family is shown schematically only.

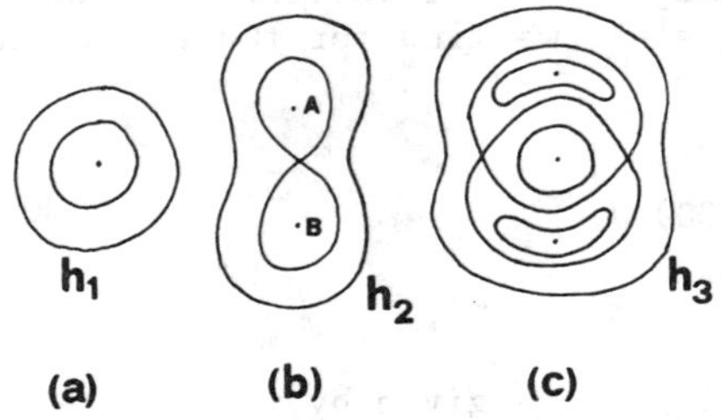

Fig.4: Surfaces of section at the energy levels shown in Fig.3.

8. THE PERTURBED MAPPING

8.1 The mapping in the unstable region $h_A \leqslant h \leqslant h_B$

Let us consider a surface of section for an energy level in the above unstable region, where the central fixed point is unstable (Fig. 4b). Let $\lambda, 1/\lambda$ be the two eigenvalues different from unity. The instability is due to the effect of Jupiter and this results in a hyperbolic mapping at the origin, in addition to the twist mapping (33) of the unperturbed problem. Consequently the perturbed mapping will be the product of these two mappings, i.e. it will be a hyperbolic twist mapping.

In order to find an explicit form of such a mapping we change first to new coordinates $x = J_1 \cos \theta_1$, $y = J_1 \sin \theta_1$, and the mapping (33) in these variables takes the form $x \to x \cos f - y \sin f$, $y \to x \sin f + y \cos f$. We perturb now this mapping to obtain the hyperbolic twist mapping

$$\begin{aligned} x &\to (x \cos f - y \sin f)s \\ y &\to (x \sin f + y \cos f)s^{-1} \end{aligned} \qquad (36)$$

It can be verified that this mapping is area-preserving in the xy plane. The value of s should be such that the central fixed point x=y=0 has the correct eigenvalue λ for every energy level h. A linear stability analysis gives the relation between s and λ:

$$S = (D \pm \sqrt{D^2 - 4})/2 \quad , \quad D = (1+\lambda^2)/(\lambda \cos f(0)). \qquad (37)$$

Note that $s=\lambda$ only when $f(0)=0$. Using numerical data for λ as a function of s and (37), we find s as a function of h:

$$s = -3817.86496\, h^2 - 13233.7820\, h - 11466.9854 \,, \qquad (38)$$

valid in the range $h_A \leqslant h \leqslant h_B$. The value of s varies between 1.0070 and 1.0019.

The invariant curves of (36) coincide qualitatively with Fig.4b. In order to have quantitative coincidence we must adjust it in such a way that the double fixed point A,B in Fig.4b has the correct eccentricity (the resonance is 3:1 in all of them) and the central fixed point the correct rotation angle f(0), for each h. We obtain finally the mapping (36) where the rotation angle f is given, as a function of h, by

$$f = \frac{2\pi}{1 - (a + Da)^{3/2}} \qquad (39)$$

where

$$Da = 158.6355084h^2 + 550.0260665h + 476.7673974, \qquad (40)$$

$$a = A_1(h_1)J_1^2 + A_2(h_1)J_1 + A_3(h_1) \tag{41}$$

with

$$\begin{aligned} h_1 &= 1.006893937h + 0.01177506 , \\ A_1 &= 4.2285h_1 + 8.3108 , \\ A_2 &= -2.50411h_1 - 5.02165 , \\ A_3 &= .689189h_1 + 1.675418. \end{aligned} \tag{42}$$

Thus, finally, we have the mapping (36), with the coefficients given by (39) - (42), which has all the basic properties of the actual system, both qualitatively and quantitatively. This is a hyperbolic twist mapping which depends on the parameter h, which is the energy of the original system.

Invariant curves obtained by the above mapping are given in the Figures 5a-b. No chaotic behaviour near the unstable fixed points was detectable (this is also the case with the actual computations). Note that in both these figures the eigenvalue of the central fixed point is $\lambda = -1.0030$ but the structure of phase space is quantitatively quite different because the rotation angle is different.

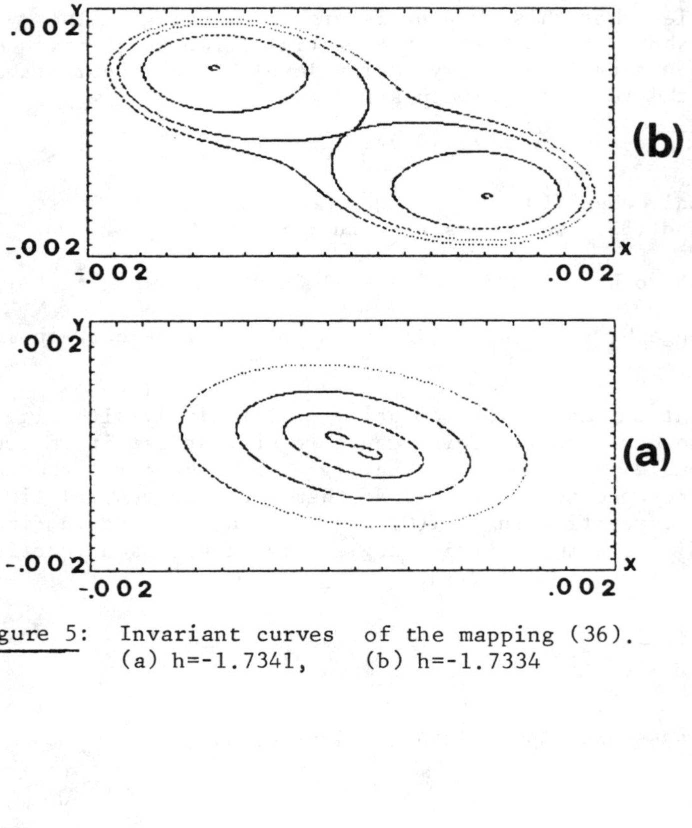

Figure 5: Invariant curves of the mapping (36).
(a) h=-1.7341, (b) h=-1.7334

8.2 Mapping in the region $h_B < h$

In this case the surface of section must have the structure of Fig.4c. The central fixed point is stable, with a ratio of frequencies n/n' smaller than 3/1, but two pairs of resonant fixed points appear, corresponding to a resonant elliptic orbit with a resonance very close to 3/1. One of these pairs is stable and the other is unstable.

We must perturb now the mapping (33) in such a way that we have the structure of Fig.4c. The technique applied in the previous case $h_A < h < h_B$ cannot be used here. In the present case we work as follows: we perturb the mapping (33),

$$J_{n+1} = J_n \; , \qquad \theta_{n+1} = \theta_n + f(J_n), \tag{43}$$

by making use of the generating function

$$F_2 = J_{n+1}\,\theta_n + R(J_{n+1}) + \varepsilon J_{n+1} \cos\left(\theta_n + \frac{\pi}{2}\right), \tag{44}$$

where $dR/dJ = f(J)$. By making use of the transformation equations

$$J_n = \partial F_2/\partial\theta_n \; , \qquad \theta_{n+1} = \partial F_2/\partial J_{n+1},$$

we obtain the area-preserving mapping

$$\begin{aligned} J_{n+1} &= J_n/(1 + \varepsilon \sin(\theta_n + \pi/2)), \\ \theta_{n+1} &= \theta_n + f(J_{n+1}) - \varepsilon \cos(\theta_n + \pi/2), \end{aligned} \tag{45}$$

where $f(J) = 2\pi/(1-a^{3/2})$, and a is a function of J and h, defined implicitly from (32). The invariant curves of the mapping (45) coincide, qualitatively, with Fig.4c. In order to have quantitative agreement with the numerical computations of the restricted problem, for $\mu = .001$, we must express ε as a function of h and also the semimajor axis a as a function of J and h. The criterion is to have the same set of resonant fixed points (i.e. the same value for the semimajor axis and the eccentricity) and the same rotation angle at the central fixed point. We obtain finally

$$\begin{aligned} a &= A_1 J^2 + A_2\, J + A_3, \\ A_1 &= 159.5451465h_1^2 + 556.908789h_1 + 486.9415535 \\ A_2 &= -18.12078322\,h_1^2 - 65.32192609h_1 - 59.46231465 \\ A_3 &= 1.502854955\,h_1^2 + 5.902659267h_1 + 6.196820767 \\ h_1 &= -.058239674h^2 + .803410937h - .165899836 \\ \varepsilon &= -6.493975034h - 11.05553936 \end{aligned} \tag{46}$$

The value of ε varies between $\varepsilon = .062$ for $h = -1.712$ and $\varepsilon = .20$ for $h = -1.7332$. The relations (46) are valid for the energy range $-1.7332 < h < -1.712$ and for $J < .01$.

Invariant curves of the mapping (45) are shown in Figs.6a,b. In

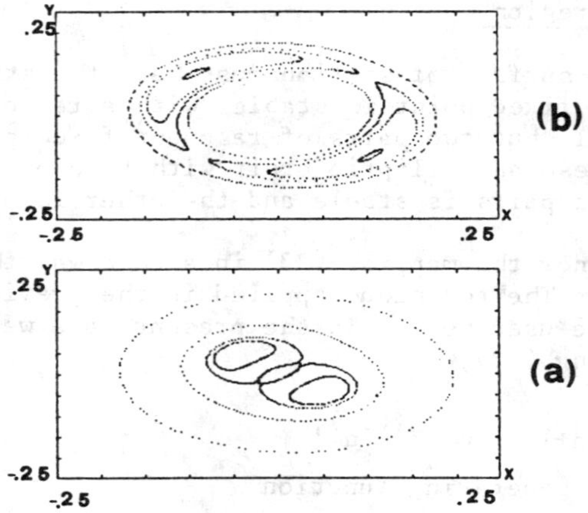

Figure 6: Invariant curves of the mapping (45) in the variables $x = \sqrt{2J_1} \cos \theta_1$, $y=\sqrt{2J_1}\sin \theta_1$. (a) h=-1.7318, (b) h=-1.712.

these figures the axes are $x=\sqrt{2J_1}\cos\theta_1$ and $y=\sqrt{2J_1}\sin\theta_1$, so that we have closed invariant curves. The mapping in the xy plane is also area-preserving. A small chaotic region appeared near the unstable fixed point at the energy h=-1.712. This corresponds to an elliptic orbit at n/n' = 3.004 with eccentricity e=.25. This region is shown in Fig.7. All points correspond to the same initial conditions and they seem to jump from one invariant curve to the other in a random way.

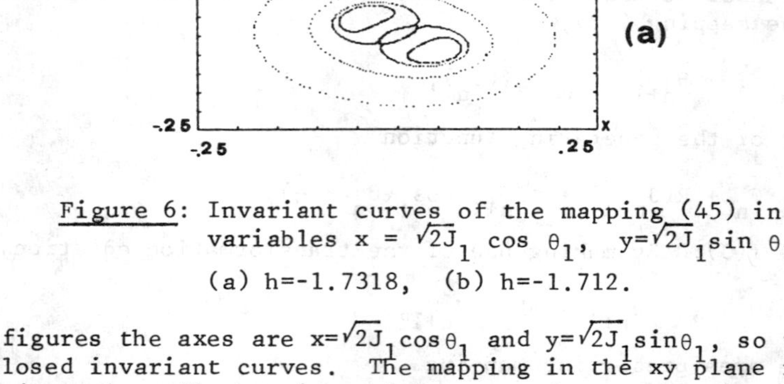

Figure 7: Detail of Figure 6(b) near the unstable fixed point. All points belong to the same initial conditions.

The region on the surface of section enclosed by the maximum invariant curve around the double stable fixed point will be called the "resonance zone".

Each point on a surface of section, both for the mapping (36) and for the mapping (45), corresponds to a set of values J_1, θ_1 which together with $\theta_2 = 0$ and the value of h determine uniquely the asteroid orbit. At this point we can compute the values of the semimajor axis and the eccentricity. These are the osculating elements of the asteroid orbit and they vary along the invariant curve (they are constant in the unperturbed mapping (30)).

9. SOME APPLICATIONS OF THE MAPPINGS (36) and (45)

We shall study in what follows the evolution of the system by considering a slow decrease of the energy. It is not the purpose of this paper to consider specific physical models, but to show the general properties of the mapping as h varies and study its behaviour for different modes of energy dissipation.

9.1 Evolution due to energy decrease only

If the effect of Jupiter is neglected, we have the unperturbed mapping (33). It is evident that as h varies, the radius J_1 remains constant. This means that for different values of h and a fixed J_1 the semimajor axis a varies, as we can see from (32) and consequently the ratio n/n' varies. This evolution is shown in Fig.8 for a starting energy h=-1.712. Each curve corresponds to an initial value of n/n', and also an initial value of the eccentricity. The computations have shown that the eccentricity remains almost constant as the energy decreases, while the value n/n' increases.

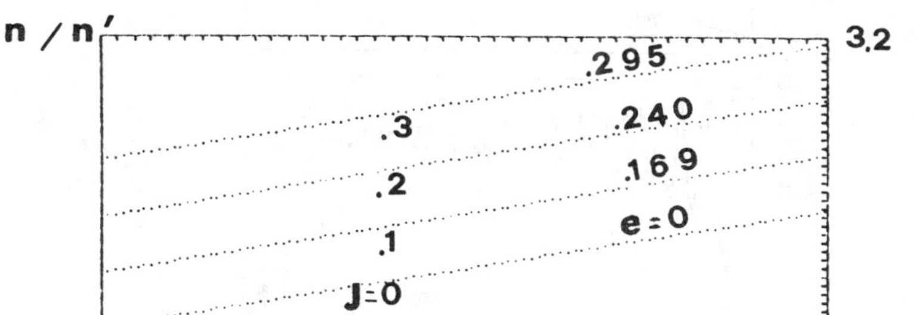

Figure 8. The evolution of n/n' for different initial conditions as the energy decreases, for $\varepsilon=0$ in (45), in the range $-1.7332<h<-1.712$. The eccentricity remains almost constant.

From Fig.8 we can deduce that a uniform initial distribution of asteroids will always remain uniform as the energy decreases.

9.2 Study of the mapping (36)

We take now into account the effect of Jupiter and consider the mapping (36) which is valid for the unstable range $h_A<h<h_B$ of Fig.3. The invariant curves of this mapping for a fixed value of h, are shown in Fig.5. The evolution of n/n' and the eccentricity as the energy h varies is very different compared to the unperturbed case, for small values of J_1, i.e. near the central fixed point. A typical evolution of an asteroid orbit starting with $J_1 \simeq 0$ (i.e. $e \simeq 0$) is shown in Figs.9a-c). We see that a jump takes place both in the ratio n/n' and in the eccentricity. This means that no asteroid orbit can be trapped near a circular resonant 3/1 orbit as it passes through the unstable region (h_A, h_B). In this way a gap is expected to appear near the 3/1 resonance, for small values of e.

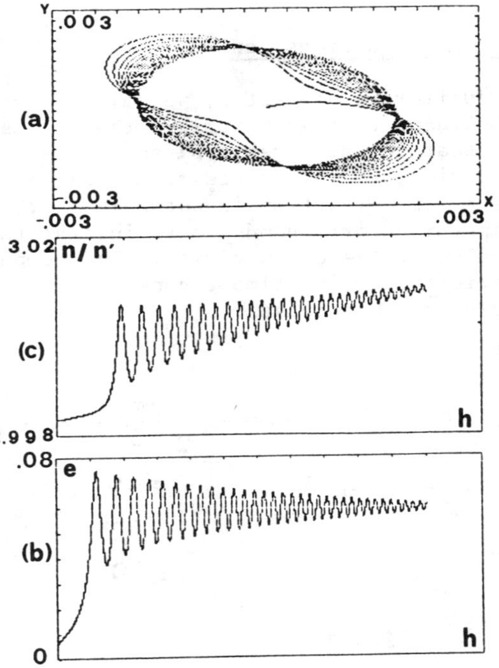

Figure 9: The mapping (36) as the energy decreases from h=-1.7332 to h=-1.7343. (a) evolution in the x-y plane, (b) evolution of n/n', (c) evolution of the eccentricity, $J_o=10^{-5}$, $\theta_o=0$.

For larger values of J_1 (and of the eccentricity) the behaviour of the perturbed mapping (36) is similar to the behaviour of the unperturbed mappings (33), because in this region of phase space the topology is the same for both mappings.

9.3 Study of the mapping (45)

We consider next asteroid orbits in the energy range $h_B<h<-1.712$ (Fig.3), perturbed by Jupiter. Their behaviour is described by the mapping (45). We shall study the evolution as the energy decreases, starting with the energy h=-1.712. We distinguish three different types of initial conditions: Outside the resonance zone, on the resonance zone and inside the resonance zone. A crucial point which determines the evolution of an orbit is the property of the mapping that as the energy decreases, the resonance zone of an energy level moves towards the origin $J_1=0$, as shown in Fig.5. This fact, together with the adiabatic invariance property described in section 3 will determine the evolution.

(a) Initial conditions outside the resonance zone: A typical evolution is shown in Figs.10a,b. Note that the invariant curves change slightly only because the enclosed area must remain constant. As a consequence, the eccentricity remains almost constant and the ratio n/n' increases slowly. This behaviour is similar to the unperturbed case (Fig.8), because at this energy range the topology of the phase space of the perturbed system is similar to that of the unperturbed problem.

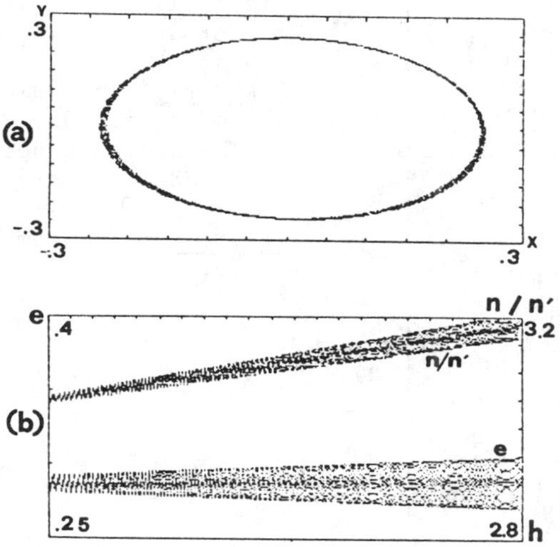

Figure 10: The same as in Fig.9 for the mapping (45) from h=-1.712 to h=-1.7332. J_o=.03, θ_o =-45°.

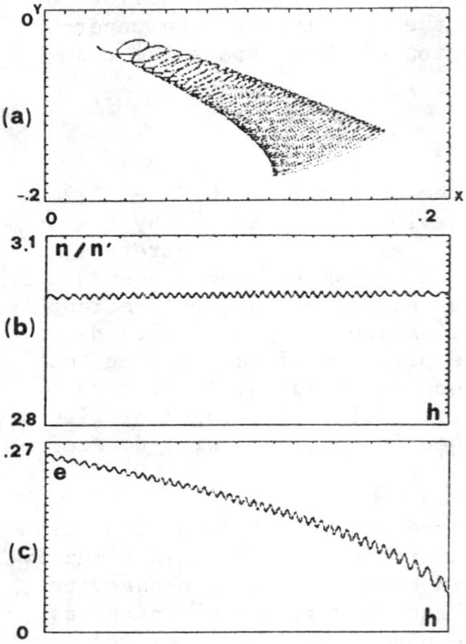

Figure 11

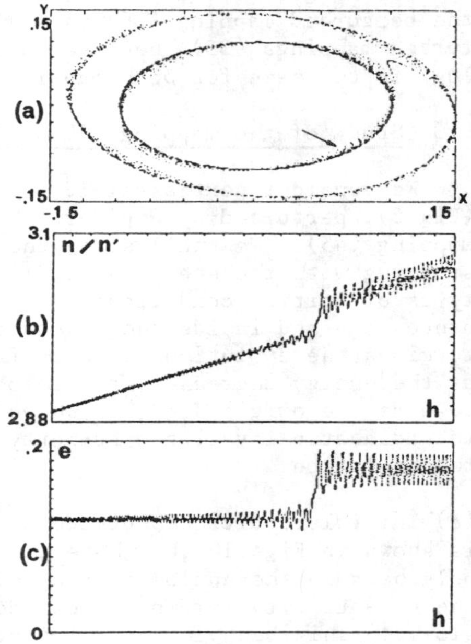

Figure 12

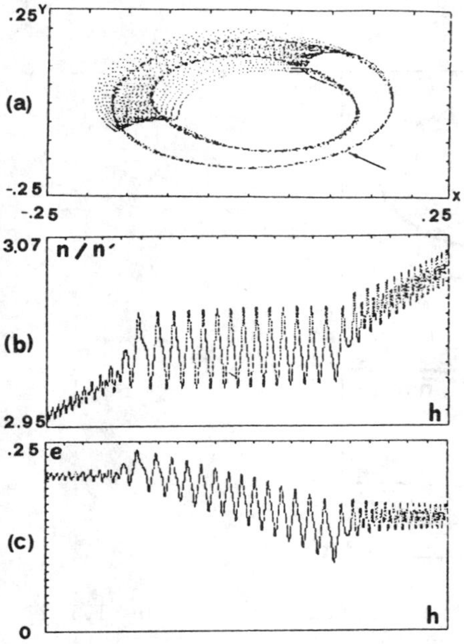

Figure 13

<u>Figure 11</u>: The same as in Fig.10 for J_o = .0216, θ_o = -45°.

<u>Figure 12</u>: The same as in Fig.10 for J_o = .005, θ_o = -45°. The starting point is indicated by an arrow.

<u>Figure 13</u>: The same as in Fig.10 for J_o = .015, θ_o = -45°.

(b) Initial conditions on the resonance zone: Due to the adiabatic invariance property, the orbit is trapped at the resonance zone and moves towards the origin as the energy decreases. If the starting point is close to the double fixed point, the orbit evolves up to $J_1 \simeq 0$, i.e. $e \simeq 0$. A typical example is given in Figs.11a-c. Note that the mean value of n/n' remains constant during the trapping at resonance.

(c) Initial conditions inside the resonance zone: The evolution of the asteroid orbit is at first similar to the unperturbed case as h decreases, i.e. it remains very close to the invariant curve of the starting point, due to the adiabatic invariance property. Since however the resonance zone moves towards the origin as h decreases, the invariant curve will touch eventually the inner boundary of the resonance zone. The evolution from this point on may be quite different, depending sensitively on the initial conditions. We may either have a jump from the inner zone to the outer zone, as shown in Fig.12, or a (temporary) trapping at the resonance zone and follow the evolution of Fig. 13. The choice between these two evolution types is a very sensitive phenomenon as is related to the chaotic region of Fig.7. We found that a small change of the initial conditions or of the rate of change of the energy, changes the evolution from that of Fig.12 to that of Fig.13. Thus we see that the small chaotic region near the unstable double fixed point at the 3:1 resonance plays an important role in the evolution of the system, in the nonconservative model. The effect of this chaotic region is negligible when h=constant, but is greatly magnified if the energy changes slowly. The more likely evolution however is that of Fig.12, since the region of phase space that corresponds to initial conditions giving rise to evolution of Fig.13 is quite small.

From all the above we see that the evolution of the system is quite complex if we pass through a resonance region as the energy decreases, compared to the smooth evolution we would have if the effect of Jupiter were neglected. This is due to the fact that the perturbation of Jupiter changes the topology of phase space. This change is not the same everywhere, but only at the resonance regions, and in particular the low order resonances. This is why the resonances play an important role in the evolution of a system. As a result of the above, a uniform initial distribution of asteroids will not remain uniform if a slow decrease of the energy takes place, but gaps are expected to appear near resonances.

The numerical computations presented in Figs.9-14 were for an energy dissipation equal to 10^{-7} normalized units of energy per iteration. It was found however that the evolution is very insensitive to this rate of decrease of energy.

REFERENCES

Arnold, V.I. and Avez, A.: 1968, "Ergodic Problems of Classical Mechanics", ch.4, Benjamin Press.
Berry, M.V.: 1978, in "Topics in Nonlinear Dynamics", S. Jorna (ed.), American Institute of Physics.
Gonzi, R., Froeschlé, Ch. and Froeschlé, Cl.: 1982, Icarus 51, 633.
Ferraz-Mello, S.: 1987, Astron. J.
Hadjidemetriou, J.D.: 1982, Celes. Mech. 27, 305.
Hadjidemetriou, J.D.: 1984, Celes. Mech. 34, 379.
Hadjidemetriou, J.D.: 1985a, in "Resonances in the Motion of Planets, Satellites and Asteroids", S.Ferraz-Mello and W. Sessin (eds.), 1-18, Univ. of São Paulo.
Hadjidemetriou, J.D.: 1985b, in "Stability of the Solar System", V. Szebehely (ed.), D. Reidel Publ. Co.
Hadjidemetriou, J.D.: 1986, JAMP 37, 776.
Hadjidemetriou, J.D. and Ichtiaroglou, S.: 1984, Astron. Astrophys. 131, 20.
Henrard, J.: 1982, Celes. Mech. 27, 3.
Henrard, J.: 1985a, in "Stability of the Solar System", V. Szebehely (ed.), D. Reidel Publ. Co., p.183.
Henrard, J.: 1985b, in "Resonances in the Motion of Planets, Satellites and Asteroids", S. Ferraz-Mello and W. Sessin (eds.), Univ. of São Paulo, p.19.
Henrard, J. and Lemaitre, A.: 1983, Icarus 55, 482.
Henrard, J. and Lemaitre, A.: 1986, Icarus 69, 266.
Henrard, J. and Lemaitre, A.: 1987, Celes. Mech. 39, 213.
Henrard, J. 1987, "The Adiabatic Invariant in Classical Mechanics", in"Dynamics Reported".
Landau, L. and Lifshitz, E.: 1973, Mechanics (3rd ed.), Pergamon Press.
Lemaitre, A.: 1984, Celes. Mech., 32, 109.
Lichtenberg, A.J. and Lieberman, M.A.: 1983, Springer-Verlag.
Murray, C.D.: 1986, Icarus 65, 70.
Roy, A.: 1982, "Orbital Motion" (2nd ed.), Adam Hilger.
Schubart, J.: 1984, Smithsonian Astrophysical Laboratory Report No.149.
Szebehely, V.: 1963, "Theory of Orbits", Academic Press.
Torbett, M. and Smoluchowski, R.: 1982, Astron. Astrophys. 111, 43.
Torbett, M. and Smoluchowski, R.: 1983, Astron. Astrophys. 127, 345.
Wisdom, J.: 1982, Astron. J. 87, 577.
Wisdom, J.: 1983, Icarus 56, 51.
Wisdom, J.: 1985, Icarus 63, 292
Yakubovich, V. and Starzhinskii, V.M.: 1975, "Linear Differential Equations with Periodic Coefficients", Vol.1, Halsted Press.

THE HIERARCHICAL RESTRICTED PROBLEM WITH APPLICATIONS

J. Pojman** and V. Szebehely***
** Research Assistant, University of Texas, Austin, TX 78712
*** Professor, R.B. Curran Chair, University of Texas, Austin, TX 78712.

ABSTRACT: The hierarchical system of restricted problems is described as the generalization of the classical restricted problem of three bodies. This is followed by a special example consisting of two interacting asteroids as influenced by the gravitational effects of the Sun and Jupiter. This binary asteroid problem is first represented in the synodic Sun-Jupiter system as a two-plus-two-body restricted problem and the existence of a generalized Jacobian integral is shown. After this a simplified and approximate system is introduced where closed surfaces of zero velocity limit the possible range of motion of the asteroids and under certain conditions allow the analytical proof of stability of the system. This stability is equivalent to establishing analytically the possibility of existence of binary asteroids and supports the presently questionable observational findings. In addition, the analytical stability investigations describe the possible behaviour of binary asteroids and assist the observers to locate them.

1. INTRODUCTION

This paper is introduced by an apology offered to those esteemed colleagues who prefer to restrict the concept of the restricted problem to its original and classical form. This form is known as the circular restricted problem of three bodies and represents one of the brilliant simplifying ideas first introduced by Euler in connection with his lunar theory (1772). A considerable variety of problems are often referred to in modern literature as modifications of the classical restricted problem such as Hill's problem, the asteroidal or astronomical restricted problem (also known as the elliptic or pseudo-restricted problem), the two fixed center problem, Lagrange's problem, the variable mass problem, etc. It is a matter of taste to derive Hill's problem starting either with the general problem of three bodies or with the classical restricted problem of three bodies. In the first case one can consider Hill's problem as a special case of the general three-body problem and using the second approach it becomes a special case of the classical restricted three-body problem (Hill 1886, Kerorkian 1962). Here the apology ends with emphasizing that termino-

A. E. Roy (ed.), Long-Term Dynamical Behaviour of Natural and Artificial N-Body Systems, 277–288.
© 1988 by Kluwer Academic Publishers.

logies do not make science but science does make terminologies.

In the following the concepts and ideas related to our generalizations of the classical restricted problem will be introduced and our terminologies will be defined.

2. THE HIERARCHICAL RESTRICTED PROBLEM

The number of participating bodies in the classical restricted problem is three. The masses of two of these bodies (often called the primaries) are sufficiently larger than the mass of the third body so that the third body has no gravitational influence on the motion of the primaries. When the question of the existence of binary asteroids was presented, the classical concept of the restricted problem had to be generalized to the motion of four bodies, two with small and two with large masses. The members of the binary plus Sun and Jupiter were the participating bodies. In the problem of binary asteroids, the two small bodies attract each other gravitationally and are attracted by the primaries. If the relative motion of a space station and a space probe is investigated, as it is often done in space mission analysis, the mutual attraction of the small masses might be neglected.

The first step in the generalization, therefore, consists of the introduction of several small bodies which might gravitationally influence each other or might not, and are always influenced by the primaries, but do not effect the motion of the primaries.

The further generalization of this model by the introduction of more than two primaries is the next significant step. As long as the number of primaries is two, their motion can be treated analytically. Furthermore, if the motion of the primaries is circular the existence of the Jacobian integral (1836) can be shown even if the number of the participating small bodies is larger than one. If the motions of the primaries are elliptic, hyperbolic or parabolic, or if their number is larger than two, the Jacobian integral does not exist any more. Note that neither the classical restricted problem nor the restricted two-plus-two body problem is integrable (Poincaré, 1982), but the existence of the Jacobian integral offers surfaces of zero velocity which allows the drawing of significant qualitative conclusions. With these remarks the first step of our process of generalization is concluded by raising the number of small bodies from one to ν and the number of the primaries from two to n, $\nu \geq 1$ and $n \geq 2$.

The next step in the process of generalization is to form a hierarchy based on the masses of the participating bodies and on their gravitational interactions. Let us add to the previous system of n primaries and ν small bodies another system of N bodies with masses sufficiently large, so that the previous primaries do not influence their motion. The new super-primaries will influence each other as well as the motions of the n primaries and of the ν small bodies. The hierarchy, of course, might also be built by adding $\bar{\nu}$ "very small" bodies (forming a system of "very small" masses) to the n large and ν small bodies. The n large interacting masses will influence the ν small bodies and the $\bar{\nu}$ very small bodies. The v small bodies will influence the motion of the $\bar{\nu}$ very small bodies but will not influence the motion of the n large bodies. The $\bar{\nu}$ very small bodies will interact with each other but will not influence the motion of the systems of n bodies and of ν bodies.

In general, therefore, we might have several categories formed by bodies influencing each other and affecting the motion of bodies belonging to other groups or affected by bodies belonging to still other groups. In this way a hierarchy of restricted problems can be established. It should be noted that the establishment of a category is not trivial and it always represents an approximation. Even in the relatively simple problem when only gravitational forces are considered, distances in addition to masses might enter the process of establishing a meaningful hierarchy of the participating bodies.

The analytical details, derivatives and evaluations of the validity of the restricted idea are presented in earlier papers (Szebehely). In what follows previous results concerning the restricted problem of 2+2 bodies will be summarized and applications to the problem of existence of binary asteroids will be presented.

3. FORMULATION OF THE RESTRICTED GRAVITATIONAL TWO PLUS TWO BODY PROBLEM.

Similarly to the classical restricted problem of three bodies a synodic system of coordinates is introduced. This system rotates around the Z axis with the primaries fixed on the X axis at points $P_1(\mu_2,0,0)$ and $P_2(\mu_2-1,0,0)$. Here $\mu_2 = m_2/(m_1+m_2)$ is the dimension-less mass parameter and m_1 and m_2 are the masses of the primaries. The convention is followed, according to which $m_1 \geq m_2$ and therefore $0 \leq \mu_2 \leq 1/2$. Conventional non-dimensional variables are used. These are obtained by dividing all distances by the fixed distance between the primaries. The non-dimensional time (t) is obtained from the actual time t* by $t = t^*n$, where n is the mean motion of the primaries. All non-dimensional masses are obtained by dividing the actual masses by the total mass of the primaries (m_1+m_2).

The equations of motion of the small bodies (asteroids) in the synodic system become

$$\ddot{X}_3 - 2\dot{Y}_3 = -\frac{\mu_1(X_3-\mu_2)}{R_{13}^3} - \frac{\mu_2(X_3+\mu_1)}{R_{23}^3} + \frac{\mu_4(X_4-X_3)}{r_{34}^3} + X_3 \qquad (1a)$$

$$\ddot{Y}_3 + 2\dot{X}_3 = \left[1 - \frac{\mu_1}{R_{13}^3} - \frac{\mu_2}{R_{23}^3}\right]Y_3 + \frac{\mu_4(Y_4-Y_3)}{r_{34}^3} \quad (1b)$$

$$\ddot{Z}_3 = \left[- \frac{\mu_1}{R_{13}^3} - \frac{\mu_2}{R_{23}^3}\right]Z_3 + \frac{\mu_4(Z_4-Z_3)}{r_{34}^3} \quad (1c)$$

$$\ddot{X}_4 - 2\dot{Y}_4 = - \frac{\mu_1(X_4-\mu_2)}{R_{14}^3} - \frac{\mu_2(X_4+\mu_1)}{R_{24}^3} - \frac{\mu_3(X_4-X_3)}{r_{34}^3} + X_4 \quad (2a)$$

$$\ddot{Y}_4 + 2\dot{X}_4 = \left[1 - \frac{\mu_1}{R_{14}^3} - \frac{\mu_2}{R_{24}^3}\right]Y_4 - \frac{\mu_3(Y_4-Y_3)}{r_{34}^3} \quad (2b)$$

$$\ddot{Z}_4 = \left[- \frac{\mu_1}{R_{14}^3} - \frac{\mu_2}{R_{24}^3}\right]Z_4 - \frac{\mu_3(Z_4-Z_3)}{r_{34}^3} \quad (2c)$$

where, X_3, Y_3, Z_3 and X_4, Y_4, Z_4 are the non-dimensional coordinates, and μ_3 and μ_4 are the non-dimensional masses of the small bodies,

$$\mu_i = \frac{\mu_i}{\mu_1+\mu_2} \quad ; \quad i=1,\dots 4,$$

$$R_{1i}^2 = (X_i-\mu_2)^2 + Y_i^2 + Z_i^2 \quad ; \quad i=3,4,$$

$$R_{2i}^2 = (X_2+\mu_1)^2 + Y_i^2 + Z_i^2 \quad ; \quad i=3,4,$$

$$R_{34}^2 = (X_4-X_3)^2 + (Y_4-Y_3)^2 + (Z_4-Z_3)^2$$

and dots denote derivations with respect to the non-dimensional time.

Note that the coupling terms (containing r_{34}) between the first and second set of equations represent the gravitational interaction of the small bodies. If these terms are omitted the above systems of equations describe the motions of two small bodies without interaction and can be used to study the motion of two space probes. Because of the presence of these coupling terms, the existence of the Jacobian integral is not obvious until it is realized that the right sides of the equations might be written as the gradients of a time-independent potential,

$$\Omega = \frac{\mu_3(X_3^2+Y_3^2)}{2} + \frac{\mu_4(X_4^2+Y_4^4)}{2} + \frac{\mu_1\mu_3}{R_{13}} + \frac{\mu_1\mu_4}{R_{14}} + \frac{\mu_2\mu_3}{R_{23}} + \frac{\mu_2\mu_4}{R_{24}} + \frac{\mu_3\mu_4}{r_{34}} \quad (3)$$

The generalized form of the Jacobian becomes

$$\mu_3 v_3^2 + \mu_4 v_4^2 = 2\Omega - C, \tag{4}$$

where C is the generalized Jacobian constant, and v_3, v_4 are the non-dimensional velocities of the small bodies.

The existence of the generalized Jacobian integral suggests its possible use to generalize the idea of zero velocity surfaces originated by Hill. As shown recently by Nobili and Milani (1988) this is not feasible for practically importsnt cases. This result is not surprising if the well-known limitations of the applicability of the surfaces of zero velocity are considered in the original formulation of the restricted problem of three bodies.

As an alternate approach, simplifications are introduced using physical approximations. In this way the practical application of the Jacobian integral becomes possible and useful, though only approximately valid results are arrived at. These approximate formulations are discussed in the next section.

4. APPROXIMATE SIMPLIFIED APPROACH

So far the motion of each minor body with respect to the synodic coordinate system has been considered. We can take a new direction by examining the motion of one minor body with respect to the other, i.e. their relative motion. This seems justifiable since the determination of those conditions under which the relative motion of the minor bodies is bounded is of prime concern. The equations describing the motion of one minor body, m_4, with respect to the other, m_3, become

$$\ddot{x}_{34} - 2\dot{y}_{34} = \left[1 - \frac{\mu_1}{R_1^3} - \frac{\mu_2}{R_2^3}\right] x_{34} - (\mu_3+\mu_4)\frac{x_{34}}{r_{34}^3} \tag{5a}$$

$$\ddot{y}_{34} + 2\dot{x}_{34} = \left[1 - \frac{\mu_1}{R_1^3} - \frac{\mu_2}{R_2^3}\right] y_{34} - (\mu_3+\mu_4)\frac{y_{34}}{r_{34}^3} \tag{5b}$$

$$\ddot{z}_{34} = \left[- \frac{\mu_1}{R_1^3} - \frac{\mu_2}{R_2^3}\right] z_{34} - (\mu_3+\mu_4)\frac{z_{34}}{r_{34}^3} \tag{5c}$$

Here x_{34}, y_{34}, z_{34} are the coordinates of m_4 in a coordinate system centred at m_3 and rotating with the mean motion of the primaries and R_1 and R_2 are the distances from the center of mass of the minor bodies to m_1 and m_2, respectively. This system of equations cannot be derived without the simplifying assumption that the minor bodies, orbiting each other in relatively close proximity, are nearly equidistant from either primary body. This is called the proximity condition, and justification can be found by considering the following. Let $R_{13} = R_{14} + \varepsilon R_{14}$ where the scalar ε is a small number if the minor bodies are in close proximity. From this we obtain

$$R^3_{13} = R^3_{14}(1 + 3\varepsilon^2 + 3\varepsilon^2 + \varepsilon^3).$$

For main belt binary asteroids, if these indeed exist, a large separation distance between the asteroids would be 20,000 km. The distance to either the Sun or Jupiter would range from 1 to 9 A.U. making $\varepsilon = O(10^{-4})$. Hence, to a first order approximation $R^3_{13} = R^3_{14} = R^3_1$, and by a similar argument $R^3_{23} = R^3_{24} = R^3_2$.

The set of equations in (5), being of only sixth-order and of fewer terms, is much simpler than the previous set. This simplicity, however, is deceiving; for it must be remembered that R_1 and R_2 are in general unknown functions of time, and more information is needed to obtain these functions.

If instead of subtracting the equations of motion for m_3 from m_4 they are added and the result simplified by applying the proximity condition, two important facts emerge. The terms representing the gravitational interaction between the minor bodies cancel, and the remaining terms can be combined. This results in the equations of motion for a single minor body in the classical circular restricted problem of three bodies:

$$\ddot{X} - 2\dot{Y} = -\frac{\mu_1(X-\mu_2)}{R_1^3} - \frac{\mu_2(X+\mu_1)}{R_2^3} + X \tag{6a}$$

$$\ddot{Y} + 2\dot{X} = \left[1 - \frac{\mu_1}{R_1^3} - \frac{\mu_2}{R_2^3}\right] Y \tag{6b}$$

$$\ddot{Z} = \left[- \frac{\mu_1}{R_1^3} - \frac{\mu_2}{R_2^3}\right] Z \tag{6c}$$

where the capital letters X, Y and Z represent the location of the center of mass of the minor bodies. Hence, when the minor bodies are close together, their center of mass moves approximately as if it were a single point mass in the circular restricted problem of three bodies.

This new formulation reveals information about the problem not apparent in the initial formulation, namely that the minor bodies are essentially orbiting their center of mass which itself moves as a single mass in the circular restricted problem of three bodies. Also, it is apparent that in the relative motion equations, the masses of the minor bodies appear as a sum.

It is conceivable that one of these minor bodies could be of "very small" mass, which constitutes a hierarchical problem in which $n=2$, $\nu=1$ and $\bar{\nu}=1$. The smaller minor body is influenced only by primaries. Such a model might be applicable to the motion of a spacecraft near an asteroid or comet.

A significant benefit of the simplified formulation results from the fact that the equations of relative motion yield an integral for some cases of practical importance. The potential function for equation (5) is given by the expression

$$\Gamma = \frac{1}{2}(x_{34}^2 + y_{34}^2) - \frac{K}{2} r_{34}^2 + \frac{(\mu_3+\mu_4)}{r_{34}} \tag{7}$$

where

$$K = \frac{\mu_1}{R_1^3} + \frac{\mu_2}{R_2^3} \tag{8}$$

Equation (7) might be used to obtain an integral of equations (5) if Γ does not show explicit time-dependence. Such is the case if the constraint that K remains invariant with time is imposed. This we call the circularity constraint for reasons which will shortly become apparent. When this constraint is met, the integral for the system of equations of relative motion becomes

$$v_{34}^2 = 2\Gamma - C_{34} \tag{9}$$

where v_{34} is the relative velocity, and C_{34} is the constant of integration.

The numerical value of K is indicative of the motion of the center of mass of the minor bodies in the synodic coordinate system. It is difficult to visualize exact solutions of the circular restricted problem of three bodies which satisfy the circularity constraint. There do exist, however, at least two such solutions represented by the triangular libration points. If the center of mass lies at either of these points, the circularity constraint is met, and the system can be integrated. It is interesting to note that for these cases the numerical value of K is unity.

We must now ask the question whether the concept of an integral of relative motion is applicable in those cases in which the circularity constraint is not met. Strictly speaking, there can be no integral of relative motion if there is any variation of K with time. However, if K is nearly constant C_{34} becomes an approximate integral of motion. The degree to which the former remains constant is the degree to which the latter is valid. Fortunately, a class of trajectories important to the problem of binary asteroids exists for which K remains approximately constant. This is the case when the center of mass orbits m_1 on a nearly circular orbit so that R_1 is approximately constant. If $\mu_2 \ll \mu_1$ (as in the case when the primaries are Jupiter and the Sun), then Jupiter's contribution to the value of K can be neglected. Note that this can be true also for orbits of high inclination. A majority of the main belt asteroids possess orbits which satisfy approximately the circularity constraint.

Given that there exists a valid or at least approximately valid integral of relative motion, Hill's method can be employed to determine regions of possible motion as in the classical circular restricted problem of three bodies. However, since the potential functions are not the same, the resulting regions are not identical.

We recall that x_{34}, y_{34}, z_{34} system is a translating and rotating coordinate system whose origin is fixed at m_3. These coordinates describe the motion of m_4 with respect to m_3, and the surface of zero velocity represents zero relative velocity. The zero velocity surfaces separate the regions where relative motion is possible from those where it is not. When the initial conditions are such that the motion is confined within the zero velocity region, the minor bodies have stability. In order to determine the shape of the surface of zero velocity, equation (9) is used. The surface of zero velocity is defined as the locus of all points which satisfy equation (9) when the velocity term vanishes. These surfaces are ellipsoids of revolution centered at m_3 and symmetric about the z_{34} axis. For actual cases of potential main belt binaries, the relative velocity term is positive inside the ellipsoid only. Hence, based on these considerations, the relative motion of main belt binary asteroids should be bounded, i.e. they should be stable in Hill's sense.

5. NUMERICAL INVESTIGATIONS

In this section some recent numerical investigations of the restricted problem of 2+2 bodies are described briefly. The investigations were aimed to test the validity of the approximate integral of relative motion and to gain insight into the three dimensional stability behaviour of the two minor bodies perturbed by the Sun and Jupiter. The system of equations given by Equations (1) and (2) was numerically integrated on a Control Data Corporation Dual Cyber 170/750 digital computer system at the University of Texas. The integration scheme employed consisted of paired, variable stepsize, seventh- and eighth-order Runge-Kutta integrators. The leading truncation error at each step was 10^{-12} which corresponds to one meter. The dimensionless integration step size varied with the distance between the minor bodies, but a typical value was 10^{-3} dimensionless time units (16 hours). Variations in the constant of integration, the generalized Jacobian integral given by Equation (4), were consistently less than 10^{-11} of its value at the initial conditions.

It is convenient to think of binary asteroid motion as two perturbed Keplerian orbits. When the two asteroids are close, they orbit each other with perturbations from the Sun and Jupiter while their center of mass orbits the Sun. The initial conditions for the asteroids are specified by two sets of orbital elements - one set for the orbit of one asteroid moving about the other and a second set for the orbit of their center of mass.

The experiments consisted of numerically integrating trajectories using four sets of initial conditions which were identical except for

for the values of a_{34}, the semimajor axis for the orbit of the minor bodies. These cases were intended to model a hypothetical main belt binary, the masses of which correspond to two spheres of diameter 50 km and 5 km, respectively, with uniform densities of 3g cm^{-3}. The non-dimensional masses of the minor bodies were thus 10^{-13} and 10^{-17}, respectively. The heliocentric orbit of the binary center of mass had the following initial Keplerian orbital elements: semimajor axis 2.8 A.U., eccentricity 0.1, inclination 15°, longitude of the ascending node 90° and argument of perihelion 90°. During the integration this orbit remained approximately constant. Only small periodic changes were observed on the semimajor axis, eccentricity, and inclination. The node and line of apsides had small secular rates of -1.6 and +2.1 degrees per century.

The initial elements for the orbit of m_4 about m_3 were: eccentricity 0.01, the inclination, longitude of ascending node, argument of periapsis and true anomaly 90°. In the four cases the semimajor axis had initial values of 3000 km, 6000 km, 8000 km, and 10000 km, respectively. Case 1, which has the smallest initial semimajor axis, proved to be stable for the entire duration of the simulation (940 yrs) during which time 3300 orbits of the minor bodies were completed. In the other cases instability occurred at increasingly earlier times as the initial value of the semimajor axis was increased. A summary for the four cases is given in the table.

Case	$(a_{34})_o$(km)	T_{34}(yrs)	Stability (yrs)	r_{zv}(k)	$2(a_{34})_o/r_{zv}$
1	3000	0.28	>940	5926	1.012
2	6000	0.80	72	11080	1.082
3	8000	1.24	4.9	13780	1.161
4	10000	1.73	3.6	16030	1.248

In this table the second column gives the initial semimajor axis for the orbit of the minor bodies. T_{34} is the period of this orbit, and the fourth column gives the duration of stability for each case.

In the fifth column r_{zv} is the radius of the nearly spherical zero velocity surface based on the initial conditions. The sixth column compared the initial value of twice the semimajor axis to this radius. It is interesting to note that the smaller this parameter, the greater the stability.

The time history of the first three orbital elements (a,e,i) for Case 1 are shown in Figures 1, 2 and 3. Time units are in Jupiter's period (11.862 yrs). It is seen that all the elements have dramatic periodic variations but do not appear to vary in a secular manner. While the magnitude of the changes in the semimajor axis are relatively small, only 80 km, the eccentricity ranges from near zero to as high as 0.98, indicating that the orbit is at times nearly rectilinear. The periapsis distance becomes as small as 46 km, making collisions between the asteroids a definite possibility. (In this simulation the asteroids were modelled as point masses). The variations in the elements seems to have a period of approximately five Jupiter years (60 yrs). It is noteworthy that this does not correspond to the synodic period between the binary center of mass and Jupiter which is (0.395 Jupiter years). This phenomenon will be investigated further.

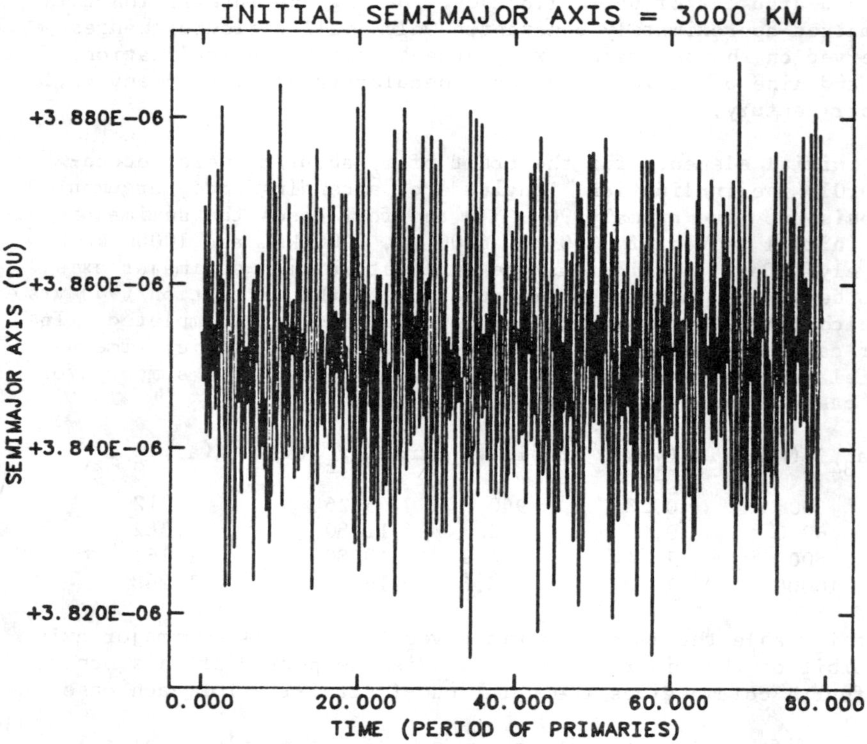

Figure 1: Semimajor Axis for Case 1

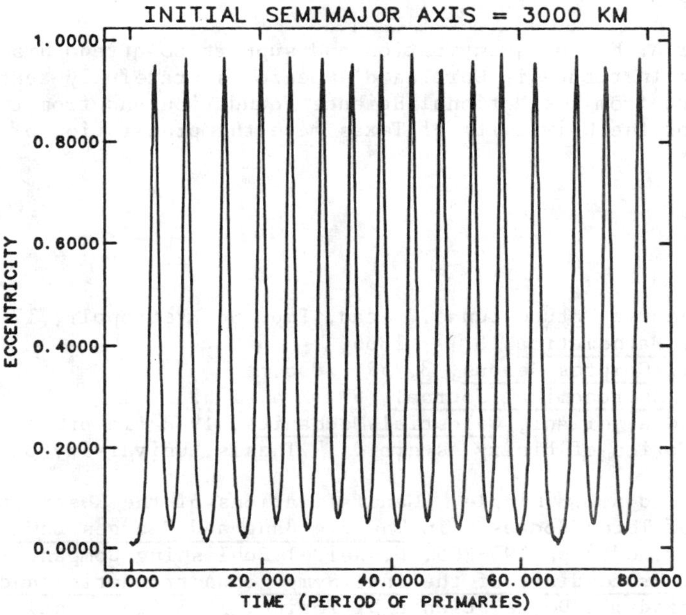

Figure 2: Eccentricity for Case 1

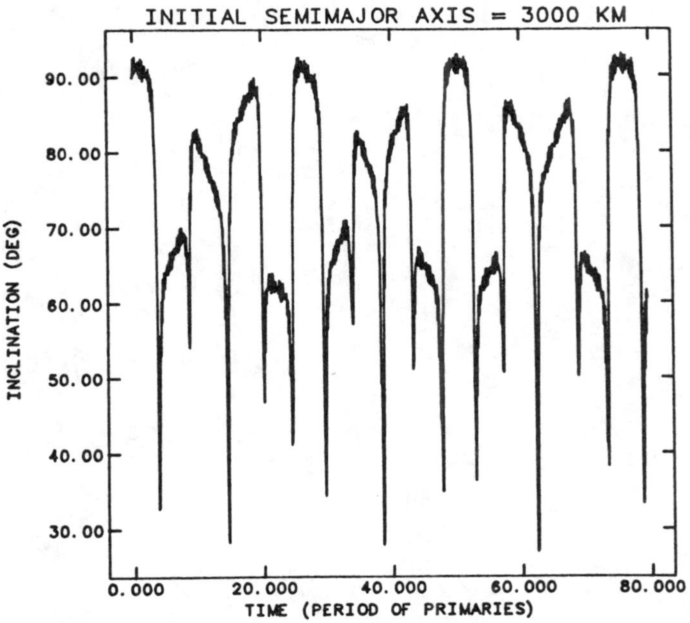

Figure 3: Inclination for Case 1

ACKNOWLEDGEMENTS

Professor A. E. Roy's invitation and support to attend his NATO Advanced Study Institute in Cortina d'Ampezzo is gratefully acknowledged. Partial support from the National Science Foundation and from the R.B. Curran Chair of the University of Texas made the preparation of this paper possible.

REFERENCES

Euler, L., "Theoria Motuum Lunae," Acad. Imp. Sc. Petropoli, 1772.
Hill, G., Acta Mathematica, 8, 1, 1886.
Jacobi, C.G.J., Comptes Rendus, 3, 59, 1836.
Kevorkian, J., Astronomical Journal, 67, 204, 1962.
Nobili, A., and A. Milani, Celestial Mechanics, 1988 (in print)
Pojman, J., "Motion of Binary Asteroids," Thesis, University of Texas, 1987.
Szebehely, V. and A.L. Whipple, "Generalizations of the Restricted Problem of Three Bodies", in The Big Bang and Georges LeMaitre (A. Berger, Ed.) pp.195-205. D. Reidel Publishing Company 1984.
Szebehely, V., Proceedings of the Int. Symp. on Astronautics and Cel. Mech., Accademia Delle Scienze Di Torino, 1988 (in print).

THE MOTIONS OF ASTEROIDS IN THE SECULAR RESONANCE ν_6: AN ANALYTICAL MODEL AND NUMERICAL CALCULATIONS

Makoto Yoshikawa
Tokyo Astronomical Observatory
University of Tokyo
Mitaka, Yokyo 181, Japan.

ABSTRACT. The motions of asteroids in the secular resonance ν_6 are investigated by a simple analytical model. By using this model, it is found that eccentricities of asteroids may become very large ($e \sim 0.8$) when asteroids are in or near this resonance. These large increases in eccentricity are also confirmed by the numerical integrations. The strongly depopulated region in the asteroidal belt, which corresponds to the position of the secular resonance ν_6, is explained well by this analytical model.

1. INTRODUCTION

The secular resonance ν_6 occurs when the longitudes of perihelion (ϖ) of an asteroid and Saturn move nearly the same rate ($\nu_6 = 26.217''$/year; Bretagnon, 1974). Williams (1969) showed the location of secular resonant surfaces in the asteroidal belt, and he found that these surfaces correspond to the regions where asteroids are depleted. Recently, Froeschlé and Scholl (1986, 1987) and Scholl and Froeschlé (1986) studied the dynamics of asteroids in secular resonances by numerical integrations, and they found that the eccentricity or inclination of asteroids becomes very large. In this paper, the dynamics of the secular resonance ν_6 is investigated by a simple analytical model and numerical integrations.

2. THE SECULAR RESONANCE ν_6

The motions of asteroids in the secular resonance ν_6 are investigated by an analytical model, in which the third degree terms of the eccentricity and inclination are taken into account (Yoshikawa, 1987). In this model, the eccentricity variations of asteroids are expressed by the Hamiltonian

$$\tilde{F}_{\nu_6} = \frac{1}{2}(b-\nu_6)e^2 + \frac{1}{4}ce^4 - d_6 e \cos(\tilde{\omega} - \tilde{\omega}_s^*),$$

where coefficient b is a function of semimajor axis (a) and inclination (I), c and d_6 are the functions of a, $\tilde{\omega}_s^*$ is longitude of perihelion

A. E. Roy (ed.), Long-Term Dynamical Behaviour of Natural and Artificial N-Body Systems, 289–293.
© 1988 by Kluwer Academic Publishers.

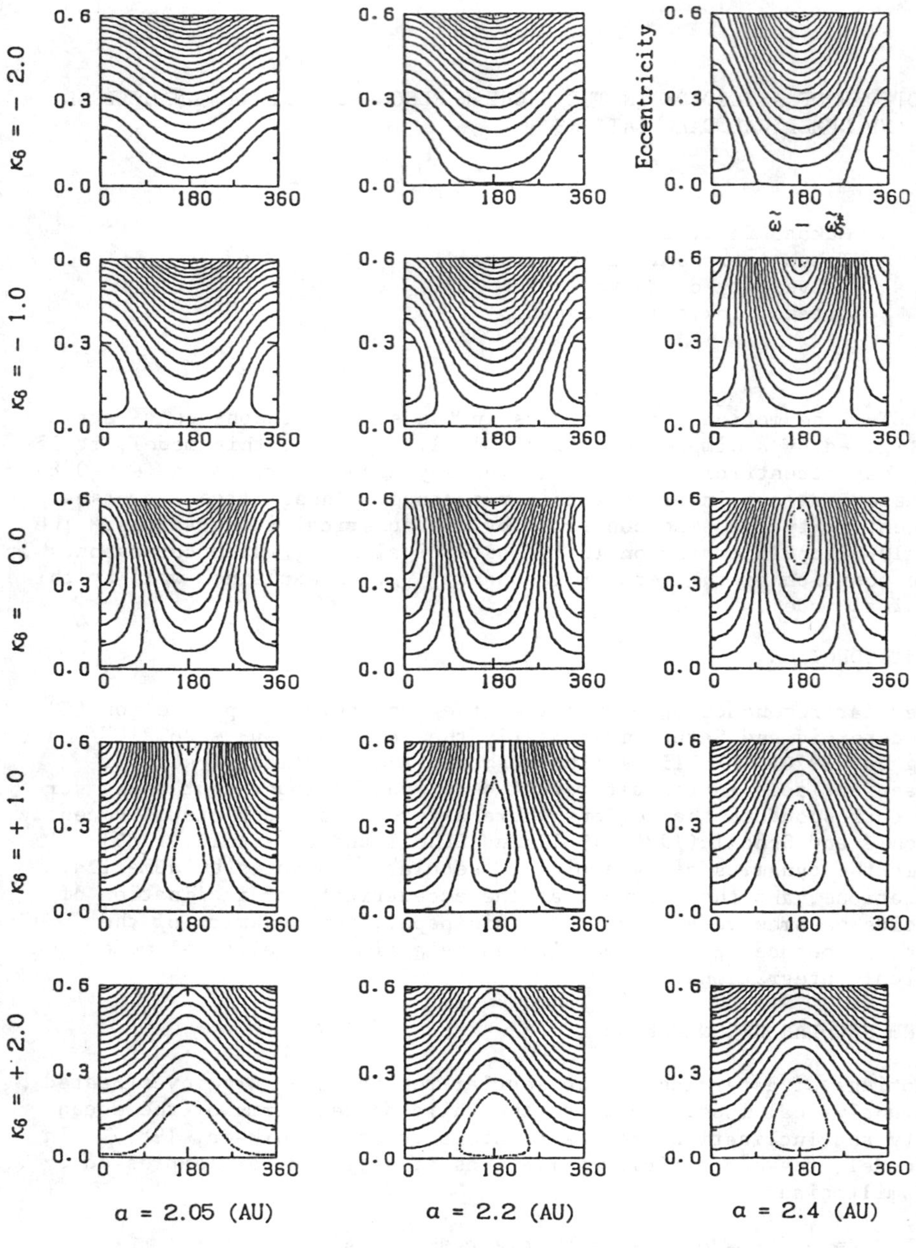

Fig.1 The diagrams of equi-Hamiltonian curves for $2.0 < a < 2.5$ and $-2.0 \leq \kappa_6 \leq 2.0$. The unit of κ_6 is (″/year).

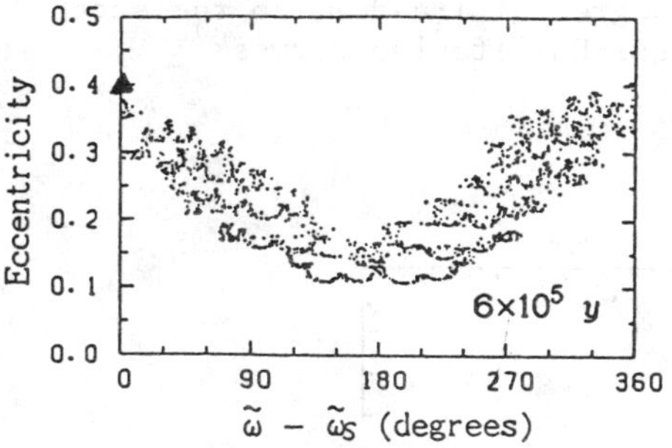

I_0 = 13° $\langle I \rangle$ = 13°.6 $\kappa_6 = -2.3$

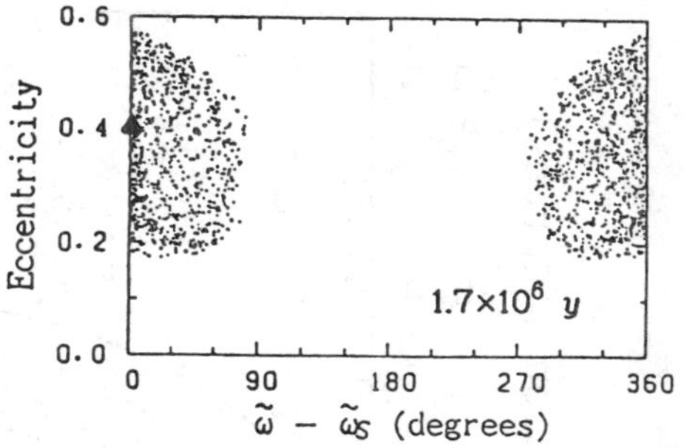

I_0 = 12° $\langle I \rangle$ = 12°.4 $\kappa_6 = -1.2$

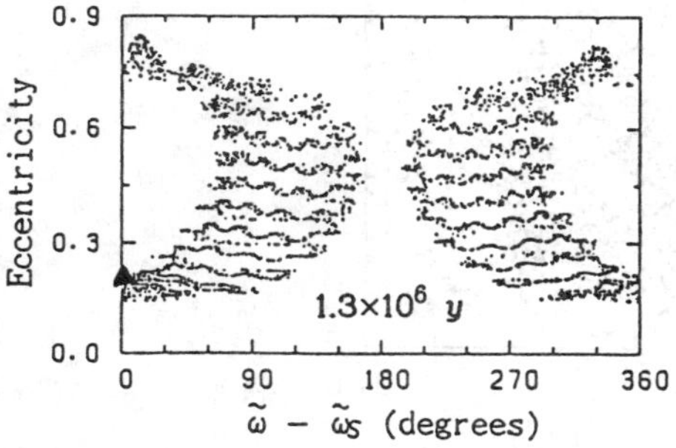

I_0 = 10° $\langle I \rangle$ = 10°.4 $\kappa_6 = 0.3$

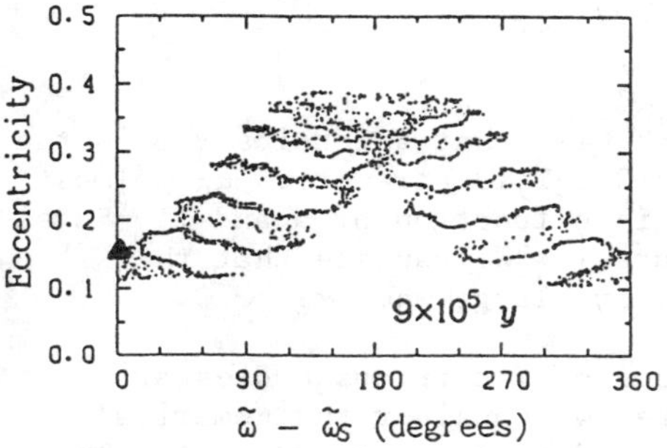

I_0 = 8° $\langle I \rangle$ = 8°.4 $\kappa_6 = 1.7$

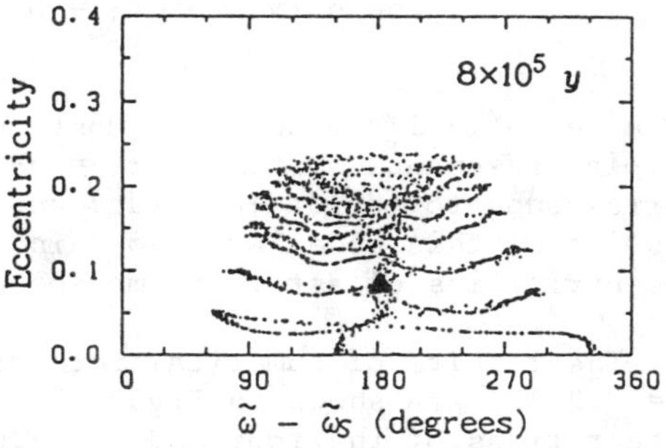

I_0 = 8° $\langle I \rangle$ = 8°.4 $\kappa_6 = 1.7$

Fig.2 The results of numerical integrations of fictitious bodies at a = 2.2(AU). I_0 and $\langle I \rangle$ are the initial and the mean value of the inclination, respectively. The initial values of e and $\tilde{\omega}-\tilde{\omega}_S$ are denoted by filled triangles.

of Saturn. This Hamiltonian is of one degree of freedom, so the eccentricity variations are represented by equi-Hamiltonian curves

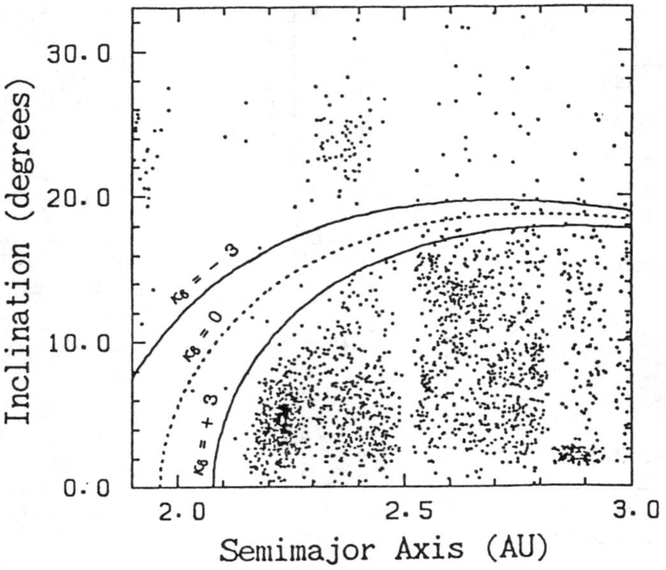

Figure 3 The osculating elements of numbered asteroids and the curves of $\kappa_6 = 0, \pm 3$ ("/year).

on a $e - (\tilde{\omega} - \tilde{\omega}_s^*)$ plane. We must notice here that asteroids are in the secular resonance ν_6 when $\kappa_6 = b - \nu_6 \sim 0$. This parameter κ_6 almost corresponds to $(d\tilde{\omega}/dt - d\tilde{\omega}_s/dt)$ and it is a function of a and I. Some results of this model are shown in Figure 1. We can see that the eccentricities of asteroids may become very large when $\kappa_6 \sim 0$.

The results of numerical integrations of fictitious bodies at a = 2.2(AU) are shown in Figure 2. Here, we consider the numerical integrations of the four body Sun-Jupiter-Saturn-Asteroid problem. The values of κ_6 in Fig.2 are calculated using values of mean inclinations. Comparing Fig.1 and Fig.2, we can find that this analytical model agrees qualitatively with the numerical integrations.

In Figure 3, the region where $|\nu_6| \lesssim 3.0$ is shown. The analytical model predicts that the eccentricity of asteroids may become large if they are in this region. This region agrees quite well with the depopulated region of asteroids especially in the inner part of the main belt. (The values κ_6 in Fig.2 are slightly different from the values in Fig.3. The reason is that values of κ_6 in Fig.2 are calculated considering only Jupiter and Saturn, although those in Fig.3 are calculated considering eight major planets).

3. CONCLUSION

The small eccentricity (0.1 $\sim$ 0.2) of asteroids increases up to about 0.8 for about one million years, if asteroids are in the secular resonance ν_6 . The analytical model in this paper agrees with the solutions of numerical integrations, and we can clearly understand the dynamics of the secular resonance ν_6 by this model. Moreover, we can explain the width of the depopulated region due to the secular resonance ν_6 in the inner part of the asteroidal belt.

Computations were carried out using the FACOM M-380R computer at the Computing Center of the Tokyo Astronomical Observatory.

REFERENCES

Bretagnon, P. : 1974, Astron. Astrophys. 30, 141-154.
Froeschlé, Ch., Scholl, H.: 1986, Astron. Astrophys. 166, 326-332.
Froeschlé, Ch., Scholl, H.: 1987, Astron. Astrophys. 179, 294-303.
Scholl, H., Froeschlé, Ch.: 1986, Astron. Astrophys. 170, 138-144.
Williams, J.G.: 1969, Ph.D. dissertation, University of California, Los Angeles.
Yoshikawa, M.: 1987, Celest. Mech. in press.

PART IV

DYNAMICS OF STARS AND GALAXIES

Institut für Reine
und Angewandte Kernphysik
Kiel

FAMILIES OF PERIODIC ORBITS IN A MODEL OF INTERACTING GALAXIES

N. Caranicolas and Ch. Vozikis
Astronomy Department
University of Thessaloniki
54006 Thessaloniki, Greece

ABSTRACT. A simple dynamical model is used for the description of the motion of a star in a binary galaxy. In this model we study the evolution of several families of periodic orbits in the primary galaxy as the perturbation caused by the companion increases.

1. INTRODUCTION

Observations show that galaxies often appear in pairs, groups or clusters. Therefore it is natural to believe that they interact in a multitude of ways with their environment.

A good example of interacting galaxies is the triple system the Galaxy, Large and Small Magellanic Clouds. The dynamical evolution of the above system have been studied in many papers (i.e. Kerr 1957, Burke 1957, Avner and King 1967, Fujimoto and Sofus 1976). On the other hand computer simulation of galactic encounters strongly support the idea that luminous bridges and tails in many multiple galaxies are the result of gravitational interaction (Toomre 1974 and references therein).

In the present paper we shall restrict ourselves to a model pair of interacting galaxies which is described in Section 2. Section 3 is devoted to the description of the characteristics of families of periodic orbits while in Section 4 we give the main conclusions of this work.

2. THE MODEL OF INTERACTING GALAXIES.

The model used consists of two galaxies. The first galaxy - hereafter the primary galaxy - is described by Richstone's potential with a softening parameter (see Binney 1982) while the second galaxy - the companion - is described by a mass point potential. The resulting potential is

A. E. Roy (ed.), Long-Term Dynamical Behaviour of Natural and Artificial N-Body Systems, 297–300.
© 1988 by Kluwer Academic Publishers.

$$V = \frac{1}{2} \ln (x^2 + ay^2 + c) - \frac{k}{[(x-R)^2+y^2]^{\frac{1}{2}}} , \quad (1)$$

where the origin of the rectangular coordinate system coincides with the centre of the primary galaxy while the companion approaches along the x-axis. R denotes the distance between the centres of the two galaxies while k, a, c are positive constants.

The above model can be used only for the study of the orbital behaviour of the primary galaxy. In what follows we shall restrict ourselves to the study of the periodic orbits in the primary galaxy as the mass point companion approaches, the two galaxies collide and finally merge. For this purpose we have chosen a series of distances R which are considered fixed for the time of the computation.

3. CHARACTERISTICS OF PERIODIC ORBITS

Figure 1 shows the characteristics of the families of periodic orbits in the (R, x) plane. Most characteristics are marked by a rational number n/m where n,m are the numbers of oscillations along the x and y axes. All periodic orbits intersect the x-axis perpendicularly while dashed and solid lines indicate unstable and stable periodic orbits respectively. Our discription begins when R=5 because for larger values of R the pattern is similar to that of the unperturbed case.

When $2.85 < R < 5$ we can see three kinds of periodic orbits. Two of them marked 1a, 1/1a are also present in the unperturbed case. The third triple periodic orbit has two branches, one stable and the other unstable.

At R = 2.85, that is a little before the collision of the two galaxies two additional periodic orbits appear. The first marked 2/2 bifurcates from the upper branch of the 1/1a characteristic and is present down to R = 1.4 while the second marked 1/2a bifurcates from the boundary line (x x x).

The dashed-dotted line in Figure 1 gives the Lagrange points of the system (1). We notice that there are two Lagrange points after the collisions of the two galaxies while there is a minimum value of R beyond which the system has no Lagrange points at all. The derivation of the coordinates of the above point is given in another paper (Caranicolas and Vozikis 1987).

The characteristics 1b, 1/2b start at the distance of collision. The corresponding periodic orbits are unstable. It is interesting to observe that the characteristic 1b is almost parallel to the line giving the Lagrange points while its end point coincides with the end point of the characteristics 1a at about r = 1.0. The characteristic

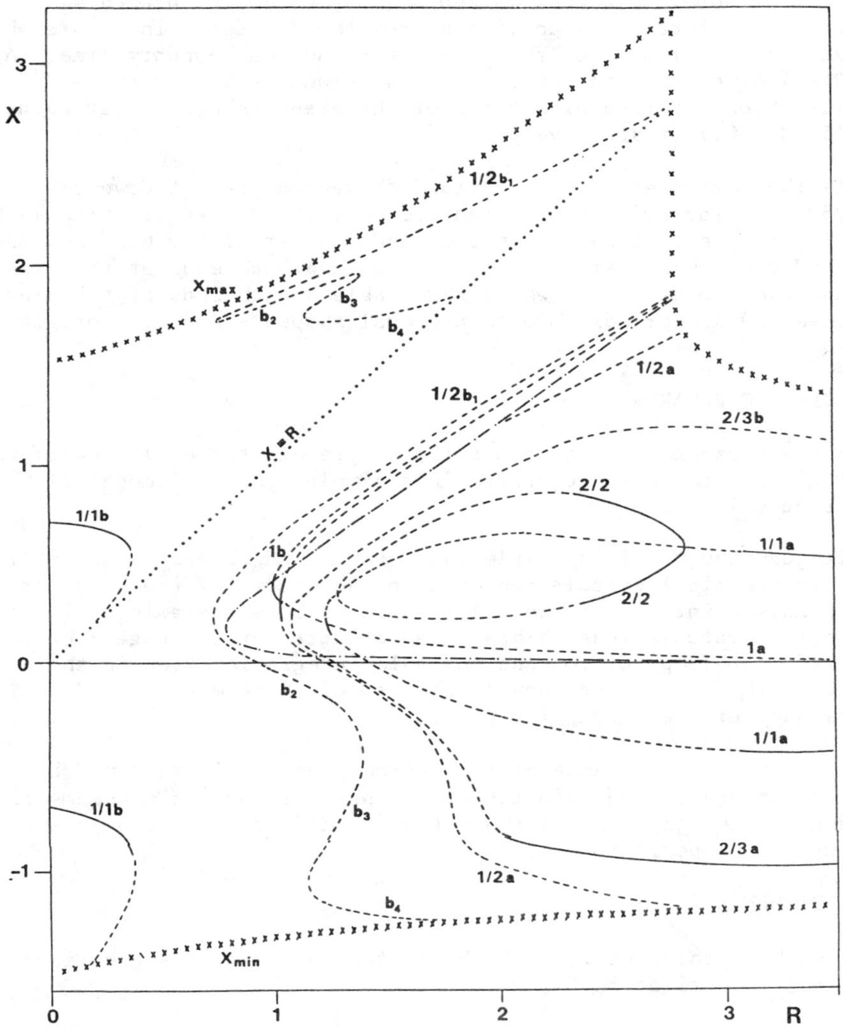

Fig. 1 Diagram giving the positions x of the periodic orbits as a function of R. The values of k , α , c and the energy constant h are 0.5, 1.777778, 0.1 and 0.1 respectively

1/2b has four double branches marked b_1, b_2, b_3, b_4. Branch b_1 appears at the distance of collision from the boundary line. Branch b_4 appears at R = 1.85 from the line x = R and the boundary line. At about R = 1.4 we have the presence of the branches b_3, b_2 the end points of which coincide with those of the branches b_4 , b_1 at about R = 1.15, R = 0.75 respectively.

All the characteristics described above are present down to R = 0.75 while for smaller R we can see only two characteristics marked 1/1b. These appear at about R = 0.36 and consist of two branches one stable and the other unstable. The unstable branch ends at the line x = R and the boundary line while the stable branch ends at the x-axis. The corresponding periodic orbits are oval shaped around the origin.

4. CONCLUDING REMARKS

In the present work we have tried to present the evolution of the characteristics of periodic orbits in a simple dynamical model of interacting galaxies.

Although the model is simple the authors had to face some difficulties in numerical calculations due to the nature of the potential near the mass point, after the collision of the two galaxies. In the numerical calculations we have used a fourth order Runge - Kutta method with double precision and variable integration step in the range 10^{-2} - 10^{-8}. The accuracy of the calculations was checked by the constancy of the energy integral.

Since the disappearance of the Lagrange points is followed by the disappearance of the majority of the periodic orbits it seems that the change in the topology of the potential affects drastically the behaviour of the periodic orbits.

ACKNOWLEDGEMENTS

We wish to thank Prof. B. Barbanis for many useful suggestions during this investigation.

REFERENCES

Avner, E. and King I. : 1967, Astron. J. 72, 650.

Binney, J. : 1982, in Morphology and Dynamics of Galaxies, L. Martinet and M. Mayor (eds.) Geneva Observatory.

Burke, B.F.: 1957, Astron. J. 62, 90.

Caranicolas, N. and Vozikis, Ch. : 1987, (preprint)

Fujimoto, M. and Sofus, Y.: 1976, Astron. Astrophys. 47, 263.

Kerr, F.J. : 1957, Astron. J. 62, 93.

Toomre, A. : 1974, in The Formation and Dynamics of Galaxies, I.A.U. Symp. No.58, ed. J.R. Shakeshaft, p.347.

QUALITATIVE CHARACTERISTICS OF DYNAMICAL SYSTEMS

G. Contopoulos
European Southern Observatory,
Garching bei München, Germany

ABSTRACT:

We compare some basic families of orbits in a number of galactic models and in the restricted three body problem. In plane rotating barred galaxies the orbits around the Lagrangian points L_4, L_5, L_1, L_2 are qualitatively similar to the orbits around L_4, L_5, L_3 of the restricted three body problem. The long period orbits are connected to the short period orbits (SPO) by an infinity of bridges. The limit of the sequence of bridges is a family of heteroclinic (in the galactic problem), or homoclinic (in the restricted problem) orbits. In various galactic models the SPO families are connected with the families around L_1, L_2 by one or more bridges. All these families end at retrograde orbits around the centre. This behaviour is compared with the restricted problem, in which the SPO families join the family around L_3, which ends at retrograde orbits around the larger primary.

In 3-dimensional systems we describe briefly the consequences of some collisions of bifurcations, like the nonuniqueness of certain families of periodic orbits. Then we examine the behaviour of nonperiodic orbits starting close to unstable periodic orbits, especially in the case of complex instability. We define a LCN (Lyapunov characteristic number) for orbits at different distances from the periodic orbit. There is a "local LCN" equal to the real part of the characteristic exponent, then LCN's appropriate for a small or large neighbourhood of the periodic orbit. We follow the change of the LCN's as certain parameters of the system vary.

I. INTRODUCTION

During the last decades an extensive study of orbits in several dynamical systems has been made. Most of these studies refer to systems of two degrees of freedom, but in recent years work has started also in systems of three degrees of freedom. Astronomical systems of this type range from the restricted three body problem to models of galaxies. However most of these studies are isolated, in the sense that they refer

A. E. Roy (ed.), Long-Term Dynamical Behaviour of Natural and Artificial N-Body Systems, 301–315.
© 1988 by Kluwer Academic Publishers.

to one particular problem, and no comparison is made with other systems. It is of interest, therefore, to show certain patterns, that are common in rather different dynamical systems.

In fact certain features of a quite wide range of systems are quite similar, and the differences are only in the details, or in the numerical values of certain quantities. Thus we may distinguish between the essential and the accidental characteristics of dynamical systems.

In the present paper we will make three qualitative descriptions:

(1) the evolution of the orbits around the Lagrangian points in a rotating galaxy in comparison with the corresponding evolution of the orbits in the restricted three body problem,

(2) the evolution of the orbits in a nonrotating galactic model in comparison with the above cases, and

(3) the evolution of complex unstable periodic orbits and of orbits close to them in a two-parameter set of 3-D galactic models.

II. THE EVOLUTION OF PERIODIC ORBITS AROUND THE LAGRANGIAN POINTS

We consider stable and unstable orbits around the Lagrangian points in two cases:

(a) A rotating galactic model that represents approximately the barred galaxy NGC 3992 (Contopoulos et al. 1987). This model has a potential

$$V=(280)^2[\ln r + E_1(0.242r)]+\frac{\varepsilon}{r^2}\exp(-0.4r)(0.79983r-1)\cos 2\theta(\mathrm{km\ s}^{-1})^2$$

for r > 1.9523 kpc, and (1)

$$V=(280)^2[\ln r+E_1(0.242r)]+\varepsilon 0.0247305537[1.72822-\cos(\frac{\pi r}{1.9523})] \times \cos 2\theta(\mathrm{km\ s}^{-1})^2$$

for r < 1.9523 kpc,
(where E_1 is the exponential integral and ε = 97422) and rotates with a constant angular velocity $\Omega_s = 52.4$ km s^{-1} kpc^{-1}. The Lagrangian points are L_4, L_5 (stable) and L_1, L_2 (unstable). L_3 represents the centre of the galaxy (Fig.1a). The results of this study are very similar to those of a different model

$$V = \frac{1}{1+(r^2+1)^{\frac{1}{2}}} + \varepsilon r^{\frac{1}{2}} (16-r)\cos 2\theta , \qquad (2)$$

discussed by Contopoulos and Papayannopoulos (1980).

(b) The restricted three-body problem with a mass ratio

$\mu = \frac{m_2}{m_1+m_2}$ in the Sun-Jupiter case ($\mu = 0.000953$), as studied by Henrard (1983)(Fig.1b). The potential is

$$V = -\left[\frac{1-\mu}{r_1} + \frac{\mu}{r_2}\right], \tag{3}$$

where r_1, r_2 are the distances of the third body from the primaries m_1, m_2 and the angular velocity of the system is $\Omega_s = 1$.

In both cases there are stable orbits around L_4, L_5, belonging to the family of short period orbits (SPO) and to the family of long period orbits (LPO) (Figs. 2a,2b). However the long period family consists of disconnected arcs joined to the short period family, by bridges, as shown schematically in Figs.2a, 2b.

Every long period arc joins a SPO described n times to a SPO described (n+1) times. The short period orbits are stable in the whole range in which they generate bridges leading to the LPO, except between the bridges of SPO described 2 and 1 times (Contopoulos 1983; Deprit and Henrard 1968, 1970). If the perturbation, ε, is small there are bridges only for n larger than a minimum n_o, up to $n \to \infty$. As ε increases the minimum n_o decreases and when it reaches 1 all the bridges $1,2,\ldots\infty$ exist. For ε somewhat larger ($\varepsilon = \varepsilon_{crit}$) the Lagrangian points L_4, L_5 become unstable. (We may say that then n_o reaches 0). For still larger ε there are no closed periodic orbits around L_4, L_5 but only orbits approaching them asymptotically. Then the sequence of SPO and LPO detaches itself from the Lagrangian points L_4, L_5. The details of this process will be described elsewhere.

A similar evolution occurs in the restricted three-body problem as the ratio μ increases. The Lagrangian points L_4, L_5 become unstable when μ goes beyond the critical value $\mu = \mu_o = 0.03852$.

The bridges joining the SPO and the LPO become thinner as n increases, and tend to the dashed line of heteroclinic (Fig.2a) or homoclinic (Fig.2b) orbits, as $n \to \infty$. The forms of these orbits are shown in Figs. 3a,3b. In Fig.3a a heteroclinic orbit starts asymptotically from an unstable periodic orbit around L_2 and ends asymptotically at an equal unstable periodic orbit around L_1. Another heteroclinic orbit starts near L_1 and ends near L_2. There is an infinity of couples of heteroclinic orbits, each couple joining two equal orbits around L_1 and L_2. In the restricted problem a homoclinic orbit starts and ends asymptotically at the same unstable periodic orbit around L_3 (Fig.3b). There is again an infinity of such orbits, one for every unstable orbit around L_3. The family of SPO becomes unstable beyond the line of heteroclinic or homoclinic orbits (Figs. 2a,2b).

The family of short period orbits around L_1 (or L_2) consists of orbits that increase in size as h (Jacobi constant) increases. They are unstable up to a certain value of h and then they become stable.

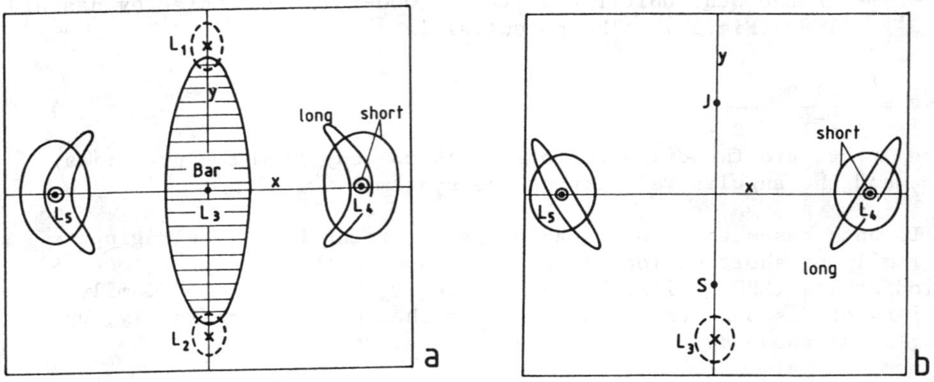

Figure 1: The Lagrangian points and the short and long period orbits around L_4 and L_5: (a) in the galactic problem and (b) in the restricted three body problem.

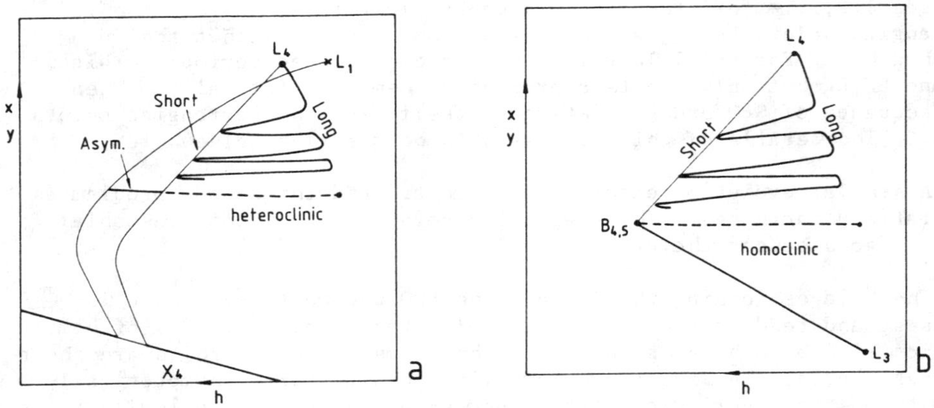

Figure 2: Characteristics of families of orbits starting near the stable and unstable Lagrangian points (schematically): (a) in the galactic problem and (b) in the restricted three body problem. In both cases the families of short and long period orbits (x vs the energy h) start at L_4 and are joined by an infinity of bridges. The limit of these bridges (----) is a family of heteroclinic or homoclinic orbits. The families starting around L_1 or L_3 (y vs the energy h) join the short period orbits through a bridge of asymmetric orbits (case (a)) or directly (orbit $B_{4,5}$, case (b)). In case (a) the families from L_1 and L_4 terminate at the retrograde family around the centre, x_4.

At the transition to stability this family is joined with the family of SPO by a bridge of stable asymmetric orbits (Fig.3a). These orbits do not surround the centre of the galaxy.

Beyond the bifurcation points with this bridge the SPO's are unstable. The instability extends up to a maximum h in the Florida model (1) and then the characteristic of the SPO turns to smaller h, as stable, and joins the family of retrograde orbits around the origin (family x_4, Fig.3a). Similarly the family around L_1 is stable beyond the bifurcation, up to another maximum h. Then its characteristic turns to smaller h, as unstable, until it joins the retrograde family x_4. The family x_4 is unstable between the points at which it joins the families SPO and L_1, and stable for smaller and larger h.

The model (2) gives a somewhat different scenario. The SPO family is unstable beyond the bifurcation of the asymmetric bridge but not all the way to its maximum h; at a point before this maximum the SPO family becomes stable, up to the maximum h. At the transition point bifurcates an unstable bridge joining the family L_1, which then becomes stable up to its own maximum h. As the families SPO and L_1 turn around towards smaller h, they are respectively unstable and stable, until they join at two points the retrograde family x_4. In this case also the family x_4 is mostly stable, with an unstable part between the points at which it joins the families SPO and L_1.

The corresponding scenario in the restricted problem is as follows. There is an unstable family around L_3 (corresponding to the family around L_1 in the galactic problem). This family joins the extensions of the SPO family at a point called $B_{4,5}$ (Fig. 2b). The orbit $B_{4,5}$ belongs to both SPO families around L_4,L_5 and to the family L_3. Beyond the point $B_{4,5}$ the family L_3 continues until it ends by joining the family of retrograde orbits around the Sun (S).

The main differences between the galactic and the restricted problem are:

(1) In the restricted problem there is an asymmetry between L_1 and L_3 (except if $\mu = 1/2$), while in the galactic problem there is a symmetry between L_1 and L_2. This explains why we have heteroclinic orbits joining L_1 and L_2, while we have homoclinic orbits around L_3 in the restricted problem.

(2) In the restricted problem the short period families around L_4, L_5 join with the family around L_3 at a common orbit. Beyond that the family L_3 continues until it joins the retrograde orbits around the Sun. On the other hand in the galactic problem the families around L_4,L_5 join with the families around L_1,L_2 through branches of asymmetric orbits, and they continue until they join separately the family of retrograde orbits around the centre.

Despite these differences the common characteristics of the galactic and restricted problems are very important. The main common behaviour is that in all cases the families of orbits around L_4, L_5 and around the unstable Lagrangian points (L_1, L_2 or L_3) are joined with each other and also with the main retrograde family around the centre, or the Sun. Furthermore the joining of the short and long period orbits and the appearance of heteroclinic or homoclinic families are in general very similar. Thus it is reasonable to assume that all rotating systems have a very similar behaviour.

III. ORBITS IN A NONROTATING GALACTIC MODEL

We have studied recently the orbits in a nonrotating galactic model of the form

$$H \equiv \frac{1}{2}(\dot{x}^2 + \dot{y}^2 + x^2 + y^2) - \varepsilon x^2 y^2 = h \qquad (4)$$

for an energy larger than the escape energy (Barbanis and Contopoulos 1987). The aim was to find the escaping orbits when the energy is larger than the escape energy.

Because of the symmetry of the problem there are 4 symmetric arcs of the curve of zero velocity (CZV)

$$\frac{1}{2}(x^2 + y^2) - \varepsilon x^2 y^2 = h \qquad (5)$$

leaving 4 open regions around the lines $x = \pm y$, through which we may have escapes. However there is an infinity of periodic orbits intersecting perpendicularly one of the axes x, or y, and orbits close to them, which do not escape. We consider now periodic orbits crossing the x-axis perpendicularly at $x > 0$. There are orbits of this type that reached the CZV at two symmetric points, for $y > 0$ and $y < 0$ (Fig.4a). The orbits of other families do not reach the CZV, but make a number n of oscillations above the x-axis before intersecting this axis a second time perpendicularly. Then these orbits form n symmetric oscillations below the x-axis and they close after a total of 2n oscillations (Fig.4a).

As $n \to \infty$ these orbits tend to homoclinic orbits starting and ending asymptotically at the x-axis (Fig.4b). The homoclinic orbits form a family (one orbit for each value of h).

We have found several infinite sequences of periodic orbits that tend to homoclinic or heteroclinic orbits. In particular there are heteroclinic orbits between unstable periodic orbits that bridge the openings of the curves of zero velocity (Fig.4c). An orbit of this type, crossing the x-axis perpendicularly, is shown in Fig.4c. These heteroclinic orbits are limits of periodic orbits that intersect the x-axis perpendicularly for $x > 0$ and make n oscillations almost parallel to each of the orbits bridging the 2 openings for $x < 0$, when $n \to \infty$. The escaping orbits are bracketed by heteroclinic orbits of this type. In fact if an orbit deviates from the heteroclinic orbit and goes be-

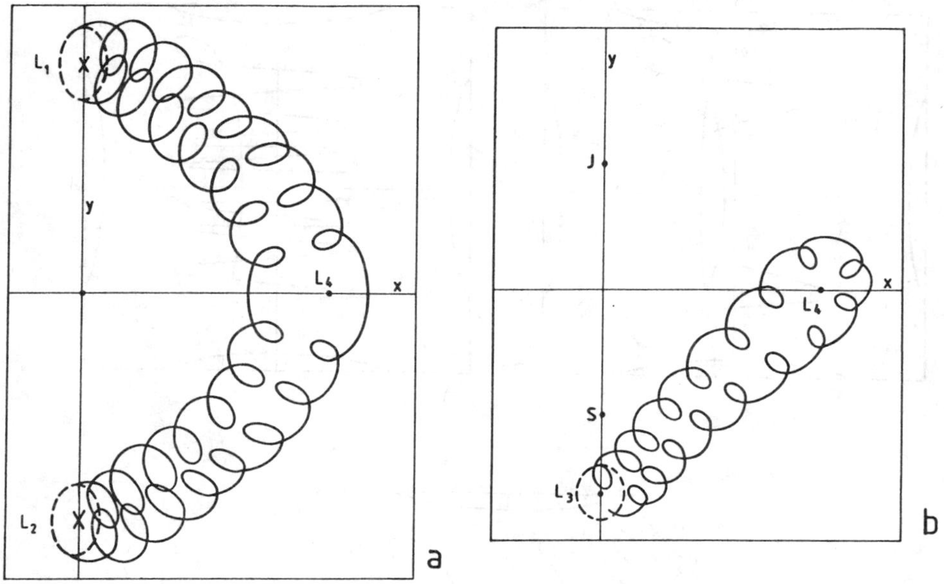

Figure 3: (a) A heteroclinic orbit in the galactic problem and (b) a homoclinic orbit in the restricted three-body problem (schematically).

yond the periodic orbit it escapes from the system.

The present model is rather different from the models considered in §II. In this case we do not have any Lagrangian points which are due to the rotation. Thus we do not have the families L_4, L_5, L_1, L_2 of the previous section. However we have sequences of infinite families of periodic orbits, and families of homoclinic, and heteroclinic orbits, which are limits of families of multiplicity n when $n \to \infty$.

IV. ORBITS IN 3-D SYSTEMS

We have studied recently the qualitative changes of the families of periodic orbits in several systems of 3 degrees of freedom. We have considered mainly Hamiltonians of the form

$$H \equiv \frac{1}{2}(\dot{x}^2 + \dot{y}^2 + \dot{z}^2 + Ax^2 + By^2 + Cz^2) - \varepsilon xz^2 - \eta yz^2 = h \qquad (6)$$

and studied the change of stability and bifurcations of various families of periodic orbits. In a first stage we keep A,B,C, and the

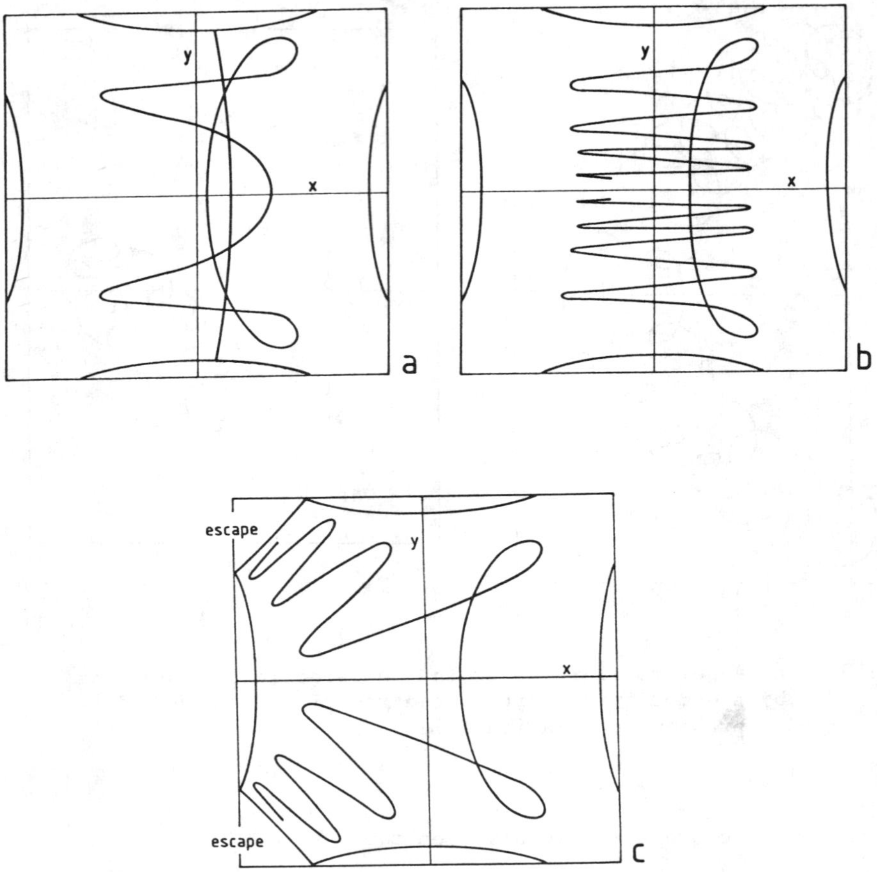

Figure 4: Orbits intersecting perpendicularly the x-axis (schematically): (a) periodic orbits, (b) a homoclinic orbit, doubly asymptotic to the periodic orbit y=0, and (c) a heteroclinic orbit asymptotic to two periodic orbits that bridge two openings of the curve of zero velocity.

energy h constant and change the parameters ε and η . Then we change also one more parameter at a time (A, and B, or C). In this way we can follow the evolution of the various families of periodic orbits in a multiparameter space.

We found that the most important qualitative changes of the system occur at various "collisions of bifurcations", when two bifurcations from one original family occur for the same values of parameters (Contopoulos 1986). Of special importance are collisions of bifurcations in limiting cases when $\varepsilon = 0$, or $\eta = 0$. Collisions of bifurcations lead, in general, to different interconnections between the various families of periodic orbits as one parameter goes beyond the collision value. E.g. before the collision (say for $A < A_{col}$) we have two separate bifurcations from the family 1a generating the families 2a and 2b, while a family 4a bifurcates from the family 2a a little further on. At the collision itself $A = A_{col}$ all these families 2a, 2b,4a bifurcate from the same point of the family 1a. Beyond the collision ($A > A_{col}$) the two bifurcations from 1a are again separate but the family 4a bifurcates now from 2b instead of 2a.

Another phenomenon related to particular collisions of bifurcations is the nonuniqueness of certain families of periodic orbits. Thus if we describe a closed path in the parameter space (ε,η) around a particular "critical point" (ε_c,η_c) and follow continuously the evolution of a particular family of periodic orbits we do not return to the original orbit, but to a different one. This phenomenon may have important consequences in the evolution of a stellar system. E.g. if a galaxy evolves in a cyclic way and returns to the original potential the stellar orbits do not need to be the same.

All the studies mentioned above refer to periodic orbits. However it is also important to study the behaviour of non-periodic orbits.

On one hand we want to know how far we can go for a stable periodic orbit and still have orbits trapped around it. The width of a trapping region gives the appropriate "weight" of the periodic orbit in constructing self-consistent models of galaxies.

On the other hand in the case of unstable orbits we want to know how fast the non-periodic orbits deviate from it. A measure of this "deviation rate" is the maximal Lyapunov characteristic number (LCN). If the deviations of an orbit from a given periodic or non-periodic orbit, are ε_o and ε at times $t = 0$ and t respectively, then

$$LCM = \lim_{t\to\infty} \chi$$

where (7)

$$\chi = \frac{\ln(\varepsilon/\varepsilon_o)}{t} \quad .$$

If the original orbit is stable then ε is bounded, therefore LCN=0. On the other hand if the whole available phase space (i.e. the phase space limited by the energy supersurface) is finite and stochastic, then the LCM is uniquely defined, independently of the initial condition (except for a set of measure zero, in which χ tends to some other Lyapunov characteristic number besides the maximal one).

In the case that the phase space is divided into a set of invariant surfaces and stochastic regions one expects that the stochastic regions communicate through Arnold diffusion, therefore the LCM is either zero (on the invariant surfaces), or has a unique positive value, as defined by Equations (7) (in the stochastic region, except for the particular set above). However it seems that Arnold diffusion is an extremely slow process, therefore the convergence of χ to a unique final value may take an extremely long time. In practice we find that χ levels off to a finite value after a long time t, but this "limiting" value may be different for various initial conditions (Contopoulos et al. 1978). Only after a much longer time should the value of χ tend to the unique final value appropriate for the whole system. We should remark here that Arnold diffusion affects also the ordered regions, because these regions also are permeated by unstable periodic orbits, followed by small stochastic regions.

If the time scale of Arnold diffusion is much longer than the time needed for a given system to change appreciably, we can ignore it altogether and calculate a LCN for a finite time, but long enough so that the value of χ is stabilized to a certain almost constant value. In such a case the LCN is different in different regions of phase space. E.g. in galactic dynamics we found that the time scale of the Arnold diffusion is much longer than the age of the Universe, therefore it can be safely ignored.

Even with these restrictions the situation is not quite clear. E.g. we have found that orbits starting close to a given unstable periodic orbit, but with slightly different initial conditions may lead to different values of χ if t is finite and not very large. Even a difference in the accuracy of calculations may give different χ's. The reason is that slightly different orbits deviate exponentially not only from the original orbit but also from each other, therefore they may go into different regions of phase space.

In particular although the stochastic regions around various unstable periodic orbits, that form the "Arnold web" (Lichtenberg and Lieberman 1983), are connected, nevertheless their thickness is variable and in some regions it may be extremely small. Therefore one or two nearby orbits may find an opening joining the stochastic regions of two different periodic orbits, while the other orbit may be reflected at some invariant surface. In this respect Arnold diffusion is similar to the "diffusion through cantori" (Bensimon and Kadanoff 1984, MacKay et al. 1984), in strongly perturbed systems of two degrees of freedom.

In order to overcome the ambiguity in the definition of a LCN we can follow two possible ways. Either we calculate the LCN's for several initial conditions over a fixed time interval and take the average (Udry and Pfenniger 1987), or we calculate some orbits close to a periodic orbit for long enough times to see the convergence of χ to a roughly constant value, which is roughly the same for nearby initial conditions. We have followed the second method (Contopoulos and Barbanis 1988) and we have found results that seem to be meaningful. Namely we can usually define a certain LCN in a finite (but small) neighbourhood of a periodic orbit, but sometimes we find a different LCN in a larger region, further away from the periodic orbit.

A particular case in which the LCN is well defined is the case of infinitesimal deviations from a periodic orbit. Such deviations can be given by the variational equations of the periodic orbit. Namely, if both ε_o are ε are infinitesimal we have

$$\varepsilon = \varepsilon_o \Sigma f(t)\, e^{\alpha t} \tag{8}$$

where α represents the characteristic exponents and f(t) are periodic functions with the period T of the periodic orbit. The characteristic exponent can be imaginary (linearly stable orbit), real (simply or doubly unstable orbit), or complex (complex unstable orbit). It is easily seen that the limit

$$\lim_{t\to\infty} \frac{\ln(\varepsilon/\varepsilon_o)}{t}$$

is equal to the maximum positive real part among the various characteristic exponents.*

In the case of stable orbits we have $\alpha_R = 0$. In the case of unstable orbits the characteristic exponents are in pairs ($\pm\alpha$, $\pm\beta$), where $\alpha > 0$ is real and β is either imaginary (simply instability), or real positive and smaller than a (double instability). In the case of complex instability the four nontrivial characteristic numbers are $\pm\alpha_R \pm i\alpha_I$ where $\alpha_R > 0$.

Thus in all cases we define a "local LCN" which is equal to

$$(LCN)_{loc} = \alpha_R . \tag{9}$$

In Figure 5 we have drawn the curves of "equal local LCN" in the parameter space (ε,η) for the Hamiltonian systems

$$H \equiv \frac{1}{2}(\dot{x}^2+\dot{y}^2+\dot{z}^2+Ax^2+By^2+Cz^2) - \varepsilon xz^2 - \eta yz^2 = h , \tag{10}$$

where the quantities A,B,C and the energy h are kept constant. This diagram refers to a particular family of periodic orbits that intersect twice perpendicularly the plane z = 0 and is called family 1a (Contopoulos and Magnenat (1985).

*If two characteristic exponents are equal then (8) contains, instead of f(t), a term $f_1(t)+tf_2(t)$, where $f_1(t)$ are periodic in t. But $\ln[f(t)]/t$ and $\ln[f_1(t)+tf_2(t)]/t$ tend to zero as $t \to \infty$, therefore these factors do not contribute to the $(LCN)_{loc}$.

This family has regions of stability (S), of simple instability (U), of double instability (DU) and of complex instability (Δ) (Broucke 1969, Contopoulos and Magnenat 1985). The local LCN is zero in the stable regions and varies continuously as we move into the unstable regions. As the complex unstable region protrudes into the stable region, the "iso-LCN" curves turn somewhat abruptly downwards into that region. However these curves are smooth, except for the boundary (LCN = 0) which has some angular points.

But the "local NCN" refers only to the neighbourhood of a particular periodic orbit. In fact for every family of periodic orbits we have a different diagram like Fig.5. If we allow a finite deviation from the original (unstable) periodic orbit, the non-periodic orbit deviates exponentially and fills a part of the stochastic region (or the whole stochastic region).

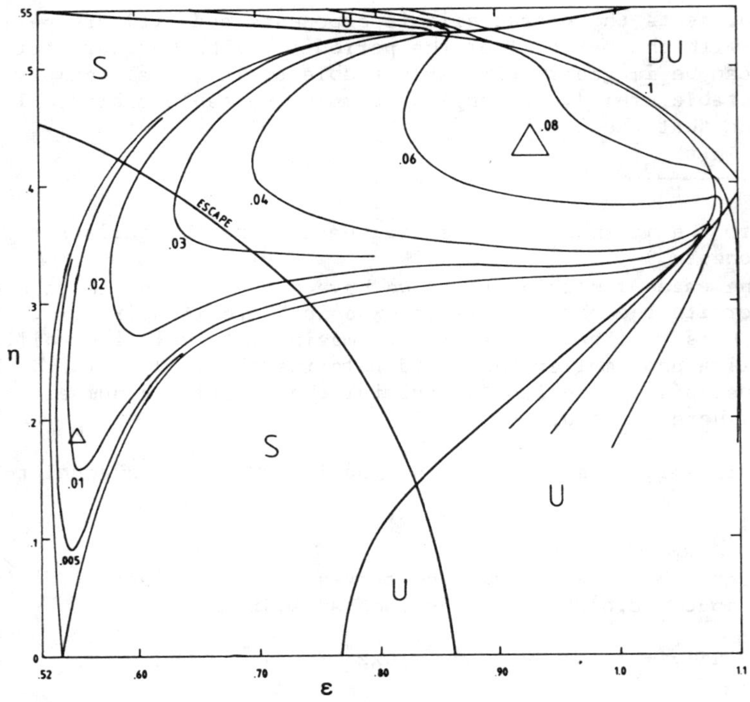

Figure 5: The curves of "equal local Lyapunov characteristic number" for the family. 1a. This family is stable (S), simply unstable (U), double unstable (DU), or complex unstable (Δ). The local LCN is zero in the stable regions.

Of special interest is the study of the regions near complex unstable periodic orbits. During the transition from a stable to a complex unstable region in the parameter space (ε,η) the stochasticity is introduced abruptly, and not through a Feigenbaum sequence of infinite period doubling bifurcations. In fact at the onset of complex instability there is no bifurcation of another family of periodic orbits, as in the other cases of transition to instability. Only quasi-period orbits bifurcate in this case (Heggie 1985, Pfenniger 1985), but this phenomenon is of a quite different nature.

Thus it is of interest to study how the LCN's change as we vary the parameters (ε,η) in the complex unstable region (Fig.6) (Contopoulos and Barbanis 1988).

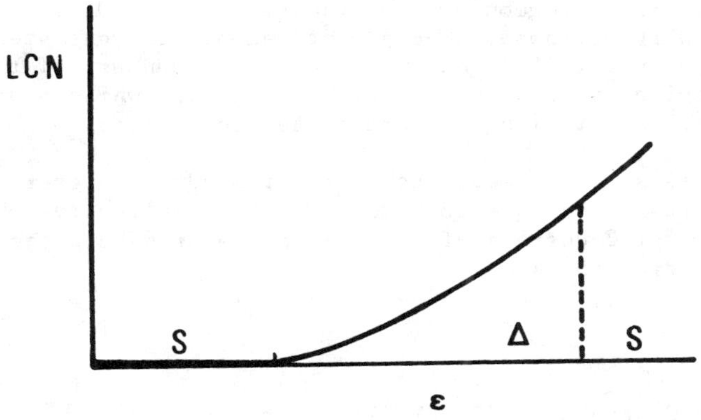

Figure 6: The variation of the (maximal) Lyapunov characteristic number (LCN) as a function of ε for fixed η, when the initial point is at a certain distance from the periodic orbit 1a (schematically). As ε increases the LCN is initially zero, when the periodic orbit is stable, then it increases in the complex unstable region (Δ). Beyond that region, although the periodic orbit is stable, the LCN for the non-periodic orbit continues to increase.

We keep $\eta=$ constant (but smaller than $\eta \simeq 3$) and increase ε. As ε crosses the boundary $S \to \Delta$, the LCN at first increases slowly. The complex unstable orbit produces some stochasticity, but this is rather small. If we move further out of the periodic orbit most orbits are ordered (quasi-periodic) and their LCN is zero. Of course we expect Arnold diffusion to take over and allow a diffusion of the orbits through the ordered region, but this will probably take a very long time, much longer than a Hubble time.

As we increase the value of ε inside the complex unstable region, the LCN increases and the stochastic region also increases. At an even larger ε the periodic orbit becomes again stable (Fig.6). This means that not only the "local LCN" is zero, but also the LCN for initial conditions in a finite small neighbourhood of the periodic orbit is zero. However if we start an orbit further away from the immediate neighbourhood of the periodic orbit the LCN is large and increases as ε increases, despite the fact that the "parent" periodic orbit is now stable.

This behaviour is typical if we scan the parameter space (ε,η) at various values of η = const with increasing ε. Thus we have the somewhat strange result that in a stable region (S) in Fig.6 we have LCN > 0. But again we have to stress that this LCN does not refer to the local effects but to a larger region outside the stable periodic orbits and their immediate neighbourhood. These conclusions are consistent with the remark of Mangenat (1982) that near the left boundary of the complex unstable region the orbits seem rather regular, while near the right boundary of this region they are quite stochastic.

Of course this subject needs much further study. However our results show that it is possible to understand the qualitative characteristics of systems of 3 degrees of freedom, as regards both the periodic and the non-periodic orbits.

REFERENCES

Barbanis, B. and Contopoulos, G.: 1987, in preparation.
Bensimon, D. and Kadanoff, L.P.: 1984, Physica 13D, 82.
Broucke, R.: 1969, Amer. Inst. Aeronaut. Astronatu. J. 7, 1003.
Contopoulos, G.: 1983, Celest. Mech. 31, 193.
Contopoulos, G.: 1986, Astron. Astrophys. 161, 244.
Contopoulos, G., and Barbanis, G.: 1988, in preparation.
Contopoulos, G., and Magnenat, P.: 1985, Celest. Mech. 37, 387.
Contopoulos, G., and Papayannopoulos, Th.: 1980, Astron. Astrophys. 92, 33.
Contopoulos, G., Giorgilli, A. and Galgani, L.: 1978, Phys. Rev.A18,1183.
Contopoulos, G., Gottesman, S.T., Hunter, J.H. Jr. and England, M.N.: 1987, preprint.
Deprit, A. and Henrard, J.: 1968, Adv. Astron. Astrophys. 6, 1.
Deprit, A. and Henrard, J.: 1970, in Giacaglia, G.E.O. "Periodic Orbits, Stability and Resonances", Reidel, p.1.
Heggie, D.C.: 1985, Celest. Mech. 35, 357.
Henrard, J.: 1983, Celest. Mech. 31, 115.
Lichtenberg, A.J. and Lieberman, M.A.: 1983, Regular and Stochastic Motion, Springer Verlag, New York.

MacKay, R.S., Meiss, J.D., and Percival, I.C.: 1984, Physica 13D, 55.
Magnenat, P.: 1982, Celest. Mech. 28, 319.
Pfenniger, D.: 1985, Astron. Astrophys. 150, 97.
Udry, S. and Pfenniger, D.: 1988, Astron. Astrophys. (in press).

ON THE GALACTIC MOTION

F.M.F. El-Sabaa & H.H. Sherief
Kuwait University
Department Of Mathematics
P.O. Box 5969
13060 Safat, Kuwait

ABSTRACT. A certain potential function is studied as a possible model for the galactic potential. Some solutions are obtained. Also the numerical study of some of the orbits is carried out.

INTRODUCTION.

The galactic motion problem is essentially one of determining the forces acting on a system which admits an integral of nonhomogeneous quadratic form in the velocities. In two dimensions and under the steady-state assumption, the problem has been studied for a long time starting with Jeans (1915) and Eddington (1915) [1,2].

The problem has been extended in the period of 1956 to 1966; by Schmidt [4], Contopoulos [5,6,7,8], Henon & Heiles [9], Ollongren [10, 11] and Hori [12].

Our paper consists of two parts: the first part is analytical study of motion for some known potential function; line surfaces of separated functions are studied and some complete solution of motion are given, while the second part is concerned with the numerical study of some periodic orbits and their stability.

1.1 The Equation of Motion

The equation of motion will be written as:

$$\dot{x} = \frac{\partial H}{\partial p_x} , \quad \dot{y} = \frac{\partial H}{\partial p_y} , \quad \dot{z} = \frac{\partial H}{\partial p_z} ,$$

$$\dot{p}_x = - \frac{\partial H}{\partial x} , \quad \dot{p}_y = - \frac{\partial H}{\partial y} , \quad \dot{p}_z = - \frac{\partial H}{\partial z}$$

where $\quad H = \frac{1}{2}(p_x^2 + p_y^2 + p_z^2) + U(x,y,z,t).$

Each star can be represented by a point in the six-dimensional phase space (x,y,z,p_x,p_y,p_z)

To solve the above equations it is required to obtain six independent integrals:

$$I_i(x,y,z,\dot{x},\dot{y},\dot{z},t) = C_i \quad , \quad i = 1,\dots,6$$

A. E. Roy (ed.), Long-Term Dynamical Behaviour of Natural and Artificial N-Body Systems, 317–327.
© 1988 by Kluwer Academic Publishers.

where each equation represents a hypersurface in the phase space and the trajectory is the intersection of the six hypersurfaces. The integrals can be found if we take the two following restrictions:

1) The potential energy U is explicitly independent of time, and then we get the integral of energy

$$I_1 = \tfrac{1}{2}(\dot{x}^2 + \dot{y}^2 + \dot{z}^2) + U(x,y,z) = C_1 = E \ .$$

2) We consider the case when the potential has axial symmetry about some fixed direction in space. Then if we take a system of cylindrical coordinates (r,θ,z) with the z-axis along the axis of symmetry, then the potential energy is $U(r,z)$.

The components of linear velocity are: $\dot{r}, r\dot{\theta}, \dot{z}$ and the Lagrangian L take the form:

$$L = \tfrac{1}{2}(\dot{r}^2 + \dot{r}^2\dot{\theta}^2 + \dot{z}^2) - U(r,z) \quad .$$

Hence the equations of motion become:

$$\frac{d^2r}{dt^2} = r\dot{\theta} - \frac{\partial U}{\partial r} \ , \quad \frac{d}{dt}(r^2\dot{\theta}) = -\frac{\partial U}{\partial r} = 0 \ , \quad \frac{d^2z}{dt^2} = -\frac{\partial U}{\partial r} \tag{1}$$

The second equation of motion gives us the angular momentum integral

$$I_2 = r^2\dot{\theta} = C_2 = h \quad . \tag{2}$$

We take a particular solution which are circles in the Galactic plane: $r = r_0$, $z = 0$, with the condition $h^2/r_0^3 = (\frac{\partial U}{\partial r})_{r=r_0}$,

which means that centrifugal force and centripetal force are equal. By substituting (2) in (1), the equations of motion reduce to two degrees of freedom, we obtain

$$\frac{d^2r}{dt^2} = -\frac{\partial U}{\partial r} \quad , \quad \frac{d^2z}{dt^2} = -\frac{\partial U}{\partial r} \tag{3}$$

where the potential function U is defined by: $U = \dfrac{h^2}{2r^2} + \phi(r,z)$, (4)

It still requires three other independent integrals to describe fully the motion of the galaxy.

It was shown that two of the required integrals are generally nonisolating. (The definition of nonisolating is such that the corresponding hypersurface consists of an infinity of sheets which usually fill the phase space densely). So that the nonisolating integrals have no physical interest.

The existence of a third integral which must be isolating is an open question and is equivalent to a third isolating integral of equation (3).

1.2 Separation of Variables

Hori [10] introduced the elliptic coordinates λ, μ such that $r = C\lambda\mu$, $0 \le \lambda < \infty$, $z = \pm C\sqrt{(1+\lambda^2)(1-\mu^2)}$, $0 \le \mu < 1$, where C is a real constant, and the transformed equation of motion can be separated if the potential function ϕ takes the form $\Phi = \frac{F(\lambda)+G(\mu)}{\lambda^2 + \mu^2}$, F and G being arbitrary functions. The Hamiltonian function H of the system (3) is

$$H = \tfrac{1}{2}(\dot{r}^2+\dot{z}^2) + \tfrac{1}{2}h^2/r^2 + \Phi(r,z) \quad (5)$$

where $$\dot{q}_i = \frac{\partial H}{\partial p_i} \quad , \quad \dot{p}_i = -\frac{\partial H}{\partial q_i} \quad , \quad i = 1,2 \quad (6)$$

To write the Hamiltonian as a functions of λ,μ,p_λ and p_μ, then we have:

$$H = \frac{1}{2C^2(\lambda^2+\mu^2)}\left[p_\lambda^2(1+\lambda^2) + p_\mu^2(1-\mu^2)\right] + \frac{h^2}{2C^2\lambda^2\mu^2} + \frac{F(\lambda)+G(\mu)}{\lambda^2 + \mu^2} , \quad (7)$$

The Hamilton-Jacobi equation of our system is $H(\lambda,\mu,\frac{\partial S}{\partial\lambda},\frac{\partial S}{\partial\mu}) = E$, where $p_\lambda = \frac{\partial S}{\partial\lambda}$, $p_\mu = \frac{\partial S}{\partial\mu}$, (8), and E is the integral of energy constant. Equation (7) becomes:

$$E = \frac{1}{2C^2(\lambda^2+\mu^2)}\left[(1+\lambda^2)(\frac{\partial S}{\partial\lambda})^2 + (1-\mu^2)(\frac{\partial S}{\partial\mu})^2\right] + \frac{h^2}{2C^2\,{}^2\,{}^2} + \frac{F(\lambda)+G(\mu)}{\lambda^2 + \mu^2} \quad (9)$$

where the complete integral S of the above equation is taken as $S = S_1(\lambda) + S_2(\mu)$. Hence equation (9) can be separated into two ordinary differential equations:

$$\lambda^2(1+\lambda^2)(\frac{dS_1}{d\lambda})^2 = 2C^2E\lambda^4 - 2C^2\lambda^2F(\lambda) + \alpha\lambda^2 - h^2$$
$$\mu^2(1-\mu^2)(\frac{dS_2}{d\mu})^2 = 2C^2E\mu^4 - 2C^2\mu^2G(\mu) - \alpha\mu^2 - h^2 \quad (10)$$

where α is an arbitrary constant. The two functions S_1 and S_2 can be taken of the form:

$$S_1 = \int\left(\frac{2C^2E\lambda^4-2C^2\lambda^2F(\lambda)+\alpha\lambda^2-h^2}{\lambda^2(1+\lambda^2)}\right)^{\frac{1}{2}} d\lambda \quad ,$$

$$S_2 = \int\left(\frac{2C^2E\mu^4-2C^2\mu^2G(\mu)-\alpha\mu^2-h^2}{\mu^2(1-\mu^2)}\right)^{\frac{1}{2}} d\mu \quad .$$

According to (8), we have:

$$\frac{\lambda\, d\,\lambda}{\sqrt{(2C^2E\lambda^4-2C^2\lambda^2F(\lambda)+\alpha\lambda^2-h^2)(1+\lambda^2)}} = \frac{dt}{C^2(\lambda^2 + \mu^2)}$$
$$\frac{\lambda\, d\,\lambda}{\sqrt{(2C^2E\mu^4-2C^2\mu^2G(\mu)-\alpha\mu^2-h^2)(1-\mu^2)}} = \frac{dt}{C^2(\lambda^2 + \mu^2)} \quad (11)$$

We introduce the artificial time τ connected with the time t by the relation $dt = C^2(\lambda^2+\mu^2)d\tau$. Hence by integrating (11), we get

$$\int_{\lambda_0}^{\lambda} \frac{\lambda\, d\lambda}{\sqrt{Q(\lambda)(1+\lambda^2)}} = \tau - \tau_0, \int_{\mu_0}^{\mu} \frac{\mu\, d\mu}{\sqrt{R(\mu)(1-\mu^2)}} = \tau - \tau_1 \qquad (12)$$

where $Q(\lambda) = 2C^2E\lambda^4 - 2C^2\lambda^2F(\lambda) + \alpha\lambda^2 - h^2$,

$R(\mu) = 2C^2E\mu^4 - 2C^2\mu^2G(\mu) - \alpha\mu^2 - h^2$,

and τ_0 , τ_1 are arbitrary constants of integration.

1.3 Phase Portrait

The equation (10) can be written as follows:

$$\begin{aligned} \lambda^2(1+\lambda^2)p_\lambda^2 &= 2C^2E\lambda^4 - 2C^2\lambda^2F(\lambda) + \alpha\lambda^2 - h^2 \\ \mu^2(1-\mu^2)p_\mu^2 &= 2C^2E\mu^4 - 2C^2\mu^2G(\mu) - \alpha\mu^2 - h^2 \end{aligned} \qquad (13)$$

Now our aim is to establish the line surfaces of the above mentioned functions in the plane $(\underline{q},\underline{p})$, where $\underline{p} = (p_1,p_2) = (p_\lambda,p_\mu)$, $\underline{q} = (q_1,q_2) = (\lambda,\mu)$ which are generalized momenta and generalized coordinates.

Let the functions $Q^*(q_1,p_1)$, $R^*(q_2,p_2)$ be defined as:

$$Q^* = 2C^2Eq_1^4 - 2C^2q_1^2F(q_1) + \alpha q_1^2 - q_1(1+q_1^2)p_1^2 - h^2 , \qquad (14)$$

$$R^* = 2C^2Eq_2^4 - 2C^2q_2^2G(q_2) - \alpha q_2^2 - q_2^2(1-q_2^2)p_2^2 - h^2 \qquad (15)$$

and consider the arbitrary functions $F(q_1)$, $G(q_2)$: $F(q_1) = a_1q_1^2+b_1$ and $G(q_2) = a_2q_2^2+b_2$.

To get the line surfaces of the first function, it is necessary to find its singular points.

Singular points of Q^* are obtained by solving the equations

$$\frac{\partial Q^*}{\partial q_1} = 8C^2(E-a_1)q_1^3 - 2(\alpha-2C^2b_1)q_1 - 2p_1^2q_1(1+q_1^2) = 0 ,$$

$$\frac{\partial Q^*}{\partial p_2} = - 2p_1q_1^2(1+q_1^2) = 0$$

The singular points of Q^* are $(0,0),(0,p),\left(\pm\sqrt{\dfrac{2C^2b_1-\alpha}{4C^2(E-a_1)}},0\right)$.

The points $(0,0)$, give no information. To investigate the type of the points

$\left(\pm\sqrt{\dfrac{2C^2b_1-\alpha}{4C^2(E-a_1)}}, 0\right)$, we put $q_1 = \pm\sqrt{\dfrac{2C^2b_1-\alpha}{4C^2(E-a_1)}} + x$, $p_1 = y$.

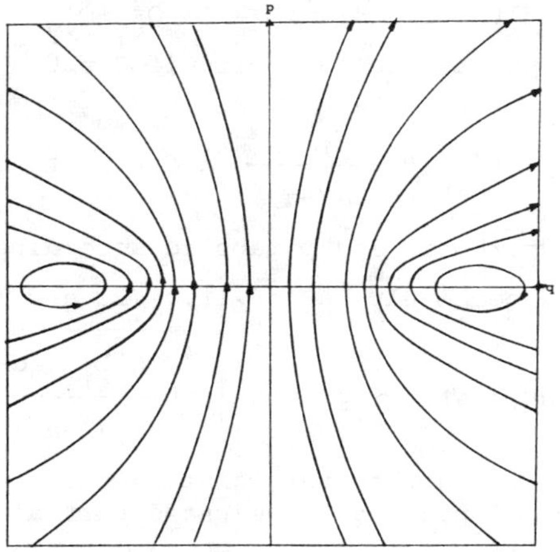

Fig. 1
Phase portrait of Q^* at the point $\left(\pm\sqrt{\frac{2C^2b_1-\alpha}{4C^2(E-a_1)}}\,,\,0\right)$

"elliptic point", where $E < a_1$.

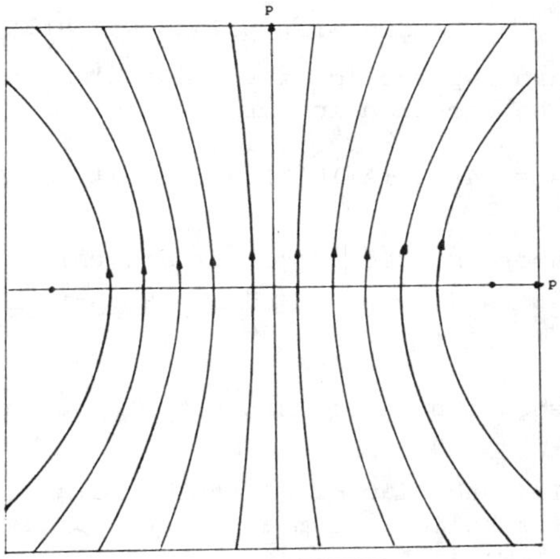

Fig. 2
Phase portrait of Q^* at the point $\left(\pm\sqrt{\frac{2C^2b_1-\alpha}{4C^2(E-a_1)}}\,,\,0\right)$

"hyperbolic point" where $E > a_1$.

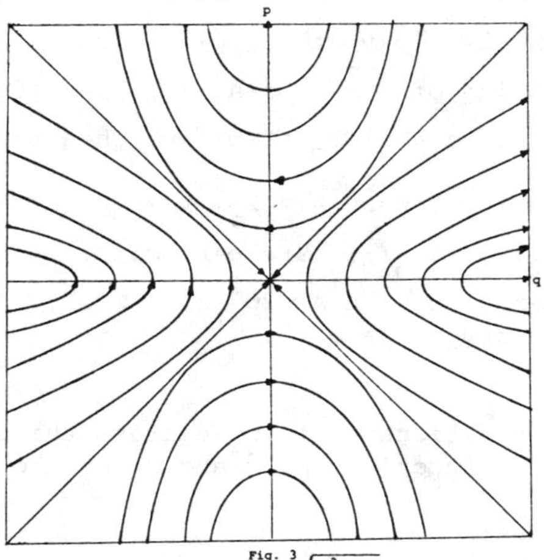

Fig. 3
Phase portrait of Q^* at the point $\left(\pm\sqrt{\frac{2C^2b_1-\alpha}{4C^2(E-a_1)}}\,,0\right) \equiv (0,0)$

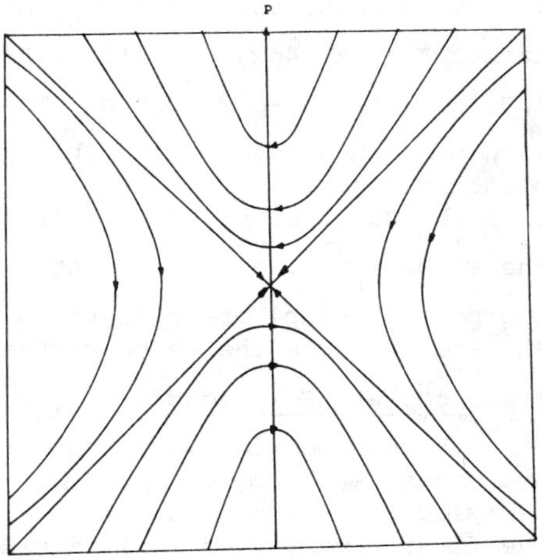

Fig 4

The singular points are elliptic if the Jacobian $J = Q^*_{xx}Q^*_{yy}-Q^*_{xy} > 0$ and hyperbolic if the Jacobian < 0. The points are parabolic if J = 0. The Jacobian in our case is

$J = -4A^2(1+A^2)[12C^2A^2(E-a) + (\alpha-2C^2b_1)]$, $A = \sqrt{\dfrac{2C^2b_1 - \alpha}{4C^2(E-a_1)}}$ and then the above mentioned points are elliptic, hyperbolic and parabolic according to $C^2 < \dfrac{\alpha}{2b_1}$, $C^2 > \dfrac{\alpha}{2b_1}$ and $C^2 = \dfrac{\alpha}{2b_1}$ respectively. Fig. (1), (2) and (3) show the phase portrait of the function Q* at the points $\left(\pm\sqrt{\dfrac{2C^2b_1 - \alpha}{4C^2(E-a_1)}}, 0\right)$.

The Jacobian of Q* at the point (0,p) = 0, and then the points (0,p) are parabolic. It means that for any value of p, we get the phase portrait shown in Fig. (4).

By the same way the function R* has the singular points (0,0), (±1,p), (±B,0), where $B = \sqrt{\dfrac{\alpha - 2C^2b_1}{4C^2(E-a_2)}}$. The points (±B,0) are hyperbolic, elliptic or parabolic according as $C^2 < \dfrac{2\alpha}{b_2}$, $C^2 > \dfrac{2\alpha}{b_2}$ or $C^2 = \dfrac{2\alpha}{b_2}$, respectively. The graphs of the phase portrait of the function R* with the above mentioned point are exactly the same as those in Figs (1), (2) and (3). To get the type of the point (±1,p), we get the function R* in the form: $R^* \simeq Ax^2+4pxy$ at (1,p), $R^* = Ax^2-2y^2-4pxy$ at (-1,p), where $A \equiv 12C^2E-12C^2a_2-\alpha-2C^2b_1+7b^2$. The Jacobian of R* at point $(1,p) = -16p^2$. If p = 0, then the point (1,p) (1,0) is parabolic, if p ≠ 0, then the point is hyperbolic.

The Jacobian of R* at $(-1,p) = 8A-16p^2$. When $A = 2p^2$ then p takes the value $\pm\sqrt{\frac{A}{2}}$, and the points $(-1,\sqrt{\frac{A}{2}})$, $(-1,-\sqrt{\frac{A}{2}})$ are parabolic. The points (-1,p) are elliptic or hyperbolic when $A > 2p^2$ or $A < 2p^2$. Figs (5) and (6) show the phase portrait at any value of p.

1.4 Complete Solutions

In this section we integrate the equations (11) and express the coordinates λ, μ by Weierstrass elliptic functions [13] and study the behaviour of the motion.

The first equation of (11) is written as:

$$I = \int \frac{\lambda \, d\lambda}{\sqrt{(\lambda^2+1)2C^2\lambda^4(E-a_1) + (\alpha-2C^2b_1)\lambda^2-h^2}}$$

Introducing the quantities K_1, K_2, K_3 such that:

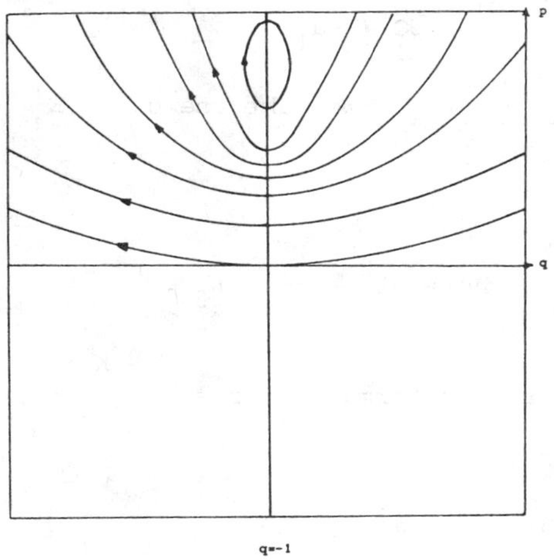

Fig 5

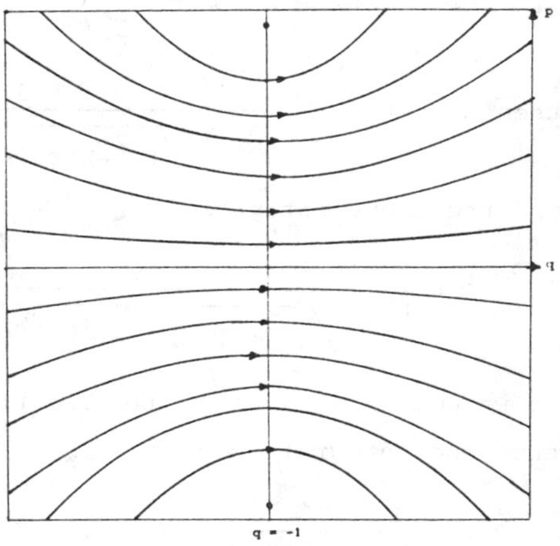

Fig. 6

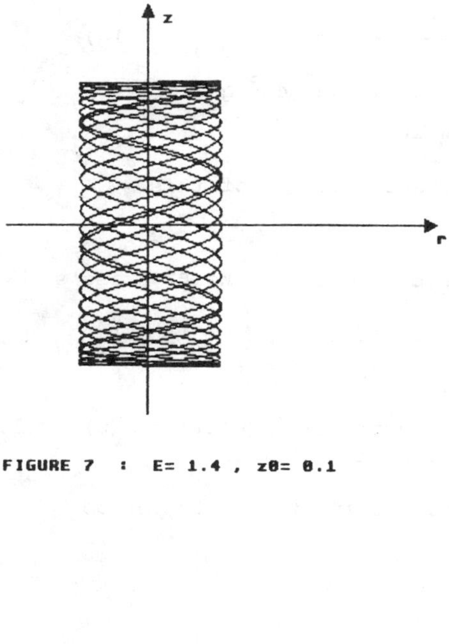

FIGURE 7 : E= 1.4 , z0= 0.1

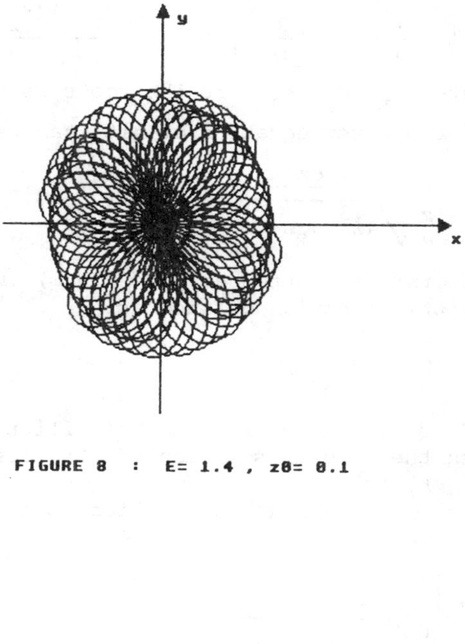

FIGURE 8 : E= 1.4 , z0= 0.1

$$K_1 = \sqrt{2C^2(E-a_1)} \quad , \quad K_2 = \frac{\alpha-2C^2b_1}{2C^2(E-a_1)} \quad , \quad K_3 = \frac{-h^2}{2C^2(E-a_1)} \tag{15}$$

then we have $I = \frac{1}{K_1}\int \frac{\lambda\, d\lambda}{\sqrt{(\lambda^2+1)(\lambda^4+K_1\lambda^2+K_2)}}$. We take the case $E > a_1$

and the above integral is

$$I = \frac{1}{K_1}\int \frac{\lambda\, d\lambda}{\sqrt{(\lambda^2+1)\ \left((\lambda^2 + \frac{k_1}{2})^2-\beta^2\right)}} \ , \text{ where } \lambda^2 = \frac{K_1^2}{4} - K_2 \ .$$

If we put $\lambda^2 = v$ and take the linear transformation $z = v - \frac{K_1+1}{3}$,
then the integral becomes:

$$I = \frac{1}{K_1}\int \frac{dz}{\sqrt{4z^3-g_2z-g_3}} \ .$$

We have the relations: $e_1+e_2+e_3 = 0$, (16)

$$-\frac{1}{4}g_2 = \left[-\frac{(K_1+1)^2}{3} + K_2 - K_1\right] = e_1e_2+e_2e_3+e_3e_1 \ , \tag{17}$$

$$-\frac{1}{4}g_3 = \left[\frac{8}{27}(K_1+1)^3 - \frac{(K_1+1)(K_2-K_1)}{3} + K_2\right] = -\ e_1e_2e_3 \ , \tag{18}$$

where e_1, e_2, e_3 are the roots of the equation $4z^3-g_2z-g_3 = 0$.
Now the first equation of (11) can be written

$$u = \int_z^\infty \frac{dz}{\sqrt{4z^3-g_2z-g_3}} \ , \ u = K_1(\tau-\tau_0).$$ The inversion of this

integral is written as $z = \rho(u,\omega)$ where $\rho(u,\omega)$ is Weierstrass's elliptic function defined as:

$$\rho(u) = \frac{1}{u^2} + g_2\frac{u^2}{20} + g_3\frac{u^4}{28} + g_2^2\frac{u^6}{1200} + \frac{3g_2g_3}{6160}u^8 + \ldots \tag{19}$$

and ω is the half period of elliptic functions: $\rho(u+2\omega) = \rho(u)$. (20)
Then the general solution of the first equation (11) is

$\lambda^2 = \frac{K_1+1}{3} + \rho(K_1(\tau-\tau_0))$, which is a periodic function with period

$$\omega = \int_0^\infty \frac{dz}{\sqrt{4z^3-g_2z-g_3}} \ .$$

The description of the motion is dependent on the roots of the equation $4z^3 - g_2 z g_3 = 0$. If two of the three roots e_1, e_2, e_3 coincide, then asymptotic motion appears, where the elliptic functions degenerate to trigonometric or hyperbolic functions. If one root in the real domain coincide with another root, then the motion makes a spiral approach to the circle corresponding to these coincident roots. If the two roots bounding the real domain coincides, then the motion is circular.

By the same way we can get the general solution of the second equation (11) and the behavior of motion according to the roots of the equation $4z_1^3 - g_2' z_1 - g_3' = 0$ with period ω_1.

When the ratio of the periods ω/ω_1 is a rational number, then the motion is periodic, where the orbits are closed after some revolutions. If the ratio ω/ω_1 is an irrational number then the motion is quasi-elliptic and the trajectory fills up some region of the phase space densely.

2.1 Periodic Orbits

The fixed point method introducing by Poincaré [14] and extended by Hénon will be used to find the periodic orbits of our system

$$\ddot{r} = \frac{h^2}{r^3} - \frac{\partial\phi}{\partial r}, \quad z = -\frac{\partial\phi}{\partial z}, \quad \text{where} \quad \phi = \frac{F(\lambda) + g(\mu)}{\lambda^2 + u^2} \tag{21}$$

The function ϕ can be expressed in terms of r, z by using the relations

$$\lambda^2 = \frac{1}{2C^2}[r^2 + z^2 - C^2 + \sqrt{(r^2+z^2-C^2)^2 + 4r^2C^2}\,]$$

$$\mu^2 = \frac{1}{2C^2}[C^2 - r^2 - z^2 + \sqrt{(r^2+z^2-C^2)^2 + 4r^2C^2}\,]$$

The system (21) possesses the integral energy $E = \frac{1}{2}(\dot{r}^2+\dot{z}^2)+\frac{h^2}{2r^2}+\phi(r,z)$.

The above equation restricts the motion, such that

$$\dot{r} = \sqrt{2E - \frac{h^2}{r^2} - 2 \quad - z^2}$$

which means that it define a three-dimensional surface of constants E and h lying within the four-dimensional phase space on which the trajectories must lie. For a given E, h the motion is specified by a curve in the space $(r,z,\dot{z})$. If the trajectory goes through a point $(z_0,\dot{z}_0)$ at time $t = t_0$, with $r = 0$, then the subsequent motion is determined, and in particular the next intersection $P_1(z_1,\dot{z}_1)$ with the same surface of section. The point P_0 is said to be mapped into P_1 which can be denoted by $TP_0 \rightarrow P_1$.

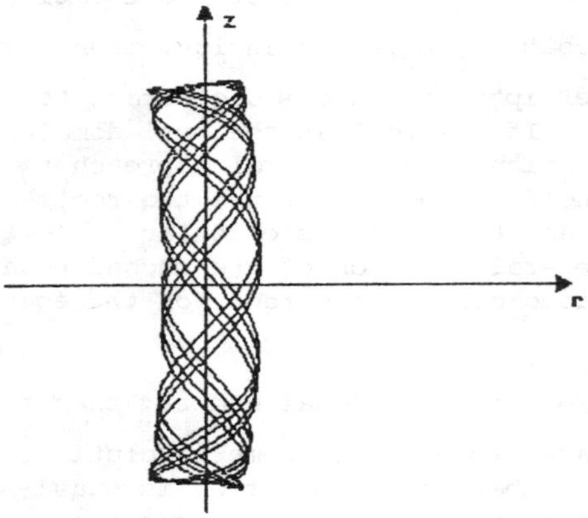

FIGURE 9 : E= 1.5 , z0= 0.5

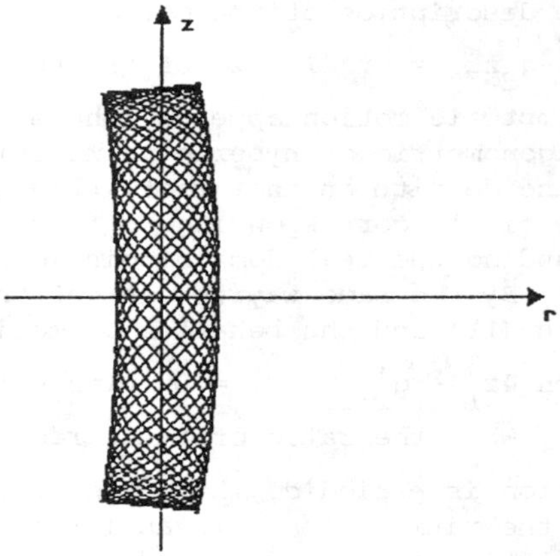

FIGURE 10 : E= 1.5 , z0= 0.5

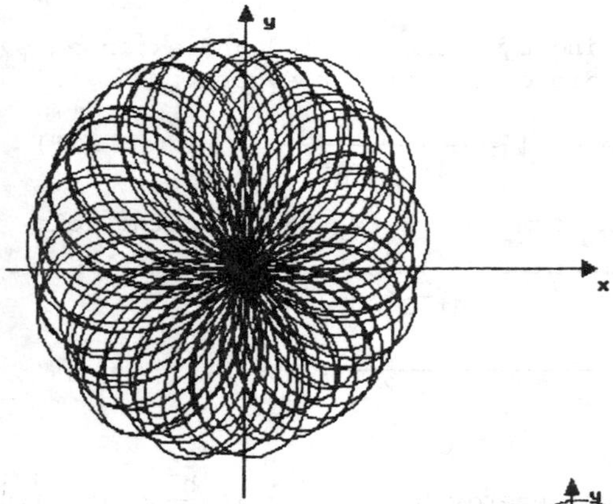

FIGURE 11 : E= 1.5 , z0=

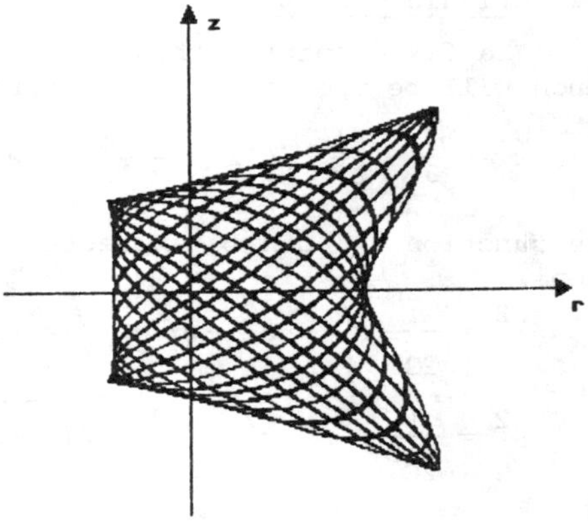

FIGURE 12 : E= 1.8 , z0= 1

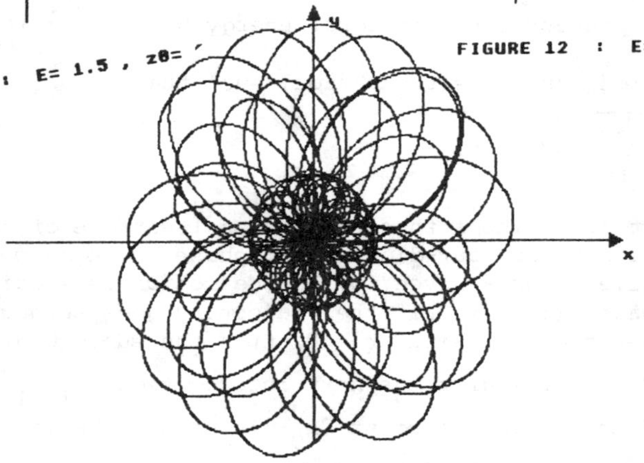

FIGURE 13 : E= 1.8 , z0= 1

It is clear that z-axis will be an axis of symmetry for the mapping, since the axis is the axis of symmetry for the equations of motion. Periodic orbits consist of a finite and discrete set of fixed points on the intersection of some surfaces. The symmetric simple periodic orbit with z-axis crosses this axis twice perpendicularly (where $z = 0$), once with $\dot{r} > 0$ and once with $\dot{r} < 0$. The periodic orbit of order m is that one that crosses the z-axis with $\dot{r} > 0$, m times and with $\dot{r} < 0$ m times. Of these 2m crossings, there is only 2 perpendicular ones.

For illustration the numerical computations were carried out for three cases. We have chosen $a_1 = 1.989$ and $a_2 = 1.346$, $b_1 = b_2 = 0$.

Case I: $E = 1.4$, $z0 = 0.1$, $\dot{z}0 = 0$.

The graph of z against r spirals around the z-axis as shown in fig. 7 going up and down. The projection on the x-y plane is shown in fig. 9 and consists of almost equal sized leafs. The moving point returns to its initial position after 33 revolutions.

Case II: $E = 1.5$, $z0 = 0.5$, $\dot{z}0 = 0$.

Again z is plotted against r, the graph was halted after 20 revolutions to show an intermediate state (fig. 9), as before the moving point spirals around the z-axis. The complete path is shown in fig. 10. The projection on the x-y plane (shown in fig. 11) consists of leafs of increasing size. The path closes after 37 revolutions.

Case III: $E = 1.8$, $z0 = 1$, $\dot{z}0 = 0$.

The graph of z against r is shown in fig. 12 with the projection on the x-y plane shown in fig. 13. The path closes after 31 revolutions.

REFERENCES

[1] Jeans, J.H.: 1915, *Monthly Notices Roy. Astrom. Soc.* 76, p. 78.

[2] Eddington, A.S.: 1915, *Monthly Notices Roy. Astron. Soc.* 76, p. 37.

[3] Oort, J.H.: 1928, *Bull. Astron. Inst. Neth.* **4**, p. 269.

[4] Schmidt, M.: 1956, *Bull. Astron. Inst. Neth.* **13**, p. 15.

[5] Contopoulos, G.: 1957, *Stockholms Obs. Ann.* No. **10**, p. 19.

[6] Contopoulos, G.: 1958, *Ibid.* No. **5**, p. 20.

[7] Contopoulos, G.: 1960, *Z. Astrophys.* **49**, p. 273.

[8] Contopoulos, G.: 1963, *Astron. J.*, Vol. **68**, No. 1, p. 1.

[9] Hénon & Heiles, C.: 1964, *Astron. J.*, Vol. **69**, No. 1, p. 73.

[10] Ollongren, A.: 1962, *Bull. Astron. Inst. Neth.*, **16**, p. 241.

[11] Ollongren, A.: 1965, *Ann. Rev. Astron. Astrophys.* **3**, p. 113.

[12] Hori, Gen'i Chiro: 1962, *Bull. Astron. Soc. Japan,* No. 4, p. 353.

[13] Whittaker, E. and Watson, T: 1943, *A course in Modern Analysis,* University Press, New York.

[14] Poincaré, H.: 1957, *Les Méthodes Nouvelles de la Mécanique Céleste,* Dover Publications, New York.

THE N-BODY PROBLEM IN STELLAR DYNAMICS

Douglas C. Heggie
University of Edinburgh
Department of Mathematics
King's Buildings
EDINBURGH EH9 3JZ, U.K.

ABSTRACT: One of the goals of stellar dynamics is the understanding of globular star clusters. In the past the large number of stars in these systems has led to the need for simplified mathematical models, but the direct simulation of a small cluster using the N-body equations should soon become possible. This paper describes some of the computational techniques which are now being developed for this purpose. One problem to which particular attention is paid is the treatment of close approaches by a small number of bodies. Another section of the paper summarises some facts and ideas about the growth of errors in large numerical N-body simulations.

1. INTRODUCTION

Several kinds of N-body problem are of importance in stellar dynamics. Binary and multiple stellar systems are described by few-body problems, while whole galaxies can be modelled as N-body systems with $N \sim 10^{11}$. This paper is concerned with N-body problems of an intermediate size, $N \sim 10^5 - 10^6$, since this is comparable with the total number of stars in a galactic globular cluster.

There are statistical methods of handling the dynamics of such systems. But progress in the use of these methods has had the effect of exposing their shortcomings, and in future greater emphasis will be placed on the direct study of the N-body equations, without further simplification of the physical model.

In this paper we shall outline some of the present approaches to the direct numerical integration of the equations of motion, i.e.

$$\ddot{\underline{r}}_i = -G \sum_{\substack{j=1 \\ j \neq i}}^{N} m_j \frac{\underline{r}_i - \underline{r}_j}{|\underline{r}_i - \underline{r}_j|^3}, \qquad i=1,\ldots,N, \qquad (1)$$

A. E. Roy (ed.), Long-Term Dynamical Behaviour of Natural and Artificial N-Body Systems, 329–347.
© 1988 by Kluwer Academic Publishers.

where $\ddot{\underline{r}}_i$ is the acceleration of the ith star, whose mass is m_i. Of course these equations are common also to celestial mechanics, but here the number of equations is much larger. Also the masses are comparable, and it turns out that the orbits are much more irregular than in most problems of celestial mechanics.

Three problems must be confronted in a numerical approach to the solution of eqs.(1). First there is the great computational effort in evaluating the right-hand sides and in following the orbits of so many stars. The second question concerns the growth of errors in the numerical solution; very little progress has been made in understanding this. Finally, the collision singularity, which is encountered when $\underline{r}_i = \underline{r}_j$ for some pair, gives rise to severe numerical difficulties in its vicinity.

Mainly because of limitations in computer time (the first problem in the previous paragraph) little progress has been made in studying systems containing more than a few thousand bodies. In this paper we shall review some of the tricks by which even this much progress has been achieved, and some recent promising developments which may soon bring within reach the goal of simulating a modest globular star cluster.

Before embarking on this survey, however, it is worth setting down some standard results from the theory of stellar systems (see, for example, Saslaw 1985). If $M = Nm$ is the total mass of the cluster (whose stars are all assumed to have the same mass m), let its potential energy be written as $-GM^2/2R$ (which defines the so-called virial radius R, which is actually the harmonic mean separation of the stars) and its kinetic energy be written as $\frac{1}{2}Mv^2$ (in which v^2 is the mean square speed of the stars). We consider systems which are nearly in "virial equilibrium', which means that

$$Mv^2 \simeq \frac{GM^2}{2R} \quad . \tag{2}$$

Then an important time scale is the <u>crossing time</u>, defined as

$$t_{cr} = 2R/v. \tag{3}$$

For many globular star clusters this is of the order of 10^6 yr. For a uniform sphere of density n it is easy to show from (2) and (3) that

$$t_{cr} \simeq (G\,m\,n)^{-\frac{1}{2}} \tag{4}$$

The literature cited in this paper refers mostly to recent reviews and articles and not always to the original sources of the ideas. However, the latter can usually be traced by consulting the bibliographies of the papers to which we shall refer.

2. EFFICIENCY

2.1 Individual time steps and the AC scheme

Suppose one attempts to integrate equations (1) numerically, using almost the simplest integration scheme on a small microcomputer. Then two problems quickly arise if N is increased. One is that storage is exhausted, but the other is the processing time taken. A rough estimate of the latter is 10 N^2sec per crossing time. If one could extrapolate this result to $N \simeq 10^6$ we would find that the simulation would take about 10^6 yr per crossing time, i.e. it would proceed at about the same rate as a real star cluster.

In order to make the simulation more efficient, the computational effort should be concentrated where it is needed. Stars moving on long-period orbits near the outside of the cluster are subject to a gravitational acceleration which fluctuates with much smaller amplitude, and on a much longer time-scale, than stars near the dense centre of the cluster. Hence in general the time step for the outer stars can be made much longer than that required for the innermost stars (Fig.1).

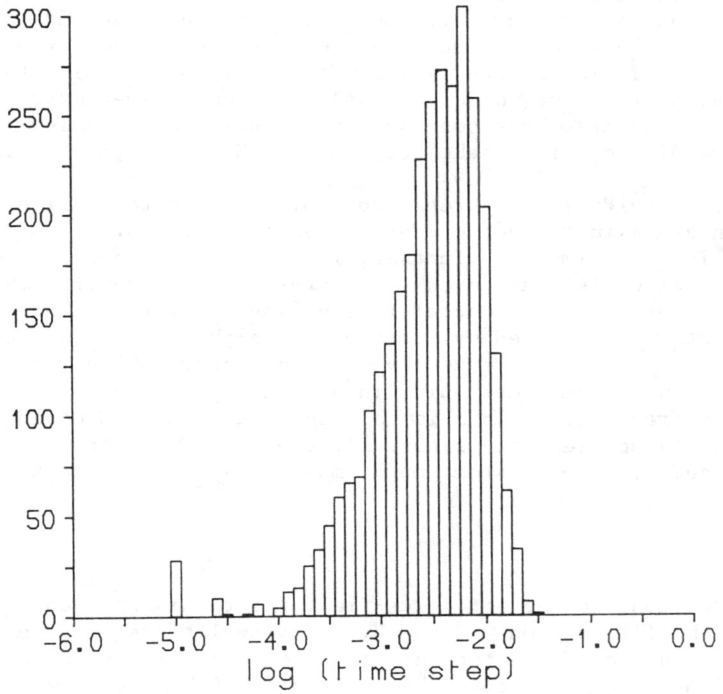

Figure 1. Histogram of time steps in a simulated cluster with N=3151. Time steps were computed using a slight modification of Aarseth's formula (Aarseth 1985, eq.(9)), except that the time steps of a number of stars were adjusted in order that

> the particle advance could be done in parallel. The distribution of stars was approximately isothermal (with density $\rho \propto r^{-2}$, where r is distance from the centre of the system) except for the core containing the innermost twenty or thirty stars, where the density was roughly uniform. The system actually represents only the innermost N stars from a larger cluster with N = 20,000. The remaining (outer) stars would generally have larger time steps than those shown. Also the inter-particle potential ϕ was 'softened' (i.e. $\phi \propto (r^2 + \varepsilon^2)^{-\frac{1}{2}}$, where r is the inter-particle separation and ε is a small constant) in order to suppress effects from close binaries. If such binaries had been allowed to form, considerably smaller time steps would almost certainly have been present.

Note that this advantage cannot be exploited if one uses an equation-solver from a standard library of mathematical subroutines, which almost always force the use of the same time step to solve all the equations of a system such as (1). Even in a purpose-built subroutine, some methods of integration are better suited to individual time steps than others. A predictor—corrector method, based on polynomial approximations to the acceleration of each star, has been preferred traditionally (Aarseth 1985). These readily allow the positions of all stars to be predicted at the time when the total force on a given star has to be calculated. A program based on these principles, and running on a conventional mainframe, might take roughly 10^{-3} N^2s per crossing time.

By use of individual time steps, most of the computational effort is expended on stars in the dense central region of the star cluster. If efficiency is to be improved further, it must be done here. One important observation is that, even for a star near the centre, the most rapid fluctuations in its acceleration (and it is those that control the time step) are caused by its nearest neighbours. Their contributions must be computed frequently, but the contribution from the other stars (which is the great bulk, in a large system) can be computed much less frequently. This is the basis of the Ahmad-Cohen scheme, which, despite its complexity, is more efficient than the basic method already discussed even for systems much smaller ($N \sim 10^2$) than a globular cluster.

2.2 Trees

The fluctuations in the net force-field due to distant bodies are small because the fluctuations caused by individual stars tend to cancel statistically. The net force they exert varies very little as the stars move about. Hence the force could be calculated with tolerable accuracy by replacing the individual stars by a suitable distribution of matter. This is the basis of 'tree codes', in which distant groups of particles are lumped in suitable ways, and the force due to the stars is replaced by the force due to the lump. The reason why these are called tree codes is that the lumping of particles is achieved by organising the information on the particles in a tree, rather than in a conventional array. Several approaches to this idea are described in Barnes and Hut (1986), Press (1986), and in some of the references in those papers.

A tree may be thought of as consisting first of a root with a fixed small number, n, of branches (e.g. n = 2 for a binary tree). At the end of each branch is a node, from which again sprout n subbranches, and so on. Information on N bodies may be accommodated within a tree containing of order log N levels of branching, and the construction of such a tree requires of order N log N operations.

In the Barnes-Hut algorithm, each node of the tree contains information on the total mass and total barycentre of all the particles associated with all the subbranches which emanate from that node. Thus the finest nodes contain information on one body each, while information at the root corresponds to the mass and barycentre of the entire cluster. When the force on a star is being evaluated, the contribution from the more distant stars may be determined by a barycentric approximation; in other words, only relatively low-level branchings (corresponding to large groups of stars) need be considered. Only for the nearest neighbours is it necessary to sum contributions from the finest subbranches.

To estimate the effort required to obtain the force on a single star, let us suppose that the barycentric approximation is adequate if $d/D < \varepsilon$, where d is the size of the clump, D is its distance, and ε is some small number. Suppose also that the clumps at the rth level of subbranching of the tree have a size d of order $2^{-r} R$, where R is a measure of the size of the whole cluster. Then clumps at this level will be suitable at distances satisfying $D > D_{min} \sim 2^{-r} R/\varepsilon$, while even coarser clumps will be adequate at distances $D > D_{max} \sim 2^{1-r} R/\varepsilon$. Hence the number of clumps to be included at the rth level is of order $(D_{max}/d)^3 \sim 8\varepsilon^{-3}$, i.e. independent of r. Since there are of order log N levels, this is how the number of contributions scales with N.

If the tree has to be constructed afresh (O(N log N) operations) before each force calculation, this method cannot compete with a direct force evaluation (O(N) operations). Therefore one of the challenges of incorporating individual time steps within a tree calculation is to economise on the evolution of the tree (whose structure may well change each time a single star is moved). For this and other reasons the suitability of tree codes for the star cluster problem is not yet established, but they have been mentioned here because they are so promising.

2.3 Parallel computation

The evaluation of the force on a star is ideally suited to vector or parallel computation. In other words with a suitable kind of computer, some or all of the (N-1) individual force contributions on a star can be computed simultaneously, rather than sequentially. The book edited by Hut and McMillan (1985) includes several papers describing ways in which this feature of the N-body equations can be exploited in a variety of computers. Much depends on the architecture of the computer.

One architecture which is not represented in the papers in this volume is the type represented by the ICL DAP (Distributed Array Processor). Until recently two such devices were in operation at Edinburgh, and the author acquired some experience in their use for N-body simulations.

A DAP consists of 4096 processors arranged in a 64 x 64 array, and can be programmed to carry out the same arithmetic operations simultaneously on all the elements of a 64 x 64 matrix of variables, or on any subset of them. Operations on vectors of length 64 (or subsets of these) and on scalars are also catered for. The processors are relatively slow: the product of two matrices takes approximately 273 μ sec, as against 1.8 μ sec for an Amdahl mainframe. On the other hand the DAP performs 4096 multiplications in the stated time, and so it can perform around 15×10^6 equivalent scalar multiplications per second.

A notable feature of the DAP is the ease of programming and running. The standard high-level programming language is a dialect of FORTRAN, with extensions to cope with vectors and matrices, and jobs may be loaded and run just as easily as on a conventional mainframe.

In transferring Aarseth's code NBODY1 (which has individual time steps, but no other major refinement) to the DAP, it was decided to arrange the bodies as a matrix. This limits N to the range $N \leq 4096$, but other arrangements would be possible and, in some circumstances, desirable; the chosen arrangement was decided on because an application to several thousand bodies was envisaged. The evaluation of the force on the body is now highly parallel; for example, the computation of one coordinate-difference between a given body and all others is a single matrix calculation. Since the evaluation of the force is so rapid, the most time-consuming part of the program is now the particle-advance, i.e. the part in which the new position and velocity of the star at the end of a time step are computed from the old values of those quantities, and values of the force at several successive times.

The particle advance is a scalar calculation, even on a device like the DAP. (Vectorising the three Cartesian components of the calculation does not lead to any dramatic improvement). Because scalar computations on the DAP are relatively so slow, it was decided to vectorise the particle calculation by forcing a number of stars to be advanced in lock step. In NBODY1 the time steps of the bodies are all different, and the next body to be advanced is found by searching through a list of the N times at which the N bodies are to be advanced. Forcing a number, n, to be advanced in lockstep means finding the next n times and bringing those forward to the earliest one. The advantage gained by vectorisation is, however, offset by two considerations: (i) n-1 stars are advanced sooner than they need be, and (ii) the force calculation is no longer parallel. Nevertheless the new strategy led to great savings for relatively small systems ($N \ll 4096$). Because of the structure of the DAP it was convenient to take n = 64. But for a star cluster this is an arbitrary number; no doubt an improved strategy could be devised in which n depends on the distribution of the time steps.

With N = 3151 the computational speed was found to be equivalent to about 7×10^6 useful floating point operations per second. This is about half the optimal figure quoted above; the loss is partly due to the fact that each matrix operation gives only 3151 equivalent scalar operations rather than 4096.

In the computation referred to the N bodies were distributed as described in Fig.1. They represent the innermost 15% (approximately) of the stars of a notional 20000-body system. The computing time per overall crossing time was estimated at 1.5 days. This would have been substantially or greatly increased had it not been for the steps which were taken to suppress the formation of very close binary stars. More details of this calculation and its astrophysical implications will be presented elsewhere.

3. ERRORS

3.1 Integrals of motion

In celestial mechanics the errors in a numerical integration can be checked in several ways: (i) comparison with an exact solution, (ii) conservation of first integrals, (iii) time reversal. The first method is not available in the star cluster problem because the highly random motions bear no resemblance to those in any exact solution of the equation of motion. The third method is never applied in practice because all except very brief integrations would fail, for reasons which will become clear in §3.2. That leaves the 10 classical integrals, and of these the most useful is the total energy.

Even here the difference between celestial mechanics and stellar dynamics is great. In celestial mechanics it is customary to use integration routines of high or moderately high order, which, with appropriate time steps, should lead to a relative error in the energy much better than $\delta E/E \sim 10^{-10}$ per orbit. In stellar dynamics the overriding need for efficiency is better served by relatively low-order schemes, and larger errors ($\delta E/E \sim 10^{-5}$ per crossing time) are tolerated.

That errors of this size do not vitiate the usefulness of the results is little more than a article of faith. For the larger systems whose simulation is now in prospect, energetic considerations suggest that a lower level of errors will be essential. A simple model of, say, a 20000-body problem (see, for example, Heggie and Ramamani 1987) shows that there are prolonged phases in which the central few percent of the stars undergo oscillations with a time scale of order 100 t_{cr}. If most of the numerical errors occurred in this central group of stars (which is plausible, since interactions are most vigorous and frequent near the centre), the energy generated by numerical errors in this time scale is several percent of the energy of this subsystem. Since the expansion is driven by a delicate exchange of energy within this subsystem, the effect of this error in the energy could be decisive. This difficulty would be still more serious for a larger system.

3.2 Exponential growth of errors

In the study of many dynamical systems an important role is played by the largest Lyapounov characteristic exponent, which measures the long-term trend in the divergence of neighbouring solutions. The same idea is applicable to N-body calculations, though to satisfy purists we must avoid referring to Lyapounov exponents, as the study is conducted over limited periods of time.

For temporarily bound triple systems, the distance δ, between neighbouring solutions may increase by a factor 10^8 - 10^{20}, or even 10^{150} in some cases (Dejonghe and Hut 1986). For N = 25 results by Standish (1968) imply magnifications around 10^3 after about 0.75 t_{cr}. The most extensive results are those of Miller (1964, 1974), who first drew attention to this question in N-body simulations. He sampled the range $4 \leqslant N \leqslant 32$, and found magnifications of around 10^5 for N = 8 after about 3 t_{cr}, and around 10^6 for N = 12 after about 2 t_{cr}. (We have used here the definition of crossing time given by eq. (3), which appears to exceed Miller's by a factor of 2).

Such results show how difficult it is to justify the accuracy of N-body simulations, even for N = 3, by time-reversal. If they can be extrapolated in time (so that, for example, a further magnification by 10^5 occurs for N = 8 between 3 t_{cr} and 6 t_{cr}), then time-reversal will usually fail completely over the lifetime of a temporarily bound triple system, or after less than 2 t_{cr} for N = 12. When the cumulative magnification over the forwards and backwards integrations exceeds the relative machine accuracy, it is extremely unlikely that the initial conditions can be recovered even approximately.

By the same token, the forwards integration is certain to be greatly in error after at most a few crossing times. Even so, N-body simulations are performed by many astrophysicists, on the grounds that statistical results, such as the mean rate of escape, will be reliable even if the detailed orbits are completely wrong. This again is little better than an article of faith. Such a point of view then calls into question the very need for accurate simulations.

The situation could be considerably clarified if more was known about the exponential divergence. For example if the bulk of the magnification were due to only a small number of particles, then it might be easier to justify the statistical reliability of the results. Apparently the only evidence on this point is provided by Standish, and it suggests that an initial perturbation on one particle takes about 1 t_{cr} to have a comparable effect on the other particles.

Another important property of this divergence is its N-dependence. The models for which it has been estimated are very small compared with those now being run. For the range of N considered by Miller he estimated that the e-folding time scale of the divergence is roughly $4 \langle s \rangle / N \langle v \rangle$, where $\langle s \rangle$ is a 'mean interparticle distance' and $\langle v \rangle$ is the r.m.s. particle velocity. Taking $\langle s \rangle \simeq R$ (cf. §1) gives a time around $2N^{-1}\, t_{cr}$. The fact that the result is so tiny if it is extrapolated to a cluster of, say, 2000 stars, underlines the need for a much better understanding of this phenomenon.

Theory should help, but at first sight does not. Gurzadyan and Savvidy (1986), by considering the equation of geodesic deviation, give an estimate which implies that the time scale, which we shall denote by t_e, is about $0.2\ N^{1/3}\ t_{cr}$, very different from Miller's. However it is not clear that what they have predicted is what he has observed. Consider a cluster of N particles in free rectilinear motion, and consider a second cluster which differs from the first at t = 0 only in the velocity of one star, which differs by $\Delta\underline{v}$. Then after time t the magnification factor is nearly $|t\Delta\underline{v}|/|\Delta\underline{v}|$, or t. (Note that Miller uses the Euclidean metric in 6N-dimensional $\underline{x},\underline{v}$ space). A roughly similar result follows if the orbits are those of an anharmonic oscillator. This prediction happens to agree quite well with Miller's results (Figs. 2 and 3 in Miller 1964) for at least the first crossing time, which is almost half the duration of the runs. Since it is not clear that the initial variation chosen by Miller for these runs was of the type we have considered, all that can be said is that there is a danger that short runs do not measure the exponential trend of the divergence that Gurzadyan and Savvidy were attempting to predict.

Possible shortcomings in the theory may also be mentioned. Those authors show that the rate of exponential divergence depends on the mean square force on a star, and for this they give an expression equivalent to the square of the force due to a single particle at the mean interparticle distance (of order $RN^{-1/3}$). Frequently, however, much closer approaches occur, and an intuitive argument (Miller 1964) suggests that the source of the divergence is to be sought in those. The following simple calculation gives an estimate of the time scale of the exponential divergence.

Consider a star on a nearly rectilinear orbit which takes it past another star to a minimum distance p. (The nearly rectilinear approximation is justified if

$$p \gg \frac{GM}{v^2} \quad , \qquad (5)$$

where m is a typical stellar mass and v is the relative speed of the stars). The duration of the encounter is of order $\frac{p}{v}$, and the maximum acceleration of the passing star is of order GM/p^2; therefore the magnitude of the change of velocity of the star is of order Gm/pv. (Exact formulae for two-body motion justify this estimate in the limit of very high eccentricity, i.e. nearly rectilinear orbits). Now suppose we consider a second system exactly like the first, except that the initial position of the passing star differs by an amount Δ_o. There will be a comparable difference in p, and therefore a difference $(Gm/p^2v)\Delta_o$ in the velocity change; as in the previous argument, this leads to a difference $(Gm/p^2v)\Delta_o t_1$ in the position of the passing star after time t_1. Hence the total difference in its position at time t_1 is $\Delta_1 \simeq \Delta_o (1 + (Gm/p^2v)t_1)$. (The extra term Δ_o arises because there would still be a difference of this magnitude at time t_1 even without an encounter).

Now in a star cluster it is relatively easy to estimate the distance of closest approach during a time t_1. The number of encounters at a distance less than p is simply the number of stars in a cylinder of radius p, length vt, whose axis is the nearly rectilinear orbit of the star under consideration; hence this number is $n\pi p^2 vt_1$, where n is the number-density of stars. The distance of closest approach can be estimated by setting this expression equal to one. Neglecting all but the closest encounter, therefore, we find that $\Delta_1 \simeq \Delta_o(1 + \pi Gmnt_1{}^2)$, provided that (5) is satisfied for the appropriate value of p. Using (4) we find that $\Delta_1 \simeq \Delta_o(1 + \pi(t_1/t_{cr})^2)$.

During each succeeding time interval of length t_1 a similar encounter will occur leading to a similar magnification, and after a time t the magnification will be given roughly by $\Delta_t \simeq \Delta_o(1 + \pi(t_1/t_{cr})^2)^{t/t_1}$. The time scale for exponential divergence is now easily estimated as $\{d(\ln \Delta_t)/dt\}^{-1}$, which gives $t_e \simeq t_1/\ln(1 + (t_1/t_{cr})^2)$. This is minimised with respect to t_1 when $t_1 \simeq t_{cr}$, and this gives a time scale $t_e \sim t_{cr}$. The result also suggests that the maximum effect arises from encounters at a distance of order $(n\pi vt_{cr})^{-\frac{1}{2}}$, and it is easily seen by use of eqs. (4) and (2) that (5) is well satisfied when N is large.

The result suggests a time scale shorter (by a factor of order $N^{1/3}$) than that of Gurzadyan and Savvidy, but of course the derivation has its own shortcomings. We have ignored the cumulative effect of numerous more distant encounters and of the rarer closer ones, which may be expected to shorten the time scale further. But we have also neglected the stochastic nature of the magnification, which might lead to partial cancellations and a longer time scale.

Readers of this book should be able to recognise a promising research topic when they see one.

4. SINGULARITIES

4.1 General considerations

The right-hand sides of equations (1) have singularities whenever $\underline{r}_i = \underline{r}_j$ ($i \neq j$). Although these singularities may never be encountered exactly in practice, great numerical difficulties can be experienced in their vicinity, i.e. during a close encounter between two or more stars. The repetitive close encounters occurring in an eccentric binary pose a particularly acute problem.

Binaries in star clusters play a crucial dynamical role (see, for example, Elson, Hut and Inagaki 1987), and so they cannot be ignored by using, say, a barycentric approximation. Their internal dynamics are essential. Up to a point the problems can be ameliorated by the reduction of time steps. But then the progress of the entire calculation becomes very slow. Furthermore, short numerous time steps accelerate the effect of rounding errors.

Rounding errors are already acute when close binaries are concerned. The force on each component is dominated by the other, i.e. $\ddot{\underline{r}}_i = - Gm_j(\underline{r}_i - \underline{r}_j)/|\underline{r}_i - \underline{r}_j|^3$ + small perturbation, but their relative coordinates $\underline{r}_i - \underline{r}_j$ are the difference of nearly equal quantities, and subject to large error. Likewise the acceleration of their barycentre is $(m_i\ddot{\underline{r}}_i + m_j\ddot{\underline{r}}_j)/(m_i + m_j)$, and this weighted sum of accelerations also involves near-cancellation of two nearly equal quantities.

Some of these problems are already greatly eased by introducing relative and barycentric coordinates, i.e. by integrating equations for $\underline{r}$ and $\underline{R}$ in place of $\underline{r}_i$ and $\underline{r}_j$, where $\underline{r} \equiv \underline{r}_i - \underline{r}_j$ and $\underline{R} = (m_i\underline{r}_i + m_j\underline{r}_j)/(m_i + m_j)$. But this does nothing to cure the large numbers of short timesteps which are needed to deal with eccentric close binaries. What is needed is some form of regularisation, in which further transformations of the dependent and independent variables are chosen in such a way as to remove the r^{-2} singularity from the equations of motion. When this is done an eccentric orbit is no harder to integrate than a circular one. The most effective method is certainly KS regularisation, whose implementation in N-body codes is described by Aarseth (1985); for a direct and appealing 'derivation' of the basic equations, see Yoshida (1982).

Even this is not enough. Interactions between a binary star and either single stars or even other binaries frequently lead to the formation of long-lived, temporarily bound triple systems. What was said above about the severity of rounding errors in close binaries is equally applicable here to close triples. Unfortunately it is not clear that the same strategy for curing the problem is applicable. Triple collisions are not regularisable in general (Siegel and Moser 1971). On the other hand it is possible to transform the variables for three-body motion ('blow-up') so that the flow is smooth at triple collision, and represented by regular differential equations (e.g. McGehee 1974). (This is not the same as regularisation, since triple collision, if it occurs, does so as the new time variable $\tau \to \infty$). Perhaps the integration of long-lived triple systems can be improved by a combination of this technique and KS regularisation (to regularise the binary collisions which can still occur). This possibility is being investigated by Dr. A. D. Gilbert.

What has in fact been done to facilitate the integration of close triples in N-body systems is not this but a different transformation of the variables which regularises collisions of two of the three pairs. If the third pair undergo a close encounter then the variables must be switched so that this becomes one of the regularised pairs.

It is possible to employ a transformation which simultaneously regularises any binary collision, and this technique (like the slightly more restricted one described above) can be extended to any number of bodies, so that quadruple or even larger systems may be handled (Heggie 1974). For application to N-body systems it is necessary to allow for perturbations of the regularised subsystem by all the other stars. The formulae for this case are presented for the first time (as far as I am aware) in the next section.

4.2 Regularisation of all binary collisions in a perturbed n-body system.

The following treatment closely parallels the concise formulation for the unperturbed case given by Mikkola (1985), and his notation, and generalises it by including the effects of perturbations by the remaining bodies (or of any other perturbation that can be represented by a potential). As special cases one can recover the case of perturbed three-body motion (Heggie 1974), and also perturbed two-body motion (i.e. KS regularisation under perturbation; cf. Aarseth (1985)). We use units in which G = 1.

The Hamiltonian for the motion of n bodies is

$$H_o = \sum_{i=1}^{n} \frac{1}{2m_i} \underline{w}_i^2 - \sum_{i=1}^{n-1} \sum_{j=i+1}^{n} \frac{m_i m_j}{r_{ij}} + R(\underline{r}_1, \ldots, \underline{r}_n, t) \tag{6}$$

where $\underline{r}_i$ ($1 \leq i \leq n$) is the Cartesian position vector of the ith body, $\underline{w}_i$ is its momentum (both with respect to an inertial frame), m_i is its mass, $r_{ij} \equiv |\underline{r}_i - \underline{r}_j|$, and R is the perturbing function, so that the perturbing acceleration on the ith body is

$$\underline{F}_i = - \frac{1}{m_i} \frac{\partial R}{\partial \underline{r}_i} .$$

Note that R may depend on t; for example, because of the motion of the perturbing bodies. Also define $M = \sum_{i=1}^{n} m_i$.

Now introduce the position vector of the barycentre $\bar{\underline{q}}$, where

$$\bar{\underline{q}} = \sum_{i=1}^{n} m_i \underline{r}_i / M, \tag{7}$$

and the $\frac{1}{2}n(n - 1)$ relative position vectors defined by

$$\underline{q}_{ij} \equiv \underline{r}_i - \underline{r}_j , \quad 1 \leq i < j \leq n, \tag{8}$$

and conjugate momenta $\bar{\underline{p}}$ and $\underline{p}_{ij}$ satisfying

$$\underline{w}_i = \frac{m_i}{M} \bar{\underline{p}} + \underline{w}_i' \tag{9}$$

where

$$\underline{w}_i' \equiv \sum_{j>i} \underline{p}_{ij} - \sum_{j<i} \underline{p}_{ji} . \tag{10}$$

(The reason for this choice is a simple extension of the argument given in Heggie 1974). If $n > 2$ the vectors $\underline{q}_{ij}$ are not independent, and so the equations (8) with (7) have many solutions, among them being the solution

$$\underline{r}_i = \left(\sum_{j=i+1}^{n} m_j \underline{q}_{ij} - \sum_{j=1}^{i-1} m_j \underline{q}_{ji} \right) /M + \bar{\underline{q}} . \tag{11}$$

Somewhat analogously, it can be seen that a solution of (10) is

$$\underline{p}_{ij} = (\underline{w}_i' - \underline{w}_j')/n . \tag{12}$$

Anyway, from (8) and (11) it follows that

$$\underline{q}_{ij} = \frac{1}{M} \left\{ \sum_{k>i} m_k \underline{q}_{ik} - \sum_{k<i} m_k \underline{q}_{ki} - \sum_{k>j} m_k \underline{q}_{jk} + \sum_{k<j} m_k \underline{q}_{kj} \right\}. \tag{13}$$

We now substitute (9) into the first term of H_o, (8) into the second term and (11) into R, obtaining the new Hamiltonian

$$H_1 = \frac{1}{2M} \bar{\underline{p}}^2 + \sum_{i=1}^{n} \frac{1}{2m_i} \underline{w}_i'^2 - \sum\sum_{i<j} \frac{m_i m_j}{q_{ij}} + R \tag{14}$$

where $\underline{w}_i'$ is given by (10) and the arguments of R are given by (11), along with t. If $n > 2$ the transformation to the new variables is not a straightforward example of a point transformation, because of the introduction of superfluous variables. Therefore we shall justify with a little care the use of the canonical equations with Hamiltonian H_1, i.e.

$$\dot{\underline{q}} = \frac{1}{M}\ \bar{\underline{p}},\ \dot{\bar{\underline{p}}} = \Sigma\, m_i\, \underline{F}_i$$

$$\dot{\underline{q}}_{ij} = \frac{1}{m_i}\ \underline{w}_i' - \frac{1}{m_j}\ \underline{w}_j' \quad ,\ \dot{\underline{p}}_{ij} = \frac{m_i m_j}{M}\ (\underline{F}_i - \underline{F}_j) - \frac{m_i m_j}{q_{ij}^3}\ \underline{q}_{ij} \tag{15}$$

where once again for w_i' is meant the expression (10). We shall show that a solution of these equations generates a solution of the canonical equations for H_o, via the transformations (8) and (11).

Suppose, then, that we are given a solution of eqs. (15), the initial values satisfying (13), and that $\underline{r}_i$ and $\underline{w}_i$ are defined by (11) and (9) respectively. We first show that (13) is satisfied at all times if (as we suppose) it is true initially. Indeed it is easily verified from (15) that

$$\dot{\underline{q}}_{ij} = \frac{1}{m}\left\{\sum_{k>i} m_k\, \dot{\underline{q}}_{ik} - \sum_{k<i} m_k\, \dot{\underline{q}}_{ki} - \sum_{k>j} m_k\, \dot{\underline{q}}_{jk} + \sum_{k<j} m_k\, \dot{\underline{q}}_{kj}\right\},$$

from which the result follows. Hence (11) implies (8). Next, differentiating (11), and using (15) to evaluate the right-hand side, we find that

$$\begin{aligned}\dot{\underline{r}}_i &= \frac{1}{M}\bar{\underline{p}} + \frac{1}{m_i}\ \underline{w}_i' - \frac{1}{m}\sum_{j=1}^{n} \underline{w}_j' \\ &= \frac{1}{m_i}\ \underline{w}_i\end{aligned} \tag{16}$$

by (9), since (10) shows that the sum vanishes. Thus $\underline{r}_i$ satisfies the correct canonical equation for the Hamiltonian H_o. Finally, differentiation of (9) and substitution for $\dot{\bar{\underline{p}}}$ and $\dot{\underline{p}}_{ij}$ from eqs.(15) yields

$$\dot{\underline{w}}_i = m_i\, \underline{F}_i - \sum_{j\neq i} \frac{m_i m_j}{|\underline{r}_i - \underline{r}_j|^3}\ (\underline{r}_i - \underline{r}_j).$$

where we have used (8). This again is the correct canonical equation for H_o.

Further steps of the regularisation are straightforward. A KS transformation $\underline{q} \to \underline{Q}$ is performed on each of the variables $\underline{q}_{ij}$, and a new independent variable τ is defined by $dt/d\tau = g$, where g is some function which behaves as q_{ij} whenever the ith and j^{th} bodies are near collision. (The choice of this function is discussed by Mikkola (1985) and Alexander (1986)). Let h_1 be the value of H_1 (or H_o). Then the appropriate Hamiltonian is

$$\Gamma = g\ (H_1 - h_1), \tag{17}$$

and Hamilton's equations for the relative motions are

$$\frac{d\underline{Q}_{ij}}{d\tau} = g \frac{\partial H_1}{\partial \underline{P}_{ij}} + (H_1 - h_1) \frac{\partial g}{\partial \underline{P}_{ij}}$$

$$\frac{d\underline{P}_{ij}}{d\tau} = -g \frac{\partial H_1}{\partial \underline{Q}_{ij}} - (H_1 - h_1) \frac{\partial g}{\partial \underline{Q}_{ij}} \quad . \tag{18}$$

Finally, let H be the part of the Hamiltonian H_1 which concerns only the unperturbed relative motion of the n bodies, i.e.

$$H = \sum_{i=1}^{n} \frac{1}{2m_i} \underline{w}_i'^2 - \sum\sum_{i<j} \frac{m_i m_j}{q_{ij}} , \tag{19}$$

and let E_o bw its value. Then the factor $(H_1 - h_1)$ in each of the above equations may be replaced by the factor $(H - E_o)$, where E_o is the energy of relative motion of the subsystem of n bodies.

Note that E_o is not obtained by evaluating H, but by integrating a differential equation. Indeed from (14) and (15) it is easy to show that

$$\frac{dE_o}{dt} = \frac{1}{M} \sum\sum_{i<j} m_i m_j \, \dot{\underline{q}}_{ij} \, (\underline{F}_i - \underline{F}_j) \; .$$

Expressed in terms of regularised variables, this becomes

$$E_o' = \frac{2}{M} \sum\sum_{i<j} m_i m_j \; \underline{Q}'^T \hat{Q}^T \; (\underline{F}_i - \underline{F}_j), \tag{20}$$

where a prime denotes differentiating with respect to τ, T denotes a transpose, and $\hat{Q}$ is the KS matrix defined, for example, in Mikkola (1985).

Now, let us denote by T and -U the two terms in the right-hand side of (19). These are, in fact, the kinetic energy of relative motion of the n bodies and their mutual potential energy. Thus

$$H = T - U$$

$$\text{and} \quad H_1 = \frac{1}{2M} \underline{\bar{p}}^2 + T - U + R. \tag{21}$$

As far as the relative motion is concerned the first term of H_1 may be dropped when evaluating the derivatives of H_1 in eqs. (18). Hence these derivatives may be replaced by derivatives of (H + R), where

$$\frac{\partial R}{\partial \underline{Q}_{ij}} = - \frac{2m_i m_j}{M} \hat{Q}^T_{ij} (\underline{F}_i - \underline{F}_j). \tag{22}$$

Hence the regularised equations of relative motion become

$$\frac{d\underline{Q}_{ij}}{d\tau} = g \frac{\partial H}{\partial \underline{P}_{ij}} + (H - E_o) \frac{\partial g}{\partial \underline{P}_{ij}},$$
$$\frac{d\underline{P}_{ij}}{d\tau} = -g \left(\frac{\partial H}{\partial \underline{Q}_{ij}} + \frac{\partial R}{\partial \underline{Q}_{ij}} \right) - (H - E_o) \frac{\partial g}{\partial \underline{Q}_{ij}}. \tag{23}$$

One final result is needed. From (9) and (12) we see that a solution for eqs. (10) is

$$\underline{p}_{ij} = \left(\underline{w}_i - \underline{w}_j - \frac{m_i - m_j}{M} \bar{\underline{p}} \right)/n,$$

and by (15) and (16) it follows that

$$\underline{p}_{ij} = (m_i (\dot{\underline{r}}_i - \dot{\underline{q}}) - m_j (\dot{\underline{r}}_j - \dot{\underline{q}}))/n \tag{24}$$

4.3 Extension of Mikkola's formulation

For convenience in applications we list here those changes that must be made to Mikkola's formulation to accommodate perturbations along the lines adopted in the previous section. We follow his specific choice for g (see below), but it would be relatively trivial to adapt the formulae to other simple choices. We shall continue to use the double suffix notation for the vectors, though Mikkola cleverly introduces a more economical labelling in many of his formulae. Mikkola breaks the computation clearly into four stages, which we discuss in turn.

The first is the conversion from coordinates and momenta in the frame of reference of the barycentre of the n bodies. The transformation given by Mikkola, i.e.

$$\underline{p}_{ij} = (m_i \dot{\underline{r}}_i - m_j \dot{\underline{r}}_j)/n$$

is still correct in the presence of external perturbation, even though the frame is not inertial. Eq.(24) is a correct generalisation to any inertial frame.

The second step of Mikkola's formulation is the introduction of KS variables $\underline{Q}_{ij}$ and their conjugate momenta. No alteration in this procedure is necessary when a perturbation is present.

The largest changes are in the third step, the evaluation of the right-hand sides of (23). We follow Mikkola in choosing for g the reciprocal of the unperturbed Lagrangian i.e. $g = L^{-1} = (T+U)^{-1}$. Then (23) can be written as

$$\frac{d\underline{Q}_{ij}}{d\tau} = \frac{\partial G}{\partial T}\frac{\partial T}{\partial \underline{P}_{ij}}$$

$$\frac{d\underline{P}_{ij}}{d\tau} = -\frac{\partial G}{\partial T}\frac{\partial T}{\partial \underline{Q}_{ij}} - \frac{\partial G}{\partial U}\frac{\partial U}{\partial \underline{Q}_{ij}} - L^{-1}\frac{\partial R}{\partial \underline{Q}_{ij}} ,$$

where

$$G = (T - U - E_o)/L. \qquad (25)$$

This differs from Mikkola's equations only in the perturbation, and formulae for L and all the partial derivatives can be taken from Mikkola without change, except for $\partial R/\partial \underline{Q}_{ij}$, which is given by (22). One new equation is needed, viz. (20). Note, incidentally, a difference in interpretation between the perturbed and unperturbed cases: in the latter G is the Hamiltonian, but in the former the Hamiltonian is Γ (eq.(17)), and G is simply the function defined in eq.(25).

Finally there is the transformation back to Cartesian coordinates in an inertial frame. As in the inverse transformation the only changes that occur arise because of the motion and displacement of the centre of mass. The position of the ith body is given by (11), and by (16), (9) and (15) its velocity is given by

$$\dot{\underline{r}}_i = \dot{\bar{\underline{q}}} + \frac{1}{m_i}\left(\sum_{j=i+1}^{n} \underline{p}_{ij} - \sum_{j=1}^{i-1} \underline{p}_{ji}\right) .$$

Again Mikkola's corresponding equation (which omits the term $\dot{\bar{\underline{q}}}$) is correct in the (non-inertial) rest-frame of the barycentre of the n bodies.

5. PROSPECTS FOR N-BODY SIMULATIONS OF STAR CLUSTERS

At present the direct simulation of even a small globular cluster is far from achievable. The simulation of 3151 stars, which was described briefly in §2, took approximately 15 CPU days for the simulation of an interval of about 10^7 yr, and so a simulation over the lifetime of a cluster would take roughly 100 years! The simulation represented only the innermost 15% of a cluster of 2×10^4 stars, and it can be estimated that a simulation of the entire system would take more like 1000 years. However, steps were taken in this simulation to suppress the formation of close binaries, and an estimate for a more realistic simulation would be higher still. Certainly there exist much faster processors than those on which this simulation was run, but not by more than a factor of 100.

From these considerations it is clear that new developments are needed before the simulation of a small globular cluster becomes possible. Efficient evaluation of forces is one area where promising developments are taking place (§2), but further improvements in processor speed are also likely to be decisive. No method will succeed without the special treatment of close approaches. In §4 we outlined the present position on this aspect of the problem, but new ideas here also may enable important progress to be made.

In view of all these problems, we may be tempted to ask whether direct simulation is the best way to solve problems in the dynamical evolution of star clusters. After all simpler methods exist (described, for example, in Heggie 1986), based on methods of non-equilibrium statistical mechanics (the Fokker-Planck equation, which is a kinetic equation) or gas dynamics (in which the stars are represented by the atoms of a gas). These methods are currently very much faster than N-body simulations; with a gas code the entire dynamical evolution of a cluster takes a few minutes, even on a standard mainframe. Undoubtedly these models will continue to be used because their speed allows many cases to be investigated. On the other hand they are highly idealised and simplified, and there are several respects in which this must be set against their advantage in speed.

(i) These methods can only include those physical processes which are well enough understood. The main one is two-body scattering (hyperbolic encounters between pairs of stars), but it is certain that 3- and 4- body scattering is crucial, and the inclusion of these processes requires much preparatory work and further simplification.

(ii) These methods are fast partly because spherical symmetry is almost always assumed, which means that rotation of the cluster is ignored. Their speed is also greatly impeded if they are made more realistic by the inclusion of a distribution of stellar masses or anisotropy of the distribution of velocities. Indeed a simplified method including rotation, binaries and a spectrum of masses soon ceases to be feasible on a conventional mainframe. By contrast an N-body integration becomes no harder the more realistic it is made, and in some respects it becomes somewhat easier.

(iii) Simplified methods can reveal the effects of only those fundamental processes which are included. The need for any essential new process is likely to be revealed only by penetrating physical insight, or in N-body systems. An example is the discovery of the dynamical importance of binaries. Their occurrence was revealed in the early N-body simulations by von Hoerner (1963), and a bold theoretical argument by Hénon (1961) at about the same time has been amply justified by subsequent work, much of it very recent. A recent discovery in large N-body simulations is the wandering of the core of the cluster, which is found to stray erratically from the geometrical centre of the system (Makino and Sugimoto 1987), whereas in all Fokker-Planck simulations the assumption of symmetry keeps the densest part of the cluster coincident with the centre.

For these reasons the goal of attempting to simulate a small globular cluster with an N-body program seems a significant one, and its feasibility in the foreseeable future is by no means out of the question. In fact it is likely that all the methods which have been brought to bear on this problem in recent years will continue to serve useful and complementary functions.

REFERENCES

Aarseth, S.J., 1985. In Brackbill, J.W. & Cohen, B.I.(eds.), Multiple Time Scales, Academic Press, New York, p.377.
Alexander, M.E., 1986. J. Comp. Phys. 64, 195.
Barnes, J. & Hut, P., 1986. Nature 324, 446.
Dejonghe, H. & Hut, P., 1986. In Hut & McMillan (1986), p.212.
Elson, R.A.W., Hut, P. & Inagaki, S., 1987. Ann. Rev. Astron. Astrophys., 25, in press.
Gürzadyan, V.G. & Savvidy, G.K., 1986. Astron. Astrophys. 160, 203.
Heggie, D.C. & Ramamani, N., 1987. 'Evolution of Star Clusters after Core Collapse',preprint.
Heggie, D.C., 1974. Celes. Mech. 10, 217.
Heggie, D.C., 1986. In Hut & McMillan (1986), p.13.
Hénon, M., 1961. Ann. d'Astrophys. 24, 369.
Hut, P., & McMillan, S.L.W., 1986. (Eds.), The Use of Supercomputers in Stellar Dynamics, Springer-Verlag, Berlin.
Makino, J. & Sugimoto, D., 1987. PASJ, in press.
McGehee, R., 1974. Invent.Math. 27, 191.
Mikkola, S., 1985. M.N.R.A.S. 215, 171.
Miller, R.H., 1964. Ap.J. 140, 250.
Miller, R.H., 1974. In D.G. Bettis (ed.), Proceedings of the Conference on Numerical Solution of Ordinary Differential Equations. Springer-Verlag, New York, p.260.
Press, W.H., 1986. In Hut & McMillan (1986), p.184.
Saslaw, W.C., 1985. Gravitational Physics of Stellar and Galactic Systems, Cambridge.
Siegel, C.L. & Moser, J.K., 1971. Lectures on Celestial Mechanics, Springer-Verlag, Berlin, p.90.
Standish, E.M., 1968. Ph.D. Thesis, Yale University.
von Hoerner, S., 1963. Z. Astrophys. 57, 47.
Yoshida, H., 1982. Celes. Mech. 28, 239.

ROTATIONALLY AND TIDALLY PERTURBED ORBITAL ELEMENTS OF CLOSE BINARY STARS

Basil Zafiropoulos
Department of Physics,
University of Patras,
Greece.

ABSTRACT. This paper summarizes some relatively recent work on the effects that rotational and tidal forces can produce on the orbital elements of a close binary system.

In the case of rotational distortion the six elements of the orbit have been expressed by means of the unperturbed true anomaly measured from the ascending node. First-order approximations are obtained for the following values of orbital inclination,

i) i = any value and   ii) i = very small.

In the case of tidal distortion the elements of the orbit have been presented as functions of the unperturbed true anomaly measured from the periastron. Our study includes the effects produced by the second, third, and fourth tidal harmonic distortions. The results obtained have been expressed by means of Hansen coefficients.

First order approximations for tidal distortion have been produced for the following cases,

i) Tides do not lag, and,
ii) Tides are lagging in longitude or in latitude.

1. INTRODUCTION

Close binary systems are bound to be rotationally and tidally distorted. The variation of light, in the case of eclipsing close binaries and the variation of the radial velocity, in the case of spectroscopic close binaries, due to deformations of one or both the components, are of great importance and must be taken into account when solving for the elements of the orbit.

In the absence of viscosity the shape of the components of the binary system is of the equilibrium form, that is, the crest of each tide is always in the direction of the radius-vector $\vec{r}$, connecting

A. E. Roy (ed.), Long-Term Dynamical Behaviour of Natural and Artificial N-Body Systems, 349–353.
© 1988 by Kluwer Academic Publishers.

the mass centres of the two stars (non-lagging tides).

In the case of stellar matter, viscosity is bound to arise from several sources-plasma, radiative, turbulent- and their joint action will make the individual bulges raised by each star on its mate to advance or lag behind the radius-vector $\vec{r}$ (lagging tides).

Problems of viscous matter in close binary systems have been considered by Zahn (1966), while a more complete study is presented by Kopal (1968a,b).

Phenomena produced by tidal lag have so far received inadequate attention, even though they are responsible for the salient processes influencing fundamentally the evolution of close binary systems, such as:

1) rectification of axes of rotation of both components to a position perpendicular to the orbital plane,
2) synchronization of their rotation and revolution, and,
3) secular circularization of their orbits.

Secular perturbations due to rotational distortion for the elements α, e and $\tilde{\omega} = \Omega + \omega$ (longitude of the periastron) are given by Kopal (1972).

2. ROTATIONAL DISTORTION

The osculating orbit of one star of mass m_1 around the centre of gravity of the other star of mass m_2 is specified by six independent parameters or elements. We choose them to be the semi-major axis α, the eccentricity e, the inclination of the orbital plane, the longitude of the ascending node Ω, the longitude of the periastron ω measured from the ascending node, and χ the mean anomaly at time t = 0, defined by

$$\chi = - nT = M - nt, \tag{1}$$

where n is the mean motion, T the time of periastron passage and M the mean anomaly.

The perturbations of the osculating elements are obtained by integrating the following equations (cf., e.g. Taff, 1985 p.314).

$$\frac{d\Omega}{du} = \frac{r^3 \sin u}{\mu p \sin i} W, \tag{2}$$

$$\frac{di}{du} = \frac{r^3 \cos u}{\mu p} W, \tag{3}$$

$$\frac{d\alpha}{du} = \frac{2\alpha^2}{\mu} \{\frac{r^2 e \sin \upsilon}{p} R + rS\}, \tag{4}$$

$$\frac{de}{du} = \frac{r^2}{\mu} \{\sin\upsilon R + \frac{r}{p} [2 \cos\upsilon + \frac{e}{2} (3 + \cos 2\upsilon)] S \}, \tag{5}$$

$$\frac{d\omega}{du} = \frac{r^2}{\mu e p} \{-p \cos\upsilon R + (p+r) \sin\upsilon S\} - \frac{d\Omega}{du}, \tag{6}$$

$$\frac{d\chi}{du} = \frac{r^2\sqrt{1-e^2}}{\mu e p} \{(p\cos\upsilon - 2er)R - (p+r) \sin\upsilon S\}, \tag{7}$$

where u and υ are the true anomalies measured from the ascending node and the periastron, respectively. R, S and W denote the components of the disturbing accelerations, p is the semi-latus rectum defined by the relation

$$r = \frac{p}{1 + e\cos\upsilon} = \frac{\alpha(1-e^2)}{1 + e\cos\upsilon}, \tag{8}$$

and $\mu = G(m_1 + m_2)$, (9)

where G stands for the gravitational constant.
Time t has been replaced by u by means of the relation

$$\frac{du}{dt} = \frac{\sqrt{\mu p}}{r^2}, \tag{10}$$

which holds good only for first-order approximations.

In order to integrate Equations (2) to (7) we need to specify the components R, S and W due to rotational distortion. After substitution for the disturbing accelerations from Kopal 1978 p.209 and the radius-vector from (8) the results given by Zafiropoulos and Kopal (1982a)have been obtained. The perturbed elements for the case of rotational distortion and very small inclination of the orbital plane are presented in a recent paper by Zafiropoulos and Zafiropoulos (1987). The correctness of the resulting expressions has been checked by means of tests.

3. TIDAL DISTORTION

In the case of non-lagging tides the components R, S and W are,

$$R = -\frac{2\mu}{r^2} \sum_{i=1}^{2} \frac{m_{3-i}}{m_i} \sum_{j=2}^{4} (j+1)(K_j)_i \left(\frac{R_i}{r}\right)^{2j+1}, \tag{11}$$

$$S = W = 0, \tag{12}$$

where $(K_j)_i$ represents constants depending on the internal structure of each star; R_i denotes the mean radius of each component. For $j = 2,3$ and 4 we obtain the second, third and fourth tidal harmonic, respectively.

The form of the components R implies that the true anomaly υ measured from the periastron must be used as independent variable. The expression which connects υ to time t has the form

$$\frac{d\upsilon}{dt} = \frac{\sqrt{\mu p}}{r^2} . \tag{13}$$

Equation (13) holds good only for first order approximations.

Substituting in Equations (2) - (7) $d\upsilon$ for du, R, S and W by means of (11) - (12) and r from (8), we obtain the first order approximations in the case of non-lagging tides:

$$\delta\Omega = \delta i = 0 \tag{14}$$

$$\delta\alpha = \alpha - \alpha_o = 4\alpha\sqrt{1-e^2}\sum_{i=1}^{2}\frac{m_{3-i}}{m_i}\sum_{j=2}^{4}(K_j)_i\left(\frac{R_i}{\alpha}\right)^{2j+1} \times$$

$$\times \sum_{j=0}^{2j+2} X_o^{-(2j+4),k}(\cos k\upsilon - 1) , \tag{15}$$

$$\delta e = e - e_o = \frac{2(1-e^2)^{3/2}}{e}\sum_{i=1}^{2}\frac{m_{3-i}}{m_i}\sum_{j=2}^{4}(K_j)_i\left(\frac{R_i}{\alpha}\right)^{2j+1} \times$$

$$\times \sum_{k=0}^{2j+2} X_o^{-(2j+4),k}(\cos k\upsilon - 1) , \tag{16}$$

$$\delta\omega = \omega - \omega_o = \frac{2\sqrt{1-e^2}}{e}\sum_{i=1}^{2}\frac{m_{3-i}}{m_i}\sum_{j=2}^{4}(j+1)(K_j)_i\left(\frac{R_i}{\alpha}\right)^{2j+1} \times$$

$$\times \sum_{k=0}^{2j+2}\frac{1}{k}\left(X_o^{-(2j+3),k-1} + X_o^{-(2j+3),k+1}\right)\sin k\upsilon , \tag{17}$$

$$\delta\chi = \chi - \chi_o = -\frac{2(1-e^2)}{e}\sum_{i=1}^{2}\frac{m_{3-i}}{m_i}\sum_{j=2}^{4}(j+1)(K_j)_i\left(\frac{R_i}{\alpha}\right)^{2j+1} \times$$

$$\times \sum_{k=0}^{2j+2}\frac{1}{k}\left(X_o^{-(2j+3),k-1} + X_o^{-(2j+3),k+1}\right)\sin k\upsilon . \tag{18}$$

The results obtained in (15) - (18) have been expressed through the zero-order Hansen coefficients $X_o^{n,m}$. The following formula permits us to write various powers of the quantity $(1 + e\cos\upsilon)^{\ell}$ in the compact form of summations by means of Hansen coefficients,

$$(1+e\cos\upsilon)^{\ell} = (1-e^2)^{\ell+1/2} \sum_{k=0}^{\ell} (2-\delta_{o,k}) X_o^{-(\ell+2),k} \cos k\upsilon, \qquad (19)$$

where the symbol $\delta_{o,k}$ stands for the Kronecker's delta. Analytic expansions for the zero-order Hansen coefficients are given by Zafiropoulos and Kopal (1982b) and Zafiropoulos and Zafiropoulos (1982).

In general the tides raised by each star on its mate are bound to lag in phase with respect to the direction of the external force. In the case of tidal lag in longitude the expressions for the perturbed orbital elements can be found in Zafiropoulos 1987a, while the osculating elements in the case of tidal lag in latitude are given by Zafiropoulos (1987b).

REFERENCES

Kopal, Z.: 1968a, Astrophys. Space Sci. 1, 74.
Kopal, Z.: 1968b, Astrophys. Space Sci. 1, 179.
Kopal, Z.: 1972, Astrophys. Space Sci. 18, 287.
Kopal, Z.: 1978, "Dynamics of Close Binary Systems", D. Reidel Publ. Co., Dordrecht, Holland.
Taff, L.G.: 1985, "Celestial Mechanics", John Wiley and Sons, New York.
Zahn, J.P.: 1966, Ann. Astrophys. 29, 313, 489 and 565.
Zafiropoulos, B. and Kopal, Z.: 1982a, Astrophys. Space Sci. 88, 331.
Zafiropoulos, B. and Kopal, Z.: 1982b, Astrophys. Space Sci. 88, 355.
Zafiropoulos, B. and Zafiropoulos, F.: 1982, Astrophys. Space Sci. 88, 401.
Zafiropoulos, B. and Zafiropoulos, F.: 1987, Astrophys. Space Sci. 136, 211.
Zafiropoulos, B.: 1987a, Astrophys. Space Sci. 136, 149.
Zafiropoulos, B.: 1978b, Astrophys. Space Sci. 137, 139.

PART V

OTHER SPECIFIC TOPICS IN DYNAMICS

ESCAPE AND CAPTURE IN THE RTBP

Antoni Benseny & Arturo Olvera
Department de Matemàtica Aplicada i Anàlisi
Universitat de Barcelona,
Spain.

ABSTRACT We study the escape and capture in the Planar and Circular Restricted Three Body Problem(RTBP) when the mass m of the secondary body is small enough. The RTBP has essentially different characteristics when m is zero or not. The non-integrability of the problem for $m \neq 0$ causes homoclinic phenomena that do not exist for $m = 0$: we show here some of the phenomena related to the escape and capture orbits when the Jacobi constant C is big enough. Otherwise, the diffusion of parabolic orbits near the collision with the secondary is much richer for $m \neq 0$: the changes of sidereal energy in paths near collision are proved to be more important in some cases than in pure collision, which is the sole possibility of change for $m = 0$. The last analysis is done using a simple model of that orbits in the RTBP made by matching arcs of Keplerian orbits on a suitable circle centered at that small body.

1. ESCAPE AND CAPTURE FOR LARGE VALUES OF C.

The RTBP for m close to zero is considered as a perturbation of the case when $m = 0$: rotating Kepler problem. The last one is completely integrable: the sidereal energy E and angular momentum K are functionally independent first integrals in involution. This implies that the phase space is folliated by 2-dimensional invariant manifolds I_{EK} involving the states with sidereal energy E and angular momentum K [1,7] : there is no possibility of diffusion between them if collisions with the secondary are not considered.

The orbits of parabolic escape and capture (i.e. $r(t) \to \infty$, $\dot{r}(t) \to 0$ $(t \to \pm \infty)$) are known to have null sidereal energy (E=0): they cannot leave the 2-dimensional manifolds I_{0K} if no collision with the secondary is given.

When $m \neq 0$ the synodical energy H (or equivalently the Jacobi integral $C = -2H = 2(K - E)$) is a first integral of the problem. Nevertheless the problem is non-integrable: the characterization E=0 for parabolic escape and capture orbits is not true at all, These orbits are

A. E. Roy (ed.), Long-Term Dynamical Behaviour of Natural and Artificial N-Body Systems, 357–362.
© 1988 by Kluwer Academic Publishers.

understood only to be the asymptotic orbits to the periodic one at the infinity (i.e. $r = \infty$, $p_r = \dot{r} = 0$, $\theta \in [0,2\pi]$, $p_\theta = C/2$) (or to the parabolic orbit Γ at the origin in McGehee variables: $q = \sqrt{2/r}$, $p = p_r$, θ, $\omega = p_\theta$). This type of orbits form respectively the unstable and stable manifolds of Γ : $W^{u,s}(\Gamma)$ [5].

For $m = 0$ and Jacobi constant C both manifolds coincide and intersect the Poincaré section Π_C of returning points ($p = 0$, $\dot{p} > 0$) in a circle P_0 of radius $q = 4/C$ centered at the origin for each value of $C > 3$. $W^u(\Gamma)$ and $W^s(\Gamma)$ do not coincide any more for $m \neq 0$. The first intersections of $W^s(\Gamma)$ and $W^u(\Gamma)$ with Π_C are two different curves ($P_{\pm 1}$) that coincide at least in two points, the intersection between P_1 and P_{-1} was proved to be nontangential if m is sufficiently small and C is sufficiently large [4].

The dominant terms in the asymptotic expansion (when $C \to \infty$) of the Fourier coefficient for the expressions $q_{\pm 1}(\theta)$ of q in $P_{\pm 1}$ have been obtained to the first order in m [8] giving:

$$q = q_{\pm 1}(\theta) = \frac{4}{C} + m \{ \frac{32}{C^5} + \frac{3.2^{17}}{C^{13}} \cos(\theta) - \frac{9.2^{12}}{C^{11}} \cos(2\theta)$$

$$+ \sum_{n \geq 3} (-1)^{n+1} \frac{n+1}{n^2} \frac{(2n-1)!!2^{3n+10}}{(2n)!! \quad C^{2n+7}} \cos(n\theta)$$

$$\mp \frac{\pi}{8C} \exp(-C^3/24)\sin(\theta)$$

$$\pm \sum_{n \geq 2} A_n C^{n-1} \exp(nC^3/24)\sin(n\theta)(1+o(1)) \}$$

$$+ O(m^2)$$

where A_n are constants.

These expressions show the transversal intersection between $P_{\pm 1}$ at $\theta = 0,\pi$. This implies the non-integrability of the problem when $m \neq 0$, but the "practical" coincidence of $P_{\pm 1}$ when C is big due to the exponential dependence of the splitting on C: this homoclinic phenomenon will be practically unobservable when C is big.

A qualitative study of the successive intersections $P_{\pm n}$ of $W^{u,s}(\Gamma)$ shows the cantorian structure of the returning points whose orbits do not escape when $t \to \pm \infty$ (i.e. stable-Lagrange returning points) [8]. A quantitative study of the escape and capture returning points near $P_{\pm 1}$ has serious difficulties, so a model for the Poincaré map associated to Π_C is required. The model takes into account only the first dominant term (when $C \to \infty$) for the splitting of $P_{\pm 1}$ and it is supposed to preserve all the essential dynamics of the problem near these curves.

The chosen model is described by

$$(\theta,t) \to (\theta',t') = (\theta+2\pi(B/t)^{3/2}, t + \sin\theta'),\ t > 0,$$

where $\rho = \varepsilon t$ means the radial distance from the returning point to $P_{\pm 1}$, θ is its angular coordinate,

$$B = \frac{C}{8\varepsilon} \quad \text{and} \quad \varepsilon = \frac{m}{4C}\exp\left(\frac{-C^3}{24}\right).$$

This map can be approximated (near different resonant values) by copies of the standard map (with different perturbation constants κ). Using the estimates for the critical values of κ given by Chirikov, Greene et al. [2,3,6]: $\kappa_{cr} \simeq 0.971636$, it is possible to infer that there are invariant curves for the model (limiting the zone of stable-Lagrange returning points) until some distance of $P_{\pm 1}$ near to

$\rho_o = \varepsilon\left(\frac{3\pi}{\kappa_{cr}}\right)^{2/5} B^{3/5}$ [8].

2. ESCAPE AND CAPTURE OF ORBITS PASSING NEAR THE SECONDARY BODY

Now we suppose that C is small enough ($|C| < 3$) so that the Hill region is connected. Some of the orbits pass close to the secondary body and eventually go to collision. The orbits of the RTBP behave like O(m) - perturbations of rotating keplerian orbits, far from the secondary, but close to it the perturbations are much stronger. We must use singular methods to describe approximately the behaviour of the orbits passing near the secondary.

In the trivial case ($m = 0$) the orbits of the rotating Kepler problem are only perturbed on eventual collisions with the secondary. The Jacobi constant does not change but we have a singular diffusion: the sidereal angular momentum K has a strong variation from K_i to $K_o = 2 - K_i$.

We define a model for orbits reaching a circle C_α of radius m^a centered on the secondary approximating the real orbits in the following way:

i) outside C_α: the orbits are arcs of rotating keplerian orbits governed only by the influence of the primary body,

ii) inside C_α: the orbits are arcs of keplerian orbits having only the attraction of the secondary body,

iii) on C_α: the orbits match in the C^1 topology,

iv) the error of this approximation must be of the same order in m.

This model works if we take $\alpha = \frac{1}{2} - \frac{\gamma}{4}$ where γ is such that $m^\gamma = 3 - C$ and assuming $\alpha > 1/3$, $\alpha + \gamma < 1$, $4\alpha + \gamma > 2$. The relative errors in position and velocity are

$O(m^{\frac{1}{2}-\frac{3\gamma}{4}})$. The keplerian orbits inside C_α are hyperbolic if $C < 3 - m^{2/3}$ [8].

In order to study the diffusion of orbits arriving to C_α we will estimate the change of sidereal angular momentum from the input keplerian arc to the disk D_α inside C_α to the output one.

We use polar synodical coordinates (R,Θ) centered at the secondary in order to describe the arcs in D_α. These arcs enter and leave D_α in points with position angles Θ_i and Θ_o related by

$$\Theta_o = \Theta_i + 2\operatorname{sgn}(K_2)\ \operatorname{arc\,cos}(c),$$

$$c = \frac{\lambda \sin^2 \Psi_i - 1}{\sqrt{1 + (\lambda^2-2\lambda) \sin^2 \Psi_i}} \quad \text{and } \lambda = m^{\alpha-1}\ v_i^2 ,$$

where Ψ_i is the angle of the input velocity (with modulus v_i) referred to the radial direction from the secondary and K_2 is the angular momentum inside D_α.

Using synodic polar coordinates (r,θ) centered at the primary, the sidereal angular momentum K in C_α has the approximate expression $K \simeq 1 + v \sin \Phi$, where Φ is the angle between the synodical velocity (with modulus v) and the radial direction from the primary.

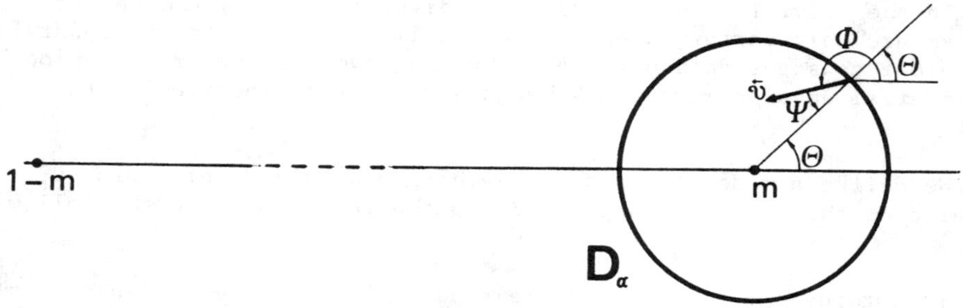

Figure 1:

We observe in figure 1 that $\Phi \simeq \pi - (\Psi+\Theta)$ on C_α.

An approximate relation between the input sidereal angular momentum K_i and the output one K_o is

$$K_o \simeq 1 + (1 - K_i)\ [\cos\Delta + \cot \Phi_i\ \sin\Delta] ,$$

where $\Delta = 2\ \operatorname{sgn}(K_2)\ \arccos(c) - 2\Psi_i$ and $v_o \simeq v_i$. The expression $\cot \Phi_i$ can be computed writing $\cot \Phi = \frac{\dot{r}}{r\dot{\theta}}$, using $\dot{\theta} = \frac{K}{r^2} - 1$ and differentiating the expression

$$r = \frac{K^2}{1 + e \cos(\theta + t - \theta_o)} .$$

For the parabolic case $e \sim 1$ close to the secondary $r \sim 1$ and $|C| < \sqrt{8}$ we have $\cot \Phi_i \simeq \sqrt{2 - K_i^2}/(K_i - 1)$. The above expression K_o is reduced to

$$K_o \simeq 1 - (K_i - 1)\frac{S^2 - 1}{S^2 + 1} + \sqrt{2 - K_i}\ \frac{S}{S^2 + 1} ,$$

defining the parameter S such that $\sin \Psi_i = Sm^{1-\alpha-\gamma}$.

From this expression we can see that $K_o \sim 2 - K_i$ for $(S \to 0)$ and $K_o \sim K_i$, $(S \to \infty)$. The maximum variation of K_o is not obtained in collision (S = 0) but for

$$S = S_{max} = \frac{-4(K_i - 1) \pm \sqrt{3K_i^2 - 8K_i + 6}}{-2\sqrt{2 - K_i^2}}$$

This means that some parabolic orbits can be transferred to elliptic or hyperbolic ones for any value of C depending strongly of the value of Ψ_i. Figure 2 shows the allowed region of the output orbits (K_o, C) for the parabolic input orbits (K_i, C), $K_i = C/2$.

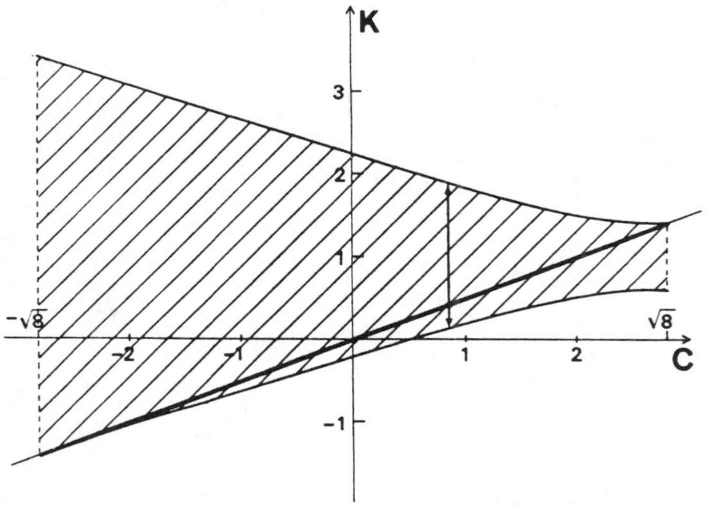

Figure 2: The parabolic input orbits $K_i = C/2$ are represented by the solid straight line. The allowed output orbits are represented by the hatched area. Hyperbolic orbits lie in the region $K_i > C/2$ and elliptic ones in the region $K_i < C/2$.

ACKNOWLEDGEMENTS

This work has been possible thanks to the Grant from PSPA-UNAM (Mexico) to the second author. The work of the first author has been supported by CAICYT Grant 3534/83C3 (Spain).

REFERENCES

[1] R. Abraham and J.E. Marsden, Foundations of Mechanics, Second Ed. Benjamin (1978).
[2] B.V. Chirikov, Phys. Reports 52 (1979) 263.
[3] J.M. Greene, J. Math. Phys. 20 (1979) 1183.
[4] J. Llibre and C. Simó, Math. Ann. 248 (1980) 153.
[5] R. McGehee, J. Diff. Eq. 14 (1973) 70.
[6] A. Olvera and C. Simó, Physica 26D (1987) 181.
[7] S. Smale, Inv. Math. 10 (1970) 305, 11 (1970) 45.
[8] A. Benseny, Contribució a l'estudi del Problema Restringit de 3 Cossos per a valors petits del paràmetre de masses. Thesis, Univ. Barcelona (1984).

A FORMULATION OF THE N-BODY PROBLEM WHERE THE VELOCITIES ARE BOUNDED

John G. Bryant
47 Avenue Feliz Faure,
75015 Paris France

INTRODUCTION

In the following, we give a brief description of how to reformulate the N-body Problem so that the velocities are always bounded, even at collision. This reformulation is more general than the usual regularization methods, since the new time variable we introduce is associated with an entirely new Hamiltonian function. As an added attraction, it is also possible to consider systems where some particles have exactly zero mass (and not simply "negligible" mass as in the restricted problem).

The construction of the new Hamiltonian is a step by step process, starting with the Kepler and Two-Body Problems. The Three and N-Body Hamiltonians are then defined by a simple (though somewhat unorthodox) procedure, just as in the classical case.

The corresponding Hamiltonian system is in practice indistinguishable from the classical one in the case of "ordinary" motions, i.e. where the velocities are small and the mutual distances large. In the vicinity of collision, however, the behaviour is radically different, since now all the velocities have a finite limit that can be determined in advance.

Finally, our point of view is essentially a classical one, but there are some interesting connections with relativistic mechanics. In particular, the constancy of the velocity of light with respect to an observer on Earth is a direct consequence of our reformulated Hamiltonian equations, when considering the zero mass case.

1. THE KEPLER PROBLEM

1.1 The classical formulation of the Kepler problem has two major drawbacks:

a) The initial velocity can be chosen arbitrarily large, and,

A. E. Roy (ed.), Long-Term Dynamical Behaviour of Natural and Artificial N-Body Systems, 363–369.
© 1988 by Kluwer Academic Publishers.

during the motion, the velocity of the particle can approach infinity by passing near collision. Although this does not happen often in practice, it is hard to accept physically.

b) The particle must have non-zero mass, which precludes a particle description of the propagation of light in a gravitational field.

As we will see, this can be remedied by introducing a new Hamiltonian for the problem. Before doing so, we recall the classical Hamiltonian function:

$$H(p_i,q_i) = \frac{p^2}{2m} - \frac{m\alpha}{r} \tag{1.1.1}$$

where: $r = \sqrt{\sum_{i=1}^{3} q_i^2}$; $p = \sqrt{\sum_{i=1}^{3} p_i^2}$; α is a positive constant:

and m is the mass ($m > 0$). The six Hamiltonian equations are written:

$$\frac{dq_i}{dt} = \frac{\partial H}{\partial p_i} = \frac{p_i}{m} \quad ; \quad \frac{dp_i}{dt} = -\frac{\partial H}{\partial q_i} = m\alpha \frac{\partial}{\partial q_i}\left(\frac{1}{r}\right) \tag{1.1.2}$$

and the elimination of the p_i/m leads to the classical system of three second-order equations in the q_i:

$$\frac{d^2q_i}{dt^2} = \alpha \frac{\partial}{\partial q_i}\left(\frac{1}{r}\right) \tag{1.1.3}$$

As is well known, the trajectories in physical space are hyperbolas, parabolas or ellipses according to the sign of H. The energy integral $H = h =$ constant yields the following relation between r and the velocity

$$v = \sqrt{\sum_{i=1}^{3} \left(\frac{dq_i}{dt}\right)^2} \quad :$$

$$\frac{v^2}{2} = \frac{\alpha}{r} + \frac{h}{m} \tag{1.1.4}$$

which shows that, for given r, v can be chosen arbitrarily large (by choosing h accordingly), and also that:

$$r \to 0 \Rightarrow v \to \infty \tag{1.1.5}$$

Finally, we note that although equations (1.1.3) no longer contain m, they cannot be used to describe the zero-mass case since it can be shown that the motion does not agree with Fermat's Principle of Least Time (which is the basis of geometrical optics).

1.2 Instead of $H(p_i,q_i)$, we now consider the new Hamiltonian function:

$$E(p_i,q_i) = c\sqrt{\frac{m_o^2c^2 + p^2}{1 + \frac{2\alpha}{c^2r}}} \qquad (1.2.1)$$

where m_o is a new mass parameter replacing m ($m_o \geq 0$), and c is identified with the velocity of light (E has therefore the dimension of energy). The corresponding system of six Hamiltonian equations is written:

$$\frac{dq_i}{d\tau} = \frac{\partial E}{\partial p_i} = \frac{p_i}{M(1 + \frac{2\alpha}{c^2r})} \quad ; \quad \frac{dp_i}{d\tau} = -\frac{\partial E}{\partial q_i} = \frac{M\alpha}{1 + \frac{2\alpha}{c^2r}} \frac{\partial}{\partial q_i} (\frac{1}{r}) \qquad (1.2.2)$$

where τ is the time variable associated with E, and where we have set $M = E/c^2$. Since M is a first integral of the motion, we can eliminate the p_i/M from (1.2.2) and obtain a system of three second order equations in the q_i:

$$(1 + \frac{2\alpha}{c^2r}) \frac{d}{d\tau} \left[(1 + \frac{2\alpha}{c^2r}) \frac{dq_i}{d\tau}\right] = \alpha \frac{\partial}{\partial q_i} (\frac{1}{r}) \qquad (1.2.3)$$

Comparison with equations (1.1.3) shows that the trajectories in physical space are identical with the classical ones (i.e. hyperbolas, parabolas and ellipses, this time according to whether M is greater, equal or less than m_o). The motion in time is only slightly different, as long as $2\alpha/c^2r \ll 1$, which is the case for ordinary (i.e. planetary) motion. M, r and the new velocity

$$w = \sqrt{\sum_{i=1}^{3} \left(\frac{dq_i}{dt}\right)^2}$$ are linked by the relation:

$$(\frac{w}{c})^2 = \frac{1}{1 + \frac{2\alpha}{c^2r}} (1 - \frac{m_o^2/M^2}{1 + \frac{2\alpha}{c^2r}}) \qquad (1.2.4)$$

which shows that w is <u>always</u> less than c, and that:

$$r \to 0 \Rightarrow w \to 0 \qquad (1.2.5)$$

Finally, if we assume $m_o = 0$, we see that the Hamiltonian function (1.2.1) as well as the Hamiltonian system (1.2.2) are still well defined, and it can in fact be shown that equations (1.2.3) agree with Fermat's principle (the refractive index $n(q_i)$ being equal to $(1+2\alpha/c^2r)^{\frac{1}{2}}$).

Our next step is to generalize the above to the Two-Body Problem, which can be done in a very straightforward way, as we shall see. Before proceeding, we note that in the case of a <u>free</u> particle (i.e. $\alpha = 0$), $E_{f.p} = c(m_o^2c^2 + p^2)^{\frac{1}{2}}$ is the Hamiltonian of a free relativistic particle, with proper mass m_o. As stated before, however, our point of view is basically a classical one, and it is in fact possible to give a classical derivation for the Hamiltonian $E(p_i, q_i)$ of (1.2.1)

2. THE TWO-BODY PROBLEM

2.1 The classical Hamiltonian of the Two-Body problem has the following form in a system of Jacobi coordinates:

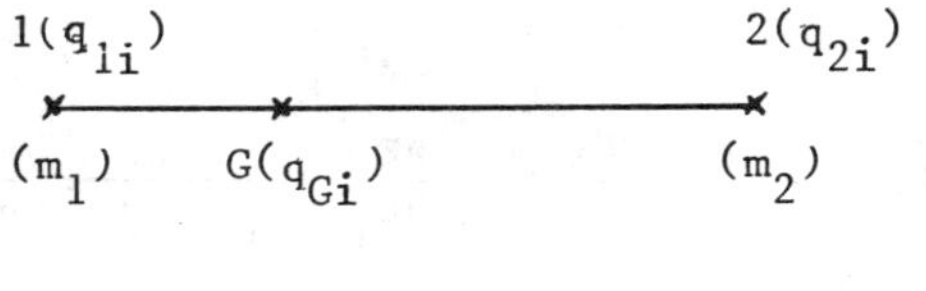

$$H(p_{Gi}, p'_i, q_{Gi}, q'_i) = \underbrace{\frac{p_G^2}{2m}}_{H_G} + \underbrace{\frac{p'^2}{2m'} - f\,\frac{mm'}{r'}}_{H'} \qquad (2.1.1)$$

where f is the gravitational constant; $m = m_1 + m_2$ and $m' = m_1m_2/m$ are the reduced masses ($m > 0$; $m' > 0$); the p_{Gi} and p'_i are the components of the total and relative momentum; and

$$r' = \sqrt{\sum_{i=1}^{3} q_i'^2} = \sqrt{\sum_{i=1}^{3} (q_{2i} - q_{1i})^2} \; ; \; p_G = \sqrt{\sum_{i=1}^{3} p_{Gi}^2} \; ; \; p' = \sqrt{\sum_{i=1}^{3} p_i'^2}$$

The advantage of Jacobi coordinates is to bring out the fundamental property that the motion separates into:

a) The motion of the center of mass G, which is that of a free particle with mass m and Hamiltonian H_G.

b) The relative motion of particle 2 with respect to particle 1, which is the Kepler motion of a particle with mass m', and Hamiltonian H'.

2.2 It follows that the Two-Body Problem has the same drawbacks as the Kepler Problem, since the relative velocity can become unbounded, and neither mass can be zero. To remedy this, it is natural to introduce a new Hamiltonian E that preserves the separability of the problem, i.e. E is the sum of a center of mass Hamiltonian E_G and a relative Hamiltonian E', but where now:

a) E_G is the modified Hamiltonian for a free particle given at the end of section 1.

b) E' is the modified Hamiltonian for the general Kepler Problem given by (1.2.1). We therefore write:

$$E = c\sqrt{m_o^2c^2 + p_G^2} + c\sqrt{\frac{m_o'^2\,c^2 + p'^2}{1 + \frac{2\alpha'}{c^2r'}}} = Mc^2 \qquad (2.2.1)$$

$$\underbrace{c\sqrt{m_o^2c^2 + p_G^2}}_{E_G = M_Gc^2} \qquad \underbrace{c\sqrt{\frac{m_o'^2\,c^2 + p'^2}{1 + \frac{2\alpha'}{c^2r'}}}}_{E' = M'c^2}$$

where m_o and m_o' are the modified reduced masses ($m_o \geq 0$; $m_o' \geq 0$) and $\alpha' \triangleq fm_o$.

2.3 The formulas of section 1, especially (1.2.4) remain valid so that the velocity w_G of G and the relative velocity w' are always less than c, and $r' \to 0 \Rightarrow w' \to 0$. Also, when $m_{2o} = 0$ ($\Rightarrow m_o' = 0$), the center of mass G coincides with particle 1, whose motion is therefore rectilinear and uniform, and thus is unaffected by 2. On the other hand, the trajectory of particle 2 is a hyperbola with respect to 1 (i.e. is deflected by 1) as in the Kepler problem, and formula (1.2.4) reduces to:

$$\left(\frac{w'}{c}\right)^2 = \frac{1}{1 + \frac{2\alpha'}{c^2r'}} \qquad (2.3.1)$$

We see that $w' \simeq c$ as long as $2\alpha'/c^2r' \ll 1$, which is the case for example on the surface of the Earth.

3. THE THREE-BODY PROBLEM

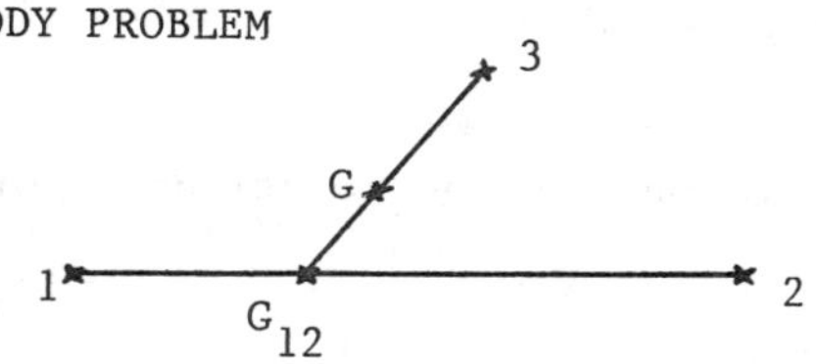

3.1 Once again using Jacobi coordinates, the Hamiltonian of the Three-Body Problem has the form $H = H_G + H' + H''$, where

$$H_G = \frac{p_G^2}{2m} \; ; \; H' = \frac{p'^2}{2m'} - f\,\frac{m'(m_1+m_2)}{r'} \; ; \; H'' = \frac{p''^2}{2m''} - fm_3\left(\frac{m_1}{r_{13}} + \frac{m_2}{r_{23}}\right)$$

(For the Four-Body Problem we would have an extra term H''', etc.) The m, m', p_G, p' have the same meaning as before; $m'' = m_3(m_1+m_2)/m$; the p_i'' are the components of the relative momentum of particle 3 with respect to G_{12} ; and

$$p'' = \sqrt{\sum_{i=1}^{3} p_i''^2}\ .$$

3.2 As before the motion of G, corresponding to H_G, separates and is that of a free particle with mass m. This is no longer the case for H' and H" since the mutual distances r_{13} and r_{23} depend on both the q_i'' (coordinates of 3 with respect to G_{12}) and q_i'. As it turns out however, all we need is the appearance of separability. This is already the case for H'. For H", we simply introduce the notation r" defined by:

$$\frac{m_1+m_2}{r''} = \frac{m_1}{r_{13}} + \frac{m_2}{r_{23}} \tag{3.2.1}$$

so that:

$$H'' = \frac{p''^2}{2m''} - f\,\frac{mm''}{r''} \tag{3.2.2}$$

(In the same way, in the Four-Body Problem, we would introduce r''' by: $(m_1+m_2+m_3)/r''' = m_1/r_{14} + m_2/r_{24} + m_3/r_{34}$, etc.) We now define a new Hamiltonian function E by:

$$E = \underbrace{c\sqrt{m_o^2c^2 + p_G^2}}_{E_G\,=\,M_Gc^2} + \underbrace{c\sqrt{\frac{m_o'^2c^2 + p'^2}{1 + \frac{2\alpha'}{c^2r'}}}}_{E'\,=\,M'c^2} + \underbrace{c\sqrt{\frac{m_o''^2c^2 + p''^2}{1 + \frac{2\alpha''}{c^2r''}}}}_{E''\,=\,M''c^2} = Mc^2$$

where m_o, m_o', m_o'' are modified reduced masses replacing m, m', m";

$\alpha' = f(m_{1o} + m_{2o})$ and $\alpha'' = fm_o$.

3.3 The corresponding set of Hamiltonian equations is written:

$$\begin{cases} \dfrac{dq_{Gi}}{d\tau} = \dfrac{p_{Gi}}{M_G} \; ; \quad \dfrac{dp_{Gi}}{d\tau} = 0 \\[2ex] \dfrac{dq_i'}{d\tau} = \dfrac{p_i'}{M'\left(1+\frac{2\alpha'}{c^2r'}\right)} \; ; \quad \dfrac{dp_i'}{d\tau} = \dfrac{M'\alpha'}{1+\frac{2\alpha'}{c^2r'}}\dfrac{\partial}{\partial q_i'}\left(\dfrac{1}{r'}\right) + \\[2ex] \qquad\qquad + \dfrac{M''\alpha''}{1+\frac{2\alpha''}{c^2r''}}\dfrac{\partial}{\partial q_i'}\left(\dfrac{1}{r''}\right) \\[2ex] \dfrac{dq_i''}{d\tau} = \dfrac{p_i''}{M''\left(1+\frac{2\alpha''}{c^2r''}\right)} \; ; \quad \dfrac{dp_i''}{d\tau} = \dfrac{M''\alpha''}{1+\frac{2\alpha''}{c^2r''}}\dfrac{\partial}{\partial q_i''}\left(\dfrac{1}{r''}\right) \end{cases} \tag{3.3.1}$$

To show that these equations are in practice indistinguishable from the classical ones, we make use of the following relations:

$$\left(\frac{w_G}{c}\right)^2 = 1 - \frac{m_o^2}{M_G^2} \quad ; \quad \left(\frac{w'}{c}\right)^2 = \frac{1}{1 + \frac{2\alpha'}{c^2 r'}}\left(1 - \frac{m_o'^2/M'^2}{1 + \frac{2\alpha'}{c^2 r'}}\right) \quad ;$$

$$\left(\frac{w''}{c}\right)^2 = \frac{1}{1 + \frac{2\alpha''}{c^2 r''}}\left(1 - \frac{m_o''^2/M''^2}{1 + \frac{2\alpha''}{c^2 r''}}\right) \qquad (3.3.2)$$

which can be solved for M_G, M' and M'' if we assume m_o, m_o' and m_o'' are non-zero:

$$M_G = \frac{m_o}{\sqrt{1 - \frac{w_G^2}{c^2}}} \quad ; \quad M' = \frac{m_o'}{\sqrt{1 + \frac{2\alpha'}{c^2 r'}}\sqrt{1 - (1 + \frac{2\alpha'}{c^2 r'})\frac{w'^2}{c^2}}} \quad ;$$

$$M'' = \frac{m_o''}{\sqrt{1 + \frac{2\alpha''}{c^2 r''}}\sqrt{1 - (1 + \frac{2\alpha''}{c^2 r''})\frac{w''^2}{c^2}}} \qquad (3.3.3)$$

This shows that so long as:

$$\frac{w_G}{c} \ll 1 \; ; \; \frac{w'}{c} \ll 1 \; ; \; \frac{w''}{c} \ll 1 \; ; \; \frac{2\alpha'}{c^2 r'} \ll 1 \; ; \; \frac{2\alpha''}{c^2 r''} \ll 1 \qquad (3.3.4)$$

we have: $M_G \simeq m_o$; $M' \simeq m_o'$; $M'' \simeq m_o''$, and system (3.3.1) is practically identical to the classical one.

3.4 Returning to formulas (3.3.2), we see that w_G, w' and w'' are always less than c, and that: $r' \to 0 \Rightarrow w' \to 0$; $r'' \to 0 \Rightarrow w'' \to 0$. We note however that the velocity of particle 3 with respect to 1 or 2 is not necessarily less than c (although it is always finite).

3.5 Finally, when $m_{2o} = 0$ ($\Rightarrow m_o' = 0$), G_{12} coincides with particle 1, and the motion of particle 2 separates, since r'' no longer depends on the q_i' . Particle 2 has the same motion with respect to 1 as in the Two-Body Problem, basically a straight line with constant velocity c (except if r' approaches zero, when it is deflected). The motion of particle 1 however is no longer Galilean, since it is a Kepler motion with respect to 3 (or a three-body motion in the case of the Four-Body Problem, etc.). This is what is actually observed on the surface of the Earth (particle 1) in its Kepler motion about the Sun (particle 3), as shown by the famous Michelson-Morley experiment.

A REPORT ON SOME NEW KAM ESTIMATES

Alessandra Celletti and Luigi Chierchia*
Forschungsinstitut für Mathematik,
ETH-Zentrum, CH-8092 Zürich

*Permanent address: Dipartimento di Matematica,
II Universita' di Roma,
00173 Rome, Italy.

In this paper we consider near-integrable Hamiltonian systems, namely, perturbations of completely integrable Hamiltonian systems. In particular, we will discuss the following example which has a direct interest in Celestial Mechanics.

Consider a rigid body on an elliptic orbit, with the spin-axis parallel to the largest principal moment of inertia and perpendicular to the orbit plane. Then the Hamiltonian of this system, under suitable assumptions (see [1]), is given by

$$H(y,z,t;\varepsilon) \equiv h(y) + \varepsilon V(x,t) \equiv \frac{y^2}{2} + \varepsilon[\cos x + \cos(x-t)], \tag{1}$$

where $(y,x,t) \in \mathbb{R} \times \mathbb{T}^2$, $T \equiv \mathbb{R}/2\pi\mathbb{Z}$ and ε is a small parameter related to the ratio of the axis of inertia of the body.

Hamilton's equations take the form:

$$\begin{cases} \dot{x} = y \\ \dot{y} = -\varepsilon V_x \end{cases}$$

and when ε is zero are trivially solved. In fact, for any initial data (y_o,x_o,t_o) the motions are either periodic or quasi-periodic:

$$(y_o,x_o,t_o) \to (y_o,x_o + \omega_o t, t_o + t), \quad \omega_o \equiv h'(\omega_o) = y_o,$$

and take place on the invariant torus $\mathcal{T}(\omega_o) = \{y_o\} \times \mathbb{T}^2$.

For ε sufficiently small, KAM theory yields that most (w.r.t. Lebesgue measure) tori $\mathcal{T}(\omega_o)$ of the unperturbed system will be deformed into tori $\mathcal{T}_\varepsilon(\omega_o)$, which are still invariant for the perturbed flow. As the magnitude of ε is increased, numerical investigations (see, e.g. [2] , [3] , [4]) suggest that these invariant surfaces break down and the motions appear to be "chaotic".

A. E. Roy (ed.), Long-Term Dynamical Behaviour of Natural and Artificial N-Body Systems, 371–375.
© 1988 by Kluwer Academic Publishers.

A relevant problem related to this picture is to try to determine the critical value of $\varepsilon, \varepsilon_c(\omega)$, at which a given KAM torus, with preassigned rotation number ω, breaks down. This problem is related to the so-called transition to stochasticity ([5],[6],[7],[8]): it is believed that, if one looks at frequencies ω between 0 and 1, there is a last torus to disappear corresponding to $\omega_o = \frac{\sqrt{5}-1}{2}$.

There are several (non-rigorous) methods to determine ε_c ([4]). For example an application of Greene's residue-criterion ([3]) indicates ([9]) that for the Hamiltonian (1)

$$\varepsilon_c(\omega_o) \equiv \varepsilon_c\left(\frac{\sqrt{5}-1}{2}\right) \simeq 0.027.$$

We discuss now a new method that allows one to give a rigorous lower bound on ε_c. An application of such method to (1) yields the stability of the golden-mean torus for:

$$|\varepsilon| \lesssim \varepsilon_r \simeq 0.015$$

so that the ratio between the numerical and the rigorous values is about 1.8 (a previous rigorous result on the same model ([10],[11]) gave a ratio of about 40). We also point out that our method can be extended to more degrees of freedom or adapted to monotone twist maps. For instance, an application to the so-called standard map:

$$\begin{cases} y' = y + \varepsilon \sin x \\ x' = x + y', \end{cases} \qquad y \in \mathbb{R},\ x \in \mathbb{T},$$

shows the existence of the golden-mean invariant curves for $|\varepsilon| \lesssim \varepsilon_r \simeq 0.65$, while the critical value is believed to be about 0.98 [3].

Going back to the Hamiltonian case, we recall that a KAM torus $\mathcal{T}(\omega)$ can be represented by the parametric equations:

$$\begin{cases} y = v(\theta,t) \\ x = \theta + u(\theta,t), \end{cases} \qquad (\theta,t) \in \mathbb{T}^2 \tag{2}$$

with u,v (depending upon ω and ε) and that the flow in the (θ,t)-coordinates is just a straight line:

$$(\theta_o, t_o) \to (\theta_o + \omega t, t_o + t).$$

Hamilton's equations and the above characterization of $\mathcal{T}(\omega)$ yield the following equation for u:

$$D^2 u + \varepsilon V_x(\theta + u, t) = 0 \quad , \quad D \equiv \omega_o \frac{\partial}{\partial \theta} + \frac{\partial}{\partial t} \ . \tag{3}$$

and the function v is related to u by

$$v \equiv \omega + Du.$$

Vice versa, given a smooth solution of (3), one obtains a KAM torus with parametric equations (2).

Now, let, as above, $\omega_o = \frac{\sqrt{5}-1}{2}$. Then:

THEOREM: For any complex ε, with $|\varepsilon| \leq 0.015$, equation (3) has a unique solution u, which is analytic in θ, t, ε, with

$$\int_{T^2} u(\theta,t;\varepsilon)\, d\theta dt = 0 .$$

The proof of this theorem is based on a KAM algorithm and contains some parts which are computer-assisted. We try here to give an idea of such proof, leaving the complete exposition to [12].

We look for a solution u, which is analytic in ε near $\varepsilon = 0$; thus we set:

$$u(\theta,t;\varepsilon) = \sum_{k=1}^{\infty} u^{(k)}(\theta,t)\varepsilon^k \equiv v_{k_o} + R_{k_o} \equiv \sum_{k=1}^{k_o} u^{(k)}(\theta,t)\varepsilon^k + R_{k_o}(\theta,t;\varepsilon), \tag{4}$$

for some k_o to be specified later.

The basic idea is now to compute "explicitly" v_{k_o} and then to apply a "superconvergent iteration", using as starting approximate-solution v_{k_o}.

To compute v_{k_o} we simply insert the series (4) into equation (3) and compare the same orders in ε, obtaining:

$$\begin{cases} D^2u^{(1)} = \sin\theta + \sin(\theta-t) \\ D^2u^{(k)} = \sum_{\ell=1}^{k-1} \frac{1}{\ell!}\left[\frac{\partial^\ell}{\partial\theta^\ell}(\sin\theta + \sin(\theta-t))\right] \sum_{\substack{k_1+\ldots+k_\ell=k-1 \\ k_i \geq 1}} u^{(k_1)} \ldots u^{(k_\ell)}, \quad k \geq 2. \end{cases} \tag{5}$$

Expanding $u^{(k)}$ in Fourier series

$$u^{(k)} = \sum_{(n,m)\in Z^2} u^{(k)}_{(n,m)} e^{i(n\theta+mt)},$$

one gets:

$$\left\{\begin{array}{l} \hat{u}^{(1)}_{(n,m)} = \dfrac{in}{(\omega n + m)^2}\, \zeta_{(n,m)} \\ \hat{u}^{(k)}_{(n,m)} = \dfrac{1}{(\omega n + m)^2} \displaystyle\sum_{\ell=1}^{k-1} \sum_{\substack{k_1+\dots+k_\ell=k-1 \\ k_i \geq 1}} \sum_{\substack{\underline{\nu}_o+\dots+\underline{\nu}_1=(n,m) \\ \nu_i \in Z^2}} \\ \qquad\qquad \zeta_{\underline{\nu}_o}\, \hat{u}^{(k_1)}_{\underline{\nu}_1} \dots u^{(k_\ell)}_{\underline{\nu}_\ell} \dfrac{(i\nu_{o1})^{\ell+1}}{\ell!} \end{array}\right. \qquad (6)$$

with $k \geq 2$ and

$$\zeta_{(n,m)} \equiv \begin{cases} \frac{1}{2} & \text{if } (n,m) = \pm(1,-1),\ \pm(1,0); \\ 0 & \text{otherwise.} \end{cases}$$

Now if we plug $v_o \equiv v_{k_o}$ in place of u in (3), we get:

$$D^2 v_o + \varepsilon V_x(\theta + v_o, t) = E_o,$$

with an "error term" E_o (which is defined by the l.h.s.). By definition of v_o one easily gets that $v_o \simeq O(\varepsilon^{k_o+1})$ and one can try to apply a KAM type of algorithm in order to get a sequence of quadratically convergent approximate solutions $\{v_j\}$, which are analytic in θ, t, ε. Following a recent scheme ([13], [14]), we construct the sequence $\{v_j\}$ as follows.

Assume that v_j and E_j satisfy

$$D^2 v_j + \varepsilon V_x(\theta + v_j, t) = E_j, \qquad (7)$$

and consider the "linearized equation"

$$D[(1 + \frac{\partial v_j}{\partial \theta})^2 Dz] = -(1 + \frac{\partial v_j}{\partial \theta}) E_j.$$

Notice that (7) implies easily that the r.h.s. has mean value zero, so that the equation can be solved provided

$$1 + \frac{\partial v_j}{\partial \theta} \neq 0 . \qquad (8)$$

Now defining

$$w_j \equiv (1 + \frac{\partial v_j}{\partial \theta})\, z, \qquad v_{j+1} = v_j + w_j,$$

one sees that v_{j+1} verifies

$$D^2 v_{j+1} + \varepsilon V_x(\theta + v_{j+1}, t) = E_{j+1},$$

and that $w_j \simeq O(E_j), E_{j+1} \simeq O(E_j^2)$.

In order to iterate this procedure so to obtain the final solution, one should check condition (8) infinitely many times. In standard KAM techniques one constructs a unique condition, which implies (8) for $j=0,1,\ldots$ However it turns out to be much more effective to control explicitly (8) for a finite number of steps and then to apply a condition which implies all the remaining ones. Analyticity in ε is a byproduct of this construction.

The role of the computer in this scheme is to "construct" v_o, using (6) and to control condition (8) for a certain number of times j_o. To prove the above theorem we took $k_o = 24$ (which implies the computation of about 10000 Fourier coefficients) and $j_o = 30$.

Finally, we mention that to cover the problem of the control of numerical errors, we performed the so-called "interval arithmetics" (see [15], [16]).

REFERENCES:

[1] Wisdom J., Peale S.J., Icarus 58, 1984.
[2] Henon M., Heiles C., Astron. J. 69, 1964.
[3] Greene J.M., J. Math. Phys. 20, 1979.
[4] Lichtenberg A.J., Lieberman M.A., Regular and Stochastic Motion (Springer, N.Y.), 1983.
[5] Escande D.F., Doveil F., J. Stat. Phys. 26, 1981.
[6] Escande D.F., Phys. Reports 121, 1985.
[7] Greene J.M., Percival I.C., Physica 3D, 1981.
[8] Percival I.C., J. of Physics A L57, 1979.
[9] Falcolini, private communication.
[10] Celletti A., Chierchia L., J. Math. Phys. 28, 1987.
[11] Celleti A., Falcolini C., Porzio A., Ann. Inst. H. Poincaré 47,1987.
[12] Celletti A., Chierchia L., in preparation.
[13] Moser J., Ann. Inst. H. Poincaré 3, 1986.
[14] Salamon D., Zehnder E., to be published.
[15] Lanford III O.E., Physica A. 124, 1984.
[16] Eckmann J.P., Koch H., Wittwer P., Memoirs AMS 47, 1984.

EXTENDED CANONICAL TRANSFORMATIONS INCREASING THE NUMBER OF VARIABLES

José M. Ferrándiz
Dpto. Mathemática Aplicada a la Técnica
E.T.S. de Ingenieros Industriales,
Valladolid

1. INTRODUCTION

Redundant variables have been successfully used in Celestial Mechanics for a long time. Here, we are concerned specially with the changes of coordinates increasing their number. Among them, the Euler parameters, K-S variables, or other sets of variables by Broucke are widely known.

In recent years, systems involving redundant variables have been also considered in a Hamiltonian framework. Typical examples are the canonical K-S variables or the Euler parameters.

In this communication we address the problem of obtaining a set of canonical variables from a given transformation which is applied to the coordinates belonging to a previous canonical set and increases their number. A partial answer to this question was given by Kurcheeva (1977). In her approach the number of coordinates is allowed to increase only in a unity and that restriction does not seem to be easily removable.

By using different ideas, we will develop below a method to construct such extended canonical sets in a straightforward way, whatever the number of added coordinates is. Moreover, we can select the new variables to satisfy desired constraints in their velocities, which may be useful in some cases.

2. CONSTRUCTION OF EXTENDED CANONICAL TRANSFORMATIONS

The derivation of the main result will be divided in several steps of progressive complexity.

1. Here, we only deal with coordinates.
   Let $x = (x_1, x_2, \ldots, x_n)$ be an n-vector of old coordinates and $Q = (Q_1, Q_2, \ldots, Q_m)$ an m-vector of corresponding new coordinates with $m > n$.

A. E. Roy (ed.), Long-Term Dynamical Behaviour of Natural and Artificial N-Body Systems, 377–383.
© 1988 by Kluwer Academic Publishers.

Let $x = x(Q)$ be a transformation sufficiently differentiable and of maximum rank on an open domain, which increases the number of variables from n to m. This transformation can be expressed in coordinates as

$$x_i = x_i(Q_\alpha) \; , \tag{1}$$

if we agree from now onwards that Latin indices run from 1 to n and Greek ones from 1 to m. Likewise, primed Latin indices appearing below will vary from n+1 to m.

If we denote $b_{i\alpha} = (\partial x_i/\partial Q_\alpha)$, the following identity holds

$$dx_i = b_{i\alpha}\, dQ_\alpha . \tag{2}$$

Let us assume that we want to determine solutions $x_i(t)$ to a given Initial Value Problem (IVP)

$$dx_i = f_i(x,t)dt \quad , \quad x_i(0) = x_i^o \; , \tag{3}$$

by inserting in the transformation (1) solutions $Q_\alpha(t)$ to a 'suitable' transformed problem in the Q-space. Such a problem cannot be unique, but that will be the case if we add m-n equations

$$b_{i'\alpha} \frac{dQ_\alpha}{dt} = 0 \tag{4}$$

to hold along the trajectories $Q_\alpha(t)$. The only requirement on the functions $b_{i'\alpha}$ of Q and t is that the matrix $B = (b_{\alpha\beta})$ must be non-singular.

Using (2) and (4) we obtain a suitable transformed problem in the form

$$dQ = B^{-1} \begin{pmatrix} dx_i \\ \\ 0 \end{pmatrix} = B^{-1} \begin{pmatrix} f_i dt \\ \\ 0 \end{pmatrix} \; , \tag{5}$$

$$Q(0) = Q^o \quad \text{such that} \quad x^o = x(Q^o).$$

Then the solutions to the new IVP (5) are transformed by (1) into the required solutions to the old IVP (3) and satisfy the constraints (4).

2. In this step, besides x we consider n variables $y = (y_1, y_2, \ldots, y_n)$. Let us replace the old problem (3) by

$$\begin{aligned} dx_i &= f_i(x,y,t)dt \\ dy_i &= g_i(x,y,t)dt \\ x_i(0) &= x_i^o \quad , \quad y_i(0) = y_i^o \, . \end{aligned} \tag{6}$$

Then, the above results still hold if the transformed problem is

$$dQ = B^{-1} \begin{pmatrix} f_i dt \\ \\ 0 \end{pmatrix} \qquad (7)$$

$$dy_i = g_i(x(Q),y,t)dt,$$

with initial values Q^o, y^o verifying $x = x(Q^o)$.

3. Let us introduce m-n auxiliary variables $y_{i'}$, and the corresponding solutions of an auxiliary problem.

$$\frac{dy_{i'}}{dt} = 0 \quad , \quad y_{i'}(0) = y^o_{i'} \quad , \qquad (8)$$

where constants y^o_i are arbitrary.

If we add the auxiliary problem (8) to the IVPs (6) and (7), the former conclusion remains true, since the corresponding solutions do not essentially differ.

The increased new problem is now 2m-dimensional in the variables Q_α, y_α. The matrix B being invertible on the considered Q-domain, we have no trouble in changing the variables y_α to P_α, defined by

$$P = B^t y \quad , \qquad (9)$$

where superscript t stands for transposition.

According to (9), the increased new problem (7+8) is immediately transformed into a new one of the form

$$\begin{aligned} dQ_\alpha &= F_\alpha(P,Q,t)dt \\ dP_\alpha &= G_\alpha(P,Q,t)dt \\ Q_\alpha(0) &= Q^o_\alpha \quad , \quad P_\alpha(0) = P^o_\alpha \quad , \end{aligned} \qquad (10)$$

the initial values of which verify

$$x^o = x(Q^o) \quad , \quad P^o = (B^t)^o y^o \ . \qquad (11)$$

Since problem (7+8) has the m-n integrals $y_{i'}$ = const., it is clear that the last rows in the inverse of the identity (9) provide the relations

$$y_{i'}(P,Q) = \text{const.}, \qquad (12)$$

which are thus integrals for system (10).

Moreover, the solution functions $Q_\alpha(t)$ to (10) do not stop satisfying the m-n selected constraints (4).

Finally, the solution of the original IVP (3) is recovered by inserting the solution of the IVP (10) into the equations of the whole transformation

$$x = x(Q) \quad , \quad y = (B^{-1})^t p \tag{13}$$

4. Let us assume now that (x,y) is a set of 2n canonical variables belonging to a Hamiltonian $H = H(x,y,t)$. We can make the canonical IVP

$$\begin{aligned} dx_i &= \frac{\partial H}{\partial y_i}\, dt \quad , \quad dy_i = -\frac{\partial H}{\partial x_i}\, dt \\ x_i(0) &= x_i^o \quad , \quad y_i(0) = y_i^o \end{aligned} \tag{14}$$

play the role of the old IVP (6) in the step 2, and apply the above transformation to obtain a new IVP with the same form as (10). We will show that such a system has indeed Hamiltonian form.

Let us define m-n differential 1-forms $w_{i'}$, be means of

$$w_{i'} = b_{i'\alpha} dQ_\alpha \quad , \tag{15}$$

so that, from (2), it follows the identity

$$\begin{pmatrix} dx_i \\ w_{i'} \end{pmatrix} = B\ dQ \tag{16}$$

On account of (16) and the definition (13) of variables p_α, we get the new identity

$$y_i dx_i + y_{i'} w_{i'} - H dt = P_\alpha dQ_\alpha - \bar{H} dt, \tag{17}$$

where $\bar{H}(Q,P,t) = H(x(Q),y(Q,P),t)$.

Taking exterior derivatives in (16), we come to the equality

$$dy_i \wedge dx_i + dy_{i'} \wedge w_{i'} + y_{i'} dw_{i'} - dH \wedge dt = dP_\alpha \wedge dQ_\alpha - d\bar{H} \wedge dt. \tag{18}$$

Now, we only need to realize that the relations $y_{i'} = y_i^o$ and $w_{i'} = 0$ hold along the solution of the problem (10), the last ones because constraints (4) are fulfilled. It suffices to choose the arbitrary constants $y_{i'}^o = 0$ to find that the equation

$$dy_i \wedge dx_i - dH \wedge dt = dP_\alpha \wedge dQ_\alpha - d\bar{H} \wedge dt \tag{19}$$

is verified on the manifold $y_{i'} = 0$ and therefore the transformation is canonical on it.

Let us remark that the transformation might be canonical on a wider manifold. That happens when some forms w_j, are exact, since one can suppress the corresponding requirement ^jy_i, = 0 in the obtaining of (19). Then, the resulting manifold would be foliated into leaves y_j, = const., and diffeomorfic solutions would lie on them.

Next, we will summarize the main results in the following

<u>Theorem</u>: Assumptions: Let $x_i(t)$, $y_i(t)$ be the solution to the Hamiltonian IVP (14); $x = x(Q)$ the transformation (1) of maximum rank on a m-dimensional region; B a non-singular mxm matrix, the terms of its n first rows verifying $b_{i\alpha} = (\partial x_i/\partial Q_\alpha)$ and the remainder being arbitrary: P_α the m variables defined in (9) which satisfy the relations

$$y_i = c_{i\alpha} P_\alpha, \tag{20}$$

with $C = (c_{\beta\alpha}) = (B^{-1})^t$; and $\bar{H}(Q,P,t) = H(x(Q),y(P,Q),t)$.

Statement: The solution of the canonical system

$$\frac{dQ_\alpha}{dt} = \frac{\partial \bar{H}}{\partial P_\alpha} \quad , \quad \frac{dP_\alpha}{dt} = - \frac{\partial \bar{H}}{\partial Q_\alpha} \tag{21}$$

for initial values Q^o, P^o such that

$$x^o = x(Q^o) \quad , \quad P^o = (B^t)^o \begin{pmatrix} y_i^o \\ 0 \end{pmatrix} \tag{22}$$

yields, by virtue of transformation (1), (20), the solution $x_i(t)$, $y_i(t)$ of the old system.

Moreover, the constraints (4) are satisfied along that solution, which lies on the manifold given by the m-n equations $y_{i'}(P,Q) = 0$. In other words, the transformation from (x,y) to (Q,P) is canonical on that manifold.

To end this section we will give an explicit expression for conditions $y_{i'} = 0$. A simple consideration about the Cramer system contained in (9) allows us to write the equivalent m-n relations

$$\begin{vmatrix} \frac{\partial x_1}{\partial Q_1} & \cdots\cdots & \frac{\partial x_1}{\partial Q_n} & \frac{\partial x_1}{\partial Q_{i'}} \\ \vdots & & \vdots & \vdots \\ \frac{\partial x_n}{\partial Q_1} & \cdots\cdots & \frac{\partial x_n}{\partial Q_n} & \frac{\partial x_n}{\partial Q_{i'}} \\ P_1 & \cdots\cdots & P_n & P_{i'} \end{vmatrix} = 0 \qquad (23)$$

Notice that they reduce to Kurcheeva's compatibility condition, if $m = n+1$.

3. APPLICATIONS

The extension of a change of coordinates increasing their number can be made in infinitely many ways by using the previous algebraic procedure. The real difficulty lies on the choice of the equations of the constraints so that the transformation has useful properties. No general rule is known although the derivations become shorter using our method, as can be seen in the following examples.

Example 1: Canonical K-S variables.

The change from Cartesian to K-S coordinates is given by $x = \Lambda(\bar{x})\bar{x}$ (Stiefel and Scheifele, p.234). On account of the identity $dx = 2\,\Lambda(\bar{x})d\bar{x}$, if we choose as additional constraint $2(\bar{x}_4\dot{\bar{x}}_1 - \bar{x}_3\dot{\bar{x}}_2 + \bar{x}_2\dot{\bar{x}}_3 - \bar{x}_1\dot{\bar{x}}_4) = 0$, our matrix B is now $2L(\bar{x})$. The new momenta are obtained through $(\bar{p}_1,\bar{p}_2,\bar{p}_3,\bar{p}_4)^t = B^t(p_1,p_2,p_3,p_4)^t$ and the resulting set is canonical on the manifold $p_4 = 0$, since the constraint is non-integrable. The inversion of the previous identity yields the equation of that manifold, which reduces to the bilinear relation $\bar{x}_4\bar{p}_1 - \bar{x}_3\bar{p}_2 + \bar{x}_2\bar{p}_3 - \bar{x}_1\bar{p}_4 = 0$.

Let us remark that no other proof is needed to establish those facts.

Example 2: A set of canonical orbital variables by Ferrandiz (1986,87). Let us change from Cartesian coordinates X_i to new coordinates x_i, z, verifying $X_i = x_i/z$ $(i = 1,2,3)$. We choose the integrable constraint $x_i\dot{x}_i = 0$. Then $\det B = |x|^2 z^{-4} \neq 0$ if $|x| \neq 0$, and the new momenta are defined through $(p_i, p_z)^t = B^t(P_i, P_4)^t$, P_i being the Cartesian ones.

The inversion of that expression gives

$$P_i = p_i z - z x_i |x|^{-2} \{(x,p)+zp_z\} , \quad P_4 = |x|^{-2}\{(x,p)+zp_z\} \tag{24}$$

Now, the transformation is canonical on the whole increased space with $z \neq 0$ and has the two integrals: $|x|^2 = \text{const.}$ and $(x,p) + zp_z = \text{const.}$ This transformation allows us to obtain equations of motion with suitable properties for numerical integration and even to reduce exactly into four harmonic oscillators the radial intermediaries by Deprit (Ferrándiz & Fernández-Ferreirós, 1987).

Other examples are available. Among them is the classical method to change the time variable in a Hamiltonian system.

ACKNOWLEDGEMENT

This work has been partially supported by D.G.A. Proy. No.CB4/85 and CAICYT Proy. No. 779-84.

REFERENCES

Ferrándiz, J.M.: 1986, ESA SP-255, pp.361-364.
——————: 1987, Submitted to Celest. Mech.
Kurcheeva, I.V.: 1977, Celest. Mech., 15, pp.353-365.
Stiefel, E. and Scheifele, G.: 1971, Linear and Regular Celestial Mechanics, Springer-Verlag, Berlin-Heidelberg-New York.

NON LOCAL SPIRALING CHARACTERISTIC CURVES

Joaquim Font
Universitat de Barcelona
Facultat de Matemàtiques
Departement Matemàtica Aplic.i Anàlisi
Gran Via 585. Barcelona 08071, Spain.

Miquel Grau
Universitat Politècnica de Catalunya
Facultat d'Informatica
Departement Matematica Aplic.II
Pau Gargallo 5, Barcelona 08028, Spain.

ABSTRACT:
In a quartic potential introduced by B. Barbanis, he found characteristic curves of symmetric periodic orbits which look like finite spirals. Considering a one parameter family of potentials including the previous one we have found also characteristic curves which are finite spirals around a finite curve. This one is built up from heteroclinic connections between a symmetrical couple of Lyapunov orbits. This allows us to explain in a very geometric way the results of B. Barbanis.

1. INTRODUCTION

We consider the Hamiltonian introduced by Barbanis (1) (a modification of the Hamiltonian of Contopoulos and Moutsoulas (4)) given by

$$H = \frac{1}{2}(p_x^2 + p_y^2 + x^2 + y^2) - xy^2 + \frac{1}{2}y^4.$$

The most important properties of this Hamiltonian are:

I) All the zero velocity curves (z.v.c.) defined by $p_x=p_y=0$ are closed and they only exist for positive energy.

II) There is a symmetry with respect to the x-axis: $H(x,y,p_x,p_y,t) = H(-x,y,p_x,-p_y,-t)$.

III) There is only one fixed point, located at the origin, with double eigenvalues $\pm i$.

IV) There is a family of symmetric periodic orbits with projection on

A. E. Roy (ed.), Long-Term Dynamical Behaviour of Natural and Artificial N-Body Systems, 385–391.
© 1988 by Kluwer Academic Publishers.

(x,y) as the one given in Fig.1. Each one of the symmetric triple periodic orbit (3-P,0) cuts y=0 in a point as A with $\dot{x}=0$. Then we characterize the orbit from the values of the energy, h, and the x coordinate of the point A. The curve obtained in the variables x and h is called the characteristic curve (C.C.). The C.C., for this family of 3-P.O., is a "finite" spiral (see Fig.9 of (1)).

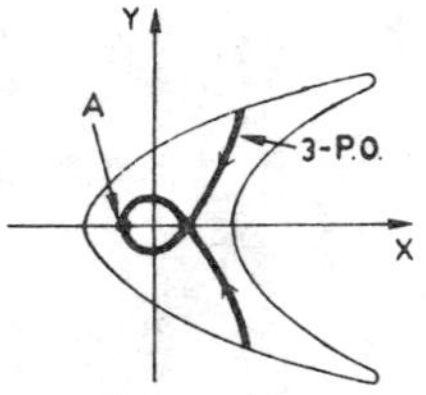

Figure 1

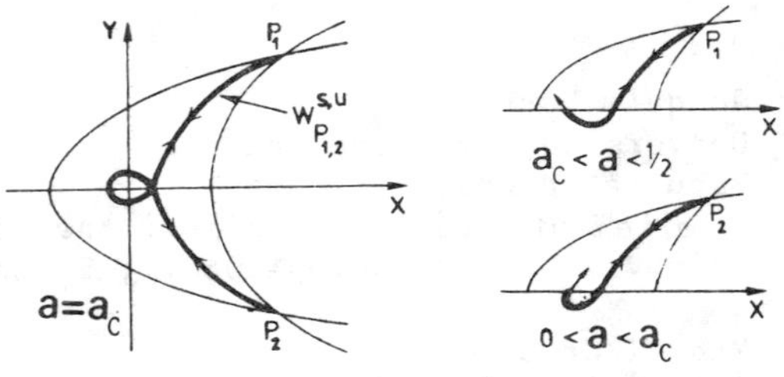

Figure 2

Our objective will be to understand the origin of this curve. We remark that Devaney (5) and also Henrard (8) have found infinite spiral C.C. going to the point which represents a homoclinic orbit to a complex-saddle fixed point. However the Barbanis' spiral is "finite" since it gives 8 revolutions inward and then it spirals outwards. Furthermore there appear some inserted bubbles in the inner part which are not connected to the main family of 3-P.O.

What we shall present here is a numerical study of the C.C. of symmetric 3-P.O. of a Hamiltonian uniparametric family which generalizes the Hamiltonian introduced by Barbanis. We give a geometrical explanation of the spiral shape of the C.C. and its evolution with respect to the parameter.

2. A FAMILY OF HAMILTONIANS

We consider the Hamiltonian family:

$$H = \frac{1}{2}(p_x^2 + p_y^2 + x^2 + y^2) - xy^2 + a\, y^4 ,$$

where a is a real number. This family has the following properties:

I) If $a \geq \frac{1}{2}$, the origin is the only fixed point.

II) If $a < \frac{1}{2}$, there are three fixed points: the origin and $P_{1,2}$ at $p_x = p_y = 0$, $x=(2-4a)^{-1}$ and $y= \pm(2-4a)^{-\frac{1}{2}}$.

The points $P_{1,2}$ are of saddle-center type, live at the level of energy $h_c = (8-16a)^{-1}$ and go to infinity when a tends to $\frac{1}{2}$.

The z.v.c. for different values of a and h are described by:

$$\frac{1}{2}(x^2+ y^2) - xy^2 + ay^4 = h ,$$

(see Fig.3, where the Hill's regions are dashed).

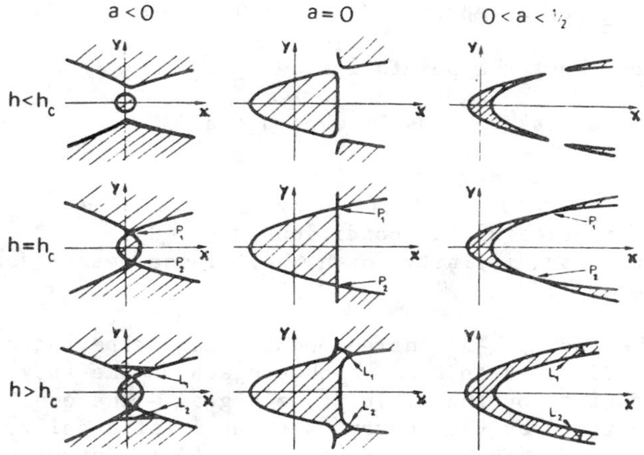

Figure 3

For different values of a there exists a double heteroclinic connection between P_1 and P_2: $a = -4.359923649...$, $a = a_c = 0.49188633722...$(see Fig.2). On the region bounded by the z.v.c. the stable and unstable manifolds of P_1 (which have the same projection on the (x,y) plane due to the symmetry) coincide with the unstable and stable ones, respectively, of P_2.

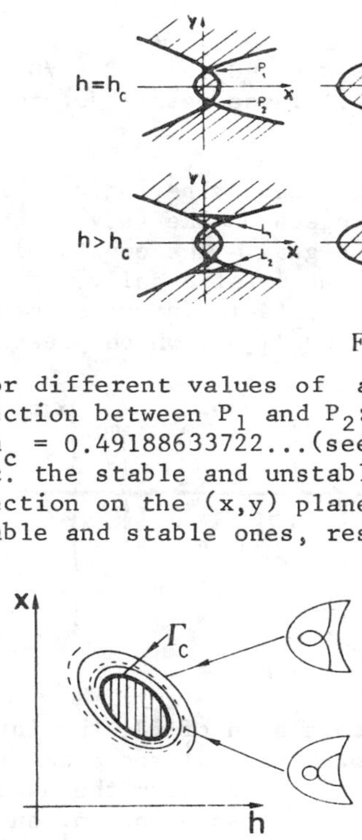

Figure 4

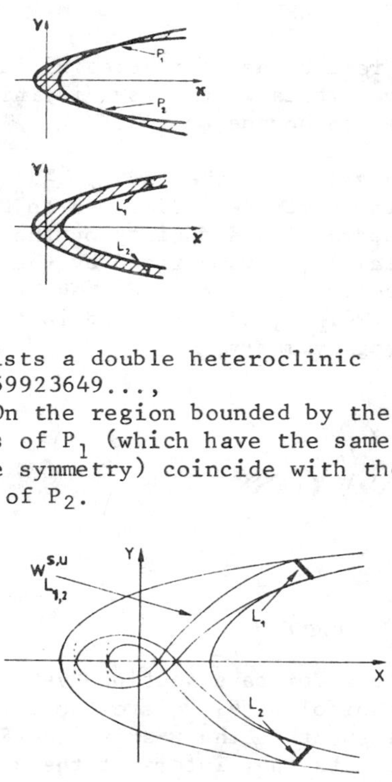

Figure 5

The C.C. of the family of symmetric 3-P.O. for $a = a_c$ consists of two spirals which approach a "big" finite curve Γ_c (see Fig.4). Several questions appeared in this new problem:

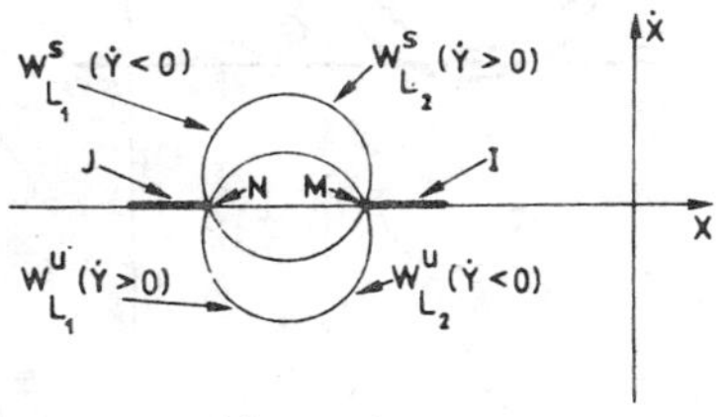

Figure 6

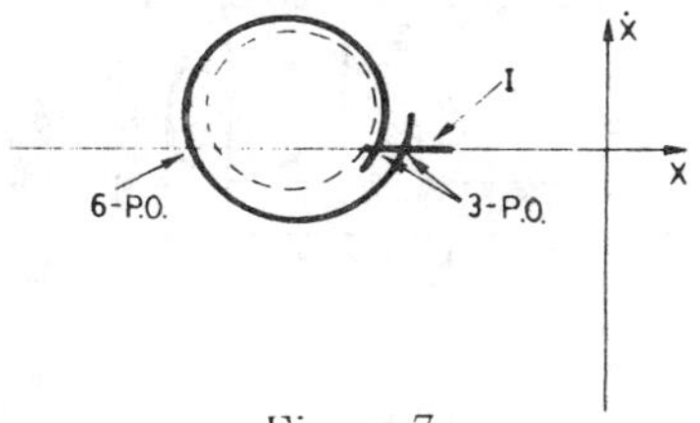

Figure 7

- What does the Γ_c curve represent?

- What can we say about the points inside Γ_c ?

- Why do the C.C. spiral towards Γ_c when $a = a_c$?

3. THE Γ_c CURVE

We remark that for some initial conditions $x_o = y_o = 0$, x_o and $\dot{y}_o \neq 0$, we have orbits which reach infinity for $h > h_c$. These points determine the region bounded by Γ_c.

The values of the energy $h > h_c$ there appear Lyapunov periodic orbits (see for instance (9)) close to $P_{1,2}$ which project on the (x,y) plane on a curve with endpoints on the z.v.c. (see Fig.5). The orbits of this family, parametrized by the energy, are, at least locally, of hyperbolic type and then they have stable and unstable invariant manifolds ($W^{s,u}_{1,2}$, also denoted in the figures by $W_L{}^{s,u}_{1,2}$) which present a strong symmetry,

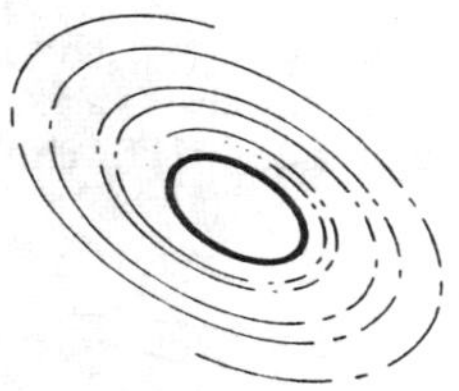

Figure 8

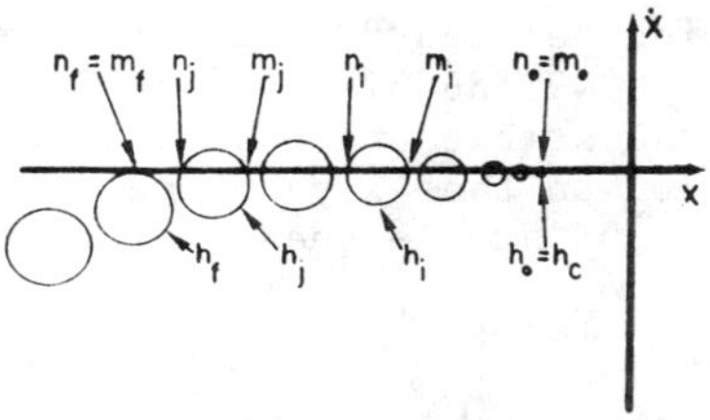

Figure 9

On the Poincaré section (y=0, with a suitable sign of $\dot{y}$), the invariant manifolds $W^{s,u}_{1,2}$ are two curves diffeomorphic to S^1 and symmetric with respect to the x-axis (see Fig.6). In accordance with the numerical results they intersect the x-axis, at most, in two points n_i and m_i for values of the energy h_i between $h_o = h_c$ and h_f; in only one point for h_o and h_f and they have no intersection for $h > h_f$ and for $h < h_o$ (see Fig.9).

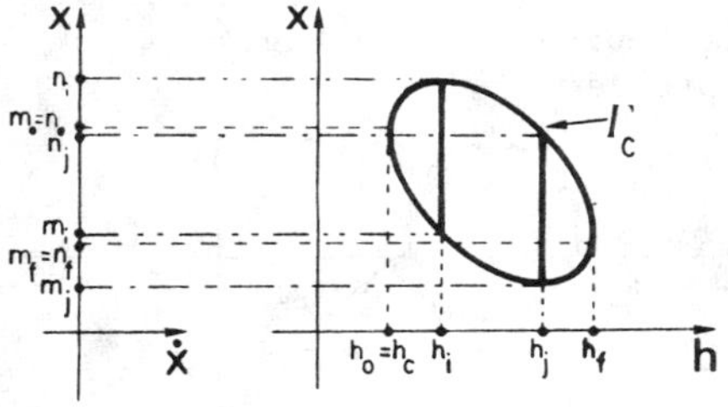

Figure 10

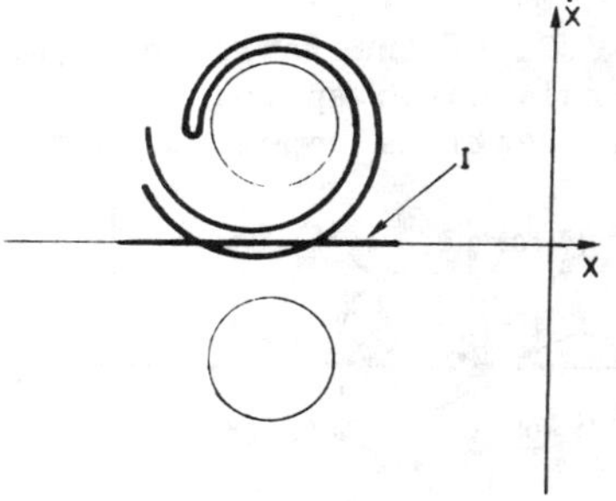

Figure 11

A point Q belonging to the (n_i,m_i) segment is inside the two invariant manifolds $W^{s,u}_{1,2}$. Therefore the flow takes Q to infinity. The boundary of this consists of the points n_i and m_i and when h ranges in $[h_o,h_f]$ it gives the closed curves Γ_c (see Fig.10).

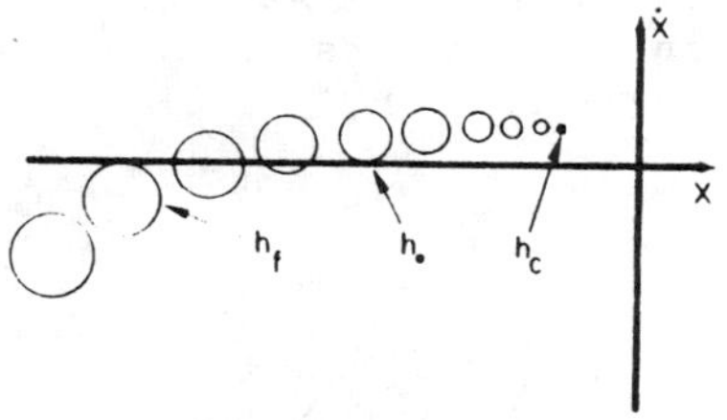

Figure 12

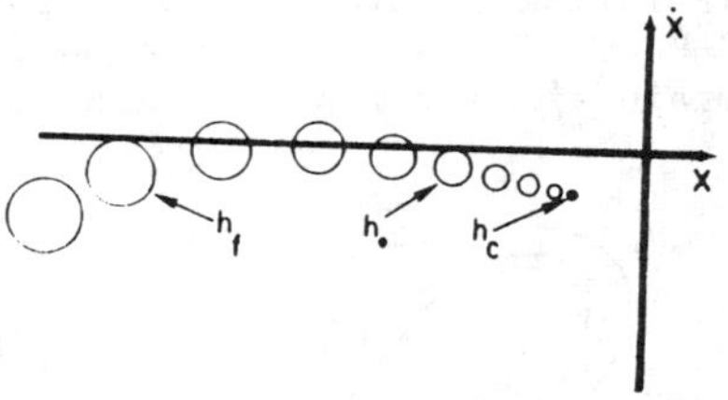

Figure 13

4. THE CHARACTERISTIC CURVE FOR $a=a_c$

Numerically we see that for $a\leq\frac{1}{2}$ and in particular for $a=a_c$ the C.C. presents 8 sequences of levels of energy $\{h_{ij}\}_{i=1,\dots,4}$, $\{k_{ij}\}_{i=1,\dots,4}$, $j\in\mathbb{N}$ so that the symmetric 3-P.O. Π_{ij} with energy $h=h_{ij}$ or k_{ij} is a parabolic orbit, using Hénon's index of stability (7). If h belongs to the interval with bounds h_{ij} and k_{ij}, $i=1,\dots,4$, Π_{ij} is an elliptic orbit and, if not, it is hyperbolic (see Fig.8). The C.C. bifurcates to 6-P.O. in each revolution and its behaviour is similar to the one described by Barbanis (1), but they exhibit infinite turns. This result is analogous to those mentioned in (3) replacing fixed points with the Lyapunov orbits.

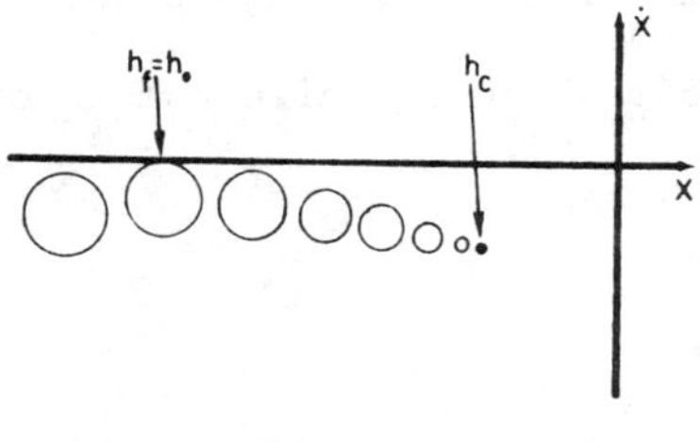

Figure 14

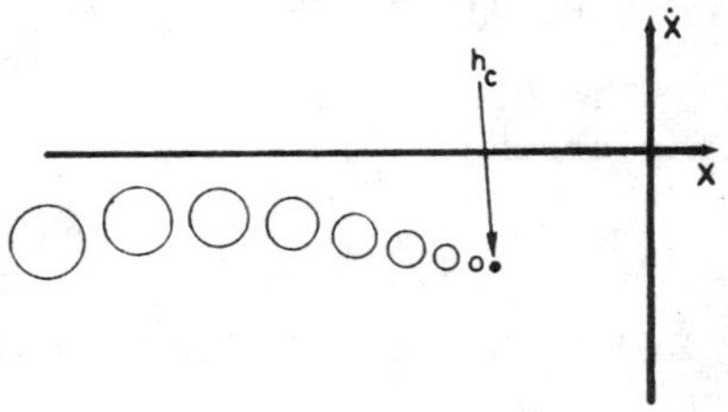

Figure 15

We find a non-local phenomenon, since the Γ_c curve cannot be completely studied by a local analysis in the neighbourhood of a point. Therefore the C.C. of the 3-P.O. cannot be described using these techniques.

The study of the behaviour of C.C. can be approached in three regions:

a) $h<h_o\approx h_c$: As the invariant manifolds of the points $P_{1,2}$ present a double heteroclinic connection, we have obtained, in their neighbourhood, the variational equations until second order and these solutions show

Institut für Reine
und Angewandte Kernphysik
Kiel

that the C.C. is a foliation which accumulates to the Γ_c curve.

b) $h_o < h < h_f$: If we fix an energy level h, the $W^{s,u}_{1,2}$ are shown in Fig.6. Take a segment I (J) ending on M(N). The image by the return map of I is a spiral close to $W^s_{1,2}$($W^u_{1,2}$ resp.)(see Fig.7 and 11)). The infinite intersection point of the spiral with the x-axis are P.O. (3-P.O. if the points are on I and 6-P.O. in any other case). We obtain similar results when we take the J segment.

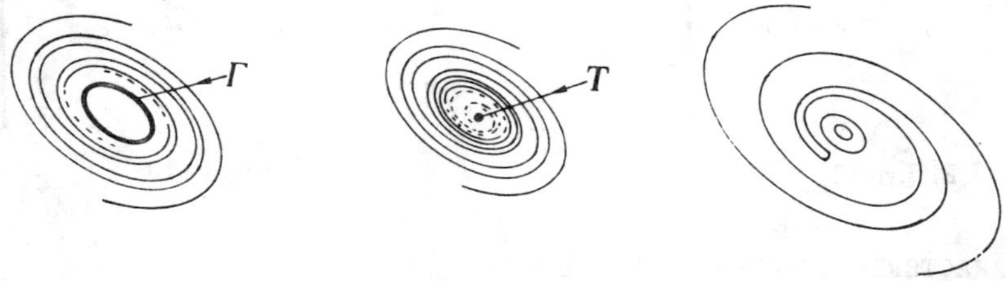

(a) (b) (c)

Figure 16

c) $h_f < h$: In this region the $W^{s,u}_{1,2}$ have not any point in common with the axis $\dot{x}=0$ (see Fig.11). As in b), the image of a segment I of the x-axis is a spiral around the $W^s_{1,2}$ ($W^u_{1,2}$ resp.), but now it cuts the segment I in a finite number of points. Therefore, the C.C. has a finite number of points in this energy level.

The previous results allow us to make the following conjecture:

The C.C. of 3-P.O. is a curve that spirals an infinite number of times around Γ_c.

5. EVOLUTION WITH RESPECT TO THE PARAMETER a

As the parameter a changes, Fig.9 undergoes some changes, which result in variations of the C.C.. This gives place to the following facts:

i) For $a \lesssim a_c$, the $W^{s,u}_{1,2}$ are shown in the Fig.12,
ii) For $a = a_c$, in Fig.9,
iii) For $a_c < a_t < \frac{1}{2}$, in Fig.13,
iv) For $a \doteq a_t$, in Fig.14, and, finally,
v) For $a_t < a$, in Fig.15.

In the cases i) and III), the C.C. are similar to the C.C. in case ii), but the points of Γ for $h=h_o$ come from a Lyapunov orbit (see Fig.16a).

For $a=a_t=0.4946439...$ we have that the curve Γ consists only of one point T, and the C.C. spirals an infinite number of times round T (see Fig.16b).

Eventually in case v), Γ vanishes. The C.C. consists of a finite spiral and a set of embedded "bubbles", and its behaviour is similar to the one described by Barbanis (1), (2). This fact has been verified numerically for a=0.495, a=0.498, a=0.540 and a=0.600. These bubbles are generally called irregular orbits, since they are not connected to the main family.

BIBLIOGRAPHY

(1) Barbanis, B.: Celestial Mechanics 36 (1985), 257.
(2) Barbanis, B.: Celestial Mechanics 39 (1986) 345.
(3) Churchill, R.C., Pecelli, G. and Rod, D.L.: Archive for Rational Mechanics and Analysis 73, (1980) 313.
(4) Contopoulos, G. and Moutsoulas, M.: Astronomical Journal 79, 1965, 817.
(5) Devaney, R.: Journal of Diff. Eq. 21 (1976) 431.
(6) Font, J. and Simó, C.: C.E.D.Y.A. Universidad de Santander (1985) (to appear).
(7) Hénon, M. and Heiles, C.: Astronomical Journal 69 (1964) 73.
(8) Henrard, J.: Celestial Mechanics 7 (1973) 449.
(9) Moser, J.K.: Communications Pure Appl. Math. XI (1958) 257.
(10) Simó, C.: Proceed. AMS Conf. on Hamiltonian Dynamical Syst., Boulder(1987) (to appear).

CONNECTANCE AND STABILITY OF LINEAR DIFFERENTIAL SYSTEMS

C. Froeschlé and J.L. Gautero
Observatoire de Nice,
BP 139, 06003 Nice cedex

ABSTRACT: Numerical experiments allow the study of the local stability of dynamical systems as a function of their connectance: when this connectance is high, this local stability is improbable. Effects of substitutability versus complementarity are also studied. Moreover, as shown by the diagonal dominance theorem, the situation is quite different for some systems resulting from economic models.

Through a Monte-Carlo method, the local stability of dynamical systems is studied as a function of a structural variable: the connectance.

The notion of connectance and its definition originates in the work of M.R. Gardner and W.R. Ashby (1). Using this variable in numerical experiments, they answer the question they formulate at the beginning of their paper: "What is the chance that a large system will be stable ?" Examples of such high-dimensional real systems may be found in biology (the human brain), technology (an airport), social sciences (an urban area) or economics (general equilibrium theory, classical gravitation) (see D. D. Siljak (2)).

First Gardner and Ashby, then R.M. May (3), and C. Froeschle (4) have studied the stability of dynamical systems, either dissipative or conservative, varying their connectance, that is the proportion of relations between their constitutive elements: connectance reaches 1 for a fully connected system, where every variable has a direct effect on every other variable. This is not the case in the real world for most high-dimensional systems: in such systems, the connectance is usually less than 1.

How does the connectance affect the local stability of a dynamical system? Of course, a Monte-Carlo exploration using numerical experiments will only give a probabilistic answer to the question.

Let us consider an element of an n-dimensional space $p = (p_1 \ldots p_n)$ and the system of differential equations

A. E. Roy (ed.), Long-Term Dynamical Behaviour of Natural and Artificial N-Body Systems, 393–403.
© 1988 by Kluwer Academic Publishers.

$$\dot{p}_k = \frac{dp_k}{dt} = E_k(p) \ , \qquad k = 1,\ldots,n \tag{1}$$

where E_k are continuously differentiable functions.

If p* is an equilibrium, we have $E_k(p^*) = 0$; thus, linearizing the system by its first order Taylor development, we get

$$\dot{p}_k = \sum_{h=1}^{n} E_{kh}(p_h - p^*) \ , \quad k = 1,\ldots,n \tag{2}$$

where

$$E_{kh} = \frac{\partial E_k(p^*)}{\partial p_k}$$

Changing the origin and taking P = p - p* gives

$$\dot{P}_k = \sum_{h=1}^{n} E_{kh} P^h \ , \qquad k = 1,\ldots,n \tag{3}$$

or, in matrix form,

$$\dot{P} = M(p^*)P \tag{4}$$

where M(p*) is the matrix of the partial derivatives of $E = (\ldots,E_k,\ldots)$ for p = p*. It is well known that the local stability of the equilibrium p* depends on the sign of the real parts of the eigenvalues of M(p*): their negativeness being a necessary and sufficient condition.

We define a binary matrix B called the interconnection matrix such that if E_{kh} is equal to zero (or weak), then $B_{kh} = 0$; else $B_{kh} = 1$.

In analogy with economic models, if $E_{kh} > 0$, with $k \neq h$, we say that the hth component exerts a gross substitutability effect upon the kth near equilibrium. Conversely, if $E_{kh} < 0$, with $k \neq h$, we shall speak of a gross complementarity effect exerted by the hth component upon the kth near equilibrium.

All the elements of the main diagonal of M(p*) are supposed to be negative, which ensures the local stability when the non-diagonal terms are zero. In this case, the matrix B is the identity matrix and the connectance C is equal to zero.

Actually, following Gardner and Ashby, we define the connectance of the system as the mean proportion of non-zero values of B_{kh} coefficients outside the main diagonal in the entries of matrix B.

We have performed two sets of numerical experiments. First, we have recovered the results of Gardner and Ashby, and by a slightly different definition of the connectance shown the robustness of their

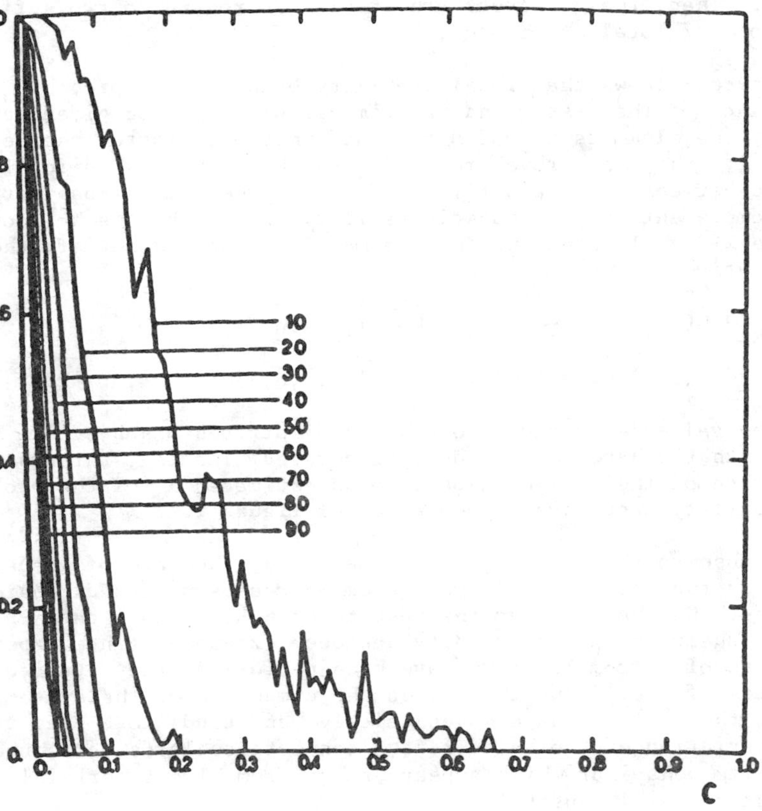

Figure 1: Variation of the estimate α of the probability of stability of a system as a function of its connectance C (its dimension varying from 10 to 90).

results. Then, these results are studied in regard to two sufficient conditions of local stability.

Figure 1 shows that local stability becomes less probable as the connectance of the system and its dimension grow. For given values of n and C, the elements E_{kk} of M are uniformly distributed between - 1.0 and - 0.1; the non-zero elements E_{kh} (with $k \neq h$) are uniformly distributed between - 1.0 and + 1.0, i.e. with the same probability for gross complementarity or substitutability exerted by the hth component over the kth. In other words, the matrix M entries satisfy the constraints

$$E_{kk} \in U(-1.0;\ -0.1)\ , \quad k = 1,\dots,n;$$
$$E_{kh} \in U(-1.0;\ +1.0)\ , \quad k = 1,\dots,n;\quad h=1,\dots,n;\quad k \neq h \qquad (5)$$

For given values of n and C, one hundred matrices M subject to the constraints (5) have been randomly generated, leading, through the computation of their eigenvalues, to an estimate α of the probability P of stability, and thus to one point of figure 1.

We observe that for a given value of α, the rate of connectance C implying the stability of the system decreases as its dimension increases. We thus confirm the results of M.R. Gardner and W.R. Ashby. However, their domain of validity has been extended, since experiments for values of n from 10 to 90 have been performed (they studied only the values of n 4, 7 and 10): when the dimension of the system increases, the curve's shape moves quickly, and tends to a step function: a weak variation of the value of the connectance leads to a huge variation of its behaviour which appears to go from almost certainly stable to almost certainly unstable.

We now generalise the notion of connectance, considering matrices where we replace the zero elements by weak interactions, i.e. by numbers close to zero. We may thus identify the influence of weak effects of the variation of the hth component upon that of the kth, dissociate them from those exerted in the same direction, but with a non-negligible amplitude, and assure that there is no discontinuity in calculations and results with the introduction of zeros in the matrix M.

Dimension 20 has been fixed for the matrix M in the numerical experiments of figure 2. The coefficients E_{kk} are uniformly distributed between -1 and -0.1. The E_{kh} which were zero for the first sequence of calculations are now uniformly distributed on segment$\{-\varepsilon;\varepsilon\}$, ε varying from 0.05 to 0.35 with 0.05 steps. The other E_{kh} satisfy the constraints (5) formerly defined. We continue to observe the phenomena which appeared in the first part of our study; this validates the notion of connectance.

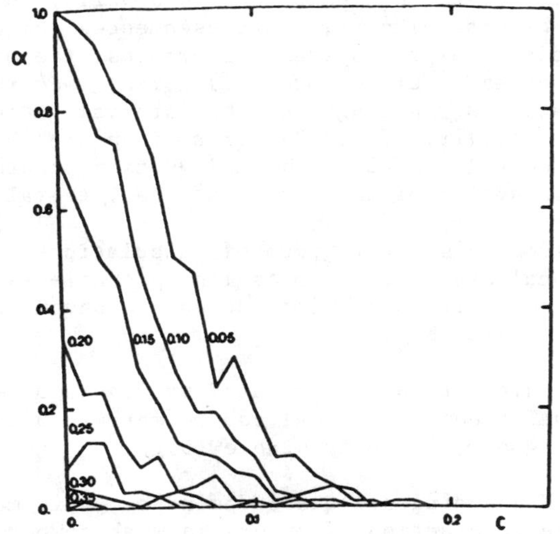

<u>Figure 2:</u> Dimension 20. Measure of the influence upon stability of weak effects, parametrized from 0.05 to 0.35.

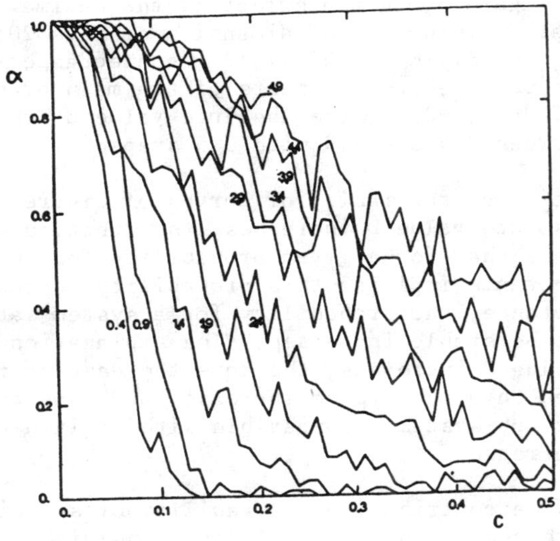

<u>Figure 3:</u> Dimension 20. Probability of stability as a function of the connectance C for various values (from 0.4 to 4.9) of the maximal feed-back.

Thus, the reaction of the kth component to a variation of the hth from equilibrium may be weak or zero, the consequences remain the same: as the rate of connectance of the system S increases, the probability of its stability decreases. Furthermore: 1) for a given rate of connectance, the greater is the length of the interval $\{-\varepsilon;\varepsilon\}$, the less probable is the stability of S; 2) for a given probability P of the stability of S, the value of C which implies this probability increases with the diminution of the length of the interval $\{-\varepsilon;\varepsilon\}$.

The conclusion from this first group of calculations (Figs.1 and 2) is as follows: local stability becomes less probable as the connectance of the system and/or its dimension increase. Beyond a given rate of C, it is very improbable.

The sequence of calculations illustrated by figure 3 refers to the quasi-dominant diagonal theorem: a real square matrix, with negative and quasi-dominant diagonal, is fully stable (5).

When the matrix M has diagonal quasi-dominance, this means that for each component the interaction of others is weak compared with its own influence aimed at the return to initial equilibrium through the damping of the original perturbation. Evidently, this concerns each of the isolated sub-systems of the global system. Actually, the magnitude of the feed-back responding to a perturbation is the variable which is more easily controlled. Therefore, we were interested, with the calculations of figure 3, in the effect of the maximal value admitted for diagonal dominance. The dimension of M is 20; the non-zero coefficients E_{kh} verify $E_{kh} \in U(-1;1)$; connectance is the proportion of the non-zero E_{kh}; the elements of the main diagonal satisfy the constraint $E_{kk} \in U(-D; -0.1)$, the absolute value of the inferior limit -D varying between 0.4 and 4.9 with 0.5 steps.

The interpretation of the family of curves of figure 3 is as follows: as the absolute value D increases, the function moves to the right, which means that for a given probability for the system to be stable, the connectance implying this probability increases; for a fixed rate of connectance, the probability for a system satisfying the above constraints to be stable increases. The explanation is that the more the absolute value D increases, the more the damping rate of perturbations exerted on a sub-system increase, and thus the more the interaction of other sub-systems, compatible with a given probability of stability, may increase.

Figure 4 shows the contribution to stability of substitutability effects compared with complementarity effects. The dimension of M remains 20. The entries E_{kk} are uniformly randomly distributed across the interval $\{-1; -0.1\}$; the non-zero elements of M outside of the main diagonal are such that $E_{kh} = X_{min} + 2U(0,1)$; and X_{min} goes from -1.8 to 0 with 0.2 steps. Thus we study the influence on the stability of the passage from a distribution where each line of M contains a mean of 20% of substitutability effects from the components h to the

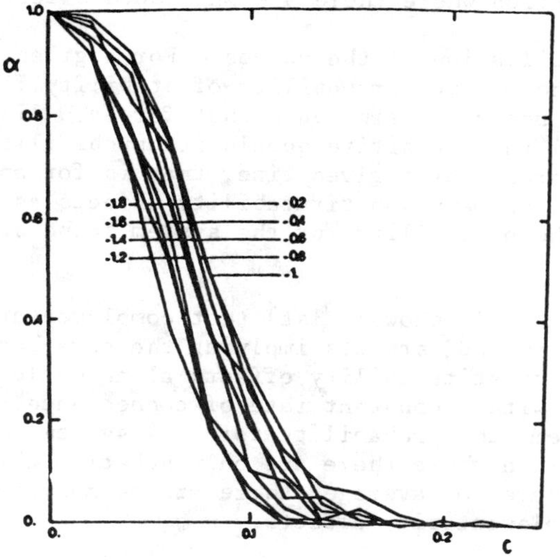

Figure 4: Dimension 20. Probability of stability as a function of C for various minimal values (from -1.8 to 0) of the non-diagonal elements.

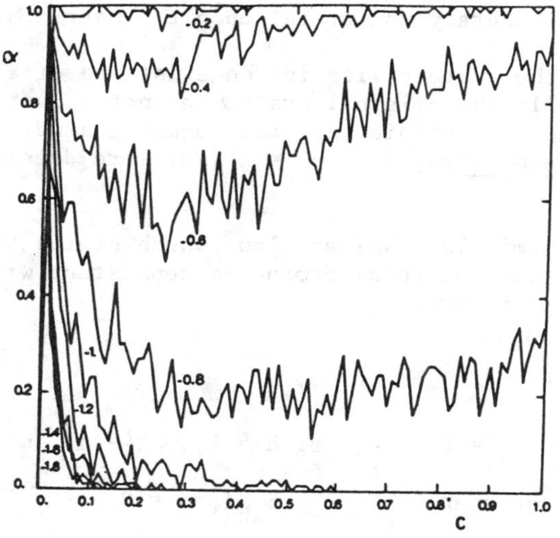

Figure 5: Dimension 20. Same as Fig.4, with the supplementary constraint (8).

component k to a situation where there are only such effects.

We observe an oscillation of the curves. For a given value of the connectance, we get the maximum probability of stability for $X_{min} = -1$, i.e. a repartition of non-zero terms such that for each line of M the probability for a term to be positive equals its probability to be negative. In other words, for a given line, that is for any component, there are, on average, as many substitutability effects as complementarity effects: the probability for the system to be stable is then maximal.

This interesting result shows: 1st) that complementarity effects do not seem to oppose the adjustments implying the come-back to initial equilibrium which the substitutability effects alone would be supposed to entail; 2nd) that with a constant rate of connectance, for a given dimension of the system, the probability for this system to be stable is smaller in a situation where there are only substitutability effects than in a situation where, on average, there are as many substitutability effects as complementarity effects.

When $X_{min} = 0$, we have: $(\forall k,h)\ E_{kh} \geq 0$; numerical calculations show that stability is very improbable (except when $C < 0.05$). In regard to the economic origins of this problem, such a result seems rather surprising: since it is known in the economic world that gross substitutability between all components entails a locally stable equilibrium. Actually we will see that the gross substitutability hypothesis imposes a supplementary constraint upon the matrix M.

The solution of the mystery lies in the existence of a <u>hidden variable</u>, since actually the original system has not n, but $l = n + 1$ differential equations. We obtain the case formerly studied by taking the lth component as <u>numeraire</u>, i.e. $p_l = 1$ (For more details about economics, see (6)).

Moreover, an economic law, Walras' law, which states that, in an economic system, the value of total demand is consistent with that of total supply, allows us to write:

$$\sum_{k=1}^{l} p_k E_k(p) = 0 \tag{6}$$

Consequently, $\sum_{k=1}^{l} p^*_k E_{kh} + E_h = 0$, for $h = 1,\dots,l-1$; when $p = p^*$, $E_h(p^*) = 0$, and then $\sum_{k=1}^{l} p^*_k E_{kh}(p^*) = 0$.

That is $\sum_{k=1}^{n} p^*_k E_{kh} = -p^*_l E_{lh} = -E_{lh}$.

If each component exert a gross substitutability effect upon the lth :

$\forall h = 1,\ldots, l-1 \quad E_{lh} > 0$

As a result $\sum_{k=1}^{n} p_k^* E_{kh} < 0$, which can be written $p_k^* E_{kh} < -p_h^* E_{kh}$. But $\forall k$, $E_{kh} < 0$, and $p_h^* > 0$; then $-p_h^* E_{kh} > 0$, and consequently we find:

$$p_h^* |E_{hh}| > \sum_{\substack{k=1 \\ k\neq h}}^{n} p_k^* |E_{kh}| \qquad (7)$$

which gives the stability of the system with the diagonal quasi-dominance theorem.

With a scaling, we may obtain: $\forall k$, $p_k^* = 1$ Thus

$$\forall h = 1,\ldots,n \quad \sum_{k=1}^{n} E_{kh}(p^*) < 0 \qquad (8)$$

Each column of the interaction matrix must verify this constraint if we suppose that all goods exert a gross substitutability effect on the lth.

The simulation corresponding to figure 5 integrates all the constraints of the matrices of figure 4, their dimension being 20, and adds the supplementary constraint (8).

We no longer obtain oscillations of curves when X_{min} goes from -1.8 to 0 as in figure 4. We may distinguish three families of curves, defining three intervals of variation for X_{min}. They correspond to the passage from a situation where, for each non-zero entry of the matrix, a gross complementarity effect is more probable than a gross substitutability effect, to a situation where the probabilities invert, reaching finally a situation where there are only effects of the latter kind in the matrix.

In the first interval, X_{min} varies from -1.8 to -1: for each non-zero entry of the matrix $P(E_{kh} < 0) > P(E_{kh} > 0)$. The shape of first five curves is the same as above: with a constant dimension, the probability of the stability of the system decreases as C increases. But now, differing from the results illustrated by figure 4, there is no oscillation of the curves: an increase in the probability for a term outside the main diagonal to be positive, $P(E_{kh} > 0)$, which happens when X_{min} goes from -1.8 to -1, tends, for a given rate of connectance, to increase the stability of S.

In the second interval, X_{min} goes from -0.8 to -0.4; in the simulation $P(E_{kh} > 0) > P(E_{kj} < 0)$: curves move upwards. With a given rate of connectance, the probability for a system to be stable increases with X_{min}. For each curve, we observe that as C increases, the probability of the stability of S first decreases, and then increases.

This evolution ends in the third interval, where X_{min} goes from -0.2 to 0. When $E_{kh} \in [-0.2; 1.8]$, the system is quasi-stable, independent of the value of C. Finally, when $E_{kh} \in [0;2]$, the system is always stable, whatever the connectance rate may be.

We may observe that the constraint $\sum_{k=1}^{n} E_{kh} < 0$ may be written

$$\sum_{\substack{k=1\\k\neq h}}^{n} E_{kh} < -E_{hh} \,.$$

If $\forall k \neq h \;\; E_{kh} \geq 0$, we have $\sum_{\substack{k=1\\k\neq h}}^{n} E_{kh} = \sum_{\substack{k=1\\k\neq h}}^{n} |E_{kh}|$.

As $E_{kh} < 0$, $-E_{hh} = |E_{hh}| > 0$. By the theorem of diagonal quasi-dominance, the equilibrium is stable if $\sum_{\substack{k=1\\k\neq h}}^{n} E_{kh} < -E_{hh}$, which is equivalent to

$\sum_{\substack{k=1\\k\neq h}}^{n} |E_{kh}| < |E_{hh}|$. When the sign of E_{kh} may be positive or negative, which is the case when X_{min} goes from -1.8 to -0.2, then $\sum_{\substack{k=1\\k\neq h}}^{n} E_{kh} \leq \sum_{\substack{k=1\\k\neq h}}^{n} |E_{kh}|$; the diagonal quasi-dominance condition may be unsatisfied, which explains why stability is not always verified - as the random simulation shows. Yet, the passage from -1.8 to -0.2, increasing the probability for each non-zero term E_{kh} to be positive, increases the probability for the system S to be stable.

We may compare the results represented in figure 5 by those of another sequence of calculations where the constraints upon the entries E_{kk} and E_{kh} of a matrix of dimension 20 are similar, but in addition E_{kk} and E_{kh} have the same sign (which happens for instance when the matrix is symmetric). In every case, the constraint (8) implies a modification of the shape of curves when X_{min} goes from -1.8 to 0. Calculations satisfying the same constraints except (8) lead to results comparable with those of figure 4: for a fixed dimension of the system, stability is very improbable.

In conclusion, we may say that our numerical experiments have shown, like those of Gardner and Ashby, that local stability seems to be improbable beyond a given rate of connectance. Moreover in the general case we have shown that for a given connectance the stability is maximum when substitutability has the same strength as complementarity. Finally, it appears through the diagonal dominance theorem that the situation is quite different for systems resulting from an economic model where each component exerts a gross substitutability effect upon numeraire: the probability of stability is higher and even increases with the connectance.

Indeed the general results displayed in figures 1,2,3,4 may change if additional constraints depending on the nature of the problem are added. For Hamiltonian systems, they are modified by the symplectic structure of the matrix M. Preliminary results show that for Hamiltonian systems, the key parameter for the stability seems to be the mean number of non-zero terms in the interconnection matrix rather than the Gardner-Ashby's definition of the connectance.

The concept of hierarchical dynamical systems developed by A. Roy (7) in the frame of celestial mechanics may also lead to interesting developments.

Since all these results have been established within the linear approximation, the introduction of non-linear terms may change the conclusions. This problem is also under investigation.

ACKNOWLEDGEMENTS: We thank A. Gilbert, for some interesting conversations about this paper.

REFERENCES

(1) M.R. Gardner, W.R. Ashby, "Connectance of Large Dynamic (Cybernetic) Systems: Critical Values for Stability", Nature Vol.228,p.784, 1970.
(2) D.D. Siljak, "Large Scale Dynamic Systems. Stability and Structure", North Holland, New York, U.S.A., 1978.
(3) R.M. May, "Will a large complex system be stable?", Nature, Vol.238, p.413,1972.
"Stability and complexity in Model Ecosystems", Princeton, New Jersey, Princeton University Press, U.S.A., 1974, 2nd ed.
(4) C. Froeschle, "Connectance of Dynamical Systems with increasing number of degrees of freedom", Phys. Rev. A, Vol.18, pp 27-281, 1 July 1978.
(5) M. McManus, "The Arrow and Hurwicz and Hahn Theorem", Palo Alto (Calif.), unpublished, c.f. J. Quirk and R. Saposnik, "Théorie de l'équilibre général et économie du bien-être", p.184, Paris, PUF,1974.
(6) C. Froeschle and A. Longhi, "Connectance et stabilité locale d'un équilibre général", Economie appliquée, XL, 1987.
(7) A.E.Roy, I.W. Walker and A.J.C. McDonald, "Studies in the stability of hierarchical dynamical systems" in "Stability of the solar system and its minor natural and artificial bodies", NATO ASI Series edited by V.G. Szebehely.

RESONANCES IN THE PLANAR ELLIPTIC RESTRICTED PROBLEM

Jacques Henrard
Facultés Universitaires N.D. de la Paix
Rempart de la Vierge, 8
5000 - Namur, Belgium

ABSTRACT. We review the recent progresses in the study of the resonances 2/1 and 3/1 in the planar elliptic restricted three body problem. These studies have been mainly motivated by their applications to the Kirkwood gaps in the asteroid belt but we shall rather emphasize here the theoretical point of view: how numerical experiments show the importance of a rather new concept in Celestial Mechanics, the chaotic behaviour of some motions, and how a perturbative approach can make us understand and even predict this kind of behaviour.

1. INTRODUCTION

The existence of the Kirkwood gaps (Kirkwood 1866) in the asteroid belt and their unmistakable association with resonances has been a puzzling problem in Celestial Mechanics for a long time. These peculiarities in the distribution of asteroids could contain some hidden clues for the cosmogonic problem of how the Solar System formed or evolved.

Numerical investigations by Torbett and Smoluchowski (1980, 1982) and Gonczi et al (1982) have suggested that the sweeping of resonances through the asteroid belt (or vice versa of asteroid through the resonance pattern) could produce gaps. This mechanism has been analysed in detail (Henrard and Lemaitre, 1983b; Lemaitre, 1984b) with the help of an extension of the adiabatic invariant theory (Henrard, 1982). A possible explanation for the resonance sweeping, the gravitational effect of the dissipation of a protosolar nebula was also investigated and showed that the dissipation of a reasonably sized nebula was sufficient to clear the gaps we now see (Henrard and Lemaitre, 1983b; Lemaitre, 1984b).

At about the same time, Wisdom (1982, 1983, 1985, 1986) indicated that maybe one did not need resonance sweeping but that Mars could serve as the "Deus ex machina" clearing up the gaps. He found evidences of widespread chaotic motion leading eventually to a large increase in eccentricity in the 3/1 resonance problem. When the eccentricity becomes larger than about 0.3, the asteroid becomes a Mars crosser and

A. E. Roy (ed.), Long-Term Dynamical Behaviour of Natural and Artificial N-Body Systems, 405–425.
© 1988 by Kluwer Academic Publishers.

can be removed from the belt by collisions or large perturbations due to close encounters.

Now if Wisdom's effect can explain the 3/1 gap just as well as resonance sweeping does, can it do the same for the 2/1 gap? This is not evident. Indeed in the case of the 2/1 resonance, the eccentricity has to reach a higher value (0.5) for the orbit to become either a Mars or Jupiter crosser. Furthermore the 2/1 resonance is stronger and may better resist the perturbations due to Jupiter eccentricity.

This question has provoked a regain of interest in the study of the dynamical systems by which the Jovian resonances can be modelled.

Among them, the averaged planar elliptic restricted problem is of particular interest. Being a system with two degrees of freedom only, it is simple enough that we can hope to understand it thoroughly by numerical or perturbative methods. On the other hand as shown by the work of Giffen (1973), Scholl and Froeschlé (1974, 1975) and Froeschlé and Scholl (1976, 1977) and more recently by Wisdom (1982, 1983, 1985, 1986), Murray and Fox (1984) and Murray (1986) it is complex enough and shows a delicate pattern of regular and chaotic motions.

It is the numerical and analytical evidences on this dynamical system in the two cases of resonances 2/1 and 3/1 that we wish to review here. The interest of this is not only its applications to the Kirkwood gaps but also the variety of phenomena encountered and of analytical or numerical tools used to describe them. These tools and the understanding of those phenomena can be useful in other problems in Celestial Mechanics or more generally in Classical Mechanics.

Of course, in this review we can only give a brief summary. We have tried to give a list of references which, without being exhaustive, is complete enough. It is hoped that this text will help the interested reader in consulting those references where he will find more detailed explanations.

2. THE AVERAGED PLANAR ELLIPTIC RESTRICTED PROBLEM IN CASE OF RESONANCE

The Hamiltonian of the restricted three body problem can be written (see for instance Szebehely, 1967)

$$H = -\frac{(1-\mu)}{2a} - \mu\left\{\frac{1}{|\vec{r}-\vec{r}'|} - \frac{\vec{r}\,|\,\vec{r}'}{r'^3}\right\} \qquad (1)$$

where $(1-\mu)$ and μ are the reduced mass of the primary and secondary, a (resp. a'), $\vec{r}$ (resp. $\vec{r}'$) the semimajor axis and position vector (relative to the primary mass) of the test particle (resp. of the secondary). We are considering here that the Hamiltonian function (1) is expressed implicitly in the usual modified Delaunay's elements

λ	=	mean longitude of the particle	$L=\sqrt{(1-\mu)a}$
$-p$	=	longitude of the pericenter of the particle	$P=L(1-\sqrt{1-e^2})$. (2)

The Hamiltonian function is also a function of the time through its dependence upon the mean longitude λ' of the secondary.

In the case where there exists a mean motion resonance

$$(i+j)n' - in \simeq 0 \tag{3}$$

between the unperturbed mean motion of the test particle $n = [(1-\mu)/a^3]^{\frac{1}{2}}$ and the mean motion of the secondary $n' = a'^{3/2}$, it is customary and useful to introduce Poincaré's resonance variables (Poincaré, 1902; Schubart, 1966)

$$\begin{aligned} \sigma &= [(i+j)\lambda' - i\lambda + jp]/j \;, & S &= P \\ \nu &= -[(i+j)\lambda' - i\lambda + j\omega']/j \;, & N &= (j/i)L+P \end{aligned} \tag{4}$$

and to average the Hamiltonian (1) over the remaining fast variable which is essentially the time. This averaging is not just a rough approximation (of the first order in μ) but can be justified in terms of transformation of coordinates from the osculating coordinates (4) to a set of "averaged" coordinates $(\bar{\sigma},\bar{\nu},\bar{S},\bar{N})$ (see for instance Message, 1966). In order to keep the notations as simple as possible, we shall drop the superscript of the averaged variables and note them also (σ,ν,S,N).

The averaging transforms the time dependent Hamiltonian (1) into a time-independent one. When expanded in powers of the eccentricities (e,e') and the difference $(a-a^*)$, where a^* is a nominal value of a close to the resonance value, this Hamiltonian reads in the case of the 2/1 resonance (i.e. with i=j=1)

$$\begin{aligned} H = H_o(N-S) + \mu\{&Ae\cos\sigma + Be'\cos\nu + Fe^2 + Ce^2\cos 2\sigma \\ &+ Gee'\cos(\sigma+\nu) + Dee'\cos(\sigma-\nu) + Ee'^2\cos 2\nu\} \\ &+ \dots \end{aligned} \tag{5}$$

where

$$H_o(x) = -\frac{j^2(1-\mu)^2}{i^2x^2} - \frac{(i+j)}{j}x + \mu\gamma\sqrt{x} \tag{6}$$

and the coefficients $\gamma,A,B,C\dots$ are numerical coefficients.

When truncated at order one in the eccentricities the Hamiltonian (5) is integrable (Sessin and Ferraz-Mello, 1984) and can be reduced to a one-degree of freedom system by a simple rotation in phase space (Wisdom, 1986; Henrard et al, 1986). This rotation is of the type

$$\sqrt{2R}\sin r = \sqrt{2S}\sin\sigma + \frac{B}{A}e'\sin\nu$$

$$\sqrt{2R}\cos r = \sqrt{2S}\cos\sigma + \frac{B}{A}e'\cos\nu$$

$$m = -\nu$$
$$M = N + Be'\sqrt{2S}\cos(\sigma-\nu) + B^2e'^2/2 \tag{7}$$

and introduces a kind of "proper eccentricity" : $\sqrt{2R}$ instead of $\sqrt{2S} \sim e$. After this transformation (5) reads

$$H = H_o(M-R) + \mu\{A\sqrt{2R}\cos r + 2F\,R + 2C\,R\cos 2r + G'e'\sqrt{2R}\cos(r+m) + D'e'\sqrt{2R}\cos(r-m) + E'e^2\cos 2m\} + \ldots \tag{8}$$

Numerical values of the coefficients may be found in Henrard and Lemaitre (1986, 1987). Notice that we have not performed here the change of scale proposed in these references in order to compare more easily the Hamiltonian (8) with the Hamiltonian for the 3/1 resonance.

In the case of the 3/1 resonance ($i=1$, $j=2$) we obtain directly

$$H = H_o(N-S) + \mu\{2FS + 2CS\cos 2\sigma + Ge'\sqrt{2S}\cos(\sigma+\nu) + De'\sqrt{2S}\cos(\sigma-\nu) + Ee'^2\cos 2\nu\} + \ldots \tag{9}$$

Where H_o is the same function as in (6) but where the numerical coefficients γ,F,C... have other numerical values which can be found in Wisdom (1985).

The expansion in powers of e' and $\sqrt{2R}$ (or $\sqrt{2S}$) in (8) and (9) have been truncated at order two. This is not always sufficient to insure a good representation of the perturbation function. Actually some of the results described later on have been computed with an expansion carried out to order 8. But this will serve our purpose here of comparing qualitatively the two main resonances in the asteroid belt.

In the restricted circular problem, when $e' = 0$, the Hamiltonian functions (5) and (9) do not depend upon the angular variable ν. The corresponding momentum N is thus a constant and both problems are reduced to a one-degree of freedom problem. The phase space of these one-degree of freedom problems is well known (Poincaré, 1902; Schubart, 1966; Message, 1966; Henrard-Lemaitre, 1983a; Lemaitre 1984 etc...).

We reproduce in Figure 1 and 2, the trajectories in the cartesian phase space ($\sqrt{2S}\cos\sigma$, $\sqrt{2S}\sin\sigma$) for typical values of the constant N. They are characterized by the apparition of homoclinic orbits (orbits asymptotic on both side to an unstable equilibrium) which isolate from the remaining part of the phase space a set of orbits (which we call the resonant orbits). This is a genuine separation, orbits on both sides of the homoclinic orbits are not continuous deformations of each other. Furthermore, it is well known, (and we shall see again evidence of it in what follows) that homoclinic orbits under perturbation do usually generate motions with very strange behaviour, the so-called chaotic motions.

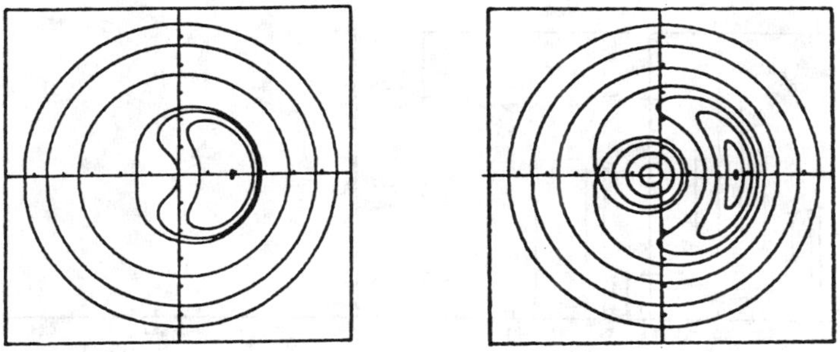

Figure 1: Typical phase portraits of the 2/1 resonance in the restricted circular problem (From Henrard-Lemaitre, 1983a, with permission).

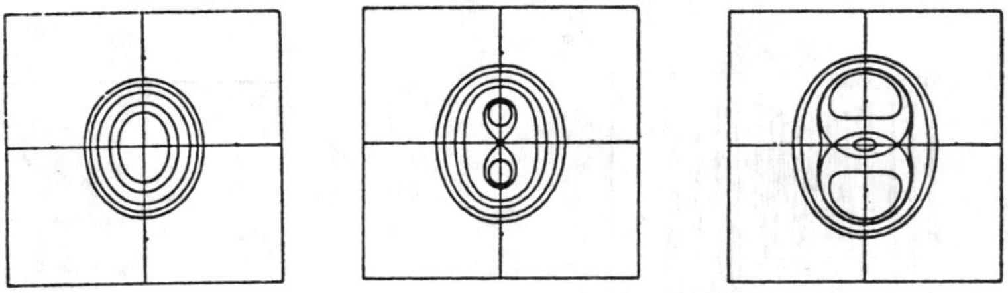

Figure 2: Typical phase portraits of the 3/1 resonance in the restricted circular problem (From Lemaitre, 1984, with permission).

3. NUMERICAL EVIDENCES IN THE 3/1 RESONANCE PROBLEM

If the resonance cases in the planar and circular restricted problem are well understood, it is not the case of the non-circular or elliptic problem, when $e' \neq 0$. The intrusion of a second non-trivial degree of freedom (ν, N) and the fact that this second degree of freedom is also in resonance (indeed the unperturbed frequency of ν has the same

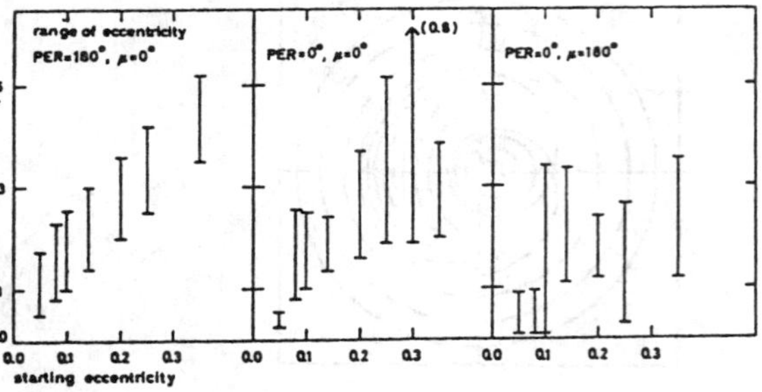

Figure 3: Excursions in eccentricity for orbits in the 3/1 resonance (From Scholl and Froeschlé, 1974 with permission).

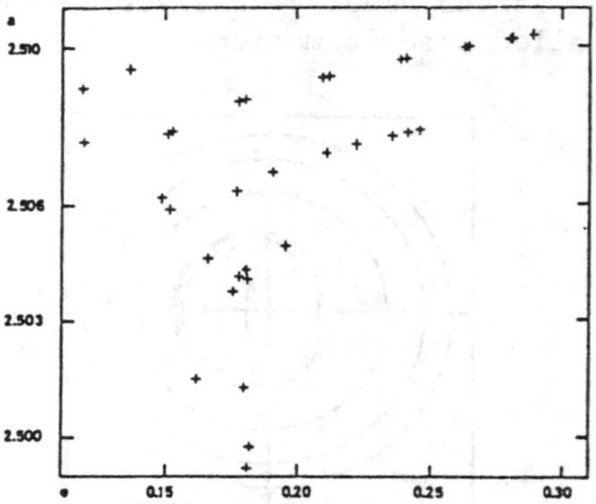

Figure 4: Surface of section for a chaotic orbit in the 3/1 resonance (From Scholl and Froeschlé, 1974 with permission).

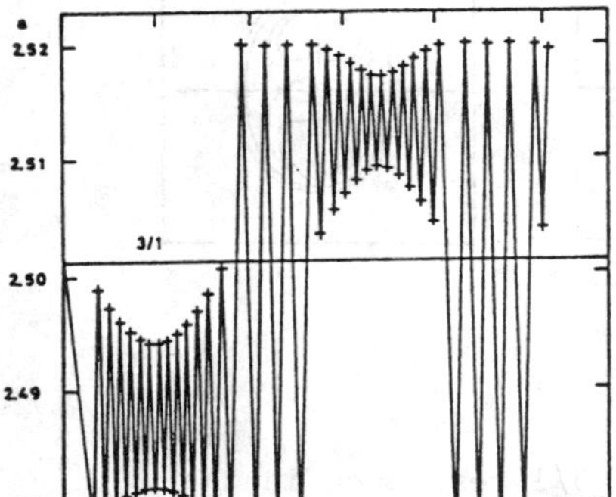

Figure 5: Variation of a with time for a "peculiar orbit" in the resonance 3/1 (From Scholl and Froeschlé, 1974 with permission)

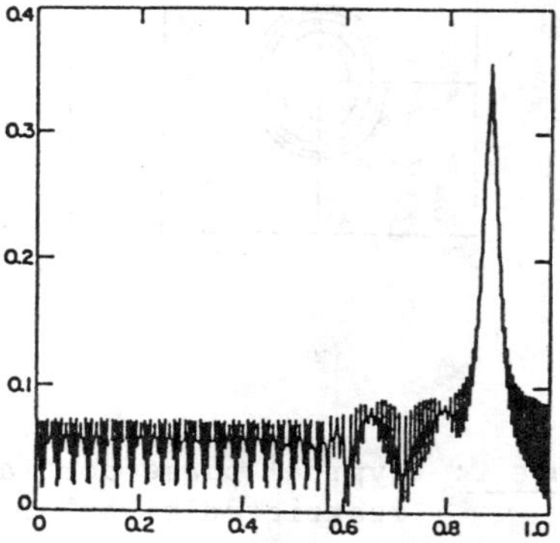

Figure 6: Variation of the eccentricity with time for a chaotic orbit. Time is measured in units of 20,000 periods of Jupiter ≃ 240,000 years (From Wisdom, 1982 with permission).

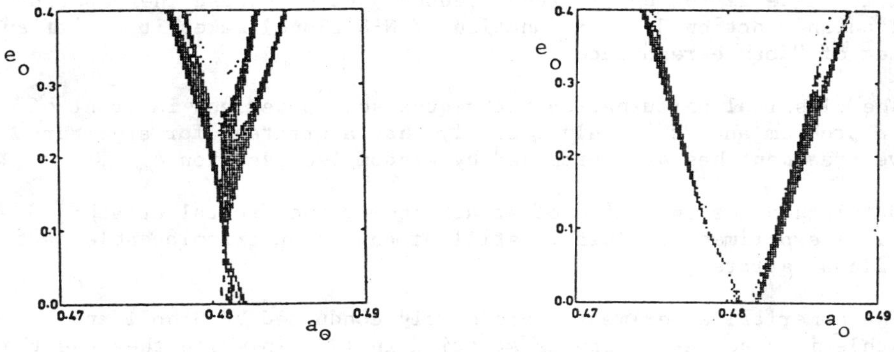

Figure 7: Initial conditions in the plane (a,e) of orbits showing a chaotic behaviour in Murray and Fox numerical experiments. Figure 7a corresponds to $(\lambda_o,\omega_o) = (\pi,\pi)$ and Figure 7b to $(\lambda_o,\omega_o) = (\pi,0)$ (From Murray and Fox, 1984, with permission).

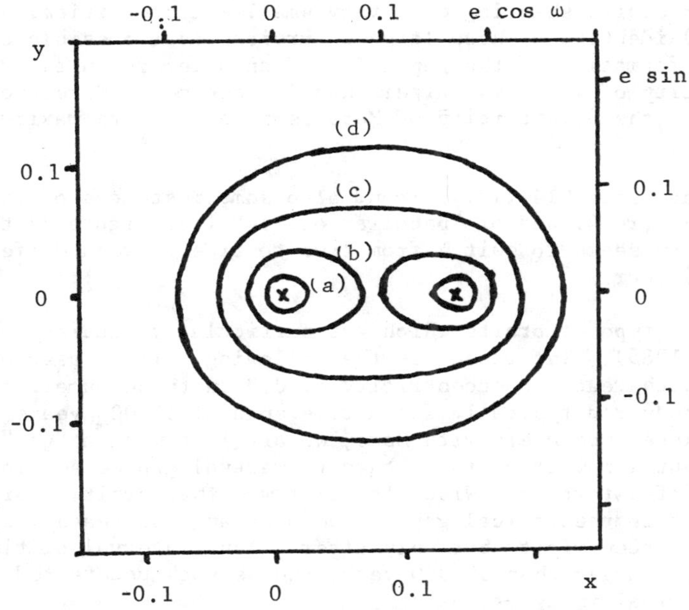

Figure 8: Level curves of the ratio Ω/μ of the two time scales the curves a,b,c,d correspond respectively to $\Omega/\mu = 5, 10, 20, 40$.

absolute value as the unperturbed frequency of σ because the unperturbed Hamiltonian function H_o is a function of N-S alone) makes it a difficult problem of "double resonance".

The classical perturbation techniques seem powerless in front of such a problem and it is only recently that a strategy for a perturbative treatment has been proposed by Wisdom (see section 4).

Until then the only line of attack seemed the frontal attack: numerical experiments. This is still of course an irreplaceable part of a global strategy.

The numerical experiments were mostly conducted by Scholl and Froeschlé in a series of papers starting in 1974 in which they use the numerical averaging procedure of Schubart (1968) and later by Wisdom and Murray using Wisdom's mapping (1982) based upon an idea of Chirikov (1979).

For the 3/1 commensurability, Scholl and Froeschlé (1974) found that most orbits present large excursions in eccentricity (see Figure 3) except for orbits starting with very small eccentricities. Later on Wisdom (1982) identified those large excursions as a possible explanation for the formation of the gap. Indeed an asteroid in the 3/1 gap, the eccentricity of which is larger than 0.3 becomes a Mars crosser (at least when the eccentricity of Mars is itself at its maximum of 0.142).

Scholl and Froeschlé (1974) found also some instances of chaotic motion (see Figure 4) and of "peculiar orbits" (see Figure 5) the motion of which seems to switch from time to time between different types of behaviour.

It is this type of orbits which was particularly studied by Wisdom (1982, 1983, 1985). The reason is the following. If we assume that the orbits which reach an eccentricity of 0.3 in these numerical experiments (conducted typically for a time span of 50,000 years) are removed by close encounters with Mars, we are left with a "gap" of the right shape but somewhat narrower than the actual gap we see in the distribution of asteroids. Wisdom's point was that orbits starting at the edges of this theoretical gap are chaotic and can reach large values of the eccentricity but only after a long interval of time typically much larger than 50,000 years and as such undetected by previous numerical integrations.

As an example and paradigm, Wisdom (1982) produced the orbit shown in Figure 6, which shows a large and sudden increase in the eccentricity only after 200,000 years. Removing both the orbits which show a chaotic behaviour (and thus are expected to eventually cross also the orbit of Mars) produces a theoretical gap which compares very well with the actual gap.

The relative importance of this chaotic zone on the edges of the gap may best be illustrated by Murray and Fox (1984) diagrams reproduced in Figure 7. These diagrams and the accompanying tables (especially Table 1) offer us a hint for the explanation of the large excursions in eccentricity of the regular orbits well inside the gap.

If we rotate Figure 7b along the a-axis and glue it below Figure 7a, we obtain a diagram the ordinate of which is this starting value of the regular variable $h = e \cos \omega$. We see then a Figure X the center point of which seems to correspond to a periodic orbit (already described by Sinclair, 1970). This periodic orbit is a member of a family parametrized by the eccentricity of Jupiter e', which starts off at $e' = 0$ on the axis $e = 0$ for a particular value of N. This is the central point in Figure 2b.

The center point of the Figure X seems to act as a kind of center of symmetry for orbits starting in the upper or lower quadrant of the Figure X. An orbit starting from the lower quadrant will after a time of about 500 Jupiter periods (see Table 1 and 2 of Murray and Fox, 1984) reach a point in the diagram symmetric with respect to the center point in the upper quadrant. As the center of symmetry is located at $h_o \sim 0.125$, if we start with $h_o \simeq -0.05$ we arrive at $h_o \sim 0.3$ (and so on... if we start at $h_o = -0.1$ we end up at $h_o \sim 0.35$). This scenario is strongly suggested by Murray and Fox (1984) numerical results and makes us understand in some way the large variations in eccentricity. Of course this value of $h = 0.125$ for the center of symmetry is directly proportional to the value of the eccentricity of Jupiter. We can expect that for smaller values of e', the displacement of the center of symmetry will be smaller and thus the variations in eccentricities.

If the numerical results just reviewed offer an interesting and fairly complete description of what happens in the case of the resonance 3/1, it does not yet offer an answer to the question why does it happen: can the regular motions be described by some perturbative approximation which would explain why Sinclair's periodic orbit acts as a kind of center of symmetry? Can the location and extent of the chaotic zone by predicted and explained by some mechanism? Why does the chaotic behaviour of orbits seem to lead to sudden jumps in the eccentricity?

4. WISDOM'S PERTURBATIVE TREATMENT OF THE 3/1 RESONANCE

The answers to some of these questions were given by Wisdom (1985). Let us consider the Hamiltonian (9) truncated at order two in the eccentricities and expand the function $H_o(N-S)$ around the value N* (the "exact resonance") for which the derivative of H_o vanishes. Let us retain only the first non-vanishing term in this expansion. Furthermore let us introduce the following canonical transformation

$$x = \sqrt{2S} \cos (\sigma+\nu) \qquad \phi = 2\nu$$
$$y = -\sqrt{2S} \sin(\sigma+\nu) \qquad \Phi = (N - S)/2 \tag{10}$$

The Hamiltonian (9) can be rewritten under the form:

$$H = \frac{1}{2}\alpha\Phi^2 - \mu A(x,y)\cos[\phi - P(x,y)] + \mu F(x^2 + y^2) + \mu Ge'x \quad (11)$$

with

$$A\sin P = 2Cxy + De'y$$
$$A\cos P = C(x^2 - y^2) + De'x + Ee'^2$$
$$\alpha = (\partial H_o/\partial N) \text{ evaluated at } N = N^* \quad (12)$$

The point of all this preparation is that when we "freeze" the (x,y) degree of freedom, the Hamiltonian (11) is the Hamiltonian of a pendulum. Of course the special form of the Hamiltonian (9) truncated at the second order is instrumental in obtaining this.

But does it make sense to "freeze" at least in a first approximation the (x,y) degree of freedom? Wisdom (1985) points out that it does if the time scale of the (x,y) motion is much larger than the time scale of the (ϕ,Φ) pendulum. The time scale of the (x,y) motion is obviously of the order of $1/\mu$ and the time scale of the pendulum can be taken as the inverse of the frequency $\Omega = \sqrt{\alpha\mu A}$ at its stable equilibrium. It is a function of the position in the (x,y) plane.

As noted by Wisdom (1985), the restoring torque A of the pendulum possesses two zeroes in the plane (x,y) situated at $(X_1 \sim 0.007, Y_1 = 0)$ and $(X_2 \sim 0.14, Y_2 = 0)$. We find that A(x,y) can be expressed easily in terms of these zeroes:

$$A^2 = C^2[(x - X_1)^2 + y^2][(x - X_2)^2 + y^2] \quad (13)$$

It is then easy to plot in the plane (x,y) the level curves of the ratio Ω/μ of the two time scales (see Figure 8). We observe that except in the immediate vicinity of the zeroes of the restoring torque μA, one of the time scales is indeed (much?) larger than the other one. To make such a statement we have of course to be ready to accept 5 as a "large" number. This is a little uncomfortable but the fact that the perturbative treatment based upon this assumption does describe at least qualitatively the numerical results makes us feel better. Indeed the time scale of the (x,y) motion is actually significantly larger than $1/\mu$ and Figure 8 underestimates the ratio of the time scales.

Having recognized the two time scales of the problem, Wisdom proceeds by introducing action-angle variables for the pendumum (ϕ,Φ) and by averaging the resulting Hamiltonian over the fast frequency.

The system is then reduced to a one-degree of freedom system in (x,y), the "averaged" slow-time scale degree of freedom. The solution of this system can then be obtained by plotting curves of constant values of H for a given value of the averaged action of the pendulum or by plotting curves of constant values of the averaged action for a given value of H. Wisdom chooses the second branch of the alternative

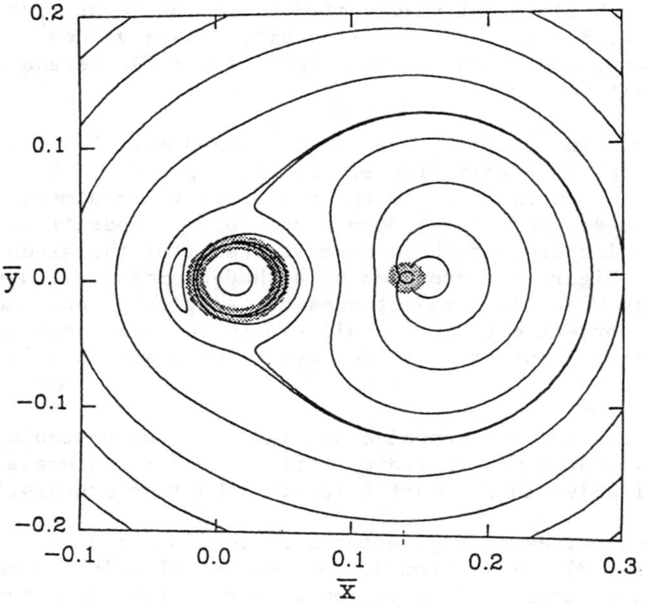

Figure 9: Trajectories in the plane $(\bar{x},\bar{y})$ of the averaged system for $H = -3.04\ 10^{-6}$. Also shown (surrounded by a shaded area) are the separatrices for this value of H. (From Wisdom, 1985 with permission).

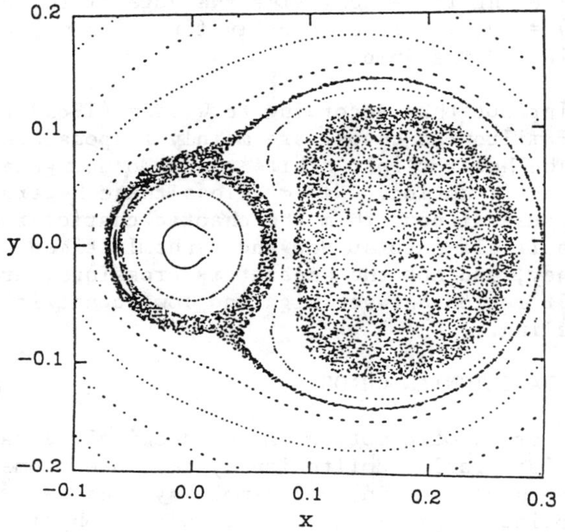

Figure 10: Surface of section for $H = -3.04\ 10^{-6}$.(From Wisdom, 1985 with permission),

and produces such plots for values of H close to the critical value corresponding to Sinclair's periodic orbit. These values of H follow closely in Murray and Fox's diagrams (see Figure 7) the chaotic X shape we have recognized there.

We reproduce in Figure 9, one of Wisdom's plots. Two remarkable things can be found in this figure. First as pointed out by Wisdom, the assumption on which is based the perturbative treatment (separation of the two time scales) breaks down close to the separatrices of the pendulum. The location of these separatrices (for the given value of H) is shown in Figure 9 surrounded by a shaded area (called the "zone of uncertainty" by Wisdom). Trajectories crossing this zone cannot be assumed to be correct but as we shall see in the next section should be assumed to dissolve into a "chaotic layer" (we shall call it the slow stochastic layer).

On the other hand we recognize on Figure 9, two unstable equilibria with their associated heteroclinic orbits. Such a feature also is expected to dissolve in a "chaotic layer" (the fast stochastic layer).

Indeed when corresponding surfaces of section of actual trajectories are plotted (see Figure 10 from Wisdom) we see clearly two chaotic regions. One (the large one) corresponds to the right hand separatrix of Figure 9. The other one blends together the left hand separatrix of Figure 9 and the heteroclinic orbits. A most interesting feature is the narrow branch of chaotic motion going up to $x = 0.3$ along one of the heteroclinic orbit. This is the path followed by the orbit shown in Figure 6 during its large jump in eccentricity. Before this jump, the orbit was following for a long time the edge of the circular chaotic zone around $(x,y) = (0,0)$. It eventually found its way along the narrow branch and experienced the jump.

We should point out that according to Wisdom (1985) the separatrix crossings (slow Hamiltonian chaos) are mainly responsible for the chaotic behaviour. The heteroclinic orbits alone do not generate a macroscopic zone of chaotic motion. Nevertheless these heteroclinic orbits are important because without them the chaotic character of the motion could not express itself into sudden and randomly timed jumps in the eccentricity. They, and not the separatrix crossings, are responsible for the random alteration between high and low eccentricity regime which was so puzzling.

5. SLOW AND FAST HAMILTONIAN CHAOS

Stochasticity or chaotic motion is the seemingly erratic wandering of orbits of non-integrable Hamiltonian systems over some part of phase space and it appears already in very simple systems. It is of interest in various disciplines: particle physics, plasma physics, celestial mechanics as it can give insight into the long term behaviour of conservative dynamical systems.

One of the best studied paradigms in this class of problems is the pendulum perturbed by a fast periodic force (Chirikov, 1979, Lichtenberg and Lieberman,1983). Let us consider for definiteness

$$H = \frac{1}{2} p^2 - [1 - \delta \cos (\frac{t}{\varepsilon})] \cos q. \tag{14}$$

If we take $t = t_o$ (modulo $2\pi\varepsilon$) as a surface of section and plot the intersection points of the trajectories of (14), we find that for most of the trajectories, the intersection points fall on regular curves which are trajectories of the autonomous system obtained by replacing the restoring torque by its mean value, except along the separatrices of this autonomous system. There we find a "stochastic layer" (we have called it the fast stochastic layer in the previous section), the width of which is exponentially small in

$$\Delta H_{mean} \sim \delta e^{-3} \exp \{- \frac{\pi}{2\varepsilon} \}. \tag{15}$$

On the other hand there has been also a great deal of interest about a seemingly unrelated topic: the effect of a slow variation of the parameters of a one-degree of freedom Hamiltonian system. This gives rise to the adiabatic invariant theory (see Henrard 1988 for a review) whereby the area enclosed by the orbit of the frozen system is an asymptotic invariant. The adiabatic invariant theory holds as long as the orbit of the frozen system is a periodic orbit but does not hold when this orbit is a saddle connection like the separatrix of the pendulum.

Let us then consider the time-dependent pendulum

$$H = \frac{1}{2} p^2 - [1 - \delta \cos(\varepsilon t)] \cos q \tag{16}$$

and the same surface of section as mentioned above. Trajectories such that the initial value of the action (the adiabatic invariant) does not belong to the interval of actions swept by the "moving separatrix" (the separatrix of the system frozen at time t, t noving from 0 to $2\pi/\varepsilon$) obey the adiabatic invariant principle. The intersection points with the surface of section fall on regular curves which are the trajectories of the autonomous system (16) frozen at $t = t_o$.

On the other hand it can be shown (Cary et al. 1986; Henrard, 1988) that when the actual orbit goes from one side (say the libration side) of the separatrix to the other, the adiabatic invariant J suffers a jump

$$\Delta J \simeq I_c - I_L - \frac{1}{\omega} \left(\frac{\partial J_c}{\partial t_c} \right) \log(2 \sin \pi x) \tag{17}$$

where I_c and I_L are the limits of the action when the orbit of the frozen system approaches the separatrix from the libration side or the circulation side and are evaluated at the crossing time t_c. The quantity ω is the time scale at the unstable equilibrium and x is a function of where the separatrix is crossed. If we consider a set of orbits the initial phase of which are uniformly distributed in [0,2π], the quantity x can be shown to be uniformly distributed in [0,1].

The first difference in (17) describes the change in the adiabatic invariant due to the change of its definition across the separatrix. Its contribution will be cancelled by the corresponding term in the subsequent jump back to the libration side. The third term introduces a small phase dependent correction. It is this last contribution which leads to chaotic motion if we consider that the phase at one jump is uncorrelated with the phase at another jump.

Hence all the orbits the initial action of which belongs to the interval of actions swept by the moving separatrix (see Figure 11) will be forced to cross periodically the separatrix and suffer periodically a (small) jump in the action. These orbits will form a "stochastic layer" (we have called it the slow stochastic layer) around the separatrix the width of which is independent of ε but proportional to δ. This stochasticity manifests itself under the form of a slow diffusion process for the action. The diffusion coefficient can be estimated and is of the order of $\varepsilon^{3/2}$ (Menyuk, 1985; Henrard, 1987).

We do not claim that the two kinds of chaotic motion we have just described are fundamentally different. But the perturbative treatments of them, their time scales and the volumes of phase space affected by them are sufficiently different that it is worth distinguishing them.

6. NUMERICAL EVIDENCES IN THE 2/1 RESONANCE PROBLEM

A systematic numerical exploration of the 2/1 resonance also was conducted by Scholl and Froeschlé (1975) and Froeschlé and Scholl (1976-1977). Their main results can be summarized as follows. The 2/1 resonance is also characterized by rather large excursions in eccentricities (see Figure 12) but not large enough for the orbits to become Mars crossers (an eccentricity of 0.5 is needed). Chaotic orbits are also found but they cannot lead to large eccentricities because they are boxed in by regular motions (see Figure 13). "Peculiar orbits" switching back and forth from a libration regime to a circulation regime (see Figure 14) were also found as in the 3/1 resonance case. We can now identify them as the generators of a slow stochastic layer.

Following Wisdom's idea, Murray (1986) mapped in detail the location of the chaotic orbits (see Figure 15). He found two distinct types of chaotic orbits:

One in the central region of the resonance is characterized by very large increases in eccentricities (see Figure 16). Actually singularities of the mapping used as a numerical integration technique were encountered when the value of e exceeded 0.9 so that individual orbits were halted if the condition $e > 0.8$ occurred and the orbit was declared "chaotic".

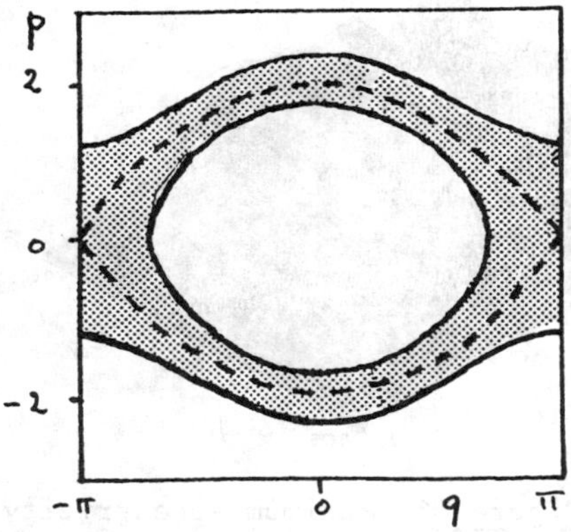

Figure 11: The slow stochastic layer for $\delta = 0.5$ and $t_o = \pi/2$. The boundaries are trajectories of the frozen pendulum which can be identified analytically.

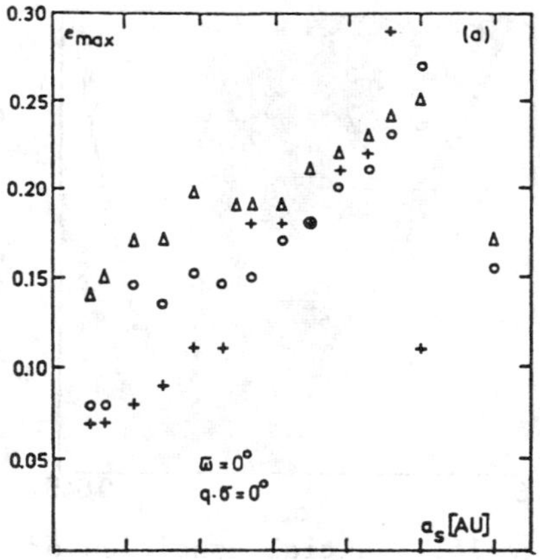

Figure 12: Maximum eccentricity reached by orbits at the 2/1 resonance starting with semi-major axis a_s and eccentricity e_s=0.04(+), 0.08 (0), 0.14(Δ) (From Scholl and Froeschlé, 1975, with permission).

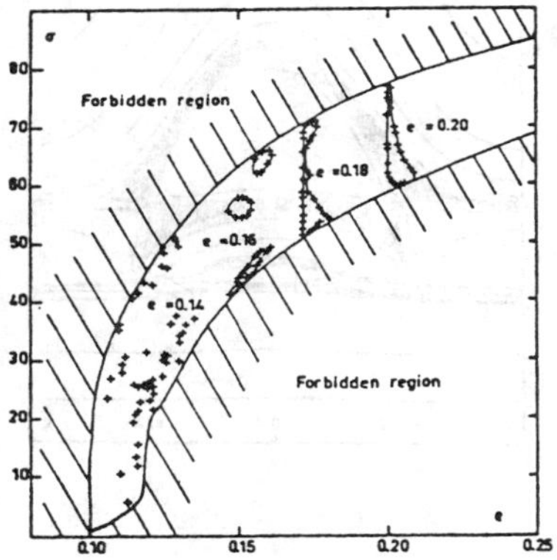

Figure 13: Surface of section at the 2/1 resonance showing a chaotic orbit (e=0.14) boxed in by regular orbits (e=0.18, e=0.20) (From Froeschlé and Scholl, 1976, with permission).

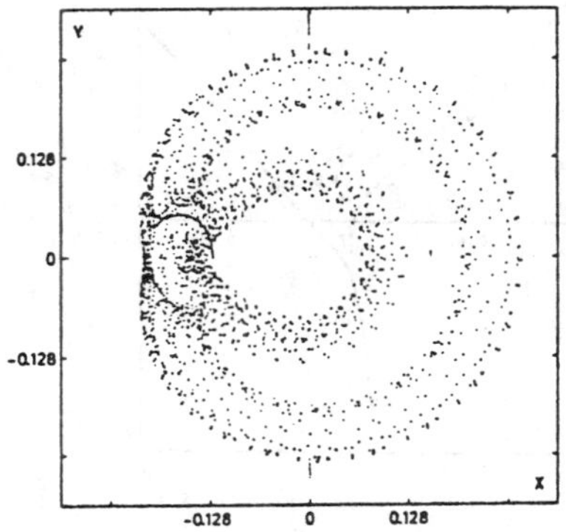

Figure 14: Peculiar orbit at the 2/1 resonance showing alternatively a libration and a circulation (From Froeschlé and Scholl, 1977 with permission).

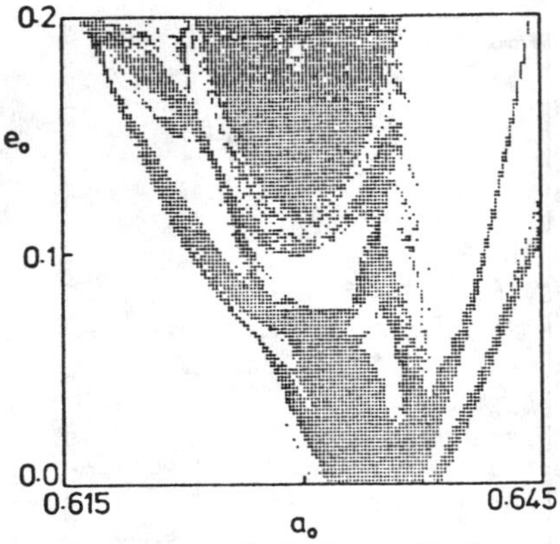

Figure 15: Initial condition of orbits showing a chaotic behaviour at the 2/1 resonance (From Murray, 1986 with permission).

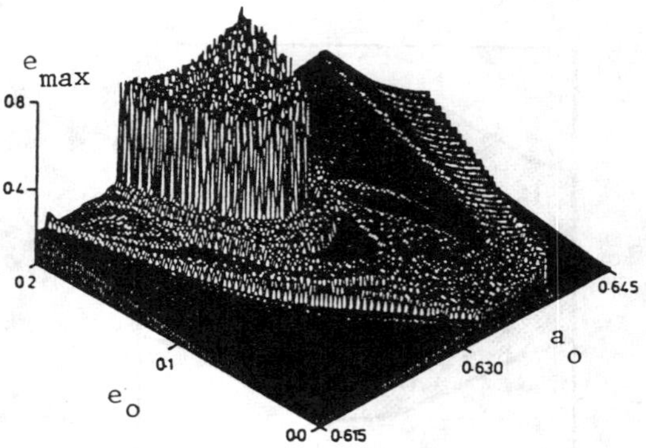

Figure 16: Maximum eccentricity achieved during 7000 Jupiter periods (From Murray, 1986 with permission).

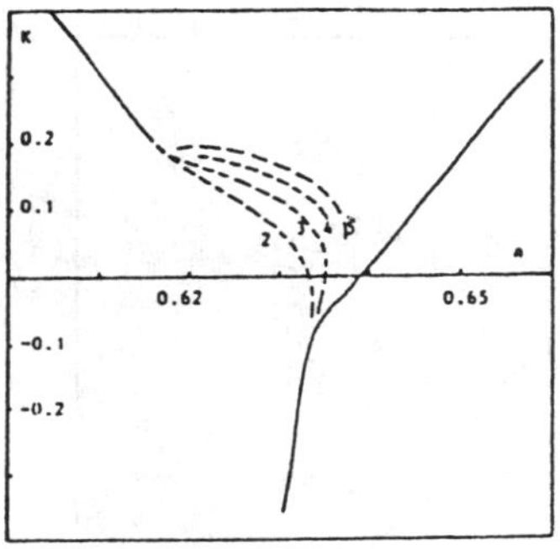

Figure 17: Level curves for the ratio ω_1/ω_2 (dashed curves). Also shown (solid curves) are initial conditions for the homoclinic orbits of the circular problem.

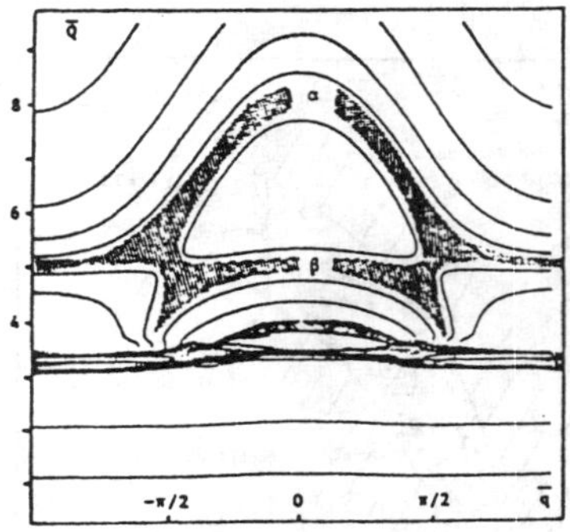

Figure 18: The phase space $(\bar{q},\bar{Q})$ for $\bar{J} = 4$ (From Henrard and Lemaitre, 1986, with permission).

The other type of orbits at the edges of the resonance and at low eccentricities did not show those high increases in eccentricity but did show genuine indications of stochasticity (i.e. a non-zero Lyapunov number).

Murray (1986) is very careful is not overemphasizing the importance of his results concerning the first type of orbits. Indeed he points out (and analyses) the poor convergence in eccentricity of the perturbation function. As the numerical technique he uses (Wisdom's mapping) can handle only a perturbation function truncated at order two, the results cannot really be trusted when the eccentricity reaches values larger than say 0.2. As a matter of fact, previous numerical integrations (without truncation of the perturbation function) by Giffen (1973) and Scholl and Froeschlé (1975) or Froeschlé and Scholl (1976) show in this region of phase space regular orbits devoid of large increase in eccentricities.

The other type of chaotic orbits seems more trustworthy. Murray's results confirm and refine the previous results of Giffen (1973) and Froeschlé and Scholl (1976). The chaotic orbit shown in Fugure 13 is one of them, starting at the location marked by a cross in Figure 15.

Murray points out also that the regular region between the two prongs of chaotic motions on the right of Figure 15, is characterized by a libration of the pericenter of the asteroid. This indication will help us in recognizing the origin of the two prongs of chaotic motions in the next section.

7. PERTURBATIVE TREATMENT OF THE 2/1 RESONANCE

Wisdom's perturbative treatment of the 3/1 resonance cannot be applied here without modifications. Indeed even when truncated at order two in the eccentricities the Hamiltonian cannot be reduced to a pendulum in one of the degrees of freedom and what is more important yet, the time scale of the motion of the pericenter is not (much) larger than the time scale of the libration of the critical argument, at least in a significant part of the domain of interest. This is principally due to the fact that the libration frequency of the critical argument is of the order of $\mu^{-2/3}$ (see Henrard and Lemaitre, 1983a) rather than $\mu^{-\frac{1}{2}}$ as in the 3/1 resonance. This is also confirmed by numerical integrations of Giffen (1973) who finds that the period of libration maintains an approximate value of 260 years but that the period of the argument of pericenter ranges from 500 years to several thousands of years for typical orbits in the domain of interest.

The two time scales are still different enough on a part of the phase space that a modification of Wisdom's perturbative treatment can be applied to the problem (Henrard and Lemaitre, 1987) but to describe the interesting features of Murray's diagram (Figure 15) a different approach must be used.

One such approach is described by Henrard and Lemaitre (1986). It is based upon the observation that the terms depending upon e' in (8) can be considered as (much?) smaller than the principal term $\sqrt{2R}$ cos r in the perturbation function because they are second order in the eccentricities and the principal term is first order. This is not true of course for the Hamiltonian (9) of the 3/1 resonance.

The following strategy may then be applied. Write the Hamiltonian (8) under the form:

$$H = H_o(M-R) + \mu H_1(M,R,r) + \mu e' H_2(M,R,m,r) \tag{18}$$

Introduce action-angle variables (J,ψ) for the one-degree of freedom problem

$$H_c = H_o(M-R) + \mu H_1(M,R,r) \tag{19}$$

describing the circular problem. This will transform the Hamiltonian H_c into

$$K_c = K_o(Q,J) \tag{20}$$

where J is the action of the one-degree of freedom problem (19) and Q = M is the momentum of the second degree of freedom. A new notation is introduced for it because the introduction of the action-angle variables (J,ψ) implies also a change of canonical variables for the second degree of freedom. This change of variables happens to be the identity for the momentum (Q = M) but not for the angular variable for which we find

$$q = m + \rho(\psi,J,M) \tag{21}$$

with a function ρ defined in Henrard-Lemaitre (1986).

Applying the transformation to H_2 also, we obtain for the Hamiltonian (18):

$$K = K_o(Q,J) + e' K_1(Q,J,q,\psi) \tag{22}$$

which is now ready for a classical perturbation treatment.

Of course this strategy implies that the action angle variables of the one-degree of freedom (19) can be actually computed, substituted in H_2 , that an averaging of (22) can be performed etc... It is certainly not obvious that formulae can be provided for doing so. Henrard and Lemaitre (1986) show how this can be done numerically. In some sense we make use of the fact that computers allow us to use as "practical", functions which are not written in terms of elementary or even special functions but for which we can provide a reasonable short algorithm.

An important factor in the application of a classical perturbation theory to (22) is the ratio of the basic frequencies:

$$\omega_1 = \frac{\partial K_o}{\partial J} \quad ; \quad \omega_2 = \frac{\partial K_o}{\partial Q} \tag{23}$$

The first one is the approximate frequency for the libration (or circulation) of the critical argument $r = \lambda - 2\lambda' + \omega$, while the second one is the approximate frequency of the angular variable $2\lambda' - \lambda$ which is equal to the frequency of the pericenter when the critical argument r is librating.

Level curves of the ratio ω_1/ω_2 of these frequencies are shown in Figure 17.

In the region corresponding to the two prongs of chaotic motion shown on the right of Figure 15, the ratio of these frequencies is large enough that an averaging over the fast angular variable (ψ) can be performed leaving us with a one-degree of freedom problem in $(\bar{q},\bar{Q})$. Level curves for this one-degree of freedom problem are shown in Figure 19 for a particular value of the constant $\bar{J}$ (unlike Wisdom - see Figure 9 - we have chosen to plot curves of constant H for a given value of the averaged action $\bar{J}$ rather than curves of constant $\bar{J}$ for a given value of H). On this diagram we have also indicated the "zone of uncertainty" surrounding the critical orbit of the (ψ,J) degree of freedom. (This is here a straight line $\bar{Q} = 3.29$) and the vicinity of heteroclinic orbits emanating from two unstable equilibria located at $\bar{Q} \sim 5$ and $\bar{q} = \pm\pi/2$. In a way very similar to what happens at the 3/1 resonance, one of the heteroclinic orbit (β) is connected with the "zone of uncertainty" and together they form a chaotic layer which corresponds in Murray's diagram (Figure 15) to the rightmost prong. The other heteroclinic orbit (α) shown itself as the left prong (still on the right of the diagram). Between these two heteroclinic orbits we find libration in $\bar{q}$ (and thus in ω) as mentioned by Murray (1986).

The larger area of chaotic motion on the left and center of Murray's diagram is due to a very different phenomenon. There as it can be seen on Figure 17, the ratio of the two main frequencies is not large and resonances (1/2, 1/3, 1/4...) are to be expected. Indeed Giffen (1973) has computed families of periodic orbits of the averaged restricted problem which present those ratios between the periods of the critical argument and the period of the pericenter. Giffen's periodic orbits fall remarkably well along the curves shown in Figure 17. (See Lemaitre-Henrard, 1988).

The structure seen in this area of Murray's diagram seems closely connected with these secondary resonances. Indeed as shown in Henrard-Lemaitre (1986) (where we have called them "super-resonance") the large island of regular motion seen on the upper left of Murray's diagram corresponds to the vicinity of stable 2/1 periodic orbits and the chaotic domain itself corresponds in part to unstable 2/1 periodic orbits and their accompanying homoclinic trajectories. This domain is apparently enlarged by a partial overlap of the 1/2, 1/3 and perhaps

1/4 secondary resonances (see Henrard and Lemaitre, 1986).

Hence most of the features found numerically by Froeschlé and Scholl or by Murray can be explained qualitatively and even for many of them quantitatively by the semi-numerical perturbative method we have outlined. The right edge of the 2/1 resonance zone has the same features than the 3/1 resonance zone but the remaining part shows a quite different behaviour. Secondary resonances seems to shape the phase space for low eccentricities and to produce a rather large domain of chaotic motion which is apparently boxed in by regular motions at higher eccentricities as indicated by Froeschlé and Scholl results.

REFERENCES

Arnold, V.I. 1978. Mathematical Methods of Classical Mechanics. Springer Verlag.

Cary, J.R., Escande, D.F., Tennyson, J.L. 1985. Adiabatic invariant changes due to separatrix crossing. Phys. Rev., A34, 4256-4275.

Chirikov, B.V. 1979. A universal instability of many-dimensional oscillator systems. Physics Reports, 52, 263-379.

Froeschlé, C. and Scholl, H. 1976. On the dynamical topology of the Kirkwood gaps. Astron. Astrophys., 48, 389-393.

Froeschlé, C. and Scholl, H. 1977. A qualitative comparison between the circular and elliptic Sun-Jupiter-Asteroid problem at commensurabilities. Astron. Astrophys., 57, 33-39.

Giffen, R. 1973. A study of commensurable motion in the asteroid belt. Astron. Astrophys., 23, 387-403.

Gonczi, R., Froeschlé, Ch. and Froeschlé, C. 1982. Poynting-Robertson drag on orbital resonances. Icarus, 51, 633-654.

Henrard, J. 1982. Capture into resonance: An extension of the use of the Adiabatic Invariant. Celest. Mech., 27, 3-22.

Henrard, J. 1987. Slow chaotic motion in Hamiltonian systems. Proc. of the 1st Belgian National Congress on Theoretical and Applied Mechanics, Brussels.

Henrard, J. 1988. The adiabatic invariant in classical mechanics. Submitted for publication in Dynamics Reported.

Henrard, J. and Lemaitre, A. 1983a. A second fundamental model for resonance. Celest. Mech., 30, 197-218.

Henrard, J. and Lemaitre, A. 1983b. A mechanism of formation for the Kirkwood's gaps. Icarus, 55, 482-494.

Henrard, J. and Lemaitre, A. 1986. A perturbation method for problems with two critical arguments. Celest. Mech., 39, 213-238.

Henrard, J. and Lemaitre, A. 1987. A perturbative treatment of the 2/1 Jovian resonance, Icarus, 69, 266-279.

Henrard, J., Lemaitre, A., Milani, A. and Murray C.D. 1986. The reducing transformation and apocentric librators. Celest. Mech., 38, 335-344.

Kirkwood, D. 1866. On the theory of meteors, Proc. A.A.A.S., 8-14.

Lemaitre, A. 1984a. High order resonance in the restricted three body problem. Celest. Mech., 32, 109-126.
Lemaitre, A. 1984b. Formation of the Kirkwood gaps in the asteroid belt. Celest. Mech., 34, 329-341.
Lemaitre, A. and Henrard, J. 1988. Chaotic motion in the 2/1 resonance. Submitted for publication.
Lichtenberg, A.J. and Lieberman, M.A. 1983. Regular and Stochastic Motion. Springer Verlag.
Menyuk, C.R. 1985. Particle motion in the field of a modulated wave. Phys. Rev., A31, 3282-3290.
Message, P.J. 1966. On nearly commensurable periods in the restricted problem of three bodies. Proc. I.A.U. Symposium 25, 197-222.
Murray, C.D. 1986. Structure of the 2/1 and 3/2 Jovian resonance. Icarus, 65, 70-82.
Murray, C.D. and Fox, K. 1984. Structure of the 3/1 Jovian resonance: A comparison of numerical methods. Icarus, 59, 221-233.
Poincaré, H. 1902. Sur les planètes du type d'Hécube. Bull. Astron. 19, 289-310.
Schubart, J. 1966. Special cases of the restricted problem of three bodies. Proc. I.A.U. Symposium 25, 187-193.
Schubart, J. 1968. Long-periods effects in the motion of Hilda-type planets. Astron. J., 73,99-103.
Scholl, H. and Froeschlé, C. 1974. Asteroidal motion at the 3/1 commensurability. Astron. Astroph. 33, 455-458.
Scholl, H. and Froeschlé, C. 1975. Asteroidal motion at the 5/2, 7/3 and 2/1 resonances. Astron. Astroph. 42, 457-463.
Sessin, W. and Ferraz-Mello, S. 1984. Motion of two planets with periods commensurable in the ratio 2/1. Celest.Mech., 32, 307-332.
Szebehely, V. 1967. Theory of Orbits. Academic Press.
Sinclair, A.T. 1970. Periodic solutions close to commensurabilities in the three body problem. M.N.R.A.S., 148, 325-351.
Torbett, M. and Smoluchowski, R. 1980. Sweeping of the Jovian resonances and the evolution of the asteroids. Icarus, 44, 722-729.
Torbett, M. and Smoluchowski, R. 1982. Motion of the Jovian commensurability resonances. Astron. Astroph. 110, 43-49.
Wisdom, J. 1982. The origin of the Kirkwood's gaps: A Mapping for asteroidal motion near the 3/1 commensurability. Astron. J., 85, 1122-1133.
Wisdom, J. 1983. Chaotic behaviour and the origin of the 3/1 Kirkwood gap. Icarus, 56, 51-74.
Wisdom, J. 1985. A perturbative treatment of motion near the 3/1 commensurability. Icarus, 63, 272-289.
Wisdom, J. 1986. Canonical solution of the two critical argument problem. Celest. Mech., 38, 175-180.

THE NEAR-RESONANCE THEOREM: ANALYSIS OF THE VICINITY OF PERIODIC SOLUTIONS OF ANALYTIC DIFFERENTIAL SYSTEMS.

C. Marchal
D.E.S. ONERA
92320 Chatillon
France

ABSTRACT. The near-resonance theorem allows us to simplify the systems of differential equations and to study the vicinity and the stability of their periodic solutions. It leads to suitable transformations that cancel all non-near-resonant terms, even those of very large orders.

The demonstration of the theorem is difficult but short and various applications are presented.

For Hamiltonian systems the near-resonance theorem leads to a strict definition of the notion of "quasi-integral", to the "generalized Birkhoff differential rotation" and to the "all-order stability" that is not necessarily a true stability but is at worse an extremely slow "Arnold diffusion". It also discloses many high-order instabilities related to resonances.

1. INTRODUCTION

Systems of differential equations appear in a very large number of models used for the description of scientific phenomena (mechanics, physics, chemistry, biology, astronomy etc...). These systems are built either empirically or according to the laws of the phenomenon of interest, laws arising from the consideration of simpler phenomena of the same nature.

In most cases, and especially in astronomy, the models are so simple and so accurate that they allow excellent long-time predictions. However most of these models are non-linear and non-integrable, fundamental questions remain open such as the question of the stability of the Solar System and many other questions implying high-order effects on very large periods of time (final evolutions, instabilities, etc.).

These difficult questions mix two causes of errors: A) The numerical integrations become inaccurate for large intervalsof time, B) The small differences between the mathematical model and the physical reality appear only after long enough periods.

A. E. Roy (ed.), Long-Term Dynamical Behaviour of Natural and Artificial N-Body Systems, 427–448.
© 1988 by Kluwer Academic Publishers.

A good understanding of these small differences, and thus the improvement of our models, require both accurate measures on long intervals and a theoretical and mathematical analysis of the solutions of differential systems.

Henri Poincaré has emphasized the great importance of periodic solutions that can be computed for all time with any given accuracy and that are very often the "backbone of the set of solutions".

The near-resonance theorem will allow to simplify the study of the vicinity of periodic solutions and to analyse accurately their stability or instability.

2. THE NEAR-RESONANCE THEOREM

2.1 The equations of motion

Let us consider an analytic differential system:

$$\begin{cases} \vec{X} = (x_1, x_2, x_3, \ldots x_n) = \text{"state vector"} & (1) \\ d\vec{X}/dt = F(\vec{X}, t) & (2) \end{cases}$$

that is:

$$dx_k/dt = f_k(\vec{X}, t) \quad ; \quad k = \{1, 2, \ldots n\} \tag{3}$$

We will assume that the analytic equations of motion (2) and (3) are either autonomous (as in the usual n-body problem) or at least periodic in terms of the time t (as in the elliptic restricted three-body problem):

$$F(\vec{X}, t) \equiv F(\vec{X}, t+P) \tag{4}$$

Let us consider now a periodic solution $\vec{X}(t)$ with the period T:

$$\forall t : d\vec{X}(t)/dt = F(\vec{X}(t), t) \tag{5}$$
$$\forall t : \vec{X}(t) \equiv \vec{X}(t+T) \tag{6}$$

If the system (1)-(3) is autonomous the period T is arbitrary, but it is a multiple of P in the general periodic case (4).

We will analyse the vicinity of the periodic solution $\vec{X}(t)$ and we will put at any time:

$$\vec{Y} = \vec{X} - \vec{X}(t) \tag{7}$$

The equation of motion of the small vector $\vec{Y}$ is then:

$$d\vec{Y}/dt = G(\vec{Y}, t) \tag{8}$$

with of course:

$$G(\vec{Y},t) \quad = \quad F(\vec{X}(t)+\vec{Y},t) - d\vec{X}(t)/dt \tag{9}$$

The analytical function of $G(\vec{Y},t)$ has the two main following properties:

$$G(\vec{0},t) = 0 \qquad ; \qquad G(\vec{Y},t) \equiv G(\vec{Y},t+T) \tag{10}$$

2.2 The classical first-order Floquet analysis

With a suitable "Jacobian matrix" M(t) the equation of motion (8) can be written:

$$d\vec{Y}/dt \quad = \quad M(t)\ \vec{Y} + O(Y^2) \tag{11}$$

The matrix M(t) has the period T and the first-order approximation is:

$$d\vec{Y}/dt \quad = \quad M(t)\vec{Y} \tag{12}$$

The integration of (12) leads to:

$$\vec{Y}(t) \quad = \quad C(t)\ \vec{Y}(0) \tag{13}$$

with:

$$C(0) = I = \text{unit matrix} \quad ; \quad dC/dt = M(t)\ C(t) \tag{14}$$

Let us call A the matrix C(T), it is the "monodromy matrix" or "transition matrix" or "Floquet matrix", its eigenvalues $\mu_1,\mu_2 \ldots \mu_n$ are the "Floquet multipliers"

(14) and the periodicity of M(t) imply:

$$\left.\begin{array}{lll} \forall t : & C(t+T) & \equiv C(t).A \\ \forall m\ \mathbb{Z} & C(mT) & \equiv A^m \\ & \det C(t) & = \exp \{\int_o^t [\text{trace } M(u)]\ du\} \neq 0 \end{array}\right\} \tag{15}$$

Let $\vec{V}_j$ an eigenvector of A:

$$A\ \vec{V}_j \quad = \quad \mu_j\ \vec{V}_j \tag{15}$$

hence, with (13) and (15), the first-order analysis gives:

$$\vec{Y}(0) = \varepsilon \vec{V}_j \quad \text{implies} \quad \vec{Y}(mT) \quad = \quad \varepsilon\, \mu_j^{\ m}\ \vec{V}_j \tag{17}$$

The classical Floquet results are then obvious: If one or several eigenvalues μ_j have a modulus larger than one the periodic solution $\vec{X}(t)$ of interest is "exponentially unstable", if all Floquet multipliers μ_j have a modulus smaller than one the periodic solution $\vec{X}(t)$ is "exponentially stable" and finally in the limit or "critical" case the largest modulus of the μ_j's is one and the stability depends upon higher-order effects.

The importance of the critical case comes from the conservative systems, e.g. the Hamiltonian systems. For these systems the product $\mu_1 \mu_2 \ldots \mu_n$ is always one and the critical case is very common, while the exponential stability is impossible.

Floquet has also shown that a suitable linear and periodic transformation $\vec{Y} \to \vec{Z}$ allows to have an autonomous first-order equation:

$$d\vec{Z}/dt = N \vec{Z} \tag{18}$$

the matrix N being a suitable "average" of M(t).

For instance if we choose $\vec{Z}(0) \equiv \vec{Y}(0)$ we can define N by:

$$\exp \{NT\} = A \tag{19}$$

Indeed then, from (13), (18) and $\vec{Z}(0) \equiv \vec{Y}(0)$:

$$\left.\begin{aligned} \vec{Y}(t) &= C(t).\vec{Y}(0) \\ \vec{Z}(t) &= \exp \{Nt\}.\ \vec{Z}(0) \end{aligned}\right\} \Rightarrow \vec{Z}(t) = \exp\{Nt\}.C^{-1}(t).\vec{Y}(t) \tag{20}$$

(15) and (19) show that the linear transformation $\vec{Y}(t) \to \vec{Z}(t)$ defined in (20) has indeed the period T.

Since det $A \neq 0$ the equation (19) has many possible solutions always of the type:

$$N = \sum_{j=o}^{n-1} a_j A^j .$$

The eigenvalues λ_j of N verify $\exp \{T\lambda_j\} = \mu_j$, their real parts are the Liapounov characteristic exponents.

We will slightly simplify the equation (18) by a suitable change of the set of axes that will "diagonalize" the matrix N, if possible, or at least "Jordanize" it and the full equations of motion of the components z_k of $\vec{z}$ will be:

$$\left.\begin{aligned} &dz_k/dt = \lambda_k z_k + \varepsilon_k z_{k+1} + g_k(z_1, z_2, \ldots z_n, t) \ ; \ k = \{1, \ldots n\} \\ &\text{with the Jordan rules on } \varepsilon_k \text{ and eigenvalues } \lambda_k: \\ &\varepsilon_n = 0 \ ; \ \lambda_k \neq \lambda_{k+1} \Rightarrow \varepsilon_k = 0 \ ; \ \lambda_k = \lambda_{k+1} \Rightarrow \varepsilon_k = \text{either 0 or 1} \end{aligned}\right\} \tag{21}$$

The analytic functions $g_k(z_1, z_2, \ldots z_n, t)$ contain the higher-order terms, they are periodic in terms of t with the period T.

2.3 Expression of the near-resonance theorem

The equations of motion (21) are equivalent to the initial equations of motion (3) but they are also much simpler: no zero-order term, autonomous and simple first-order terms after a suitable linear and periodic transformation from $\vec{X}$ to $\vec{Z}$.

The near-resonance theorem uses the same idea for the higher-order terms: a suitable periodic transformation $\vec{Z} \to \vec{U}$ will allow us to simplify them considerably.

We will decompose the analytic functions $g_k(z_1,\ldots z_n,t)$ of (21) into power series of the z_j and Fourier series of the time t:

$$g_h(z_1,\ldots z_n,t) = \sum_{\vec{\alpha},m} G_h(\vec{\alpha},m) z_1^{\alpha_1} z_2^{\alpha_2} \ldots z_n^{\alpha_n} \exp\{m\omega_o it\} \tag{22}$$

with:

$$\left.\begin{array}{l} \vec{\alpha} = (\alpha_1,\alpha_2,\ldots\alpha_n) \ ; \ \alpha_1,\alpha_2,\ldots,\alpha_n \in \mathbb{N} \ ; \ \alpha_1+\alpha_2+\ldots+\alpha_n \geq 2 \\ m \in \mathbb{Z} \ : \ G_k(\vec{\alpha},m) \in \mathbb{C} \ ; \ \omega_o = 2\Pi/T \end{array}\right\} \tag{23}$$

The near-resonance theorem is then the following:

For any positive scalar $\varepsilon_{\vec{\alpha}}$, even very small, there exists an analytic transformation $\vec{Z} \to \vec{U}$ such that

I) $\vec{Z}$ and $\vec{U}$ are equivalent:

$$\vec{U} - \vec{Z} = O(Z^2) = O(U^2) \tag{24}$$

II) The transformation $\vec{Z} \to \vec{U}$ may depend on t and is periodic in terms of t:

$$\vec{U}(\vec{Z},t) \equiv \vec{U}(\vec{Z},t+T) \tag{25}$$

III) The equations of motion of $\vec{U}$ have much less terms:

$$du_k/dt = \lambda_k u_k + \varepsilon_k u_{k+1} + j_k(u_1, u_2,\ldots,u_n,t) \ : \ k = \{1,\ldots n\} \tag{26}$$

The λ_k's and ε_k's are those of (21) while the power series j_k have neither zero-order term nor first-order term, are periodic in terms of t, are convergent in a small neighbourhood of the origin and verify the following:

The power-Fourier series $j_k(u_1,\ldots u_n,t)$ have only "near-resonant" terms, i.e. terms

$$J_k(\vec{\alpha},m) u_1^{\alpha_1} u_2^{\alpha_2} \ldots u_n^{\alpha_n} \exp\{m\omega_o it\}$$

for which

$$\left| \lambda_k - m\omega_o i - \sum_{q=1}^{n} \alpha_q \lambda_q \right| \leq \varepsilon (\alpha_1 + \alpha_2 + \ldots + \alpha_n) \qquad (27)$$

2.4 First applications of the near-resonance theorem

We will see the demonstration of the near-resonance theorem in section 3 and as a first application let us consider the following:

Let us assume the indices of the eigenvalues λ_k to be chosen so that:

$$R_e(\lambda_1) \leq R_e(\lambda_2) \leq R_e(\lambda_3) \leq \ldots\ldots \leq R_e(\lambda_n) \qquad (28)$$

and let us define the integers n_-, n_o, n_+ by:

$$\left.\begin{array}{l} R_e(\lambda_{n_-}) < 0 \quad ; \quad R_e(\lambda_{(1+n_-)}) = 0 \quad ; \quad R_e(\lambda_{(n_o+n_-)}) = 0 \ ; \\ R_e(\lambda_{(1+n_o+n_-)}) > 0 \quad ; \quad n_- + n_o + n_+ = n \end{array}\right\} \qquad (29)$$

The corresponding vectors $\vec{U}_-$, $\vec{U}_o$, $\vec{U}_+$ are defined by:

$$\left.\begin{array}{lll} \vec{U}_- & = & (u_1, u_2, \ldots, u_{n-}) \\ \vec{U}_o & = & (u_{(1+n_-)}, \ldots u_{(n_o+n_-)}) \\ \vec{U}_+ & = & (u_{(1+n_o+n_-)}, \ldots, u_n) \\ \vec{U} & = & (\vec{U}_-, \vec{U}_o, \vec{U}_+) \end{array}\right\} \qquad (30)$$

In these conditions, if we choose ε smaller than $R_e(\lambda_{(1+n_o+n_-)})$, the equations (26) and the conditions (27) imply that the n_+ dimensional manifold $(0,0,\vec{U}_+)$ is <u>invariant</u>, it is the "unstable subset" of the periodic solution $\vec{X}(t)$ of interest.

Inside that unstable subset the integration of equations of motion (26) is simple and, if ε is sufficiently small, this integration leads to:

$$u_k = P_k(t).\exp\{\lambda_k t\} \quad ; \quad k = \{1+n_o+n_-, \ldots, n\} \qquad (31)$$

with $P_k(t)$ being polynomials of the time t.

Similar properties are true for the "stable subset" $(\vec{U}_-,0,0)$ when ε is smaller than $R_e(-\lambda_{n-})$.

2.5 Another example of simplification: the Hartman theorem.

The Hartman theorem [Ref.1] specifies that the higher-order terms can disappear completely through a suitable analytic transformation $\vec{Z} \rightarrow \vec{U}$ if the following conditions are satisfied:

I) The equations of motion (21) are autonomous.
II) There exist two suitable positive scalars ε and μ such that:

$$\left.\begin{array}{l} \alpha_1, \alpha_2, \ldots, \alpha_n \in \mathbb{N} \\ \alpha_1 + \alpha_2 + \ldots + \alpha_n \geq 2 \\ 1 \leq k \leq n \end{array}\right\} \text{ imply: } \left| \lambda_k - \sum_{q=1}^{n} \alpha_q \lambda_q \right| > \varepsilon (\alpha_1 + \alpha_2 + \ldots + \alpha_n)^{-\mu} \quad (32)$$

When applicable this theorem is of course more interesting than the near-resonance theorem; it leads to the easily integrable equations of motion:

$$du_k/dt = \lambda_k u_k + \varepsilon_k u_{k+1} \quad (33)$$

with solutions similar to those of (31).

Unfortunately the Hartman theorem is rarely applicable. It is especially not applicable in the "critical case" and even when one Liapounov characteristic exponent $R_e(\lambda_k)$ is zero. Indeed then $\bar{\lambda}_k$ is another eigenvalue and is opposite to λ_k which leads to the following contradiction of (32):

$$\lambda_k - (2\lambda_k + \bar{\lambda}_k) = 0 \quad (34)$$

3. DEMONSTRATION OF THE NEAR-RESONANCE THEOREM

3.1 The formal development

We will prove the near-resonance theorem only under the restriction that all ε_k of (21) are zero (e.g. all eigenvalues λ_k are different). For the case of multiple eigenvalues and ε_k equal to one the demonstration is more intricate but causes no real conceptual difficulties (see Ref.2, appendix 3).

Let us define the analytic transformation $\vec{Z} \rightarrow \vec{U}$ by the following:

$$\left.\begin{array}{l} z_k = u_k + \varphi_k(u_1, u_2, \ldots, u_n, t) \\ \text{with: } \varphi_k(\vec{U}, t) = O(U^2) \quad ; \quad \varphi_k(\vec{U}, t) \equiv \varphi_k(\vec{U}, t+T) \end{array}\right\} k = (1, 2, \ldots, n) \quad (35)$$

As the functions g_k of (21), (22) and j_k of (26) the functions

φ_k will be decomposed into power and Fourier series:

$$\left.\begin{array}{l}\varphi_k(u_1, u_2,\ldots u_n,t) = \sum_{\vec{\alpha},m} F_k(\vec{\alpha},m)u_1^{\alpha_1} u_2^{\alpha_2} \ldots u_n^{\alpha_n} \exp\{m\omega_o it\} \; ; \\ \qquad k = (1,2,\ldots,n) \\ \text{with:} \\ \vec{\alpha} = (\alpha_1,\ldots,\alpha_n) \; ; \; \alpha_1,\alpha_2,\ldots\alpha_n \in \mathbb{N} \; ; \; \alpha_1+\alpha_2+\ldots+\alpha_n = \text{degree} \geq 2 \; ; \\ m \in \mathbb{Z} \; : \; F_k(\vec{\alpha},m) \in \mathbb{C} \; ; \; \omega_o = 2\Pi/T \end{array}\right\} \quad (36)$$

The relations between the functions g_k, j_k, φ_k are obtained directly from (21), (26) and (35) with the assumption all $\varepsilon_k = 0$:

$$d(u_k+\varphi_k)/dt = dz_k/dt \quad ; \quad k = \{1,2,\ldots,n\} \qquad (37)$$

that is:

$$\lambda_k u_k + j_k(\vec{U},t) + \sum_{q=1}^{n} \frac{\partial\varphi_k}{\partial u_q}(\lambda_q u_q + j_q) + \frac{\partial\varphi_k}{\partial t} = \lambda_k z_k + g_k(\vec{Z},t) =$$

$$= \lambda_k(u_k+\varphi_k) + g_k(\vec{Z},t) \qquad (38)$$

hence:

$$j_n(\vec{U},t) + \sum_{q=1}^{n} \lambda_q u_q \frac{\partial\varphi_k}{\partial u_q} + \frac{\partial\varphi_k}{\partial t} - \lambda_k\varphi_k(\vec{U},t) = g_k(\vec{Z},t) - \sum_{q=1}^{n} j_q \frac{\partial\varphi_k}{\partial u_q} \qquad (39)$$

The equations (35) and (39) and the knowledge of the power-Fourier series $g_k(\vec{Z},t)$ allow a formal development of the power-Fourier series of the functions j_k and φ_k. Indeed if these functions are known up to the degree D-1 the right-hand member of (39) is known up to the degree D in terms of $u_1, u_2,\ldots,u_n,t$.

Let us then assume that, in the right-hand member of (39), the coefficient of the term $u_1^{\alpha_1} \ldots u_n^{\alpha_n} \exp\{m\omega_o it\}$ is $B_k(\vec{\alpha},m)$, this equation (39) gives then:

$$J_k(\vec{\alpha},m) + \left(\sum_{q=1}^{n} \alpha_q\lambda_q + m\omega_o i - \lambda_k\right) F_k(\vec{\alpha},m) = B_k(\vec{\alpha}, m) \qquad (40)$$

Our purpose is to obtain j_k functions as simple as possible and thus to erase as many $J_k(\vec{\alpha},m)$ coefficients as possible. This is of course impossible, if $B_k(\vec{\alpha},m) \neq 0$ and $\sum_{q=1}^{n} \alpha_q \lambda_q + m\omega_o i - \lambda_k = 0$

and it is dangerous for the convergence if this latter quantity is too small; we are thus led to the following rule of choice of J_k and F_k :

$$\left\{\begin{array}{l} J_k(\vec{\alpha},m) = B_k(\vec{\alpha},m) \text{ and } F_k(\vec{\alpha},m) = 0 \quad \text{if the near-resonance condition (27) is satisfied} \\ J_k(\vec{\alpha},m) = 0 \text{ and } F_k(\vec{\alpha},m) = B_k(\vec{\alpha},m)/\left(\sum_{q=1}^{n} \alpha_q\lambda_q+m\omega_o i-\lambda_k\right) \\ \quad \text{if the condition (27) is not satisfied} \end{array}\right. \qquad (41)$$

The rule of choice (41) completes the construction of the formal power-Fourier series $j_k(\vec{U},t)$ and $\varphi_k(\vec{U},t)$ that are then uniquely defined and the only remaining question is of the convergence of these series at least for sufficiently small $\vec{U}$.

3.2 The convergence

The functions $g_k(\vec{Z},t)$ are analytic for all t in the vicinity of $\vec{Z} = 0$ and this is equivalent to the following limits on their development given in (22), (23):

$$\left.\begin{array}{l} \exists a,b,c \text{ such that: } 0 < a;\quad 0 < b;\quad 0 < c < 1 \quad \text{and} \\ \forall \vec{\alpha},m : \quad |G_k(\vec{\alpha},m)| \leq a.b^{(\alpha_1+\alpha_2+\ldots+\alpha_n)}\,.\, c^{|m|} \end{array}\right\} \quad (42)$$

Notice that:

$$1 + \sum_{m=1}^{\infty} c^m \,(\exp\{mi\theta\} + \exp\{-mi\theta\}) = (1-c^2)/(1+c^2-2c\cos\theta) \quad (43)$$

Hence, the n power-Fourier series $g_k(\vec{Z},t)$ are "majorized" by the power-Fourier series of the function $g(\vec{Z},t)$ given, with suitable c_1 and c_2 by:

$$\left.\begin{array}{l} g(\vec{Z},t) = \dfrac{c_1\, Z^2}{(1-c_1 Z)(1-c_2\cos\omega_o t)} \\ \text{with: } \quad 0 < c_1;\quad 0 < c_2 = 2c/(1+c^2) < 1;\quad Z = z_1 + z_2 + \ldots + z_n \end{array}\right\} \quad (44)$$

By definition a power-Fourier series as $g(\vec{Z},t)$ "majorizes" a power-Fourier series $g_k(\vec{Z},t)$ if and only if for all $\vec{\alpha}$,m:

$$G(\vec{\alpha}\ m) \geq |G_k(\vec{\alpha},m)| \quad (45)$$

The majorizing series has then only real non-negative coefficients and its radius of convergence is smaller than or equal to that of the majorized series.

Let us also define the power-Fourier series $h_k(\vec{U},t)$ and their coefficients $H_k(\vec{\alpha},m)$ by:

$$\left.\begin{array}{l} j_k(\vec{U},t) = \displaystyle\sum_{q=1}^{n} u_q \frac{\partial h_k(\vec{U},t)}{\partial u_q} \\ J_k(\vec{\alpha},m) = (\alpha_1 + \alpha_2 + \ldots + \alpha_n)\, H_k(\vec{\alpha},m) \end{array}\right\} \quad k=(1,2,\ldots n) \quad (46)$$

For a given k the power-Fourier series $j_k(\vec{U},t)$ and $h_k(\vec{U},t)$ have the same radius of convergence.

Let us now consider the near-resonance condition (27), the equation (39) and the rule of choice (41).

These 3 relations are used for the step-by-step computation of the coefficients $F_k(\vec{\alpha},m)$ and $J_k(\vec{\alpha},m)$ (and then $H_k(\vec{\alpha},m)$) and we can observe by a direct comparison that for all k the three functions $h_k(\vec{U},t)$, $\varepsilon\,\varphi_k(\vec{U},t)$ and $h_k(\vec{U},t)+\varepsilon\,\varphi_k(\vec{U},t)$ are majorized by the power-Fourier series $\varepsilon\,\psi(\vec{U},t)$ defined by the following:

$$\varepsilon \sum_{q=1}^{n} \frac{\partial\psi}{\partial u_q} \cdot u_q = \frac{c_1\, Z^2}{(1-c_1 Z)(1-c_2 \cos \omega_o t)} + \varepsilon \sum_{q=1}^{n} \sum_{p=1}^{n} \frac{\partial\psi}{\partial u_q} \cdot \frac{\partial\psi}{\partial u_p} \cdot u_p \tag{47}$$

$$\left.\begin{aligned} Z &= z_1 + z_2 + \dots + z_n \\ z_k &= u_k + \psi(\vec{U},t) \end{aligned}\right\} \tag{48}$$

Indeed, at each step of the analysis, the right-hand member of (47) majorizes that of (39) and their left hand members reinforce this effect.

It is then sufficient to study the convergence of the power-Fourier series $\psi(\vec{U},t)$ defined in (47), (48) and, since all coefficients are real and nonnegative, it is sufficient to study the case when $u_1 = u_2 = u_3 = \dots = u_n$.

We will call u this common value and, since then $Z = n(u+\psi)$, we only need to study the convergence of the two-parameter function $\psi(u,t)$ defined by:

$$\varepsilon\, u \frac{\partial\psi}{\partial u} = \frac{c_1\, n^2 (u+\psi)^2}{[1-c_1\, n(u+\psi)][1-c_2 \cos \omega_o t]} + \varepsilon\, u \left(\frac{\partial\psi}{\partial u}\right)^2 \tag{49}$$

$\psi(u,t)$ is majorized by $u.\,\frac{\partial\psi}{\partial u}$ and they are both majorized by $F(u,t)$ defined by:

$$\varepsilon\, F = \frac{c_1\, n^2 (u+F)^2}{[1-c_1\, n(u+F)]\,[1-c_2 \cos \omega_o t]} + \varepsilon \frac{F^2}{u} \tag{50}$$

The function $F(u,t)$ is given by an <u>algebraic</u> equation and can be developed in power-Fourier series with a positive radius of convergence: it always converges when

$$|u| \leqslant \varepsilon(1-c_2)/2c_1\, n(4n + \varepsilon - \varepsilon\, c_2)\ .$$

This limit can be improved in several ways, but always goes to zero with ε and this forbids us from using $\varepsilon = 0$ in the near-resonance condition (27).

All functions of the demonstration, i.e. j_k, φ_k, h_k, ψ have at least the same radius of convergence and this completes the demonstration of the near-resonance theorem.

3.3 Improvements?

We cannot use $\varepsilon = 0$ in the condition (27) of the near-resonance terms; can we at least use an improved condition as for instance:

$$|\lambda_k - m\,\omega_o\, i - \sum_{q=1}^{n} \alpha_q \lambda_q| \lesssim \varepsilon \quad ? \tag{51}$$

The answer is "yes" in some cases (for instance when the Hartman conditions of section 2.5 are satisfied). But it is "no" in the following simple system of differential equations:

$$dx/dt = x^2 \quad ; \quad dy/dt = y + x^2 \tag{52}$$

The answer is also "no" in the following Hamiltonian problem:

$$H = p_1 q_1 + p_2^2 q_1 + p_2^2 q_2 \tag{53}$$

With the condition (51) these two cases lead to diverging series φ_k in the analysis of sections 3.1 and 3.2.

However Joerg Waldvogel, Carlès Simò and Joaquim Font think that it is perhaps possible to improve the near-resonance theorem up to the condition (51) when all eigenvalues λ_k are non-zero.*

If this is true, the n_o-dimensional manifold $(0,\vec{U}_o,0)$ of section 2.4 would be invariant for $\overset{o}{\varepsilon}$ smaller than $\inf\{ R_e(\lambda_{(n_o+n_+ +1)};\ -R_e(\lambda_{n-})\}$.

This manifold would be the "central subset" of the periodic solution of interest and considerable simplifications would occur.

Notice that for systems (52) and (53) the "central subset" is non-analytic and then our analytic procedure cannot converge if we try to put this central subset into the manifold $(0,\vec{U}_o,0)$.

* Or, perhaps, when the "central subset" has no motion asymptotical to $\vec{Z} = 0$.

4. THE NEAR-RESONANCE THEOREM AND THE ANALYTIC HAMILTONIAN SYSTEMS.

The analytic Hamiltonian systems are classical and very useful differential systems.

The usual Hamiltonian notations are the following:

$$\vec{p} = (p_1, p_2, \ldots, p_r) \quad \vec{q} = (q_1, q_2, \ldots, q_r) \quad \text{The state-vector and its conjugate sometimes called "action-angle variables"} \tag{54}$$

$$r = \text{"number of degrees of freedom"} \tag{55}$$

$$H = H(\vec{p},\vec{q},t) = \text{analytic Hamiltonian function} \tag{56}$$

We will consider the autonomous case ($H = H(\vec{p},\vec{q})$) and the periodic case ($H(\vec{p},\vec{q},t) \equiv H[\vec{p},\vec{q},(t+P)]$).

The usual Hamiltonian equations are the following with $\vec{p}' = d\vec{p}/dt$; $\vec{q}' = d\vec{q}/dt$:

$$\vec{p}' = -\partial H/\partial \vec{q} \quad ; \quad \vec{q}' = \partial H/\partial \vec{p} \tag{57}$$

We will analyse the vicinity of a periodic solution $\vec{p}(t)$, $\vec{q}(t)$ of period T (that period T is a multiple of P in the non-autonomous case).

In the analytic Hamiltonian problems the eigenvalues λ_k are pairs of opposite and all the simplifications of the sections 2.1 to 3.1 can be obtained by canonical transformations with the only exception of the "Jordanization" (21) for which the canonical rules for the ε_k are slightly different and will be given thereafter.

These canonical transformations allow to keep the Hamiltonian form of the problem.

4.1 Reduction to the origin.

The transformation (7) becomes:

$$\overrightarrow{p_a} = \vec{p} - \vec{p}(t) \quad ; \quad \overrightarrow{q_a} = \vec{q} - \vec{q}(t) \tag{58}$$

with the new Hamiltonian $H_a(\overrightarrow{p_a},\overrightarrow{q_a},t)$:

$$H_a(\overrightarrow{p_a},\overrightarrow{q_a},t) = H(\vec{p},\vec{q},t) - \overrightarrow{p_a}\overrightarrow{q'}(t) + \overrightarrow{q_a}.\overrightarrow{p'}(t) - H[\vec{p}(t),\vec{q}(t),t] \tag{59}$$

The Hamiltonian $H_a(\overrightarrow{p_a},\overrightarrow{q_a},t)$ has neither zero-order term nor first-order term; it has the period T:

$$H_a(\overrightarrow{p_a},\overrightarrow{q_a},t) \equiv H_a[p_a,q_a,(t+T)] \tag{60}$$

It will be decomposed into terms of degree n:

$$H_a(\vec{p}_a,\vec{q}_a,t) \equiv \sum_{n=2}^{\infty} H_{an}(\vec{p}_a,\vec{q}_a,t) \tag{61}$$

4.2 The Floquet transformation

This second simplification given in section 2.2 is a linear and periodic transformation that leads to a Hamiltonian $H_b(\vec{p}_b,\vec{q}_b,t)$ with autonomous second-order terms H_{b2} :

$$\begin{pmatrix}\vec{p}_a\\ \vec{q}_a\end{pmatrix} = R(t)\begin{pmatrix}\vec{p}_b\\ \vec{q}_b\end{pmatrix} \quad ; \quad R(t) \equiv R(t+T) \tag{62}$$

The $2r \times 2r$ matrix $R(t)$ is "symplectic" for all t and the higher-order terms are given by:

$$n \geq 3 \text{ implies } \quad H_{bn}(\vec{p}_b,\vec{q}_b,t) \equiv H_{an}(\vec{p}_a,\vec{q}_a,t) \tag{63}$$

The simplest expression of this transformation is given by the method of the generatrix function S with mixed parameters:

$$\left.\begin{aligned} &S(\vec{p}_a,\vec{q}_a,t,H_b) = t\,H_b + S_1(\vec{p}_a,\vec{q}_b,t) \\ &\vec{p}_b = \partial S/\partial q_b \quad ; \quad \vec{q}_a = \partial S/\partial \vec{p}_a \quad ; \quad H_a = \partial S/\partial t \end{aligned}\right\} \tag{64}$$

This method gives canonical transformations for C_2-functions S_1 and uses here an homogeneous function S_1 of degree two in $\vec{p}_a$ and $\vec{q}_b$ in order to obtain the linear transformation (62) and the identities (63).

In Hamiltonian problems the simple equations (21) don't always correspond to a canonical transformation, but a similar system can always be obtained with the following Hamiltonian $H_b(\vec{p}_b,\vec{q}_b,t)$:

$$\left.\begin{aligned} &H_b = \sum_{n=2}^{\infty} H_{bn} \\ &H_{b2} = \sum_{k=1}^{r} (\lambda_k\, p_{bk}\, q_{bk} + 0.5\,\delta_k\, q_{bk}^2) + \sum_{k=1}^{r-1} \varepsilon_k\, p_{b,k+1}\, q_{bk} \end{aligned}\right\} \tag{65}$$

with rules generalizing the Jordan rules:

$$\left.\begin{aligned} &\varepsilon_r = 0 \;\; ; \;\; \lambda_k \neq \lambda_{k+1} \Rightarrow \varepsilon_k = 0 \;\; ; \;\; \lambda_k = \lambda_{k+1} \Rightarrow \varepsilon_k = \text{either } 0 \text{ or } 1 \\ &\lambda_k \text{ and/or } \varepsilon_k \neq 0 \Rightarrow \delta_k = 0 \;\; ; \;\; \lambda_k = \varepsilon_k = 0 \Rightarrow \delta_k = \text{either } 0 \text{ or } 1 \end{aligned}\right\} \tag{66}$$

Hence:

$$\left.\begin{aligned} dp_{bk}/dt &= -\,\partial H_b/\partial q_{bk} = -\,\lambda_k\, p_{bk} - \varepsilon_k\, p_{b,k+1} - \delta_k q_{bk} - \partial \sum_{n=3}^{\infty} H_{bn}/\partial q_{bk} \\ dq_{bk}/dt &= \partial H_b/\partial p_{bk} = \lambda_k\, q_{bk} + \varepsilon_{k-1}\, q_{b,k-1} + \partial \sum_{n=3}^{\infty} H_{bn}/\partial p_{bk} \end{aligned}\right\} \qquad (67)$$

The equations (67) show that the 2r eigenvalues of the system are $\pm\lambda_1, \pm\lambda_2, \pm\lambda_3, \ldots \pm\lambda_r$, they are pairs of opposite. The last terms are the higher-order terms, they are periodic with the period T.

4.3 The near-resonance theorem for analytic Hamiltonian systems.

The near-resonance theorem helps to simplify the higher-order terms of (65) and (67), terms that can be written:

$$H_b = H_{b2} + \sum_{\vec{a},\vec{b},m} K_b(\vec{a},\vec{b},m)\, p_{b1}^{a_1} \cdots p_{br}^{a_r}\, q_{b1}^{b_1} \cdots q_{br}^{b_r} \exp\{m\omega_o it\} \qquad (68)$$

with:

$$\vec{a} = (a_1, a_2, \ldots, a_r)\ ;\ \vec{b} = (b_1, b_2, \ldots, b_r)\ ;\quad a_1, a_2, \ldots, a_r, b_1, \ldots, b_r \in \mathbb{N}$$

$$a_1 + a_2 + \ldots + a_r + b_1 + \ldots + b_r \geqslant 3\ ;\ m \in \mathbb{Z}\ ; \qquad (69)$$

$$K_b(\vec{a}, \vec{b}, m) \in \mathbb{C}\ ;\quad \omega_o = 2\pi/T$$

The near-resonance theorem for analytic Hamiltonian systems is then the following:

For any positive scalar ε, even very small, there exists an analytic and canonical transformation $(\vec{p}_b, \vec{q}_b) \to (\vec{p}_u, \vec{q}_u)$ such that:

I) $(\vec{p}_b, \vec{q}_b)$ and $(\vec{p}_u, \vec{q}_u)$ are equivalent:

$$\vec{p}_u - \vec{p}_b\ ,\quad \vec{q}_u - \vec{q}_b = 0\,(p_b^2, q_b^2) \qquad (70)$$

II) The transformation $(\vec{p}_b, \vec{q}_b) \to (\vec{p}_u, \vec{q}_u)$ may depend on t and is periodic in terms of t:

$$\vec{p}_u(\vec{p}_b, \vec{q}_b, t) \equiv \vec{p}_u(\vec{p}_b, \vec{q}_b, t+T);\quad \vec{q}_u(\vec{p}_b, \vec{q}_b, t) \equiv \vec{q}_u(\vec{p}_b, \vec{q}_b, t+T) \qquad (71)$$

III) The Hamiltonian $H_u(\vec{p}_u, \vec{q}_u, t)$ has much fewer terms:

$$H_u(\vec{p}_u, \vec{q}_u, t) = H_{u2}(\vec{p}_u, \vec{q}_u) + \sum_{\vec{a},\vec{b},m} K_u(\vec{a},\vec{b},m) \times$$
$$\times\, p_{u1}\, p_{u2} \cdots p_{ur}\, q_{u1} \cdots q_{ur} \exp\{m\,\omega_o it\} \qquad (72)$$

Indeed the K_u terms satisfy the near-resonance conditions:

$$\left.\begin{array}{l}\sum_{k=1}^{r} (a_k + b_k) = D \geq 3 \\ \left| m\omega_o i + \sum_{k=1}^{r} \lambda_k (b_k - a_k) \right| \leq \varepsilon D\end{array}\right\} \quad (73)$$

IV) The series of (72) is convergent for all t in a neighbourhood of the origin and the expression of H_{u2} is identical to that of H_{b2} given in (65) with the same λ_k's, δ_k's and ε_k's:

$$H_{u2} = \sum_{k=1}^{r} (\lambda_k p_{uk} q_{uk} + 0.5\delta_k q_{uk}^2) + \sum_{k=1}^{r-1} \varepsilon_k p_{u,k+1} q_{uk} \quad (74)$$

4.4 Critical and regular Hamiltonian systems.

The near-resonance theorem has many applications as those developed in the section 2.4.

Some very interesting applications are particular to the case of "critical and regular analytic Hamiltonian systems".

A Hamiltonian system is "regular" if all δ_k and ε_k of (65), (66) and (74) are zero. It is always the case if all eigenvalues $\pm\lambda_k$ are different.

A differential system is "critical" if its largest Liapounov characteristic exponent $R_e(\lambda_k)$ is zero and in the Hamiltonian case this implies that all Liapounov characteristic exponents $R_e(\lambda_k)$ are zero since the eigenvalues are pairs of opposite numbers.

Most usual differential systems for which the first-order analysis does not provide an answer to the question of stability are critical and regular Hamiltonian systems.

The "reduction to the origin" of section 4.1 is a real canonical transformation. Let us assume that we use again only real canonical transformations for the Floquet transformation of section 4.2; what is then the simplest possible autonomous function H_{b2}?

This function is:

$$H_{b2} = \sum_{k=1}^{r} \frac{\omega_k}{2} (p_{bk}^2 + q_{bk}^2) \quad (75)$$

The 2r eigenvalues $\pm\lambda_k$ are $\pm i\omega_k$ and we will call "pulsations" the r real scalars ω_k.

The corresponding first-order differential equations are then:

$$dp_{bk}/dt = -\omega_k q_{bk} \quad ; \quad dq_{bk}/dt = \omega_k p_{bk} \quad ; \quad k = \{1,2,\ldots r\} \tag{76}$$

Let us put:

$$z_{bk} = p_{bk} + iq_{bk} \quad ; \quad \bar{z}_{bk} = p_{bk} - iq_{bk} \quad ; \quad k = \{1,2,\ldots r\} \tag{77}$$

The first-order equations (76) are then equivalent to:

$$dz_{bk}/dt = i\omega_k z_{bk} \quad ; \quad d\bar{z}_{bk}/dt = -i\omega_k \bar{z}_{bk} \tag{78}$$

and the 2r parameters z_{bk}, $\bar{z}_{bk}$ play the role of the parameter z_k in (21) and p_{bk}, q_{bk} in (67).

Hence, with these new notations, the near-resonance theorem leads to real parameters $\vec{p}_N, \vec{q}_N$ equivalent to the new $\vec{p}_b$, $\vec{q}_b$ and to Hamiltonians $H_N(\vec{p}_N, \vec{q}_N, t)$ such that:

$$p_{Nk} + i\, q_{Nk} = z_{Nk} \quad ; \quad p_{Nk} - i\, q_{Nk} = \bar{z}_{Nk} \tag{79}$$

$$H_N(\vec{p}_N, \vec{q}_N, t) = \sum_{k=1}^{r} \frac{\omega_k}{2}(p_{Nk}^2 + q_{Nk}^2) + \sum_{(\vec{\alpha},\vec{\beta},m)} K_N(\vec{\alpha},\vec{\beta}, m) \times$$

$$\times\, z_{N1}^{\alpha_1} \ldots z_{Nr}^{\alpha_r}\, \bar{z}_{N1}^{\beta_1} \ldots \bar{z}_{Nr}^{\beta_r} \exp\{m\omega_o it\} \tag{80}$$

with reality conditions on coefficients K_N:

$$K_N(\vec{\beta},\vec{\alpha}, -m) = \bar{K}_N(\vec{\alpha},\vec{\beta}, m) \tag{81}$$

and with the near-resonance conditions:

$$\left.\begin{aligned} &\vec{\alpha} = (\alpha_1, \alpha_2, \ldots, \alpha_r) \; ; \; \vec{\beta} = (\beta_1, \beta_2, \ldots, \beta_r) \; ; \\ &\alpha_1, \alpha_2 \ldots \alpha_r, \beta_1 \ldots \beta_r \in \mathbb{N}; \sum_{k=1}^{r} (\alpha_k + \beta_k) = D \text{ (degree)} \geq 3 \\ &\left| m\omega_o + \sum_{k=1}^{r} (\alpha_k - \beta_k)\, \omega_k \right| \leq \varepsilon D \end{aligned}\right\} \tag{82}$$

Let us recall that m is an arbitrary integer ($m \in \mathbb{Z}$) while ω_o is the pulsation of the period T (that is $\omega_o = 2\Pi/T$).

We can notice the two following properties:

I) $\sum_{k=1}^{r} \frac{\omega_k}{2} (p_{Nk}^2 + q_{Nk}^2)$ can be written $\sum_{k=1}^{r} \frac{\omega_k}{2} z_{Nk} \bar{z}_{Nk}$,

the Hamiltonian H_N can then be considered as $H_N(\vec{z}_N, \vec{\bar{z}}_N, t)$ with the obvious notations:

$$\vec{z}_N = z_{N1}, z_{N2}, \ldots, z_{Nr} \quad ; \quad \vec{\bar{z}}_N = \bar{z}_{N1}, \bar{z}_{N2}, \ldots, \bar{z}_{Nr} \tag{83}$$

and with the simple following "formal" differential equations:

$$dz_{Nk}/dt = 2i\ \partial H_N/\partial \bar{z}_{Nk} \quad ; \quad d\bar{z}_{Nk}/dt = -2i\ \partial H_N/\partial z_{Nk}; \ k = \{1,\ldots,r\} \tag{84}$$

(the "formal partial derivative" $\partial(z^\alpha \bar{z}^\beta)/\partial z$ is $\alpha z^{\alpha-1}\ \bar{z}^\beta$).

II) Consider the terms with a degree D smaller than or equal to the integer N; their number is finite and for sufficiently small ε only the strictly resonant terms remain, i.e. the terms for which:

$$m\omega_o + \sum_{k=1}^{r} (\alpha_k - \beta_k)\omega_k = 0 \tag{85}$$

Thus, for a given ε, we will define the integer N and the Hamiltonian $H_N(\vec{p}_N, \vec{q}_N, t)$ such that the first non-strictly resonant term be of degree N + 1.

Notice that $D \leq N$ implies not only (85) but also "$K_N(\vec{\alpha}, \vec{\beta}, m)$ is independent of N". These coefficients K_N with $D \leq N$ are the generalization of the "Birkhoff invariant coefficients" of Hamiltonian systems with one or two degrees of freedom.

The comparison of $\vec{p}_N, \vec{q}_N$ and $\overrightarrow{p_{N+1}}$, $\overrightarrow{q_{N+1}}$ shows that their differences are of Nth-order while H_N and H_{n+1} have a difference of $(N+1)^{th}$-order and an identical expression up to the degree N.

4.5 The quasi-integrals. The generalized Birkhoff differential rotation.

Let us assume that the (r+1) pulsations $\omega_o, \omega_1, \ldots \omega_r$ have no rational relations, i.e. that:

$$\text{with:} \quad \left. \begin{array}{l} \{\vec{c}\,\vec{\omega} = 0 \ ; \ \vec{c} \neq 0 \quad \text{is impossible}\} \\ \vec{\omega} = (\omega_o, \omega_1, \ldots \omega_r) \ ; \ \vec{c} = (c_o, c_1, \ldots, c_r) \\ c_o, c_1, \ldots c_r \in \mathbb{Z} \ ; \ \vec{c}\,\vec{\omega} = \sum_{k=0}^{r} c_k \omega_k \end{array} \right\} \tag{86}$$

Hence the relation (85) implies:

$$m=0 \quad ; \quad \alpha_k = \beta_k \quad ; \quad k = \{1,2,\ldots r\} \tag{87}$$

and, up to the order N, the Hamiltonian $H_N(\vec{p}_N, \vec{q}_N, t)$ or $H_N(\vec{z}_N, \vec{\bar{z}}_N, t)$ has only strictly resonant terms and is only a function of the r products $z_{Nk}\,\bar{z}_{Nk}$, that is the r sums $(p_{Nk}^2 + q_{Nk}^2)$.

Let us define the real non negative quantities w_{Nk} by:

$$w_{Nk} = z_{Nk}\ \overline{z}_{Nk} = p_{Nk}^2 + q_{Nk}^2 \quad ; \quad k = \{1,2,\ldots r\} \tag{88}$$

hence, with $\vec{w}_N = (w_{N1}, w_{N2}, \ldots W_{nr})$:

$$H_N = H_N(\vec{w}_N) + (N+1)^{th} \text{ - order} \tag{89}$$

and, with (84), for $k = (1,2,\ldots r)$:

$$\left.\begin{aligned} d\, z_{Nk}/dt &= 2i\ z_{Nk}\,.\ \partial H_N/\partial w_{Nk} + N^{th}\text{-order} \\ d\, \overline{z}_{Nk}/dt &= -2i\ \overline{z}_{Nk}\,.\ \partial H_N/\partial w_{Nk} + N^{th}\text{-order} \\ d\, w_{Nk}/dt &= (N+1)^{th}\text{-order} \end{aligned}\right\} \tag{90}$$

We will call "quasi-integral" <u>a sequence</u> as $w_{3k}, w_{4k}, w_{5k} \ldots$ with the following properties:

$$\forall\ N\ ;\ (w_{(N+1)k} - w_{Nk}) \text{ and } dw_{Nk}/dt \text{ are of } (N+1)^{th}\text{-order} \tag{91}$$

If this sequence converges its limit is a true integral of motion, however in most cases these sequences are diverging.

The "angular velocities" of z_{Nk}'s are $2\ \partial H_N/\partial w_{Nk} + (N-1)^{th}$-order, they are also quasi-integrals (with the main terms equal to the pulsations ω_k) and this leads to the "generalized Birkhoff differential rotation" of figure one.

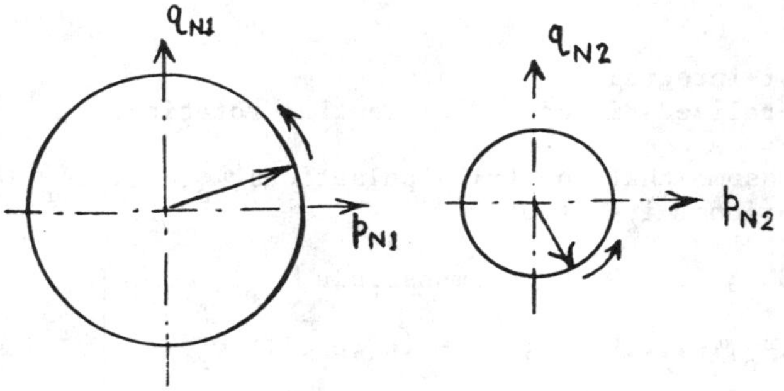

Figure 1. The generalized Birkhoff differential rotation.

The r vectors $\overrightarrow{(p_{Nk}, q_{Nk})}$ have constant moduli and are rotating in their complex z_{Nk} planes with constant rates of rotation, rates that are only functions of the r moduli.

This picture is true to any order and leads to the "all-order stability" that is not necessarily a true stability but implies at worse an "Arnold diffusion".

4.6 Other quasi-integrals

What happens if the (r+1) pulsations $\omega_o, \omega_1 \ldots \omega_r$ have one or several rational relations? For instance if they have s independent rational relations:

$$\left.\begin{array}{l} \vec{c}_j \; \vec{\omega} = 0 \quad ; \quad \vec{c}_j = c_{jo}, c_{j1} \ldots c_{jr} \quad ; \quad j = \{1,2,\ldots s\} \\ \text{The } \vec{c}_j \text{ vectors are independent and have integer components} \end{array}\right\} \qquad (92)$$

In this case the w_{Nk} of the previous section are generally no longer quasi-integrals but (r-s) independent linear functions of them remain quasi-integrals.

Let us look for the vectors $\vec{e}$ verifying the following:

$$\left.\begin{array}{l} \vec{e} = (0, e_1, e_2, \ldots e_r) \\ \forall \; j = \{1,2,\ldots s\} \; : \; \vec{c}_j \; \vec{e} = 0 \end{array}\right\} \qquad (93)$$

The number of independent $\vec{e}$ vectors is (r-s) and it is easy to verify with (84) and (85) that the functions

$$s_N = \sum_{k=1}^{r} e_k \, w_{Nk} \quad \text{are quasi-integrals.}$$

If all components $e_1, e_2, \ldots, e_r$ can be chosen positive the quasi-integral s_N will only allow bounded variations of the real and non-negative quantities w_{Nk} and the system will have another type of "all order stability". This analysis emphasizes the importance of the signs of the r pulsations $\omega_1, \omega_2, \ldots \omega_r$.

Another simple and very interesting quasi-integral is the following.

Let us define the real scalars P_{Nk}, Q_{Nk} and the complex scalars $Z_{Nk}, \overline{Z}_{Nk}$ by:

$$\left.\begin{array}{l} Z_{Nk} = P_{Nk} + i \, Q_{Nk} = z_{Nk} \exp(-i\omega_k t) \\ \overline{Z}_{Nk} = P_{Nk} - i \, Q_{Nk} = \overline{z}_{Nk} \exp(i\omega_k t) \end{array}\right\} \qquad (94)$$

The transformation $(\vec{p}_N, \vec{q}_N)$ to $(\vec{P}_N, \vec{Q}_N)$ is canonical and leads to the following Hamiltonian

$\mathcal{H}_N(\vec{P}_N, \vec{Q}_N, t)$ or $\mathcal{H}_N(\vec{Z}_N, \vec{\overline{Z}}_N, t)$:

$$\mathcal{H}_N(\vec{P}_N, \vec{Q}_N, t) = H_N(\vec{p}_N, \vec{q}_N, t) - \sum_{k=1}^{r} \frac{\omega_k}{2} (p_{Nk}^2 + q_{Nk}^2) \qquad (95)$$

The new Hamiltonian $\mathcal{H}_N(\vec{P}_N, \vec{Q}_N, t)$ has no more second-order terms but keeps the higher-order terms of $H_N(\vec{p}_N, \vec{q}_N, t)$; it is no more periodic in terms of t but, because of (85), it is autonomous up to the Nth-order.

The autonomous Hamiltonians are usual integrals of motion and the derivative of $\mathcal{H}_N(\vec{P}_N,\vec{Q}_N,t)$ is of $(N+1)^{th}$-order, hence the sequence of Hamiltonians $\mathcal{H}_N$ is a quasi-integral.

For sufficiently large N (for $N \geq M$) the main parts of the Hamiltonians $\mathcal{H}_N(\vec{P}_N,\vec{Q}_N\ t)$ are the same, they are a real autonomous and homogeneous polynomial $h_M(\vec{P}_N,\vec{Q}_N)$ of degree M.

The problem of the motion with the Hamiltonian $k_M(\vec{P}_N,\vec{Q}_N)$ is the "main problem" of higher-order effects, many of its characteristics also belong to the general problem. For instance if the main problem has asymptotic solutions the general problem has similar asymptotic solutions and is unstable.

In most problems M is equal to 4 because the terms in $z_{4k}^2\ \bar{z}_{4k}^2$ or in $z_{4k}\ \bar{z}_{4k}\ z_{4j}\ \bar{z}_{4j}$ satisfy always the resonance relation (85).

If M is infinite, i.e. if for all N the Hamiltonian $\mathcal{H}_N$ is of $(N+1)^{th}$-order, the quasi-integral $\mathcal{H}_N$ disappears but fortunately all P_{Nk} and Q_{Nk} are then quasi-integrals.

The (r-s) quasi-integrals $s_N = \Sigma\, e_k\, w_{Nk}$ and the quasi-integral $\mathcal{H}_N$ allow very often to conclude to the "all-order stability" (the "main problem" has then of course no asymptotic solutions). However there remain many unsolved cases, even if the conditions of existence of Arnold tori [Ref.3] imply the "all-order stability".

5. THE SIX MAIN TYPES OF STABILITY AND INSTABILITY.

We have already met the six following types of future evolution for the set of neighbouring solutions of the periodic solution of interest.

A) The exponential instability.

For instance: $dx/dt = \lambda x \quad ; \quad Re(\lambda) > 0$

B) The power-n instability.

For instance: $dx/dt = x^n$

C) The Arnold diffusion

D) The non-asymptotical stability

E) The power-n stability

For instance: $dx/dt = -x^n$

F) The exponential stability

For instance: $dx/dt = \lambda x \quad ; \quad Re(\lambda) < 0$

The types A,B,C are unstable and the types D,E,F are stable but they have very large differences.

The first-order analysis leads either to the type A or to the types B,C,D,E (critical types) or to the type F.

The types E and F are impossible for Hamiltonian systems and even for conservative systems (i.e. systems for which the sum of Liapounov characteristic exponents $Re(\lambda_k)$ is zero).

Periodic solutions with a "main problem" having asymptotic solutions (in the past) are of type B; they have a power-n instability with $n = M - 1$.

The types C and D correspond to the usual "all-order stability"; they are very difficult to discriminate. However physically the Arnold diffusion is not very different from stability and the time necessary to go from the infinitesimal distance ε to some bounded distance D is of the order of $Log(1/\varepsilon)$ in the type A, of the order of $\varepsilon^{(1-n)}$ in the type B and larger than all powers of $(1/\varepsilon)$ in the type C (the Arnold example [Ref 3] gives $\exp\{1/\sqrt{\varepsilon}\}$). In these conditions extremely large durations are easy to obtain and what would be the meaning of an instability that would require millions of years to appear?

For Hamiltonian problems it seems that the difference between the types C and D is the following:

If the periodic solution of interest is "all-order stable" and if the integrals of motion of the Hamiltonian system and its eventual decomposability allow us to reduce it either to integrable systems or to systems with one degree of freedom or to autonomous systems with two degrees of freedom, the type is the stable type D; if not it seems to be always the type C of Arnold diffusion.

Let us close this section by a beautiful symmetry hypothesis:

It seems that the periodic solutions of conservative problems have always the same type for the future evolution of the set of neighbouring solutions and for its past evolution. This can easily be verified at least for the types A and D (while then the types E and F are impossible).

6. AN EXTENSION OF THE NEAR-RESONANCE THEOREM

The demonstration of section 3 has been done for analytic equations of motion (21) with the analytic developments (22), (23) in power-Fourier series.

However the analyticity with respect to the time is not required for the demonstration and to the factor $1/(1-c_2\cos\omega_o t)$ of (44) we can substitute any converging Fourier series of $\omega_0 t$ with real and positive coefficients.

This extension allow to consider initial equations of motion (1)-(3) analytic in terms of $\vec{X}$ for all t and C_1 in terms of $(\vec{X},t)$ in some vicinity of the periodic solution of interest.

7. CONCLUSION

The near resonance theorem can be applied to the periodic solutions of analytic differential systems; it allows us to simplify very much the analysis of their stability and their vicinity, especially in the stable and unstable subsets.

This theorem has a particular version for Hamiltonian systems and allows a detailed analysis of critical and regular Hamiltonian systems with the emergence of the notion of quasi-integral and the classification of the different stable or unstable evolutions.

It would be interesting to improve the criteria of classification, to apply the theorem to concrete problems, to analyse the case of critical and singular Hamiltonians and to extend the theorem to quasi-periodic solutions.

REFERENCES

1 Siegel, C.L. "Über die Normalform analytischer Differentialgleichungen in der Nähe einer Gleichgewichtslösung". Nachr. Akad. Wiss. Göttingen, Math-Phys.Kl-p.21-30 1952.

2 Marchal, C. "The three-body problem, Appendix 3" Elsevier Scientific Publishing, to appear. 1988.

3 Arnold V.I., Avez A. "Problèmes ergodiques de la Méchanique Classique" Gauthier-Villars, Paris, 1967.

HOMOCLINIC AND HETEROCLINIC SOLUTIONS OF THE RTBP JOINING THE TRIANGULAR EQUILIBRIUM POINTS

Josep Masdemont
Departament de Matemàtica Aplicada,
E.T.S.E.I.B. Universitat Politecnica de Cataluyna,
Diagonal, 647, Spain.

INTRODUCTION

From Strömgren's time it has been known that some families of periodic orbits of the restricted three body problem (RTBP) end at an "orbit" formed by a pair of heteroclinic orbits connecting the two triangular equilibrium points.

In fact, and for the value of the mass parameter equal to one half, Strömgren computed five symmetric heteroclinic orbits ("asymptotic periodic orbits" according to the classical nomenclature) some of whose combinations by pairs are natural endings of well-known families of symmetric periodic orbits (see [4]).

In this way, looking at the evolution of periodic orbits, appeared Strömgren's conjecture (1935) according to which, a class of doubly asymptotic orbits are limit members of families of periodic orbits. In 1967 the conjecture seemed to be confirmed itself by the numerical experiments done by Danby, Szebehely, Nacozy and Van Flandern among others (see [1], [5], [6]), and in 1972 Henrard examined the conjecture in the more general framework of analytic Hamiltonian systems of two degrees of freedom. He gave sufficient conditions involving transversality (see [3]). Further results of Devaney prove that the phenomena occurs stably in both Hamiltonian and reversible systems (see [2]).

Our aim is to give a method for finding these orbits, and the classification of them.

METHOD OF WORK

(a) Classification of the orbits

The orbits have been classified according to the number of cuts that they have with $y = 0$.

We have	$2K+1$	Heteroclinic orbits	$K = 0,1,2,\ldots.$
	$2K$	Homoclinic orbits	

A. E. Roy (ed.), Long-Term Dynamical Behaviour of Natural and Artificial N-Body Systems, 449–455.
© 1988 by Kluwer Academic Publishers.

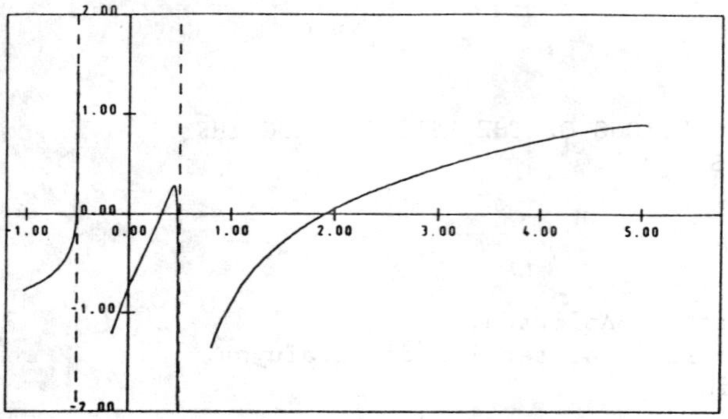

Figure 1

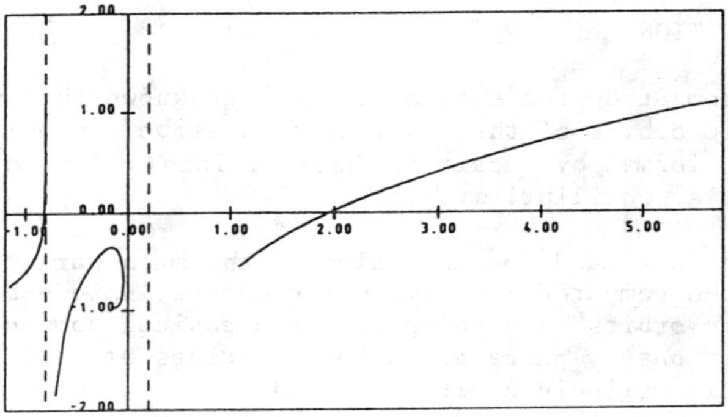

Figure 2

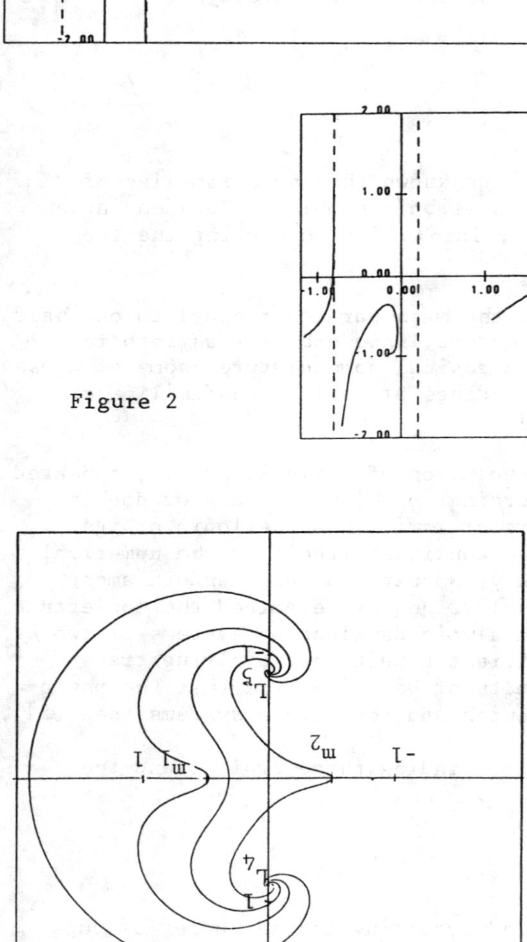

Figure 3

If some orbit is tangent with this axis, the cut is counted with its multiplicity.

(b) Globalization of the manifold

It is well known that dim $W^s(L_4)$ = dim $W^u(L_4) = 2$ where W^s and W^u denote the stable and unstable manifold respectively, and the orbits of these manifolds spiral around L_4. Looking towards a global view of the manifold, this suggests a good set of initial conditions taking x_0, y_0 in a small circle around L_4 and computing $\dot{x}_0, \dot{y}_0$ in such a way that $(x_0, y_0, \dot{x}_0, \dot{y}_0) \in W^u(L_4)$ (or $W^s(L_4)$). Then we follow the orbit until it cuts the x-axis; at this point we plot x and $\dot{x}$ in a graphic. This graphic determines the manifold, because at that point y=0 and $\dot{y}$ follows from the Jacobi's integral $\dot{x}^2 + \dot{y}^2 = 2\Omega - 3$.

Also because of the well-known symmetry of the vector field

$$(x,y,\dot{x},\dot{y},t) \xrightarrow{R} (x,-y,-\dot{x},\dot{y},-t).$$

We have that the unstable and stable manifolds of L_5 can be obtained by the symmetry of the L_4 opposite ones:

$$W^u(L_5) = R(W^s(L_4)) \qquad W^s(L_5) = R(W^u(L_4))$$

and so we need only $W^u(L_4)$ and $W^s(L_4)$.

Also by the symmetry, $\dot{x}=0$, $y=0$ implies symmetric heteroclinic orbit.

GRAPHICS OF THE MANIFOLDS AT THE FIRST CUT

Figs. 1, 2 represent the unstable manifolds at the first cut for $\mu = 0.5$ and $\mu = 0.2$ respectively.

The graphic has two kinds of discontinuities, one near the primaries, due to the collision orbits, and a second one due to the non-global character of y=0 as a surface of section, then any tangent orbit produces a discontinuity on the representation of the manifold with regard to close initial conditions. Anyhow both facts do not affect at all our purposes.

As we have noted before, when the graphic cuts $\dot{x} = 0$ we have an 1-heteroclinic orbit, in this way when $\mu = 0.5$ we have four 1 - heteroclinic orbits whilst $\mu = 0.2$ we have only two. These orbits are displayed in Figures 3, 4.

Also, the shape of the graphics gives us a clear view of the evolution with μ.

When we plot the manifold for $\mu = 0.1$, though the main shape remains, the graphic is essentially different to the previous one (see

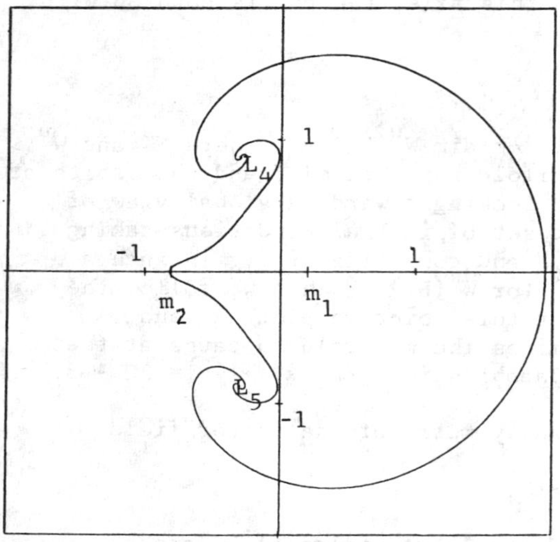

Figure 4

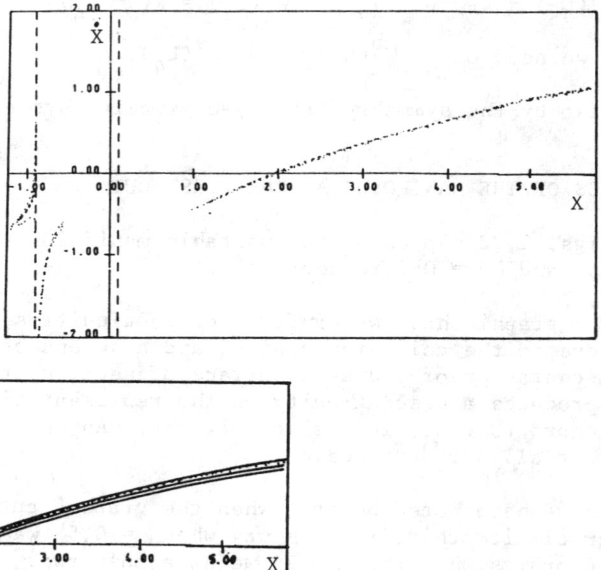

Figure 5

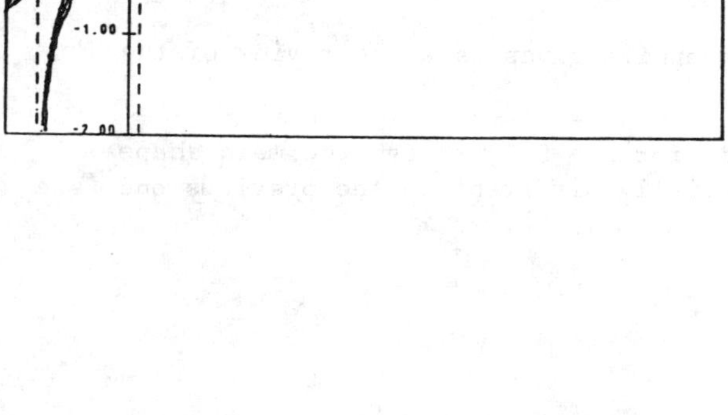

Figure 6

Figure 5).

We have a sequence of graphics tending to the main shape, and this implies an infinite number of 1-heteroclinic orbits close by, when y=0 to the "natural ones"inherited from the previous μ. In Figure 6 we can see this interpretation.

This fact is due to the existence of a 0-homoclinic orbit because the unstable and stable manifolds of L_4 intersect transversally before they cross the x-axis (see Figures 7, 8, 9,10).

GRAPHICS OF THE MANIFOLDS AT THE SECOND CUT

Looking at the unstable manifold at the second cut for $\mu = 0.5$ (see Figure 11), we find an interesting property, in some ways like the graphics of $\mu = 0.1$ that we have seen before.

The graphics have two parts well differentiated, a "proper one" and an"accumulation one" around $W^u(L_5)$ at first cut because as we can see it is the symmetrical of $W^s(L_4)$ at first cut (Figure 12). This part of the graphic is formed taking initial conditions on the unstable manifold of L_4 close to the 1-heteroclinic orbits from L_4 to L_5. These orbits after doing n-rounds around L_5 reproduce an approximation of the unstable manifold of L_5.

About the two-homoclinic orbits from L_4 tp L_5 which are exactly the intersection of the graphics represented in Figures 11, 12. Leaving aside some isolated ones, we have an infinite number of them with shape very similar to a pair of 1-heteroclinic orbits, one from L_4 to L_5 and the other from L_5 to L_4 (the limits found by Strömgren).

CONCLUSIONS

Proposition 1. For a fixed value of μ, the existence of a pair of non-degenerate 1-heteroclinic orbits, one from L_5 to L_4 and the other from L_4 to L_5, implies the existence of an infinite number of 2k-homoclinic (L_4 to L_4 and L_5 to L_5) and (2k+1)-heteroclinic (L_4 to L_5 and L_5 to L_4) orbits for k = 1,2,3,...

Proposition 2. Under the same hypothesis of proposition 1, the existence of a 0-homoclinic non-degenerate orbit ($L_4 \longrightarrow L_4$ or $L_5 \longrightarrow L_5$) implies the existence of an infinite number of 1-heteroclinic orbits $L_4 \longrightarrow L_4$ and $L_5 \longrightarrow L_4$.

Proposition 3. For the mass ratios explored, the number of 1-heteroclinic orbits is given in the following table:

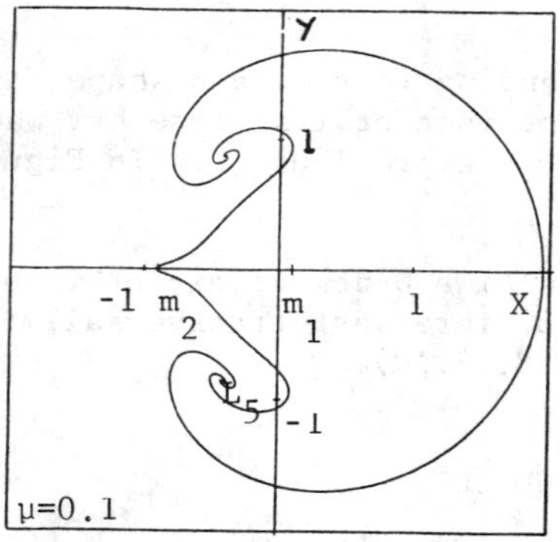

Figure 7

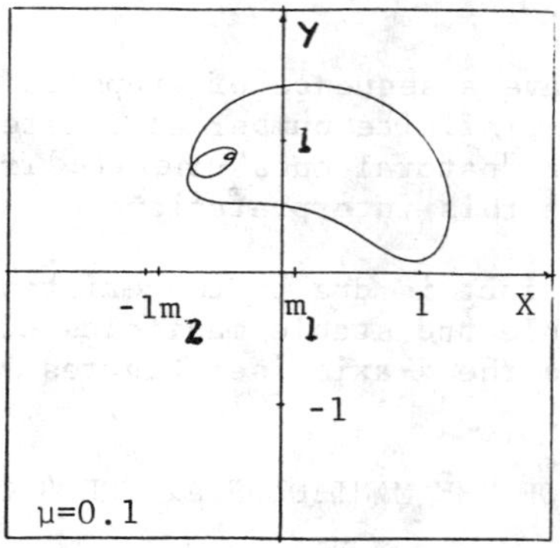

Figure 8

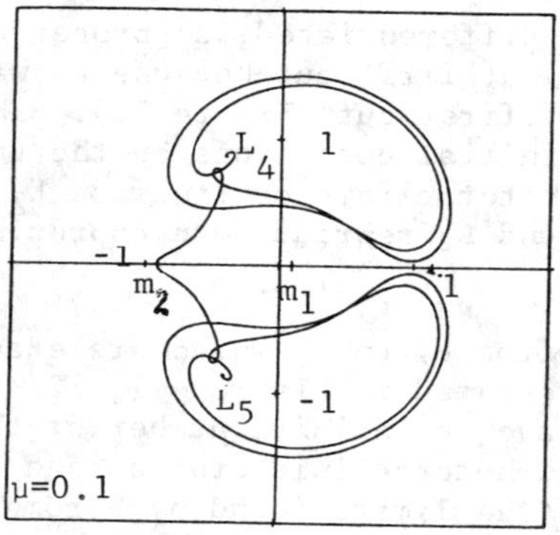

Figure 9

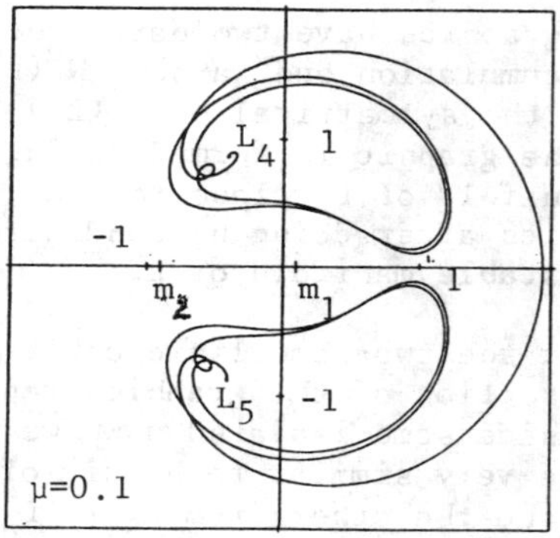

Figure 10

Figure 11

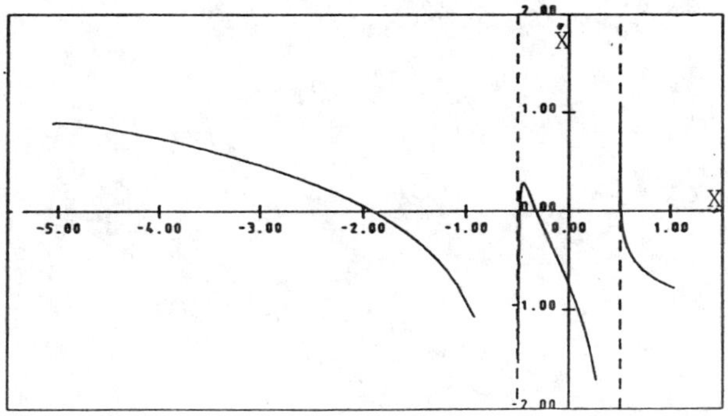

Figure 12

μ	$W^u(L_5) \cap W^s(L_4)$	$W^s(L_5) \cap W^u(L_4)$
0.5	4	4
0.4	3	4
0.3	2	2
0.2	2	2
0.1	∞	∞

Remark. Proposition 1 and proposition 2 are based in the fact that the unstable and stable manifolds cut themselves transversally, and on the understanding that at each step there is an infinite number of them involving the shape of the previous ones.

REFERENCES

[1] Danby, J.M.A.: Orbits in the Copenhagen Problem Asymptotic at L_4, and their Genealogy. The Astronomical J. 72, (1967), 198-201

[2] Devaney, R.L.: Blue Sky Catastrophes in Reversible and Hamiltonian Systems. Indiana Univ. Math. J. 26 (1977), 247-263.

[3] Henrard, J.: Proof of a Conjecture of E. Strömgren. Cel. Mechanics 7,(1973), 449-457.

[4] Strömgren, E.: Connaissance actuelle des orbites dans le problème des trois corps. Copenhagen Obs. Publ. 100 (1935).

[5] Szebehely, V. and Nacozy, P. A class of E. Strömgren's Direct Orbits in the Restricted Problem. The Astronomical J. 72 (1967), 184-190.

[6] Szebehely, V. and Van Flandern, T. : A Family of Retrograde Orbits around the triangular Equilibrium Points. The Astronomical J. 72 (1967), 373-379.

Institut für Reine
und Angewandte Kernphysik
Kiel

[illegible]

REFERENCES

[1] [illegible]

[2] [illegible]

[3] [illegible]

[4] [illegible]

[5] [illegible]

[6] [illegible]

Institut für Reine
und Angewandte Kernphysik
Kiel

DOUBLE COLLISION ORBITS AND SECOND SPECIES SOLUTIONS IN THE RESTRICTED THREE-BODY PROBLEM

M. Ollé
Departament de Matemàtica Aplicada
ETSEIB. Diagonal, 647
08028 Barcelona, Spain

ABSTRACT: The characteristic curves of double collision orbits for the circular and elliptic restricted three-body problem are given in the case of $\mu=0$. A numerical exploration of the characteristic curves of symmetric SSS for $\mu>0$, but small, in the circular RTBP is presented. The evolution of a particular family of symmetric SSS is described when varying the period.

INTRODUCTION

This communication is devoted to the study of some kinds of periodic orbits in the plane restricted three-body problem in the circular and elliptic cases.

This study is divided in two parts. In the first one, we describe the characteristic curves of double collision orbits (for $\mu=0$) and we give a proposition about the number of double collision orbits that exist, in the synodical system, when varying the energy $h = -1/(2a)$.

In the second part, we study the symmetric periodic second species solutions (SSSS) of the circular RTBP and we give the characteristic curves.

They are computed numerically, for small values of the mass parameter, stating at those orbits computed analytically for the case $\mu=0$.

1. DOUBLE COLLISION ORBITS

Let us study the orbits described by a particle P_3, which collides with both primaries P_1 and P_2 at two different epochs.

(i) We want to obtain the characteristic curves of these orbits and to do so, we consider the motion in the sidereal plane. We assume that P_3 describes a periodic rectilinear solution (e=1) and it begins its motion at the pericenter or apocenter of its orbit (according to ε"

A. E. Roy (ed.), Long-Term Dynamical Behaviour of Natural and Artificial N-Body Systems, 457–464.
© 1988 by Kluwer Academic Publishers.

equal to +1 or -1) while P_2 (without mass) describes an elliptic orbit (of eccentricity e_p) around P_1. We consider the cases $\varepsilon_p = \pm 1$ (P_2 begins at the pericenter or apocenter), so we have four initial configurations (see Fig.1).

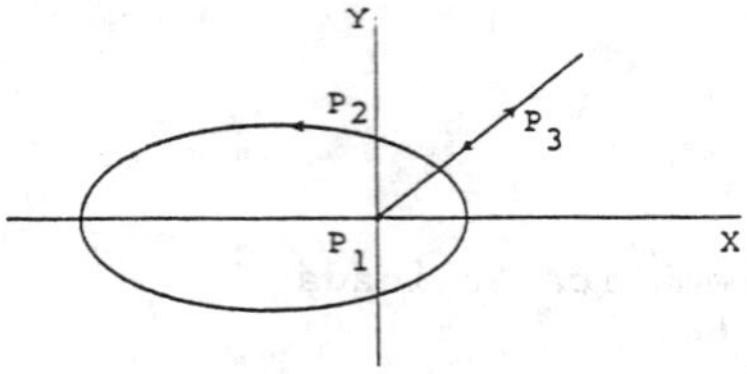

Figure 1: Rectilinear solutions in the sidereal plane (α=ctant. is the angle between the rectilinear solution and the X-axis).

When P_3 collides with P_2, we denote by τ(true anomaly of P_2), η (eccentric anomaly of P_3) and we get ([3])

$$\{2\,\mathrm{artg}(K_p^{-\varepsilon_p}\,\mathrm{tg}\,\tfrac{\tau}{2}) - \varepsilon_p e_p \sin[2\,\mathrm{artg}(K_p^{-\varepsilon_p}\,\mathrm{tg}\,\tfrac{\tau}{2})]\}$$

$$\times\,(1-\varepsilon'' \cos\eta)^{3/2} = \frac{(1-e_p^2)(\eta - e'' \sin\eta)}{(1+\varepsilon_p e_p \cos\tau)^{3/2}} \tag{1}$$

where $K_p = \left(\frac{1+e_p}{1-e_p}\right)^{\frac{1}{2}}$

The above formula is an implicit relation between τ and η. We have solved it numerically, and we show the diagram of solutions for $e_p = 0.5$ (see Fig.2). For the circular case, we obtain the characteristic equation of double collision orbits just by doing $e_p=0$ in (1). In the diagram of solutions, $\eta < 2\pi$ if $\varepsilon''=1$, and $\eta<3\pi$ if $\varepsilon'' = 1$, i.e., we want P_3 to collide with P_1 and P_2 and both collisions must occur before P_3 collides with P_1 again.

(ii) Let us now discuss how many double collision orbits exist, in the synodical system, when varying the value of the sidereal energy, $h = -1/(2a)$ or equivalently the value of the semiaxis a.

In the circular case, we know the following result (see [9]).

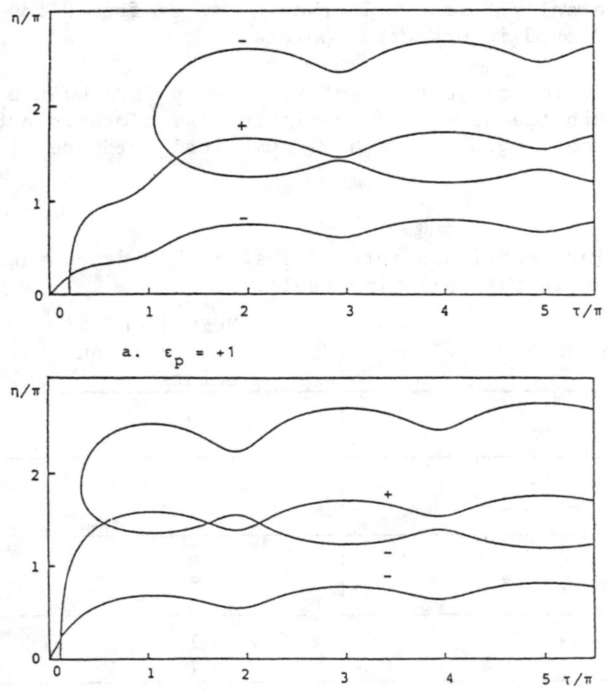

b. $\varepsilon_p = -1$

Figure 2: Diagram of solutions of equation (1) for $e_p = 0.5$. Each point represents a double collision orbit.

THEOREM: If $h<-1$, there exist no double collision orbits. If $h=-1$ there exist two double collision orbits. If $-1<h<0$ there exist two or four double collision orbits.

In general, for the circular case as well as for the elliptic one, there will be an even number of orbits, because if τ, η is a solution, so will $-\tau$, $-\eta$ (the symmetrical solution respect the synodical x-axis).

We have obtained a similar result for the elliptic case by studying the intersection of the curves defined by

$$\begin{aligned} \sigma - \varepsilon_p e_p \sin\sigma &= a^{3/2}(\eta - \varepsilon'' \sin\eta) \\ a(1 - \varepsilon'' \cos\eta) &= 1 - e_p \cos\sigma \end{aligned} \qquad (2)$$

when a value of a is fixed (σ is the value of the eccentric anomaly of the primaries when P_3 collides with P_2). For $\varepsilon''=1$, we have 0,1,2 intersection points if equations (2) don't intersect, intersect tangentially or transversally. If $\varepsilon''=-1$, then η varies from 0 to 3π and there may be 0, 1, 2 or 3 intersection points.

We have followed the tangential intersection points when e_p varies from 0 to 1, to obtain the number of double collision orbits according to the value of a (see Fig.3) so we have numerical evidence of the following

PROPOSITION

The number of intersection points of system (2), depending on the semiaxis a, is given in the following table:

ε_p	Semiaxis a	ε''	Number of intersection points
	$0<a<\frac{1-e_p}{2}$	1 -1	0 0
	$a=\frac{1-e_p}{2}$	1 -1	0 1
	$\frac{1-e_p}{2}<a<a^{++}$	1 -1	0 1
1	$a=a^{++}$	1 -1	1 1
	$a^{++}<a<a^{+-}$	+1 -1	2 1
	$a=a^{+-}$	+1 -1	2 2
	$a^{+-}<a$	+1 -1	2 3
	$0<a<a^{--}$	1 -1	0 0
	$a=a^{--}$	1 -1	0 1
-1	$a^{--}<a<a^{-+}$	1 -1	0 2
	$a=a^{-+}$	1 -1	1 2
	$a^{-+}<a<\frac{1+e_p}{2}$	1 -1	2 2
	$\frac{1+e_p}{2}\leq a$	1 -1	2 3

where a^{++}, a^{+-}, a^{-+}, a^{--} are those values of a for which there is a tangential intersection.

For each intersection point, we obtain two double collision orbits (plus the two symmetrical ones).

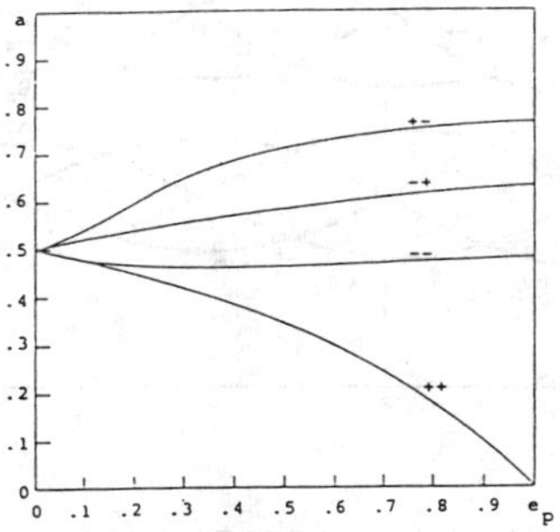

Figure 3: Tangential intersections of system (2). The pairs of signs represent those of ε_p and ε''.

2. SYMMETRIC SECOND SPECIES SOLUTIONS (SSSS)

According to Guillaume [4], a second species solution of the restricted problem is, roughly speaking, a solution which passes too close to the Moon to be considered as a weakly perturbed Keplerian orbit around the Earth.

The concept of second species solution was introduced by Poincaré (1899) and several authors have been analysing them. Let us summarize some of what it is known in the circular and elliptic RTBP.

In the circular restricted problem, Hénon [5] and Brjuno [1] studied the limiting case $\mu=0$, i.e., the orbits with consecutive collisions. For $\mu>0$, small, some numerical research has been done by Perko [7], Brjuno [2], and the analytical proof of their existence was given by Perko [8], using matching theory, and Henrard [6] (his proof is based on topological equivalence between differential systems in the neighbourhood of an equilibrium).

At present, we have done a systematic numerical extension from the case $\mu=0$ to $\mu > 0$ small, and we shall describe the characteristic curves of SSSS.

For the elliptic restricted problem, the characteristic curves of orbits with consecutive collisions ($\mu=0$) are described in [3].

As we mentioned above, our goal was to find the characteristic curves of SSSS for μ small, as an extension of the limiting case $\mu=0$ (see Fig.4).

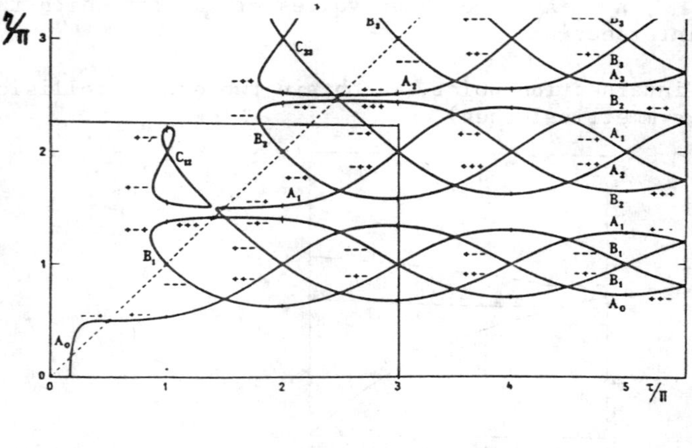

Figure 4: Hénon's diagram for orbits with consecutive collisions ($e_p = 0$).

So, we extended the families A_0, A_1, B_1, B_2, C_{12} for the value of half period less than 3π and we show the diagram of these curves for $\mu=10^{-6}$ in different variables T (half period) and C (Jacobi Constant) - Fig.5a -, X (the value of the synodical x-axis at the perpendicular crossing far from the smaller primary)and C - Fig.5b -. Each point (X,C), in principle, represents a SSSS of the circular restricted problem.

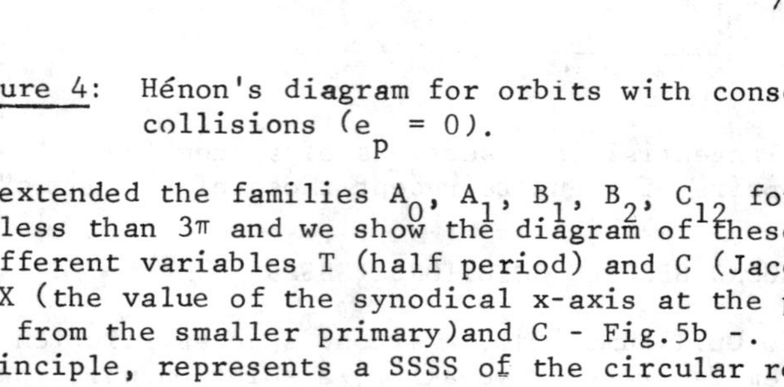

Figure 5a

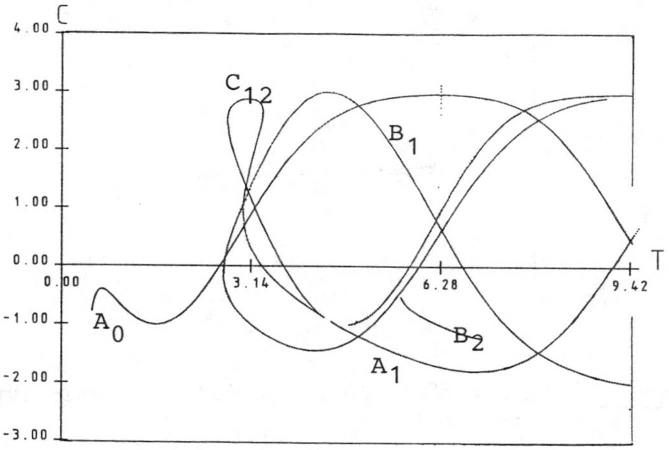

Figure 5b

<u>Figure 5</u>: Characteristic curves of SSSS ($\mu=10^{-6}$, $e_p=0$).

In particular, we have studied family A_0 with more detail (see Figure 6) and we can observe that it seems that there is a bifurcation near T = 2π .

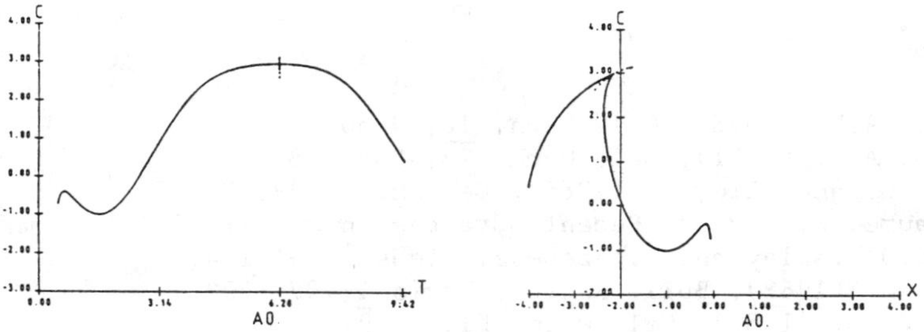

<u>Figure 6:</u> Family A_0 of SSSS ($\mu=10^{-6}$).

This behaviour is well revealed in Figure 7. There, we have plotted variables T and LC (the Levi-Civita coordinate u or v) and we see how the second perpendicular crossing (the one near the smaller primary p_2) goes further near T = 2π.

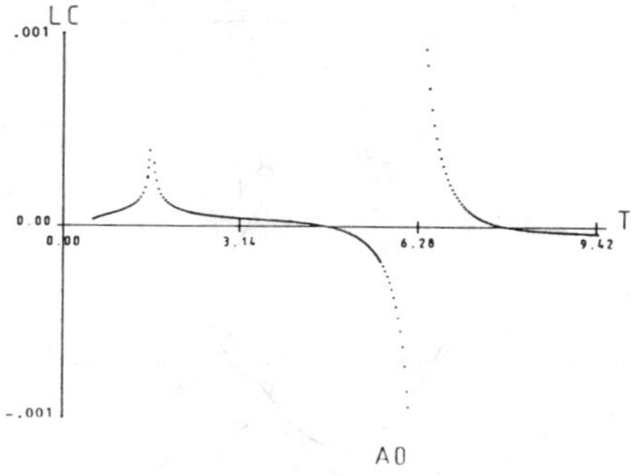

Figure 7: Family A_0. T(half period), LC (Levi-Civita coordinate).

Finally, if we follow the evolution of the SSSS when increasing T, the same behaviour is observed from π to 2π, 2π to 3π and so on. That is, from $j\pi$ (j odd) on, SSSS exist and the distance to P_2 becomes smaller up to $T = (j + \frac{1}{2})\pi$ where there is a collision orbit, then that distance begins to increase again and around $T = (j+1)\pi$, it seems there doesn't exist SSSS as an extension of orbits with consecutive collision ($\mu=0$).

BIBLIOGRAPHY

[1] Brjuno, A.P. (1978), Cel. Mech. 18, 9-50.
[2] Brjuno, A.P. (1981), Cel. Mech. 24, 255-268.
[3] Gómez, G. and Olle, M. (1986), Cel. Mech. 39, 33-55.
[4] Guillaume, P. (1973), Recent advances in Dynamical Astronomy. B.O. Tapley and V. Szebehely (eds.), 161-174.
[5] Hénon, M. (1968), Bull. Astron. Paris 1, 377-402.
[6] Henrard, J. (1980), Cel. Mech. 21, 83-97.
[7] Perko, L.M. (1965), Ph. D. Dissertation.
[8] Perko, L.M. (1974), Siam J. Appl. Math. 27, 200-237.
[9] Piñol, C. (1983), Master thesis (Universitat Autonoma Barcelona).

BIFURCATIONS AND INSTABILITIES IN THE RESTRICTED THREE-BODY PROBLEM

A. D. Pinotsis
Department of Astronomy
University of Athens
Greece

1. INTRODUCTION

We have studied the bifurcations of the main families of periodic orbits in the restricted three-body problem in the case of equal masses (μ=0.5). These are the families a,b,c,f,h,g,i,k,l and n (the families a,f,g are symmetric to b,h,i, respectively). We have studied these families near the critical points (Hénon 1965) b1,b2,b3,$b_2$6,$b_2$7 for the family b (where the subscript 2 means double periodic orbits), c1,c2,c5 for the family c, h5,$h_2$6,$h_2$9 for the family h, i3 for the family i, k2 for the family k, ℓ1,ℓ2, for the family ℓ and n2, $n_2$3,$n_2$4 for the family n, and we found the families that bifurcate from the above at the critical points.

2. BIFURCATIONS

We have found, first, the symmetric bifurcations of the main families of simple, double, and quadruple periodic orbits (Contopoulos and Pinotsis 1984, Pinotsis 1986, 1987). In some cases we have found indications that there are infinite period doubling bifurcations with bifurcation ratio $\delta \approx 8.72$, according to the Feigenbaum scenario. All the bifurcated families have in general large unstable regions.

We then considered the sequence of simple, double and quadruple non-symmetric periodic orbits that start at the points b1,c1,ℓ2 and $n_2$4 (Hénon 1965) and Z1, Bas, Nas (Pinotsis 1986, 1987). In all cases (except at the point $n_2$4, where the bifurcating family is unstable) we have found the Feigenbaum phenomenon.

The evolution and termination of various bifurcating families follow an interesting pattern. We consider, as an example, the sequence of bifurcations starting at the point b1 of the family b. The family b is composed of simple symmetric periodic orbits starting around the Lagrangian point L_1. The bifurcating family consists of non-symmetric simple periodic orbits (Figure 1). This new family is the short period family (SPO). Its characteristic has a second branch outside Figure 1. As C increases, the SPO family becomes unstable with Hénon stability parameter <u>a</u> = -1.

A. E. Roy (ed.), Long-Term Dynamical Behaviour of Natural and Artificial N-Body Systems, 465–469.
© 1988 by Kluwer Academic Publishers.

At this point (first bifurcation point), there is a new bifurcating family of double non-symmetric periodic orbits, with two branches. This bifurcation is followed by a cascade of periodic doubling bifurcations, forming a Feigenbaum sequence.

We continued the characteristic of the SPO family, looking for inverse Feigenbaum sequences. As C increases, the characteristic of the family continues further to the right and downwards, while the stability parameter a of the family continuously decreases. The value of a reaches large negative values and then it starts to increase again. At the point (x = -1.61536592, C = 3.0538105) there is a second bifurcation point, and the SPO family becomes stable again, with a = -1. A new family of non-symmetric periodic orbits bifurcates from the SPO family, towards smaller values of the Jacobi constant C, with two branches and double period. As C decreases, this new family becomes unstable and we found again the usual Feigenbaum cascade of period-doubling bifurcations (Figure 1). However the bifurcating families do not join the corresponding families generated on the right of the first bifurcation point.

As C increases beyond the second bifurcation point the SPO family is stable up to the first turning point, where the family loses its stability, with a = +1 and the family has a maximum C. Following the family beyond the first turning point (as C decreases) the instability of the family continuously increases and reaches a maximum value. Then the instability parameter a starts to decrease and the family becomes stable again, at the second turning point, with a = +1. The characteristic of the family has there a minimum C (Figure 1). After the second turning

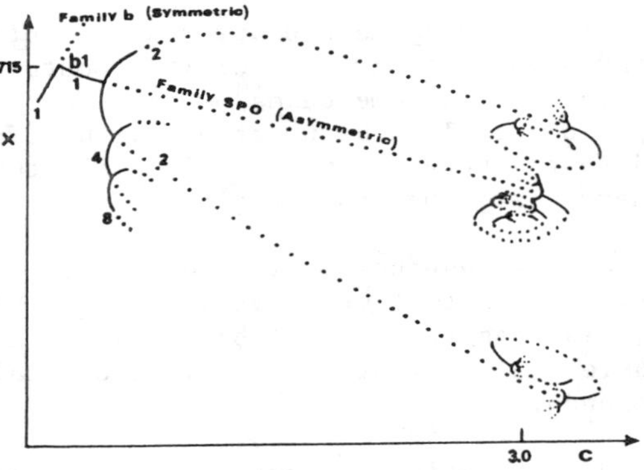

Figure 1: Bifurcations of the families of period 1,2,4,8, starting at the point B1, in the (x-C) plane, (schematically). (———)stable, (.....)unstable orbits.

point, the family has a small stable segment. At the point x = -1.6262116296, C = 2.9931158177) the family loses its stability, with a = -1 (third bifurcation point). We found here again the Feigenbaum scenario. After this point, for C increasing, the orbits become more and more unstable with a < -1, but after a while the parameter a starts to increase again and the family becomes stable after a goes through a = -1. At this fourth bifurcation point there bifurcates a new stable family of double non-symmetric periodic orbits. After this point the family has a stable segment up to a third turning point. At this turning point we have a maximum C and $\underline{z}$ = +1; then the SPO family becomes unstable again, as C increases, and so on.

It seems that the SPO characteristic spirals around a terminating point that corresponds to an asymptotic orbit, which can be considered as a periodic orbit with infinite period. The same phenomenon occurs in the other branch of the family. As one approaches asymptotically the terminating point, one passes through infinite changes from stability to instability and vice versa, while each bifurcation point generates infinite Feigenbaum cascades. The termination point is approximately at $x \approx -1.625$, C = 3.000. Namely, the first branch of the SPO family spirals around the point ($x \approx -1.625$, C = 3.00), while the second branch spirals around the point ($x \approx -0.684$, C = 3.000). Similar spirals have been found by Contopoulos (1988) along the short period family (SPO) in a galactic - type potential. Contopoulos found that such spirals are generated when the Lagrangian points L_4, L_5 become unstable. In our case the spirals should appear as soon as μ becomes larger than μ_1 .

Figure 2 shows a periodic orbit of the family b close to the bifurcating point b1 and three periodic orbits of the SPO family, near the first turning point, the third turning point and the fourth turning point correspondingly. Near the first turning point a cusp appears which turns to a loop. As C varies, the size of the loop of the orbit grows and encloses the triangular libration point L_4. Near the fourth turning point a new inner loop appears surrounding the Lagrangian point L_4 and so on. Thus, as we approach the terminating point we have a growing number of self-similar loops, one inside the other. The limiting orbit will be an infinite period spiral orbit.

We followed also the evolution of the family of double periodic orbits, generated at the first bifurcation point, beyond the bifurcation 2⟶4. We have found similar bifurcation behaviour, as in the case of the SPO family (Figure 1). Namely, as C increases, the two branches of the family extend to the right and are above and below the characteristic of the SPO family. So, the two branches pass outside the spirals of the SPO family. The instability of this family reaches a large negative value and then starts to increase continuously and at the second bifurcation point the family becomes stable again with $\underline{a}$ = -1. There bifurcates a new stable family of quadruple periodic orbits with two branches, towards smaller C, which becomes later unstable, and so on. Beyond the second bifurcation point the family of double periodic orbits has a stable part up to the first turning point, where it loses its stability (with $\underline{a}$ = +1 and the family has a maximum C), and so on. So, as C varies, its two branches spiral around two different points at C = 3.000, $x_{up} \approx -1.66$ or $x_{low} \approx -1.51$, for the upper or lower branch correspondingly, while each branch has infinite successive stable and unstable segments. We have here also self-similar orbits, which grow loops surrounding the Lagrangian point L_4 .

Similar phenomena, i.e. similar bifurcation behaviour, we have found in the cases of the bifurcation points ℓ2 and c1. As C varies, the orbits

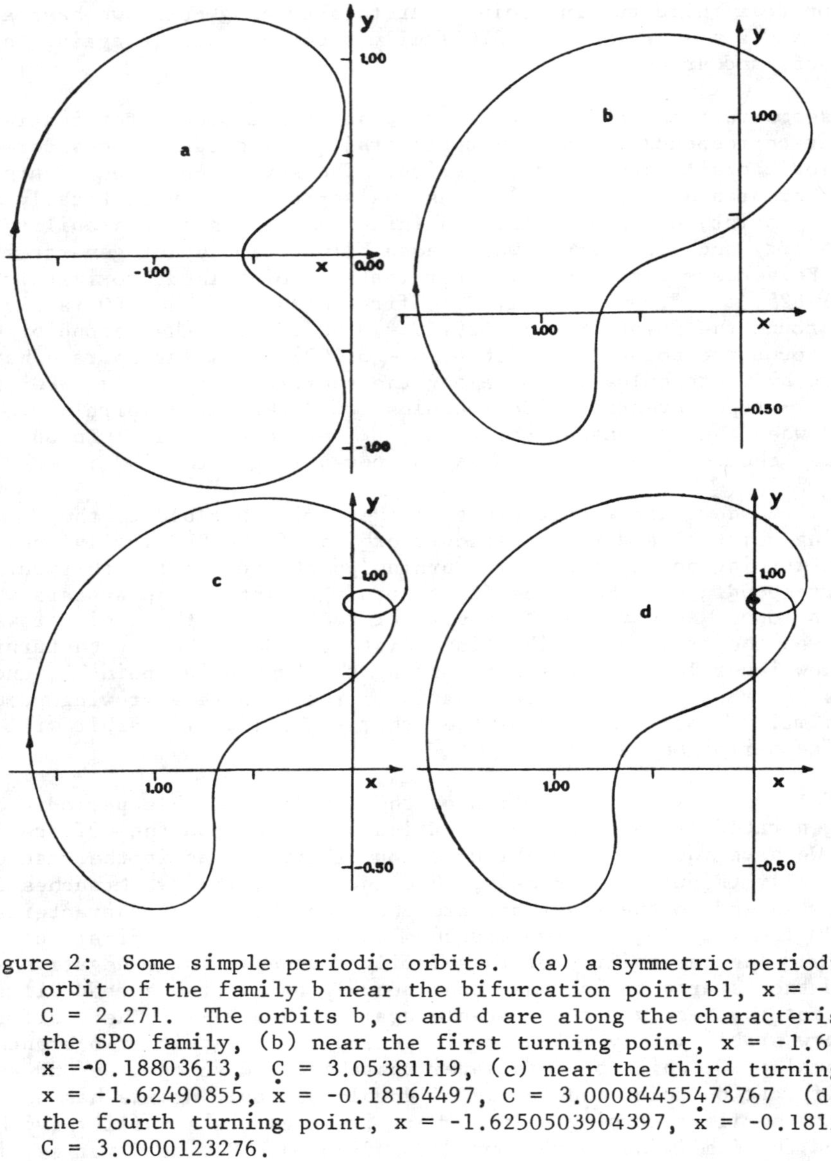

Figure 2: Some simple periodic orbits. (a) a symmetric periodic orbit of the family b near the bifurcation point b1, $x = -1.715557$, $C = 2.271$. The orbits b, c and d are along the characteristic of the SPO family, (b) near the first turning point, $x = -1.61534262$, $\dot{x} = -0.18803613$, $C = 3.05381119$, (c) near the third turning point, $x = -1.62490855$, $\dot{x} = -0.18164497$, $C = 3.00084455473767$ (d) near the fourth turning point, $x = -1.6250503904397$, $\dot{x} = -0.1815406045172$, $C = 3.0000123276$.

of the main families form new inner loops continuously, surrounding the Lagrangian point L_4 in the case of the bifurcation point ℓ2 and the Lagrangian point L_5 in the case of the bifurcation point c1.

We remark that every bifurcating family produces the same phenomenon, i.e. spirals and Feigenbaum sequences. Therefore we have infinite spirals and each one of them generates infinite new inner spirals and so on. Each member of these spirals reproduces a basic pattern. The maximum absolute value of the instability parameter a of each family is very large and becomes larger as the multiplicity of the family increases (of the order of 10^6 or larger). Therefore the appearance of infinite unstable families, whose characteristics spiral around different points is a new mechanism that introduces chaos.

ACKNOWLEDGEMENTS

I wish to thank Professor G. Contopoulos for his useful comments.

REFERENCES

Contopoulos, G.: 1988, in R. Dvorak (ed) "Long term evolution of planetary systems" (in press).
Contopoulos, G., Pinotsis, A.: 1984, Astron. Astrophys. 133, 49.
Hénon, M.: 1965, Ann. Astrophys. 28, 992.
Pinotsis, A.: 1986, Astron. Astrophys. 159, 231.
Pinotsis, A.: 1987, Astron. Astrophys. 174, 317.

THREE DIMENSIONAL EQUATIONS OF SZEBEHELY OF THE INVERSE PROBLEM AND FRENET REFERENCE FRAME

François Puel
Observatoire de Besançon

I. INTRODUCTION

In 1974 Szebehely obtained a linear first order partial differential equation for the force-function which gives rise to a one parameter family of planar orbits with a given total energy.

This equation is at the origin of many works on the inverse problem, and his explicit solution using Joukovsky's formula was studied by Broucke and Lass (1977), Molnar (1981), Puel (1985).

This equation was generalised to a two parameter family of tridimensional orbits by Erdi (1982), Bozis (1983), Puel (1984).

In 1983 Varadi and Erdi obtained by modern geometrical methods the solution of this system of two partial linear equations.

We attempted to resolve the same problem with elementary methods, to study the cases they neglect and to separate more clearly the geometrical terms and the energy dependant terms.

In most cases the formula of Joukovsky can not be extended to the three dimensional case. However there is a reference frame in which the equations are particularly simple and in which their signification is obvious, the frame of Frenet.

We define

- $\vec{t}$: the unit vector tangent to the trajectories
- $\vec{n}$: the principal normal towards the concavity
- $\vec{b}$: the binormal
- 1/R: the curvature of the trajectories

All we need is the classical result

$$\vec{t} \times \text{curl} \frac{n}{R} = - \frac{n}{R}$$

A. E. Roy (ed.), *Long-Term Dynamical Behaviour of Natural and Artificial N-Body Systems*, 471–475.
© 1988 by Kluwer Academic Publishers.

II FROM SZEBEHELY'S EQUATIONS TO THE INTRINSIC EQUATIONS

Let the two parameter known family of trajectories be defined implicitly by

$$f(M) = c \quad , \qquad g(M) = d \tag{1}$$

The energy of the corresponding motions is

$$h(M) = h(c,d)$$

The force function U is solution of Szebehely's equations (Broucke 1980, Erdi 1982, Puel 1984 a,b)

$$\begin{cases} \vec{\text{grad}}\, f.\ \vec{\text{grad}}\, U + 2\,(U + h)\, W_f = 0 \\ \vec{\text{grad}}\, g.\ \vec{\text{grad}}\, U + 2\,(U + h)\, W_g = 0 \end{cases} \tag{2}$$

with

$$W_f = \vec{t}.\ (\partial^2 f)\ \vec{t} = (\vec{\text{grad}}\, f, \vec{t}, \vec{\text{curl}}\, t)$$

$$W_g = \ldots$$

The system can be written in the form

$$\begin{cases} \left(\dfrac{\vec{\text{grad}}\, f}{W_f} - \dfrac{\vec{\text{grad}}\, g}{W_g}\right).\vec{\text{grad}}\, U = 0 \\ \dfrac{1}{2}\left(\dfrac{\vec{\text{grad}}\, f}{W_f} + \dfrac{\vec{\text{grad}}\, g}{W_g}\right).\ \vec{\text{grad}}\, U + 2\,(U + h) = 0 \end{cases}$$

We know (Puel 1984 a) that

$$W_f = \frac{\|\text{grad}\, f\|}{R_f} \quad , \qquad W_g = \frac{\|\text{grad}\, g\|}{R_g}$$

where $1/R_f$ and $1/R_g$ are the normal curvatures of the surfaces $f = cst$ and $g = cst$ along their intersections.

Using Meusnier's theorem, we establish that the vector

$$\frac{\vec{\text{grad}}\, f}{W_f} - \frac{\vec{\text{grad}}\, g}{W_g}$$

is parallel to $\vec{b}$, and that the vector

$$\frac{1}{2}\left(\frac{\vec{\text{grad}}\, f}{W_f} + \frac{\vec{\text{grad}}\, g}{W_g}\right)$$

can be expressed as $- R\,\vec{n} + \lambda\vec{b}$ (Needless to precise the value of λ).

Thus Szebehely's equations are equivalent to the intrinsic equations of motion

$$\begin{cases} R\ \vec{n}.\vec{grad}\ U = 2\ (U + h) \\ \vec{b}.\vec{grad}\ U = 0 \end{cases} \qquad (3)$$

We can express $\vec{t}$, $\vec{b}/R$, $\vec{n}/R$ as functions of

$\vec{grad}\ f$, $\vec{grad}\ g$, W_f , W_g :

$$\vec{t} = \frac{\vec{grad}\ f\ x\ \vec{grad}\ g}{\|grad\ f\ x\ grad\ g\|},\ \frac{\vec{b}}{R} = \frac{W_g\ \vec{grad}\ f - W_f\ \vec{grad}\ g}{\|grad\ f\ x\ grad\ g\|}\ \ldots\ldots$$

III. DERIVATION OF THE SOLUTION

We start from the intrinsic equation of motion

$$\vec{grad}\ U = U'_s\ \vec{t} + 2\ (U + h)\ \frac{\vec{n}}{R} \qquad (4)$$

where we note U'_s for $\vec{t}.\vec{grad}\ U$

Writing $\text{curl}\ \vec{grad}\ U = \vec{0}$ and projecting on $\vec{t}$, we obtain

$$A\ U'_s = -2B\ (U+h) + 2\ \frac{\vec{b}}{R}.\vec{grad}\ h \qquad (5)$$

with

$$\begin{cases} A = \vec{t}.\ \text{curl}\ \vec{t} \\ B = \vec{t}.\ \text{curl}\ \frac{\vec{n}}{R} = -\ \text{div}\ (\frac{\vec{b}}{R}) \end{cases}$$

We assume $A \neq 0$ then

$$\vec{grad}\ U = -2(U+h)\ \{\frac{B}{A}\ \vec{t} - \frac{\vec{n}}{R}\} + 2D(h)\vec{t} \qquad (6)$$

with

$$D(h) = \frac{\vec{b}}{AR}\ .\ \vec{grad}\ h$$

We write again $\text{curl}\ \vec{grad}\ U = \vec{0}$
We obtain after some calculations

$$(U+h)\text{curl}\ \{\frac{B}{A}\vec{t} - \frac{\vec{n}}{R}\} = (U+h)\vec{V}_o\ \ x\ \vec{t} = \vec{V}_1(h)\ \ x\ \vec{t} \qquad (7)$$

With

$$\vec{V}_o = \vec{grad}\ (\frac{B}{A}) - \frac{1}{A}\ \text{curl}\ \vec{t}\ x\ \text{curl}\ \frac{\vec{n}}{R}$$

$$\vec{V}_1(H) = \text{grad}\ (D(h)) - 3D(h)\ \frac{\vec{n}}{R} - \frac{B}{A}\ \vec{grad}\ h$$

Now we have U as a function of h and its derivatives by projection on the vectors $\vec{n}$ or $\vec{b}$.

We see that h is not arbitrary: we have

$h'_s=0$(of course) and

$$\vec{V}_1(h).\text{curl }\{\frac{B}{A}\vec{t} - \frac{\vec{n}}{R}\} = 0 \qquad (8)$$

IV SOME PARTICULAR CASES

1 - Previously, we excluded the case A = 0, that is the case of trajectories orthogonal to a one parameter family of surfaces.

In this case B = 0 (B = div (A t)...) and from eq.(5)

$$\vec{b}.\vec{\text{grad}}\, h = 0$$

In fact in this case we obtain the solution U by means of a tridimensional generalisation of Joukovsky's formula.

2 - The second important case is when

$$\text{curl }\{\frac{B}{A}\vec{t} - \frac{\vec{n}}{R}\} = \vec{0}$$

In this case h is a solution of the system

$$h'_s = 0 \quad , \quad \vec{V}_1(h) \times \vec{t} = \vec{0}$$

The inverse problem has solutions even if $\vec{\text{grad}}\, h = \vec{0}$, ($h = h_o$) and we obtain $U + h_o$, up to an arbitrary constant factor, by the equation

$$\vec{\text{grad}}\,(\ln(U + h_o)) = -2\{\frac{B}{A}\vec{t} - \frac{\vec{n}}{R}\}$$

In order to have the general solution with $\vec{\text{grad}}\, h \neq \vec{0}$ we add a particular solution of this inhomogeneous problem and the general solution of the problem with h = 0.

V. QUESTIONS OF COMPATIBILITY, CONCLUSION

In order to obtain the solution we assumed the force field that generate the given orbits to be conservative and we assumed we knew the corresponding energy h(M).

Yet we saw h(M) is not arbitrary (eq.8).

If we do not know H(M), we can derive Szebehely's equations and consider the system

$$\begin{cases} \vec{b}.\vec{\text{grad}}\, U = 0 \\ \vec{t}.\vec{\text{grad}}\,(R\, \vec{n}.\vec{\text{grad}}\, U - 2U) = 0 \end{cases} \qquad (9)$$

Thus we get one first order and one second order equation in U.

Equivalently we can consider a system of three first order equations in two unknown functions U and h.

$$\begin{cases} \vec{t}.\vec{grad}\ h = 0 \\ \vec{b}.\vec{grad}\ U = 0 \\ R\ \frac{\vec{n}}{2}\ .\vec{grad}\ U = U + h \end{cases} \qquad (10)$$

Bozis and Nakhla (1986) pointed out that an arbitrary two parameter family of three dimensional trajectories was generally incompatible with a conservative force field (i.e. generally the preceding systems have only solutions U and h such that U + h = 0).

Yet, we know two cases where trajectories are compatible with a conservative force field:

The case A = 0 (trajectories orthogonal to a family of surfaces) and the case

$$\text{curl}\ \{\frac{B}{A}\ \vec{t} - \frac{\vec{n}}{R}\} = \vec{0}$$

It would be interesting to express explicitly the general geometric condition of compatibility.

But perhaps do we have to give first a geometrical or physical (hydrodynamical ?) interpretation for the vector

$$\{\frac{B}{A}\ \vec{t} - \frac{\vec{n}}{R}\ \}$$

because it is very important for the discussion ?

REFERENCES

Bozis, G.: 1983, Celest. Mech. 31, 43.
Bozis, G. and Nakhla, A.: 1986, Celest. Mech. 38, 357.
Broucke, R. and Lass, H.: 1977, Celest. Mech. 16, 215.
Erdi, B. : 1982, Celest. Mech. 28, 209.
Molnar, S.: 1981, Celest. Mech. 25, 81.
Puel, F.: 1984a, Celest. Mech. 32, 209.
Puel, F.: 1984b, Celest. Mech. 32, 349.
Puel, F.: 1985, NATO ASI Series C, Ed. V. Szebehely, 154, 401.
Szebehely, V.: 1974, Rend. Sem. Fac. Sc. Univ. Cagliari. XLIV, suppl.31.
Varadi, F. and Erdi, B.: 1983, Celest. Mech. 30, 395.

FOURIER TRANSFORM IN ASTRODYNAMICS

John F. Segrest
Sparta, Inc., Huntsville, Al. USA

Joseph J. F. Liu
HQ Air Force Space Command, Colorado Springs, Co.,USA

Victor G. Szebehely
University of Texas, Austin, Tx., USA.

ABSTRACT

One of the fundamental properties of astrodynamics is the almost periodicity exhibited by the perturbed motion of natural and artificial satellites. For near Earth artificial satellites, in particular, this property has been demonstrated analytically by many researchers and their general perturbation theories are widely used by practitioners in the field of astrodynamics. The analytical representations of the nearly periodic perturbed motion are approximate solutions obtained by using perturbation techniques or transformation theories. These solutions generally can be obtained by first introducing a proper transformation to the equations of motion, e.g.,

$$\begin{aligned} x_i &= \bar{x}_i + \varepsilon P_{1i}(\bar{x}_m;\bar{y}_n) + \varepsilon^2 P_{2i}(\bar{x}_m;\bar{y}_n) + O(\varepsilon^3), \\ y_j &= \bar{y}_j + \varepsilon Q_{1j}(\bar{x}_m;\bar{y}_n) + \varepsilon^2 Q_{2j}(\bar{x}_m;\bar{y}_n) + O(\varepsilon^3). \end{aligned} \tag{1}$$

In the above equations, x and y stand for the original slow and fast variables, respectively, while the bar-elements represent the corresponding transformed variables. The perturbing terms are identified by the small parameter ε. The P's and Q's are the periodic variations having amplitudes which are functions of slow variables, the x's, and frequencies which are functions of fast variables, the y's. Following the substitution (or transformation), the equations of motion are significantly simplified. The transformed equations of motion can then be integrated analytically or integrated numerically with a much larger step size to obtain an approximate solution.

With this as background information, let us look into the theory of the Fourier transform. Historically, the Fourier transform and its digital version, the Fast Fourier Transform (FFT), have been widely used

A. E. Roy (ed.), Long-Term Dynamical Behaviour of Natural and Artificial N-Body Systems, 477–479.
© 1988 by Kluwer Academic Publishers.

in communications, radar and other fields dealing with functions or signals with definitive periodicities such as those discussed in the above equations. Application of FFT has become more popular with continued advances in high speed digital computers. Although in these fields the frequency is traditionally measured in cycles per second while the periodicity in astrodynamics is usually measured in revolutions per day, the mathematical concept is the same. With this in mind, the authors examined the feasibility of using the FFT to represent the numerically simulated orbital motion by identifying the magnitudes, frequencies and phase angles of the significant periodic variations. Once these characteristics are determined, the periodic variations are removed from the input data and the so-called "smoothed" variations are generated which may be numerically fitted with well-chosen functions of polynomials using least squares or other techniques. By retaining the meaningful periodic and secular variations, an approximated analytical representation of a given orbit may be accurately and efficiently provided without the use of a perturbation method. This numerically generated representation may then be used in various scientific and engineering analyses and in other applications without repetitious use of the special perturbations theory.

One additional factor motivating this study is the fact that for some difficult orbits, perturbation methods either generate slowly converging solutions or the orbital configurations invalidate the application of such methods. This study employs FFT and the state-of-the-art techniques in Signal Processing Techniques (SPT) to identify and remove the significant frequencies and to simplify the numerical analysis procedures.

A near circular orbit including only the perturbations due to the oblateness of the Earth is chosen as an example for an initial concept. The data base was generated by a special perturbations program and analyzed using FFT and SPT in the frequency domain. The first order periodic variations are successfully identified and represented while nonperiodic variations are also fitted numerically. The numerically determined analytical representation has the form

$$z_i = \sum_{i=0}^{q} a_{ik} t^k + \sum_{j=1}^{m} A_{ij} \cos (2\pi\nu_{ij} t + \phi_{ij}), \tag{2}$$

where the z's are a complete set of the six orbital elements. The a's and A's are the fitted coefficients of the secular variations and the amplitudes of the periodic variations, respectively. The ν's and φ's are the determined frequencies and the associated phase angles, respectively. The values of q and m are determined based on the required accuracy as well as the characteristics of the input data base. It should be noted that this analytical representation is given as a function of time explicitly and it should be interpreted differently with respect to those obtained by perturbation techniques (Eqs.1). This representation applies to one particular orbit only.

Although there are noted differences between this numerically determined analytical representation and the analytical solutions discussed earlier, the results obtained for the test case were compared with known first-order analytical solutions and demonstrated excellent agreement. Detailed procedures and numerical results for the near circular orbit are presented and discussed in this paper. The exercise shows the feasibility of the numerical technique and indicates some promise for further application of the Fourier transform in astrodynamics.

However, for more eccentric orbits, preliminary analyses show difficulties which take the form of numerous significant components in the spectra which require extensive computer time. The authors are presently examining the use of the true anomaly as the alternate independent variable for the study and expect that significant improvements may be realised. Additional studies concerning the generation of the data base (e.g., the frequency of data points, the choice of the independent variables, etc.) have been initiated. Other issues to be addressed are the length of the chosen data interval required to provide accuracy and optimal efficiency, and the necessary steps to further improve overall computer efficiency and limitations.

REFERENCES

[1] Segrest, J.F., Liu, J.F., and Szebehely, V.G., "Orbital Motion in the Frequency Domain", AIAA Paper 86-2003, AIAA/AAS Astrodynamics Conference, Williamsburg, VA, 18-20, 1986, pp 13-20, AIAA CP867.

[2] Rabiner, L.R. and Gold, B., Theory and Application of Digital Processing, Prentice-Hall, Inc., Englewood Cliff, N.J., 1974.

[3] Brigham, E.O., The Fast Fourier Transform, Prentice-Hall, Inc., Englewood Cliff, N.J., 1974.

[4] Oppenheim, A.V. and Schafer, T.W., Digital Signal Processing, Prentice-Hall, Inc., Englewood Cliff, N.J., 1975.

ESTIMATES OF THE ERROR IN NORMAL FORMS OF HAMILTONIAN SYSTEMS. APPLICATIONS TO EFFECTIVE STABILITY AND EXAMPLES

Carles Simó
Departament de Matemàtica Aplicada i Anàlisi
Universitat de Barcelona, Gran via 585,
Barcelona 08007, Spain.

ABSTRACT:

In this work we consider an n-degrees of freedom Hamiltonian system near an elliptic equilibrium point. The system is transformed to normal form with respect to a given modulus and up to a given order. Using standard techniques developed in [7,8] a lower estimate of the domain of convergence of the transformation as well as an upper estimate of the remainder are obtained. Then the notion of effective stability is introduced: We construct an initial ball (in a suitable norm) such that, after a given big time interval, a point starting in the ball still belongs to a slightly bigger ball. In other words, we bound the effect of diffusion on the motion. An optimization of the size of the ball with respect to several free parameters is performed. Then a simple application is done to a 3 degrees of freedom third degree Hamiltonian and, finally, to the effective stability of the $L_{4,5}$ points in the spatial Restricted Three Body Problem (RTBP), working out, explicitly, the Sun-Jupiter case.

1. INTRODUCTION

This paper contains an expository summary of work done in a joint project carried out at the Universities of Milano and Barcelona by L. Galgani and A. Giorgilli (Milano) and A. Delshams, E. Fontich and myself (at Barcelona) plus some improvements and additional technicalities. I refer to [8] for most of the general technical details. Further improvements following a different approach can be found in [21].

First we recall some facts about the stability of equilibrium points of Hamiltonian systems (in the sense of Lyapunov):

(a) If some of the eigenvalues of the linearized equations at the equilibrium point have non-zero real part, then the equilibrium is unstable.
(b) If all the eigenvalues are imaginary and different from zero (which

A. E. Roy (ed.), Long-Term Dynamical Behaviour of Natural and Artificial N-Body Systems, 481–503.
© 1988 by Kluwer Academic Publishers.

is a degenerate case) and the quadratic part of the Hamiltonian is definite, then the equilibrium point is stable [20].

(c) If all the eigenvalues are imaginary and different from zero, and the quadratic part of the Hamiltonian is indefinite we know that, under generic conditions, there are KAM tori [1,19,10] . There are then two relevant questions:
   c1) Up to which distance of the equilibrium point can we assure the existence of invariant tori?
   c2) What is the effect of the existence of the tori on the stability of the equilibrium point?

(d) I turn first to c2). Using either the KAM theorem for Hamiltonian systems of two degrees of freedom or the Moser twist theorem [15] for the related Poincaré section map, there are lots of results concerning lower bounds of the allowed perturbations based on careful analytical estimates as [19,10] or even computer assisted rigorous bounds [11,3,18] as well as upper bounds [14,12,17] for general or specific examples. I apologize for not including here a complete list of those contributions.

(e) For more than two degrees of freedom the possible existence of Arnol'd diffusion [2] prevents us from using KAM theory to give results concerning stability. Furthermore Arnol'd diffusion has been proved to be generic in the set of analytical Hamiltonians, at least at a level of microdiffusion [4] and some examples [5] and the general belief seem to confirm this fact.

(f) Taking for granted that we are faced with Arnol'd diffusion, the estimates of Nekhorosev [16] allow us to have bounds for the velocity of diffusion. However, to my knowledge, those bounds are pessimistic for specific examples and somewhat difficult to implement for practical purposes.

Then we follow another strategy having in mind that for systems of more than two degrees of freedom we cannot, in general, ask for stability of the equilibrium points. We can ask whether the mathematical instability is relevant or not for practical purposes. As a guiding example consider the $L_{4,5}$ points in the spatial RTBP, having in mind the application to the Sun-Jupiter case, for example. If we are able to prove the existence of some not too small ball around the equilibrium point such that a point initially in this ball stays in a somewhat larger ball for a time interval larger than the age of the solar system, then we do not care about instability. Hence the objectives of this paper are:

i) Given a Hamiltonian near an elliptic equilibrium point, a $\mathbb{Z}$-modulus and some order, to estimate the region of convergence of the normal form with respect to this $\mathbb{Z}$-modulus and up to this order.

ii) To bound the remainder of the Hamiltonian when the change to normal form is performed.

iii) To bound the deformation in the passage from old to new variables and in the inverse transformation.

iv) Given a suitable norm, to bound from below the time required to increase the norm by a given ratio.

(v) Given a time interval, to obtain the size of the ball such that the system remains on a slightly larger ball during all that time (effective stability ball).

vi) To give a simple application to a 3 degrees of freedom third degree Hamiltonian.

vii) To give a more complex application to the neighbourhood of $L_{4,5}$ in the spatial RTBP.

One should remark that there are many points which contribute to the fact that the estimates given here are still quite pessimistic:

1) The effect of the small divisors on the generating function, and hence on the remainder and on the transformation, is strong only on a few terms, but we bound it as if all the terms were affected by the smallest divisor at a given order. This can be improved by computing effectively the normal form up to some high order (as it is done in [21] for the Sun-Jupiter $L_{4,5}$ points). The present paper is free from effective computations of normal forms, but the nice effect of computing it to order 4 (which is very easy) is also discussed.
2) The norm used for the homogeneous polynomials in the generating function, in the transformation of coordinates and in the remainder gives easily a bound of those terms on a given ball, but that norm can be modified to a sharper one.
3) The bound of the norms of the homogeneous terms both in the remainder and in the deformations are rough. They can be improved (very much) if the generating function to a given order is explicitly known, and (a little) if some inequalities are studied more carefully.
4) The cancellations of terms in the remainder and in the transformations are not taken into account (and it is difficult to do it).

The introduction is closed with some notation and definitions.

Definition 1.1. The following norms are used for vectors:

(a) If $v = (q,p) \in \mathbb{R}^{2n}$ then $|v| = \max_j (q_j^2+p_j^2)^{\frac{1}{2}}$.

(b) If $w = (x,y) \in \mathbb{C}^{2n}$ then $|w| = \max_j (|x_j|^2+ |y_j|^2)^{\frac{1}{2}}$, where $|x_j|$ is the modulus x_j in $\mathbb{C}$.

Definition 1.2. Let E_k be the space of homogeneous polynomials of degree k in $(x,y) \in \mathbb{C}^{2n}$ with complex coefficients. Let $E = \oplus_k E_k$ (formal power series). If $f \in E$ we put $f = \sum_k f_k$ with $f_k \in E_k$.

For $f \in E_k$, if $f = \Sigma f_{\ell,m} x^\ell y^m$, $\ell,m \in \mathbb{Z}_+^n$, $|\ell+m| = k$, $|\ell! = \Sigma \ell_j$, $x^\ell = \prod_{j=1}^n x_j^{\ell_j}$, we define the norm of f as $\|f\| = \Sigma |f_{\ell,m}|$.

Definition 1.3. For $f,g \in E$ we define the Poisson bracket $\{f,g\}$ by $\sum_{j=1}^{n} \left(\frac{\partial f}{\partial x_j}\frac{\partial g}{\partial y_j} - \frac{\partial f}{\partial y_j}\frac{\partial g}{\partial x_j}\right)$. Then it is easily checked that if $f \in E_k$, $g \in E_{k'}$ one has $\|\{f,g\}\| \leq kk'.\|f\|.\|g\|$. For $g \in E_{k'}$ we denote by L_g the linear operator $L_g : E_k \to E_{k+k'-2}$ given by $L_g f = \{g,f\}$ for $f \in E_k$.

Let K be an analytical Hamiltonian that can be written as

$$K(q,p) = \frac{1}{2} \sum_{j=1}^{n} \omega_j (q_j^2 + p_j^2) + \ldots .$$

Define a linear canonical change S by $x_j = (q_j - ip_j)/\sqrt{2}$, $y_j = (-iq_j + p_j)/\sqrt{2}$, for $j=1,\ldots,n$, which has norm 1 using the norms of l_1^1.

Then the complex transformed Hamiltonian, H, is of the form

$$H(x,y) = i \sum_{j=1}^{n} \omega_j x_j y_j + \ldots = \sum_{k=2} H_k = \sum h_{\ell,m} x^\ell y^m \text{ with } |\ell+m| \geq 2.$$

Definition 1.4. We call M_ω the resonance modulus associated to H_2 defined by $M_\omega = \{\nu \in \mathbb{Z}^n, \nu.\omega = 0\}$, where the dot $\cdot$ denotes the scalar product. If $M_\omega = 0$ then H is said to be non-resonant at 0.

Definition 1.5. Given a $\mathbb{Z}$-modulus $M \supset M_\omega$, H is said to be in normal form up to order r with respect to M if it has the form $H = Z^{(r)} + R^{(r)}$, where

$$Z^{(r)} = \sum_{|\ell+m| \leq r,\ \ell-m \in M} a_{\ell,m} x^\ell y^m \quad , \quad R^{(r)} = \sum_{|\ell+m| > r} a_{\ell,m} x^\ell y^m .$$

We shall meet the equation $L_{H_2} G + Z = F$ for the unknowns G,Z for a given $F \in E_k$. We shall need

$$Z \in E_k^1 = \{f \in E_k : f = \sum_{|\ell+m| = k,\ \ell-m \in M} f_{\ell,m} x^\ell y^m\} \text{ for a given modulus}$$

$M \supset M_\omega$. Then we shall take $Z = \Pi_1(F)$, where Π_1 is the projection of E_k on E_k^1, and we can take $G_{\ell,m} = iF_{\ell,m}/\omega.(\ell-m)$ for $\ell-m \notin M$, $G^k_{\ell,m} = 0$ for $\ell-m \in M$.

§2. SUMMARY OF RESULTS CONCERNING ESTIMATES ON THE NORMAL FORM AND THE REMAINDER

In this section we proceed to present the basic results concerning the construction of the generating function of the canonical transformation to normal form up to a given order, the estimates on the normal form and, finally, the estimates on the remainder of the Hamiltonian. Full details are given in [8].

a) <u>The canonical transformation</u>

<u>Definition 2.1</u>. Given $G \in E$, $G = \sum_{k\geq 3} G_k$, that will be used as generating function, we define $T_G: E \to E$ as follows: if $f \in E$, $f = \sum_{k\geq 1} f_k$, then

$T_G f = \sum_{k\geq 1} F_k$, where $F_k = \sum_{l=1}^{k} f_{\ell,k-1}$ and $f_{\ell,0} = f_\ell$,

$f_{\ell,k} = \sum_{m=1}^{k} \frac{m}{k} L_{G_{2+m}} f_{\ell,k-m}$. Hence $F_k \in E_k$.

<u>Lemma 2.2</u>. Given G as before with $\|G_k\| \leq a^{k-3}b$ for $k \geq 3$, $a,b > 0$ and $f_\ell \in E_\ell$, then

$$\|f_{\ell,k}\| \leq C_{\ell,k} \|f_\ell\| \text{ for } \ell,k \geq 1, \text{ where } C_{\ell,k} = 3b \binom{\ell+k-1}{k}(3b+\tfrac{8}{3}a)^{k-1}.$$

Sketch of the proof: From 1.3 and the abounds on $\|G_k\|$ we have

$$\|f_{\ell,1}\| \leq 3\ell b \|f_\ell\|,$$

$$\|f_{\ell,k}\| \leq \frac{b}{k} \sum_{j=1}^{k-1} j(2+j)(\ell+k-j)a^{j-1} \|f_{\ell,k-j}\| + (2+k)\ell a^{k-1} b \|f_\ell\| \text{ for } k\geq 2.$$

Introduce $B_{\ell,k}$, $\ell,k \geq 1$ instead of $\|f_{\ell,k}\|$, 1 instead of $\|f_\ell\|$ and replace by equality. Comparing the expressions of $B_{\ell,k}$, $B_{\ell,k-1}$ one has

$B_{\ell,k} \leq \frac{\ell+k-1}{k}(3b+\frac{8a}{3})B_{\ell,k-1}$. One finds $B_{\ell,k} \leq C_{\ell,k}$ and this proves the Lemma.

<u>Proposition 2.3</u>. Let $G = \sum_{k\geq 3} G_k$, $f = \sum_{k\geq 1} f_k$, $\|G_k\| \leq a^{k-3}b$, $\|f_k\| \leq c^{k-1}d$, with $a,b,c,d > 0$, and $T_G f = F = \sum_{k\geq 1} F_k$, $R^{(r)} = \sum_{k>r} F_k$. Then:

(1) $\|F_k\| \leq d\,(3b+8a/3+c)^{k-1}$,

(2) $T^k_G f$ is convergent in any polydisc $D_R = \{(x,y) \in \mathbb{C}^{2n}, |(x,y)| \leq R\}$ with $R < R^*$, where $R^* = (3b+8a/3+c)^{-1}$.

(3) If for functions $|\cdot|$ denotes the supremum norm, then in D_R one has
$|T_G f| \leq (1-R/R^*)^{-1} Rd$, $|R^{(r)}| \leq (R/R^*)^r (1-R/R^*)^{-1} Rd$.

The proof is easily obtained from Lemma 2.2.

We define a (canonical) transformation from X,Y to x,y by $x_j = T_G X_j$, $y_j = T_G Y_j$, for $j=1,\dots,n$. Given $f \in E$, $f=f(x,y)$ we consider $F(X,Y) = f(T_G X, T_G Y)$. By the properties of the Lie series (see for instance [9]) one has $F(X,Y) = T_G f(X,Y)$. This is not only formal because of the absolute convergence given by 2.3.

<u>Proposition 2.4.</u> Let G as before and R* defined with any $c > 0$. Consider D_R with $R^*/R > \hat{\rho} > 1$ and let $(x,y) = T_G(X,Y)$. Then for $(X,Y) \in D_R$ one has

$$(1-\delta)\ |(X,Y)| \leq |(x,y)| \leq (1+\delta)\ |(X,Y)|,$$

where $\delta = (1+8a/(9b))^{-1}\ \sqrt{2}/(\rho-1)$.

This is simply obtained as in 2.3 where instead of f we use any one of the coordinate functions. The proposition 2.4 measures the maximum of the deformation introduced by the change of variables.

(b) <u>Construction of the generating function</u>

Till now we have supposed that the generating function, G, is known. However we should use the generating function that converts the Hamiltonian to normal form. We need a polynomial generating function $G^{(r)} = \sum_{k=3,\dots,r} G_k$. From the equation $T_{G^{(r)}} H = Z^{(r)} + R^{(r)}$ one should determine the unknowns $G^{(r)}$, $Z^{(r)}$ recursively.

This can be done using different formulations. The one given here is suitable for the bound desired [6].

<u>Proposition 2.5</u>. Let $H = \sum_{k\geq 2} H_k$, $H_2(x,y) = i \sum_{j=1}^{n} \omega_j x_j y_j$ and $M \supset M_\omega$. Then there exists a generating function $G^{(r)} = \sum_{k=3}^{r} G_k$ such that $T_{G^{(r)}}H$ is in normal form up to order r with respect to M. If the normal form is written as $Z^{(r)} = \sum_{k=2}^{r} Z_k$ one has

$$Z_2 = H_2, \text{ and } L_{H_2} G_k + Z_k = F_k \quad \text{for } k\geq 3, \text{ where } F_3 = H_3 \text{ and}$$

$$F_k = \sum_{m=1}^{k=3} \frac{m}{k-2} L_{G_{m+2}} Z_{k-m} + \sum_{m=1}^{k-2} \frac{m}{k-2} H_{m+2,k-m-2}, \quad \text{for } k \geq 4.$$

Sketch for low order terms: We have $T_G H = \sum_{k\geq 2} \sum_{m=0}^{k-2} H_{m+2,k-m-2}$.

Obviously $Z_2 = H_2$. From $Z_3 = H_{2,1} + H_{3,0} = L_{G_3} H_{2,0} + H_{3,0}$ we have $L_{H_2} G_3 + Z_3 = H_{3,0} = H_3$. For k=4 one has $Z_4 = H_{2,2} + H_{3,1} + H_{4,0} = \frac{1}{2} L_{G_3} H_{2,1} + L_{G_4} H_{2,0} + L_{G_3} H_{3,0} + H_{4,0}$.

Using $H_{2,1} = Z_3 - H_{3,0}$ we get

$$L_{H_2}G_4 + Z_4 = -\frac{1}{2}L_{G_3}H_{3,0} + \frac{1}{2}L_{G_3}Z_3 + L_{G_3}H_{3,0}+H_{4,0} = \frac{1}{2}L_{G_3}Z_3 + \frac{1}{2}H_{3,1}+H_{4,0},$$

as desired. Some manipulation of terms and rearrangement proves the general case.

(c) Estimates on the generating function and on the normal form

We assume, as before, $H = \sum_{k\geq 2} H_k$ with $\|H_k\| \leq c^{k-2}d$ for $k \geq 2$, $c,d > 0$. However, solving $L_{H_2}G_k+Z_k = F_k$ introduces a finite number of small denominators that should be bounded from below. If $M\supset M_\omega$ then there exist $\alpha_r > 0$ such that $|\nu\cdot\omega| \geq \alpha_r$ for $\nu \in \mathbb{Z}^n \setminus M$ and $|\nu| \leq r$.

Proposition 2.6. Let H as before with the bound $\|H_k\| \leq c^{k-2}d$ and $|\nu\cdot\omega| \leq \alpha_r$ as described. Then the generating function $G^{(r)} = \sum_{k=3}^{r} G_k$ for the transformation to normal form with respect to M up to order r satisfies the bound

$$\|G_k\| \leq \frac{cd}{\alpha_r}[6c(1+\frac{d}{\alpha_r})]^{k-3}(k-2)! \quad \text{for } k=3,\dots,r.$$

Simpler expressions are given by

$$\|G_k\| \leq \frac{cd}{\alpha_r}[6sc(1+\frac{d}{\alpha_r})]^{k-3} \quad \text{with } s = ((r-2)!)^{1/(r-3)} \qquad \text{and}$$

$$\|G_k\| \leq \frac{3cd}{\alpha_r}\left[\frac{12k}{5}c(1+\frac{d}{\alpha_r})\right]^{k-3}.$$

Sketch of the proof:

It follows easily $\|G_k\| \leq \frac{1}{\alpha_r}\|F_k\|$ and $\|Z_k\| \leq \|F_k\|$. Hence we should only produce an estimate of $\|F_k\|$. We look for positive numbers $\eta_1,\dots,\eta_{r-2}$; $\theta_{\ell,0},\dots,\theta_{\ell,r-\ell-2}$ for $0 \leq \ell \leq r-2$, such that $\|H_{2+\ell,k}\| \leq \theta_{\ell,k}cd$ and $\|F_{2+k}\| \leq \eta_k cd$. As $H_{\ell,0} = H_\ell$ and $F_3=H_3$ we can take $\theta_{\ell,0}=c^{\ell-1}$, $\eta_1=1$. From the definitions of F_k and $H_{\cdot,\cdot}$ we have two coupled recursive inequalities:

$$\theta_{\ell,k} \leq \frac{cd}{k\alpha_r}\sum_{j=1}^{k} j(2+j)(2+\ell+k-j)\eta_j\theta_{\ell,k-j}, \quad \text{for } k\geq 1,\ \ell\geq 0,$$

$$\eta_k \leq \frac{cd}{k\alpha_r}\sum_{j=1}^{r-1} j(2+j)(2+k-j)\eta_j\eta_{k-j} + \frac{1}{k}\sum_{j=1}^{k} j\,\theta_{j,k-j}, \quad \text{for } k \geq 2.$$

Let $a_{\ell,k}$, b_k be defined by $a_{\ell,0} = 1$, $b_1 = 1$ and

$$a_{\ell,k} = \frac{1}{k} \sum_{j=1}^{k} j(2+j)(2+\ell+k-j) b_j\, a_{\ell,k-j} \quad , \quad k \geq 1, \quad \ell \geq 0,$$

$$b_k = \frac{1}{k} \sum_{j=1}^{k-1} j(2+j)(2+k-j)\, b_j\, b_{k-j} + \frac{1}{k} a_{1,k-1} \quad , \quad k \geq 2.$$

To end the proof of the proposition we need two lemmas.

Lemma 2.7. For $\ell \geq 0$, $k \geq 1$ we have $\Theta_{\ell,k} \leq c_\ell c_2^{k-1} c^{\ell+k-1} a_{\ell,k}$, $\eta_k \leq c_2^{k-1} c^{k-1} b_k$, where $c_1 = d/\alpha_r$ and $c_2 = c_1 + 1$.

The proof of 2.7 requires several steps. In each one induction and rearrangement of terms should be used. We list the main steps:

i) $b_s = a_{1,s-1}$ for $s \geq 1$.

ii) $a_{1,k} \geq 3(k+1)(k+2) a_{1,k-1}/k$ for $k \geq 2$.

iii) $a_{\ell,s}/a_{\ell-1,s} \leq (\ell+s+1)/(\ell+1)$ for $\ell \geq 1$, $s \geq 0$.

iv) $a_{1,k-1} \geq \sum_{j=2,\dots,k} j a_{j,k}$.

Now, introducing $\bar{\Theta}_{\ell,k} = c_1 c_2^{k-1} c^{\ell+k-1} a_{\ell,k}$, $\bar{\eta}_k = c_2^{k-1} c^{k-1} b_k$ and using the properties of $a_{\ell,k}$, b_k one has $\bar{\Theta}_{\ell,k} \geq \Theta_{\ell,k}$ and $\bar{\eta}_k \geq \eta_k$ as desired.

At this point, and before ending the proof of the proposition, we remember that we should bound the functions F_k, that is, we should bound η_k. This is equivalent, according to 2.7, to bound b_k. We have

$$b_1 = 1 \;, \quad b_k = \frac{1}{k-1} \sum_{j=1}^{k-1} j(2+j)(2+k-j) b_j\, b_{k-j} \quad \text{for } k > 1.$$

Lemma 2.8. For $k \geq 1$ we have $b_k \leq 6^{k-1}\, k!$.
To prove 2.8 we note that for $k \leq 8$ the inequality is checked by direct computation. Assume it is true for $j=1,\dots,k$. Then, assuming k even (the case k odd is done in a similar way) and $m=k/2$, and putting together symmetrical terms (i.e. 1 and k, 2 and k-1, etc.) we have

$$b_{k+1} = 6^{k-1} \frac{k+1}{k} \sum_{j=1}^{m} \ell_j \;, \text{ where } \ell_j = (2+j)(3+k-j) j!(k+1-j)! \;.$$

Define $\bar{\ell}_j = \ell_j/\ell_{j+1}$. One checks $\bar{\ell}_j > 1$ for $1 \leq j \leq m-1$, and $\ell_1 > 3k/8$, $\ell_2 \geq k/4$ if $k \geq 4$. Then $b_{k+1}/6^{k-1}$ is bounded by $3(k+1)!(k+2)$ $[1+ \frac{8}{3k}(1+ \frac{4}{k}(\frac{k}{2} - 2))]/k$. Hence, to end 2.8, we require

$(k+2)[\,1+8(3k-8)/(3k^2)]/k \leq 2$, which holds for $k \geq 9$.

To end the proof of 2.6 we use $\|G_k\| \leq \frac{1}{\alpha_r}\|F_k\| \leq \frac{cd}{\alpha_r}\eta_{r-2}$
$\leq \frac{cd}{\alpha_r}(c_2c)^{k-3}6^{k-3}(k-2)!$.

The second and third bounds in 2.6 are obtained by bounding $(k-2)!$ by $((r-2)!)^{(k-3)/(r-3)}$ and by $3(k/2.5)^{k-3}$, respectively, for $k=3,\ldots,r$. which are easily established by induction and use of the Stirling's formula.

From the estimates on the generating function and using $a=\frac{12rc}{r}(1+\frac{d}{\alpha_r})$, $b=3cd/\alpha_r$, as simpler expressions, we can use 2.3 to obtain the

Theorem 2.9. Let $H = \sum_{k\geq 2} H_k = i\sum_{j=1}^{n}\omega_j x_j y_j + \ldots$ an n-degrees of freedom Hamiltonian with $\|H_k\| \leq dc^{k-2}$ for $k\geq 3$. Let $M\supset M_\omega$ as defined in 1.4 and assume $|\nu.\omega| \geq \alpha_r > 0$ $\forall \nu\in \mathbb{Z}^n\setminus M$, $|\nu| \leq r$, for a given positive integer $r\geq 3$.
Then $T_{G(r)}H = Z^{(r)} + R^{(r)}$, in normal form up to order r with respect to M, is convergent in any polydisk D_R of radius $R < R^*_r$ with $R^*_r = [(9+6.4r)\frac{d}{\alpha_r} + (1+6.4r)]^{-1}c^{-1}$. Furthermore, in D_R one has
$|R^{(r)}(X,Y)| \leq \frac{d}{c}R(R/R^*_r)^r(1-R/R^*_r)^{-1}$.

§3. BOUNDS FOR EFFECTIVE STABILITY

Till now we have obtained a Hamiltonian in the "new" variables $(X^{(r)}, Y^{(r)})$: $\tilde{H}(X^{(r)},Y^{(r)})$ or, going back to real variables $(Q^{(r)},P^{(r)})$ we have $\tilde{K}(Q^{(r)},P^{(r)})$, a real Hamiltonian.

Define action variables by $I_j^{(r)} = iX_j^{(r)}Y_j^{(r)} - \frac{1}{2}(Q_j^{(r)^2} + P_j^{(r)^2})$, non-negative for real $Q_j^{(r)}$, $P_j^{(r)}$. Let $I^{(r)} = \max_j |I_j^{(r)}|$. It is clear that $(X^{(r)}, Y^{(r)}) \in D_R$ implies $I^{(r)} \leq R^2/2$.

We wish to bound the variation of the size of the ball $I^{(r)} \leq R^2/2$ with respect to time. To this goal we bound $|\dot{I}^{(r)}|$ for the non-resonant case, and a suitable linear combination of the $I_j^{(r)}$ for the resonant one. Finally we perform optimization with respect to the order of normalization.

Non resonant case:
Proposition 3.1. In the hypotheses of 2.9 and for $M = \{0\}$ one has

$$|\dot{I}^{(r)}| \leq \frac{d}{c}R(R/R^*_r)^r(r+1-rR/R^*_r)(1-R/R^*_r)^{-2}.$$

Indeed, as the normal form (Birkhoff normal form) $Z^{(r)}=Z^{(r)}(I_1^{(r)},\ldots,I_n^{(r)})$ is integrable, the contribution to $\dot{I}^{(r)}$ is zero.

If $h_{k,\ell} X^{(r)^k} Y^{(r)^\ell}$ is a term in $R^{(r)}$ its contribution to $\dot{I}_j^{(r)}$ is $i(\ell_j - k_j) h_{k,\ell} X^{(r)^k} Y^{(r)^\ell}$. From the bounds on the terms of the remainder the proposition follows.

The resonant case:

If $M \neq \{0\}$ we should use a resonant modulus M. However, despite the fact that resonance is not generic, it can be useful to handle the non-resonant case as a resonant one. One can have quasi-resonances, i.e., it can exist $\nu^o \in \mathbb{Z}^n$ with $0 < |\nu^o| \leq r$ such that $|\nu^o . \omega|$ is small compared with $|\nu . \omega|$ for $0 < |\nu| \leq r$, ν independent of ν^o (see the RTBP example). In this way $\alpha_r = \alpha_r(M)$ is larger and the constants improve. It remains to see when and how the velocity of diffusion can be bounded in the resonant case.

We have again $\dot{I}_j^{(r)} = \{I_j^{(r)}, Z^{(r)}\} + \{I_j^{(r)}, R^{(r)}\}$ but now, in general, $\{I_j^{(r)}, Z^{(r)}\}$ is different from zero.

Let $\phi = \sum_{j=1}^{n} \gamma_j I_j^{(r)}$. If we select $\gamma \in M^{\perp}$ $(\gamma = (\gamma_1, \ldots, \gamma_n)^T)$ then $\{\phi, Z^{(r)}\} = 0$ and $\dot{\phi} = \sum_{j=1}^{n} \gamma_j \{I_j^{(r)}, R^{(r)}\}$, where $\{I_j^{(r)}, R^{(r)}\}$ can be bounded as in the non-resonant case. However, to bound the velocity of diffusion, we should select one or several combinations ϕ that ensure some positive definiteness. This leads to introduce admissible linear combinations.

Definition 3.2. For a given $\mathbb{Z}$-modulus M, let $\bar{\gamma} = \{\gamma^{(1)}, \ldots, \gamma^{(s)}\}$ with $s \leq \operatorname{codim} M$ be a finite set of n-vectors. We say that $\bar{\gamma}$ is admissible if the following conditions hold:

i) $\gamma^{(j)} \in M^{\perp}$; ii) $\gamma_m^{(j)} \geq 0$ for $j=1,\ldots,s$ and $m=1,\ldots,n$;

iii) $\min_m \{\gamma_m^{(j)} : \gamma_m^{(j)} \neq 0\} = 1$ for $j=1,\ldots,s$; iv) For any $m=1,\ldots,n$ there exists some $j \in \{1,\ldots,s\}$ such that $\gamma_m^{(j)} \geq 1$.

Definition 3.3. Let Γ be the set of admissible $\bar{\gamma}$. Assume $\Gamma = \Gamma(M)$ non empty (which is not necessarily true).
Define $C_1 = C_1(M) = \min_{\bar{\gamma} \in \Gamma} \{ \max_j \{ \sum_{m=1}^{n} \gamma_m^{(j)} \} \}$.

Examples: 1) $M = \{p.(1,0,-1)^T, \ p \in \mathbb{Z}\} \Rightarrow C_1 = 2$;

2) $M = \{p.(1,1,0)^T, \ p \in \mathbb{Z}\} \Rightarrow \Gamma = \phi$;

3) $M = \{p.(1,0,-1,0)^T + q.(0,1,-1,-1)^T, \ p,q \in \mathbb{Z}\} \Rightarrow C_1 = 3$:

4) $M = \{0\} \Rightarrow C_1 = 1$.

Let $\Gamma(M) \neq \phi$, $\bar{\gamma} \in \Gamma$ such that gives the minimum C_1. Then we define $\phi_j^{(r)}(I) = \sum_{m=1}^{n} \gamma_m^{(j)} I_m^{(r)}$ and $\phi^{(r)}(I) = \max_{1\leq j\leq s} |\phi_j^{(r)}(I)|$.

We note that $I^{(r)} \leq \phi^{(r)}(I) \leq C_1 I^{(r)}$ for $I^{(r)} \in \mathbb{R}_+$.

Proposition 3.4. In the hypotheses of 3.1 but with $M \neq \{0\}$ such that $\Gamma(M) \neq \phi$ one has in D_R the bound

$$|\dot{\phi}^{(r)}| \leq C_1 \frac{d}{c} R\, (R/R^*_r)^r\, (r+1-rR/R^*_r)^{-2} .$$

Now we can minimize the bound of $|\dot{\phi}^{(r)}|$ with respect to r. To work analytically we use $R^*_r \geq \alpha_r/(Nr)$, where N is an upper bound of $[(\frac{9}{r} + \frac{32}{5})d + (\frac{1}{r} + \frac{32}{5})\, \alpha_r]\, c$. For instance, if no more information is available, for $r \geq 3$ and $\alpha_r < 1$ one can take $N = (47d + 34)\, c/5$. We also use the bound $\alpha_r \geq C_2/r^{m+1}$ with C_2, m depending on M and ω. In this way we reach the

Theorem 3.5 Let $K(q,p) = \frac{1}{2} \sum_{j=1}^{n} \omega_j(q_j^2+p_j^2)+ \ldots$ an n-degrees of freedom real analytical Hamiltonian, and $H(x,y) = \sum_{k\geq 2} H_k = i \sum_{j=1}^{n} \omega_j x_j y_j + \ldots$ the transformed Hamiltonian to complex variables using the change S. Suppose $\| H_k \| \leq c^{k-2}$ for $k \geq 3$. Let $M \supset M_\omega$ and for any $r \geq 3$ assume $|\nu.\omega| \geq \alpha_r > 0$ $\forall\, \nu \in \mathbb{Z}^n \setminus M, |\nu| \leq r$, and $\alpha_r \geq C_2/r^{m-1}$ with $C_2 > 0$. $m \geq 1$. Denote

$$\bar{\delta} = \frac{\sqrt{2}}{3(e^m-1)} , \quad C_3 = C_1 \frac{4d}{3c} (1-e^{-m})^{-2}, \quad \hat{R} = (1+\bar{\delta})C_2/N , \quad \sigma_o = \frac{1+\bar{\delta}}{1-\bar{\delta}} ,$$

with the constants C_1 and N defined as before. Then, if $(q,p)(0) \in D_{R_o}$ one has $(q,p)(t) \in D^1_{\sigma R_o}$ $\forall |t| \leq T$ with

$$T = T_o \exp \{m \left[\frac{1}{e} (\frac{\hat{R}}{\sigma R_o})^{1/m}\right]\} ,$$ (here [ ] denoting the integer part).

where $T_o = R_o \dfrac{\sigma^2 - C_1 \sigma_o^2}{2\sigma\, \sigma_o} e(C_3(1-\bar{\delta})(\frac{\hat{R}}{\sigma R_o})^{1/m})^{-1}$, provided

$$\sigma > \sqrt{C_1}\, \sigma_o , \quad R_o \leq \frac{\hat{R}}{\sigma (3e)^m} .$$

The proof is done by minimizing $|\dot{\phi}^{(r)}|$ with respect to r in an approximate way. Several radii are involved:

(a) The radius of convergence R^*_r for $(X^{(r)},Y^{(r)})$;

(b) The initial radius, R_o, for (x,y) or, equivalently, for (q,p);

(c) The initial radius, R_1, for $(X^{(r)},Y^{(r)})$ given by $R_1=R_o/(1-\bar{\delta})$;

(d) The radius $R_1\sqrt{C_1}$ due to the use of $\phi^{(r)}$ instead of $I_o^{(r)}$;

(e) The final radius, R_2, in the new variables $X^{(r)}$, $Y^{(r)}$ due to the diffusion from $R_1\sqrt{C_1^2}$ to R_2 during the time T;

(f) The final radius $R_f = \sigma R_o = R_2(1+\bar{\delta})$, due to the deformation going back to the (x,y) or (q,p) variables.

The result is of Nekhorosev type, but with all the constants "easily" computable.

Now we pass to the determination of a region of "effective stability".

As we are interested in all the points on a ball near the equilibrium point and not only in a very big part of them, as in the KAM theory, in principle there is no way to avoid Arnol'd diffusion. Hence we coined the name effective stability to denote a moderate increase in the radius of a ball containing the point in the phase space for a very long time.

Now consider the time interval, T, given and look for a radius, R_o, such that, at the end of the time interval we still are in a ball of radius σR_o. Both radii are considered in the $(q,p)\in\mathbb{R}^{2n}$ variables. The computations are done with a radius $R=R_2$ in the $(X^{(r)},Y^{(r)})$ variables such that starting at the ball of radius $R_1=R/\alpha$ in the same variables we still are on the ball of radius R at the end. The passage to (q,p) variables introduces the factors $1/(1-\bar{\delta})=R_1/T_o$ and $1+\bar{\delta} = \sigma R_o/R_2$.

One obtains a relation like $|\dot{\phi}^{(r)}{}_{opt}.T| \leq R^2(\alpha^2-C_1)/(2\alpha^2)$. In this case, given T we look for an optimal r which maximizes R. The result, using the previous notations, is given in the next Theorem.

<u>Theorem 3.6.</u> In the conditions of 3.5, using $\sigma_o,\bar{\delta},\hat{R},C_2$ as before and $T>0,\sigma>\sqrt{C_1}\sigma_o$, a point in D_{R_o} remains in $D_{\sigma R_o}$ for $|t|<T$ provided

$$R_o \leq \frac{\hat{R}}{\sigma}\left(\frac{C_4}{2}\, r_{opt}^{-m-1}\right)^{1/(r_{opt}-1)} . r_{opt}^{-m},$$

where

$$C_4 = \frac{cC_2(\sigma^2-C_1\sigma_o^2)}{2NdC_1\sigma^2\ T} \quad \text{and} \quad r_{opt} = \left[-\frac{1}{m}\ln\left(\frac{C_4}{2}\right)\right].$$

§4. A SIMPLE EXAMPLE

Dealing with specific examples allows us to use the available information more carefully.

(1) One can use, for moderate values of r, the value of α_r obtained by direct computation as $\min|\nu\cdot\omega|$, for $\nu \in \mathbb{Z}^n \setminus M$, $|\nu| \leq r$.

(2) For a value of r one can compute the value of R such that
$$|\dot{\phi}^{(r)}.T| = R^2 \frac{\alpha^2 - C_1}{2\alpha^2} .$$
Then the optimization is done by trial and error with respect to r.

(3) Given r and the related value of α_r the convergence radius R^*_r is obtained from the parameters c,d such that $\| H_k \| \leq c^{k-2}d$. However several choices of c and d are available. One can also use this freedom to improve the bounds.

We take first the following easy example: $K = \frac{1}{2}(q_1^2+p_1^2)+\frac{1}{2}\beta(q_2^2+p_2^2)+\frac{1}{2}(q_3^2+p_3^2) + K_3$, where to make the estimates very simple, we select $-\beta$ = golden mean = $(\sqrt{5}-1)/2$, and K_3 is a homogeneous polynomial of degree 3 in q_j, p_j, j=1,2,3, such that $\|K_3\| = 1/\sqrt{8}$. Using the change S we obtain H_3^j with $\| H_3 \| = 1$. Obviously the only condition for c and d is $cd \geq 1$.

r	α_r
3,4	0.236068
5,6,7	0.145898
8,...,12	0.090170
13,...,20	0.055728
21,...,33	0.034442
34,...,54	0.021286
etc.	

Table 4.1

For $r \geq 3$ one has the table 4.1 and a rough approximation allows us to write $\alpha_r \geq 0.7082/r$.

As $M = \{p.(1,0,-1)^T,\ p \in \mathbb{Z}\}$, we have $C_1=2$.

We take as data the values $\alpha=1.1\sqrt{2}$, $T=10^{10}$. For R^*_r we obtain $R^*_r = ((9+6.4r)/\alpha_r + c(1+6.4r))^{-1}$. Let us introduce $\rho = R^*_r/R$. Then the equation for R can be written as

$$\rho^{1-r}(r+1-\frac{r}{\rho})(1-\rho^{-1})^{-2} = \frac{R^*_r\, c(\alpha^2-C_1)}{2dT\alpha^2 C_1} .$$

Given r the r.h.s. is known. Solving for ρ allows us to compute R. Finally we can obtain δ and $R_o = R(1-\delta)/\alpha$, while $\sigma = \alpha(1+\delta)/(1-\delta)$.

After some attempts the suitable values seem to be c=2.5, d=0.4. Then, the optimum value of R is obtained for r=20. For this one gets $R = 55.56\cdot 10^{-6}$. Using $b = cd/\alpha_r$, $a = 6sc(1+d/\alpha_r)$ the value $\delta = 0.00533$ is obtained from $\delta = (1+8a/(9b))^{-1}\sqrt{2}/(\rho-1)$. Finally $R_o=35.5\cdot 10^{-6}$ and $\sigma = 1.5723$.

A direct application of Theorem 3.6 with m=2, $C_2=0.7082$, c=2.5,

$\sigma = 1.7$, $T = 10^{10}$ produces the more conservative value $R_o = 4.84\cdot 10^{-6}$ ($r_{opt} = 14$). However, m=1, $C_2 = 0.034442$ and $N = 7.25$, expecting r close to 30, gives in fact $r_{opt} = 30$ and $R_o = 21.0\cdot 10^{-6}$, not far from the more elaborated bound.

§5. AN APPLICATION TO THE $L_{4,5}$ POINTS FOR THE SPATIAL RESTRICTED THREE BODY PROBLEM

In the RTBP L_4 and L_5 are stable in the planar case for the mass parameter $\mu \in (0,\mu_1)$ unless $\mu = \mu_2$ or $\mu = \mu_3$ [13] where $\mu_j = (1 - (1 - 16j^2/(27(j^2+1)^2))^{\frac{1}{2}})2$ is the value for which the j to 1 resonance is produced.

The Hamiltonian of the spatial RTBP centered at L_4 is cartesian coordinates is

$$J = \frac{1}{2}(p_x^2+p_y^2+p_z^2) - (xp_y - yp_x) - x(\mu - \frac{1}{2}) - y\sqrt{3}/2 - (1-\mu)/r_1 - \mu/r_2,$$

where $r_1^2 = 1-x+\sqrt{3}y+x^2+y^2+z^2$, $r_2^2 = 1+x+\sqrt{3}y+x^2+y^2+z^2$. The expression for L_5 is similar.

Hence $J = \sum_{k\geq 2} J_k$ with $J_2 = \frac{1}{2}(p_x^2+p_y^2) - xp_y + yp_x + x^2/8 - 5y^2/8 - axy + (p_z^2 + z^2)/2$ where $a = -(1-2\mu)\sqrt{27}/4$ and $J_k = r^k P_k((x-\sqrt{3}y)/2r)(1-\mu) + r^k P_k((-x-\sqrt{3}y)/2r)\mu$, for $k \geq 3$ and where P_k denotes the Legendre polynomials, with $r^2 = x^2+y^2+z^2$.

If the eigenvalues at L_4 are denoted by $\lambda = i\omega$ we obtain the characteristic equation $(\omega^4 - \omega^2 + 27/16 - a^2)(\omega^2-1) = 0$. The value $\omega = \omega_3 = 1$ is related to the vertical mode. The remaining ω_1, ω_2 are real and different for $\mu \in (0,\mu_1)$. We call ω_1 the one such that $\omega_1^2 > \frac{1}{2}$ (then $\omega_2^2 < \frac{1}{2}$). Let $U = \mathrm{Hess}(J_2)$. Introduce the vectors

$$e_j = (a, -\frac{3}{4} - \omega_j^2, \frac{3}{4} - \omega_j^2, a)^T \quad\text{and}\quad f_j = (2\omega_j, 0, a\omega_j, (\frac{5}{4} - \omega_j^2)\omega_j)^T.$$

The complex vectors $e_j \pm i f_j$, j=1,2 are the eigenvalues of

$$\begin{pmatrix} 0 & I_2 \\ -I_2 & 0 \end{pmatrix} .U.$$

Let $m_j = \omega_j(2\omega_j^4+\omega_j^2/2-3/4)$ and $e_j' = e_j m_j^{-\frac{1}{2}}$, $f_j' = f_j m_j^{-\frac{1}{2}}$. Then the matrix $M_o = (e_1', e_2', f_1', f_2')$ is symplectic and one has $M_o^T U M_o = \mathrm{dia}(\omega_1, \omega_2, \omega_1, \omega_2)$. From the reality of M_o one has $m_j > 0$ and therefore $\omega_1 > 0$, $\omega_2 < 0$.

The symplectic change $(x,y,p_x,p_y)^T = M_o(q_1,q_2,p_1,p_2)^T$, $z=q_3$, $p_z=p_3$, produces a new Hamiltonian K with $K_2 = \frac{1}{2}\sum_{j=1}^{3}\omega_j(q_j^2+p_j^2)$. The fact that K_2 is indefinite prevents of using it as a Lyapunov function and to obtain stability from this.

Making the change S we have new coordinates (x_j, y_j), j=1,2,3 and a new Hamiltonian H. We should estimate $\|H_k\|$, $k\geq 3$. It is enough to bound the norm of

$$r^k P_k((x-\sqrt{3}y)/2r) = \sum_{j=0}^{[k/2]} a_j^{(k)} \left(\frac{x-\sqrt{3}y}{2}\right)^{k-2j} (x^2+y^2+z^2)^j ,$$

where $a_j^{(k)}$ are the coefficients appearing in P_k.

We express $(x - \sqrt{3}y)/2$ and r^2 in the new variables (x_j, y_j), j=1,2,3, getting

$$\| Q_1 \| = \sqrt{2} \sum_{j=1}^{2} \frac{1}{\sqrt{m_j}} \left(\left(\frac{a}{2} + \frac{\sqrt{27}}{8} + \frac{\sqrt{3}}{2}\omega_j^2\right)^2 + \omega_j^2\right)^{\frac{1}{2}} ,$$

$$\| Q_2 \| = \sum_{j=1}^{2} \frac{1}{m_j} \left\{\left[\left(2a^2 - \frac{9}{8} - \frac{3}{2}\omega_j^2\right)^2 + 16a^2\omega_j^2\right]^{\frac{1}{2}} + 2a^2 - \frac{9}{8} + \frac{13}{2}\omega_j^2 \right\} + 2 +$$

$$\frac{2}{\sqrt{m_1 m_2}}\left\{\left[(3 - 4\omega_1\omega_2)^2 + 4a^2(\omega_1+\omega_2)^2\right]^{\frac{1}{2}} + \left[(3+4\omega_1\omega_2)^2 + 4a^2(\omega_1-\omega_2)^2\right]^{\frac{1}{2}}\right\},$$

where Q_1 and Q_2 are the polynomials of degree 1 and 2, respectively, in the new variables.

Let $\eta_j = \|Q_j\|$, j=1,2. Then one has $\|H_k\| \leq \sum_{j=0}^{k/2} |a_j^{(k)}| \eta_1^{k-2j} \eta_2^j$.
Using the properties of the Legendre polynomials we get

$\|H_k\| \leq \eta_2^{k/2} P_k(i\gamma)i^{-k}$, where $\gamma = \eta_1\eta_2^{-\frac{1}{2}}$. Now let $C_k = P_k(i\gamma)i^{-k}$, $\rho = \gamma + (1+\gamma^2)^{\frac{1}{2}}$ and $\Lambda > 0$. Then if $C_{j-1} \leq \Lambda\rho^{j-1}$, $C_j \leq \Lambda\rho^j$, for some j, one has (using induction and the recurrence formulas $C_k \leq \Lambda\rho^k$ for all k greater than j.

We are interested in bounding C_j for $j \geq 3$ by an expression of the form $\Lambda\rho^j$. We look for a value, γ_o say, of γ such that $C_4/C_3 = \gamma_o + (1+\gamma_o^2)^{\frac{1}{2}}$, recalling that $C_3(\gamma) = \gamma(5\gamma^2+3)/2$ and $C_4(\gamma) = (35\gamma^4 + 30\gamma^2 + 3)/8$. It is easily found that γ_o^4 is the positive root of $175\gamma^8 + 340\gamma^6 + 210\gamma^4 + 36\gamma^2 - 9$ (roughly, $\gamma_o = 0.3602255$). Then we can take $\Lambda = C_3\rho^{-3}$ if $\gamma \geq \gamma_o$ and $\gamma = C_4\rho^{-4}$ if $\gamma < \gamma_o$.

Hence we obtain $\|H_k\| \leq h_k = \eta_2^{k/2}\Lambda\rho^k$ $\forall$ $k \geq 3$, where Λ and ρ depend on $\gamma = \eta_1\eta_2^{-\frac{1}{2}}$ as described above. We remark that $h_k = c^{k-2}d$ with $c = \rho\eta_2^{\frac{1}{2}}$, $d = \Lambda c^2$ and we have obtained a bound of the desired type.

In [8] we display the values of R_o for $\mu \in (0,\mu_1)$ with step 10^{-6} without considering any resonance, i.e., $M = \{0\}$, and some results for the specific example of the Sun-Jupiter case, i.e., $\mu \simeq 1048.355^{-1} \simeq 0.95387536.10^{-3}$. Now we wish to study this case more carefully and to improve the value of R_o in §6.

The eigenvalues are roughly $\omega_1 = 0.996757525$, $\omega_2 = -0.080463876$, $\omega_3 = 1$. So we are close to the 1:1 resonance between ω_1 and ω_3, and we are led to choose $M = \{p.(1,0,-1)^T , p \in \mathbb{Z}\}$. The values of r for which α_r changes (α_r is non-increasing with respect to r) are all the odd numbers between 1 and 33, then 42, 48, 93,... . The minimum of α_r, for a given r, is generated for the triples (s,1,-s) for s=0 to 5 (r=1 to 11); (s+1,12,-s) for s=0 to 10 (r = 13 to 33), then (14,-13,-15) for r = 42, (7,37,-4) for r = 48, etc. For example $\alpha_{17} = \alpha_{18} \simeq 0.0247061$, $\alpha_{19} = \alpha_{20} \simeq 0.0214636$, $\alpha_{33} = \ldots = \alpha_{41} \simeq 0.001234$, etc.

For the M chosen we have $C_1 = 2$. We select for α and T the values $\alpha = 1.1\sqrt{2}$ and $T = 10^{10}$, and we do the optimization with respect to r as described in §4, 2).

Note that as 2Π units of time equals the period of revolution of Jupiter around the Sun, the time interval 10^{10}units is of the order of magnitude of the estimated age of the Universe. Certainly this can be considered as an "effective stability".

Computing c and d for $\mu = 0.95387536\cdot 10^{-3}$ we obtain $c \simeq 13.673658$, $d \simeq 46.705790$. The optimal value of r turns out to be r = 20. Then we obtain $R^*_r = 2.4521\cdot 10^{-7}$ and, successively $\rho = 11.114325$, $R = 2.20625\cdot 10^{-8}$ and $\delta = 0.0091815$. Hence $R_o = 1.40521\cdot 10^{-8}$and $\sigma = 1.584466$.

To compare with a rougher application of Theorem 3.6 we can select m=1 and hence $C_2 = \alpha_r$ (we shall try for different values of r). Using $\sigma = 2.6$ and again $T \equiv 10^{10}$, with N as defined after 3.4, with the previous values of c and d we obtain $\bar{\delta} = 0.2743465$, $\sigma_o = 1.7561362$ and $\hat{R} = (1+\bar{\delta})\alpha_r/\{c[(\frac{9}{r} + 6.4)d + (\frac{1}{r} + 6.4)\alpha_r]\}$. By trial and error with respect to r the optimum is found again for r=20, for which one has $R_o = 1.01360\cdot 10^{-8}$.

§6. IMPROVEMENT OF THE BOUNDS

Till now we have used only a small amount of the specific characteristics of the Hamiltonian of the spatial RTBP around L_4. We wish to improve the bounds obtained by using them.

a) First of all it is possible to compute explicitly $\|H_k\|$ for some values of $k \geq 3$. To this end we write

$$J_m = \sum_{j=0}^{[m/2]} \sum_{j_1=0}^{m-2j} \sum_{i_1=0}^{j} \sum_{i_2=0}^{j-i_1} \frac{(-1)^{j+j_1} 3^{j_1/2} (2m-2j)! [1-\mu+\mu(-1)^{m-j_1}]}{4^{m-j}(m-j)!(m-2j-j_1)! j_1! i_1! (j-i_1-i_2)! i_2!} x^{k_1} y^{k_2} z^{k_3}$$

with $k_1=m-2i_1-2i_2$, $k_2 = 2i_2 + j_1$, $k_3 = 2i_1$, and where we have used the expression of $P_m(\cdot)$ and the binomial formula.

If we call $\bar{S}$ the matrix giving the transformation S^{-1} which expresses (q_j,p_j), j=1,2 in terms of (x_j,y_j), j=1,2, and put $A=M_o\bar{S}$ we have $(x,y,p_x,p_y)^T = A(x_1,x_2,y_1,y_2)^T$. In particular $x = a_{11}x_1+a_{12}x_2+a_{13}y_1+a_{14}y_2$ and we have a similar expression for y, where a_{ij} are (complex) elements of A. Substituting the expressions of x and y in J_m we obtain for J_m the formula

$$J_m = \Sigma\, \text{coef}_{k_1,k_2,k_3,k_4,k_5} x_1^{k_1} x_2^{k_2} y_1^{k_3} y_2^{k_4} z^{2k_5}$$

with $k_1+k_2+k_3+k_4+2k_5 = m$, $k_j \geq 0$. To do this computation in an efficient way we start with the couple (x,y) expressed in terms of x_1, x_2, y_1, y_2. Then, multiplying the couple by x and the last element by y we obtain the triple (x^2,xy,y^2) in terms of x_1, x_2, y_1, y_2. Performing this step again and again we obtain successively the expressions for $x^k, x^{k-1}y, x^{k-2}y^2, \ldots, xy^{k-1}, y^k$ in terms of x_1,x_2,y_1,y_2. All the products of homogeneous polynomials of degree k in x_1,x_2,y_1,y_2 by a homogeneous polynomial of degree 1 in the same variables are easily done if they are stored in lexicographical order.

Finally we should substitute the (even) powers of z in terms of x_3,y_3. However, if we are only interested in $\|J_m\|$ it is enough to use the formula $\|J_m\| \leq \Sigma|\,\text{coef}_{k_1,k_2,k_3,k_4,k_5}| \cdot 2^{k_5}$. The number of terms in $J_3,J_4,\ldots,J_m$ expressed in the variables x_j,y_j, j=1,2,3, but taking into account the symmetry in the variable z is given by

$$NT(m)= \frac{1}{5760}[4m^6+84m^5+730m^4+3360m^3+8536m^2+11046m-98235+(-1)^m(90m+315)].$$

For instance, for m=30 one obtains 980406 terms. Generically all of them are different from zero. From now on we call H_m the homogeneous polynomial J_m expressed in the (x_j,y_j) variables. Table 6.1 gives $\|H_m\|$ for m=3,...,30.

We wish to bound $\|H_k\|$ by dc^{k-2} for $k \geq 3$. For $k \geq 30$ we know the Table 6.1, and for k > 30 we can use the bound found in §5: $c \simeq 13.673658$, $d \simeq 46.705790$. From this one has $cd \geq 254.4538$. If d is chosen small (and this will produce good estimates of R_o) one has $c^{k-2}d \geq 13.673658^{k-2} \cdot 46.70579$ and the bound is correct for all $k \geq 3$.

In fact it is enough to take $d \leq \| H_3\|^2/\| H_4\| \simeq 9.438057$.

b) If $h_{k\ell} x_1^{k_1} x_2^{k_2} x_3^{k_3} y_1^{\ell_1} y_2^{\ell_2} y_3^{\ell_3}$ is one of the generic terms in H, we have, due to the symmetry, that $k_3+\ell_3$ is even. This property is preserved in the terms, Z_n, of the normal form and on the G_n of the generating function. In a term of Z_n one should have $\ell_1-k_1-(\ell_3-k_3) = 0$ and $\ell_2=k_2$. Hence $Z_n \equiv 0$ if n is odd. Furthermore, when we construct G_n from F_n only the divisors of the type $|j_1\omega_1+j_2\omega_2+j_3|$ with $0 \leq j_1$, $|j_2| \leq n-j_1$, $|j_3| \leq n-j_1 - |j_2|$, j_3 even, and with $j_1+j_3 \neq 0$ or $j_2 \neq 0$, should be considered. This produces an improvement on the small divisors found. Table 6.1 contains also the smallest divisors for every order in the range 3 ÷30 under these conditions, that we denote by $\bar{\alpha}_r$.

c) The explicit knowledge of $H_{j,n-j}$, Z_n, G_n for n from 3 up to some order r will produce a substantial improvement in the bounds (see [21]). However this implies a big computing task. But nice bounds at low cost can be obtained just computing $G_3, G_4, H_{3,1}$ and Z_4. It is enough to solve $L_{H_2} G_3 = H_3$ and $L_{H_2} G_4 + Z_4 = \frac{1}{2}\{G_3,H_3\} + H_4$. Doing this one obtains

$$\| G_3\| = 439.204192,\ \| G_4\| = 2456.598718,$$

$$\| Z_4\| = 0.786519 \text{ and } \| H_{3,1}\| = 11345.9069 .$$

Then the recurrence

$$\|F_k\| \leq \sum_{m=1}^{k-3} \frac{m}{k-2}(2+m)(k-m)\|G_{2+m}\|\cdot\|Z_{k-m}\| + \sum_{m=1}^{k-2}\frac{m}{k-2}\|H_{2+m,k-m-2}\|,$$

where

$$\|H_{2+m,k-m-2}\| \leq \sum_{\ell=1}^{k-m-2} \frac{\ell}{k-m-2}(2+\ell)(k-\ell)\|G_{2+\ell}\|\cdot\|H_{2+m,k-m-2-\ell}\|,$$

can be used, and the estimates $\| Z_k\| = 0$ if k is odd, $\| Z_k\| \leq \|F_k\|$ if k is even, and $\|G_k\| \leq \|F_k\| /\bar{\alpha}_k$ for $k > 4$ are obtained. Table 6.1 displays also the values of the bounds for $\|G_k\|$ obtained in this way.

Suppose that we compute the normal form to some order r. Then $G_k=0$ for $k>r$, and values of a and b can be obtained such that $\|G_k\| \leq ba^{k-3}$ for $k = 3,\ldots,r$. From table 6.1 it is seen that one should take $b \geq 439.204192$ and then, if b is close to this limit value, one has $a \geq (\| G_r\| /b)^{1/(r-3)}$, at least for $r \leq 30$. We remark that one should increase b very much to produce a noticeable effect on a and, hence, a will be always close to

$$(\| G_r\| / \| G_3\|)^{1/(r-3)} .$$

m	$\|H_m\|$	$100\ \bar{\alpha}_m$	$\|G_m\|$
3	2.544539 E2	8.046388	4.392042 E2
4	6.860182 E3	16.092775	2.456599 E3
5	3.407255 E4	7.397893	4.948209 E8
6	7.745497 E5	15.444280	3.578154 E12
7	4.512199 E6	7.397893	6.896063 E16
8	9.150039 E7	15.444280	4.433915 E20
9	5.960107 E8	6.749398	8.991450 E24
10	1.115414 E10	14.795785	5.819965 E28
11	7.901692 E10	6.749398	1.153897 E33
12	1.385529 E12	11.165489	9.815977 E36
13	1.044966 E13	3.119102	3.858744 E41
14	1.737046 E14	4.927286	6.919481 E45
15	1.381681 E15	3.119102	2.186881 E50
16	2.205341 E16	4.278791	4.437747 E54
17	1.834808 E17	2.470607	1.703367 E59
18	2.822827 E18	4.278791	3.383525 E63
19	2.431040 E19	2.470607	1.285727 E68
20	3.620791 E20	3.630296	2.989459 E72
21	3.219483 E21	1.822112	1.499504 E77
22	4.679693 E22	3.630296	3.427354 E81
23	4.281961 E23	1.822112	1.708372 E86
24	6.078868 E24	2.981801	4.731537 E90
25	5.682722 E25	1.173617	3.569921 E95
26	7.892353 E26	2.981801	9.733789 E99
27	7.535499 E27	1.159690	7.396961 E104
28	1.030276 E29	2.333306	2.567999 E109
29	1.003511 E30	0.525122	4.204969 E114
30	1.350113 E31	2.333306	1.438958 E119

Table 6.1

d) Now we describe how to obtain the region of "effective stability". Given $C_1=2$, α and T, for any triple of values of r,c,a one computes d and b as shown in a) and c) respectively. Then R* is obtained directly of 2.3.2) as $R^* = (3b+8a/3+c)^{-1}$. Before proceeding with the computations we give some improvement of previous results, in particular propositions 2.3 and 3.4.

<u>Proposition 6.2.</u> Under the hypotheses of 2.3 and using the same notation one has

$$\|F_k\| \leq d[3b(3b+8a/3+c)^{k-1} + 8ac^{k-1}/3]/(3b+8a/3).$$

Proof:

$$\| F_k \| \leq \sum_{j=1}^{k} C_{j,k-j} c^{j-1} d \leq 3db \sum_{j=1}^{k-1} \binom{k-1}{j-1} (3b+8a/3)^{k-j-1} + dc^{k-1} =$$

$$\frac{3db}{N} \sum_{j=1}^{k} \binom{k-1}{j-1} N^{k-j} c^{j-1} + dc^{k-1}(1 - 3b/N),$$

where N stands for 3b+8a/3. Hence $\| F_k \| \leq \frac{d}{N}\, [3b(N+c)^{k-1}+8ac^{k-1}/3\ ]$ as desired.

Proposition 6.3. Under the hypotheses of 3.4 and using the same notation one has

$$|\dot{\phi}^{(r)}| \leq C_1 \{3bdR \frac{(r+1-rR/R^*)(R/R^*)^r}{(1-R/R^*)^2} + \frac{8}{3}\, adR \frac{(r+1-rcR)(cR)^r}{(1-cR)^2} \}/(c(3b+\frac{8}{3}\, a)).$$

Proof: Using N as in 6.2, R* and the result of 6.2 and taking into account the shift in the exponent f_k in 2.3 and $\|H_k\| \leq c^{k-2}d$, we get

$$|\dot{\phi}^{(r)}| \leq \frac{C_1}{c} \sum_{k>r} k \frac{dR}{N}\, [3b(\frac{R}{R^*})^{k-1} + \frac{8}{3}a(cR)^{k-1}] =$$

$$C_1\ \{3bdR \sum_{k>r} k(\frac{R}{R^*})^{k-1} + \frac{8}{3}\, adR \sum_{k>r} k(cR)^{k-1}\ \}/cN,$$

and the result follows.

Now, using the bound $|\dot{\phi}^{(r)}|_b$ given by 6.3 we rewrite the condition derived before 3.6 as

$|\dot{\phi}^{(r)}|_b \cdot T \leq R^2(\alpha^2-C_1)/2\alpha^2$. Introducing $z = R/R^*$ (= ρ^{-1}) and $\hat{\alpha} = \alpha/\sqrt{C_1}$ one has

$$z = \{\frac{R^*(\hat{\alpha}^2-1)c(3b+8a/3)}{2^2\, TC_1 d} \Big/ \Big[3b\ \frac{r+1-rz}{(1-z)^2} + \frac{8}{3}a\ \frac{(r+1-rczR^*)(cR^*)^r}{(1-czR^*)^2}\ \}\Big]^{\frac{1}{r-1}} ,$$

which can be solved by iteration. We remark that in practice cR* is very small.

Carrying out all those computations we optimize with respect to the choices of r,c,a. In our problem the best result is obtained, when $\alpha = 1.1\sqrt{2}$ and $T = 10^{10}$, for r=28, a = 17077.2, c = 3814.2. Then $R = 0.691478.10^{-5}$ and using $\delta = \sqrt{2}/((1+8a/(9b))(z^{-1}-1))$ one has $\delta = 0.0214506$. Hence $R_o = 0.434964.10^{-5}$ and $\sigma = 1.623836$. This is a big improvement compared with the results in §5.

e) Going back to the physical variables.

We have obtained that for $(q,p) \in D_{R_o}$ in the norm of 1.1 we still are confined in the ball $D_{\alpha R_o}$ after a time interval T. However we can ask for the physical meaning of the ball D_{R_o}. To this goal we recall that going back to the physical variables x,y,z,p_x,p_y,p_z we should use the change given by the matrix M_o. We are interested on the projection of this ball of the phase space on the (x,y) plane.

Let $x=m_{11}q_1+m_{12}q_2+m_{13}p_1+m_{14}p_2$, $y=m_{21}q_1+m_{22}q_1+m_{23}p_1+m_{24}p_2$ the change due to M_o, concerning only the x,y physical variables. Let (ρ,θ) the polar coordinates in the (x,y) plane. Keeping θ fixed we ask for the maximum value of ρ. Introducing $c_j=m_{1j}\cos\theta+m_{2j}\sin\theta$, $a_j=m_{1j}\sin\theta-m_{2j}\cos\theta$ and $w=(q_1,q_2,p_1,p_2)^T$, we look for

$$\rho = \max\{\sum_{j=1}^{4} c_jw_j : \sum_{j=1}^{4} a_jw_j = 0,\ w_1^2+w_3^2 \leq 1,\ w_2^2+w_4^2 \leq 1\}.$$

Using duality this is equivalent to

$$-\min\{-\sum_{j=1}^{4} c_jw_j : \sum_{j=1}^{4} a_jw_j = 0,\ w_1^2+w_3^2 \leq 1,\ w_2^2+w_4^2 \leq 1\} =$$

$$-\max_{\lambda\in\mathbb{R}}\min\{\sum_{j=1}^{4}(\lambda a_j - c_j)w_j : w_1^2+w_3^2 \leq 1,\ w_2^2+w_4^2 \leq 1\} =$$

$$\min_{\lambda\in\mathbb{R}}\max\{\sum_{j=1}^{4}(c_j-\lambda a_j)w_j : w_1^2+w_3^2 \leq 1,\ w_2^2+w_4^2 \leq 1\} =$$

$$\min_{\lambda\in\mathbb{R}}\Big[\max\{(c_1-\lambda a_1)w_1+(c_3-\lambda a_3)w_3 : w_1^2+w_3^2 \leq 1\} +$$

$$\max\{(c_2-\lambda a_2)w_2+(c_4-\lambda a_4)w_4 : w_2^2+w_4^2 \leq 1\} = \min_{\lambda\in\mathbb{R}}\phi(\lambda),$$

where

$$\phi(\lambda) = \sqrt{A_1-2B_1\lambda+C_1\lambda^2} + \sqrt{A_2-2B_2\lambda+C_2\lambda^2} \text{ and } A_j=c_j^2+c_{j+2}^2,$$

$$B_j = a_jc_j+a_{j+2}c_{j+2},\quad C_j = a_j^2+a_{j+2}^2,\quad j=1,2.$$

It is easily seen that the minimum occurs for the only λ in the segment $\langle B_1/C_1, B_2/C_2\rangle$ with $\phi'(\lambda)=0$, and the last equation is equivalent to a quartic equation. From the explicit expression of A_j, B_j, C_j it is also clear that they only depend on μ and $\bar\theta = 2\theta$. That is, by symmetry, the same value of ρ is found for θ and $\theta+\pi$.

The curve described by $\rho=\rho(\theta)$ is the projection of the unitary ball D_1 in the (q_j,p_j) variables going back to x,y,z,p_x,p_y,p_z and projecting on the plane (x,y). This curve is close (but not exactly equal) to an ellipse. It is seen that the extreme values of ρ as a function of θ are obtained when $\lambda=0$. The major and minor axes are given by

$$\hat{d}_{\pm} = 2 \sum_{i=1}^{2} \{\frac{1}{m_i} [\omega_i^4 + \frac{9}{4}\omega_i^2 + \frac{9}{8} \pm (\frac{\omega_i^2}{2} + \frac{3}{8})(9 - 4\omega_1^2\omega_2^2)^{\frac{1}{2}}]\}^{\frac{1}{2}}$$

in terms of the frequencies ω_1, ω_2 and the auxiliary functions m_1 introduced in §5. The angle θ_o between the x axis and the major axis of the curve is given by $\tan(2\theta_o) = -4a/3$, where $a = -(1-2\mu)\sqrt{27}/4$. For the value of μ of the Sun-Jupiter system we obtain $\theta_o = 0.523185$, close to $\Pi/6$. The major and minor axes are $\hat{a}_+ = 16.329149$ and $\hat{a}_- = 2.670875$. These values are converted to kilometers multiplying by the semimajor axis of Jupiter equal to 5.20256 AU × $1.496.10^8$ km/AU. Then the projection of D_1 on (x,y) gives a curve close to an ellipse of axes $1.270902.10^{10}$ km and $0.207875.10^{10}$ km. For the ball D_{R_o} with $R_o = 0.434964.10^{-5}$ these figures amount to 55280 km and 9842 km. This is certainly a very small neighbourhood, but these estimates can be improved by means of the effective computation of G_n to high order (in fact, by a factor greater than 100, see [21]).

ACKNOWLEDGEMENTS.

This work has been partially supported by a NATO Grant RG.85/0197 and a CAICYT Grant 3534/83C3. The author is indebted to his colleagues Giorgilli, Delshams, Fontich and Galgani for many helpful discussions and hints. He is also grateful to Dr. J. E. Martinez Legaz for the hints concerning duality in section 6.e), to the computing centre of the Universitat Autònoma de Barcelona for free use of the VAX 8800 computer and, very specially, to the organizers of the NATO ASI at Cortina d'Ampezzo for making it possible to present these results.

REFERENCES

[1] Arnol'd, V.I., Russian Math. Surveys 18 (1963), 85.

[2] Arnol'd, V.I., Avez, A., Ergodic Problems of Classical Mechanics, Benjamin (1968).

[3] Celletti, A., Chierchia, L., J. Math. Phys. 28, (1987), 2078.

[4] Delshams, A., Por qué la difusión de Arnol'd aparece genéricamente en los sistemas hamiltonianos con más de 2 grados de libertad, Ph.D. Thesis, U.de Barcelona (1983)

[5] Froeschlé, C., Scheidecker, J.P., in Transformations ponctuelles et leurs applications, Editions du C.N.R.S. (1976), 227.

[6] Giorgilli, A., Galgani, L., Cel. Mech. 17 (1978), 267.

[7] Giorgilli, A., Galgani, L., Cel. Mech. 37 (1985), 95.

[8] Giorgilli, A., Delshams, A., Fontich, E., Galgani, L., Simo, C., Effective Stability for a Hamiltonian System near an Elliptic Equilibrium Point, with an Application to the Restricted Three Body Problem, preprint (1987).

[9] Grobner, W., Die Lie-Reihen und Ihre Anwendungen, VEB Deutscher Verlag der Wissenschaften (1967).

[10] Herman, M., Sur les courbes invariantes par les difféomorphismes de l'anneau, Astérisque 144 (1986).
[11] Llave, R. de la, J. Math. Phys. 24 (1983), 2118.
[12] Mackay, R., Percival, I., Commun. Math. Phys. 98 (1985), 469.
[13] Markeev, A.P., PMM 33 (1969), 112.
[14] Mather, J., Ergod. Th. and Dynam. Sys. 4 (1984), 301.
[15] Moser, J., Nach. Akad. Wiss. Gottingen, Math. Phys. Kl. (1962),1.
[16] Nekhorosev, N.N., Russian Math. Surveys 32 (1977), 1.
[17] Olvera, A., Simó, C., Physica 26D (1987), 181.
[18] Rana, D., Proof of Accurate Upper and Lower Bounds to Stability Domains in Small Denominator Problems, Ph.D. Thesis, U. of Princeton (1987).
[19] Rüssmann, H., in Lecture Notes in Math. 1007, Springer Verlag (1983), 677.
[20] Siegel, C.L., Moser, J., Lectures in Celestial Mechanics, Springer Verlag (1971).
[21] Simó, C., Effective Stability for the Triangular Points in the Spatial RTBP, preprint (1988).

PASSAGES NEAR TRIPLE COLLISION

Antoni Susin
Department de Matematica Aplicada, E.T.S.E.I.B.
Universitat Politècnica de Catalunya,
Diagonal 647, Barcelona, Spain.

ABSTRACT: To study passages near triple collision, in the planar three-body problem, it is important to know if it is possible to approach to and escape from collision in triangular configurations. This is related to connecting orbits between Lagrange's points in the non-rotating triple collision manifold. The masses for which these connections exist are obtained numerically. The methods and results are described and some consequences are derived.

1. THE PLANAR THREE-BODY PROBLEM

First we consider the three-body problem in $\mathbb{R}^2$. Let p_k, $k = 1,2,3$, be point particles in $\mathbb{R}^2$ with positions q_k, momenta p_k and masses m_k. In vectorial notation, let q, p be two 6-dimensional vectors (q_1, q_2, q_3), (p_1, p_2, p_3). Then the 3-body problem is governed by the Hamiltonian function $H(p,q) = \frac{1}{2} p.M^{-1} p - U(q)$ where $M = \text{diag}(m_1, m_1, m_2, m_2, m_3, m_3)$ and the potential function U(q) is defined by

$$U(q) = \frac{m_1 m_2}{|q_1-q_2|} + \frac{m_1 m_3}{|q_1-q_3|} + \frac{m_2 m_3}{|q_2-q_3|} .$$

The Hamiltonian equations are $\dot{q} = M^{-1}p$, $\dot{p} = \nabla U(q)$, where $\cdot = \frac{d}{dt}$.

We will assume that the center of mass remains at the origin, and we have the two relations:

$$m_1 q_1 + m_2 q_2 + m_3 q_3 = 0$$

$$p_1 + p_2 + p_3 = 0.$$

2. BLOW UP AND SCALING TIME

Now, following McGehee (1974) we introduce a blow up and scaling time method as follows. Let $r^2 = I$ where I is the moment of inertia,

A. E. Roy (ed.), Long-Term Dynamical Behaviour of Natural and Artificial N-Body Systems, 505–512.
© 1988 by Kluwer Academic Publishers.

defined by

$$I = \sum_{i=1}^{3} m_i |q_i|^2.$$

Since the center of mass is at the origin, triple collision occurs if and only if r=0. We introduce new variables $s = q.r^{-1}$, $z = p.r^{\frac{1}{2}}$ and new time $t = r^{3/2}.\bar{t}$.

The normalized position variables satisfy

$$\langle s, Ms \rangle = 1 \ , \tag{1}$$

and the center of masses integrals give

$$\begin{aligned} m_1 s_1 + m_2 s_2 + m_3 s_3 &= 0 \\ z_1 + z_2 + z_3 &= 0 \end{aligned} \tag{2}$$

Let V be the inner product between s and z, V = <s,z> and denote by ' the derivation with respect to $\bar{t}$.

Then one finds this system of differential equations

$$\begin{aligned} r' &= r.V \\ s' &= M^{-1}.z - V.s \\ z' &= \nabla U(s) + \frac{1}{2} V.z \end{aligned} \tag{3}$$

The levels of the energy H(p,q) = h have the expression:

$$\frac{1}{2} \langle z, M^{-1} z \rangle - U(s) = r.h \ , \tag{4}$$

and for solutions with angular momentum w = <p,Dq> where D is a block diagonal matrix

$$D = \mathrm{diag}(J,J,J) \quad \text{and} \quad J = \begin{pmatrix} 0 & 1 \\ -1 & 0 \end{pmatrix}$$

$$\delta = \langle s, Dz \rangle = r^{-\frac{1}{2}} \ . \ w. \tag{5}$$

3. THE COLLISION MANIFOLD

We define the triple collision manifold C as a subset of the (r,s,z) -space $\mathbb{R}^{13}$ defined by (1)-(2)-(4)-(5) and with r=0, w=0. This manifold is 5-dimensional and we consider a submanifold N of C introduced by Waldvogel (1979) and called the non-rotating planar triple collision manifold which is 4-dimensional. We define N imposing the additional condition δ = <s,Dz> = 0 since, in C, δ is no longer a constant of motion. However, from the first equation of (3) we have $\delta' = -\frac{1}{2} V.\delta$; hence the set δ=0 forms an invariant manifold of C.

4. CRITICAL POINTS

There are five critical points in C in each one of the subsets with $V > 0$ and $V < 0$ and all of them lie on N. These critical points agree with the five classical central configurations, the two equilateral triangular configurations (Lagrange's points) denoted by E_+ or E_-, when the triangle is upward or downward, and three collinear configurations (Euler's points) denoted by C_i, i=1,2,3 depending on which mass is in the middle of the other two.

Sundman's inequality $V' \geq r.h + \frac{1}{2}\frac{w^2}{r}$ describes the behaviour of V. As a consequence V always increases and it is constant at the critical points. Therefore, we can represent the critical points as a function of V as shown in Fig.1.

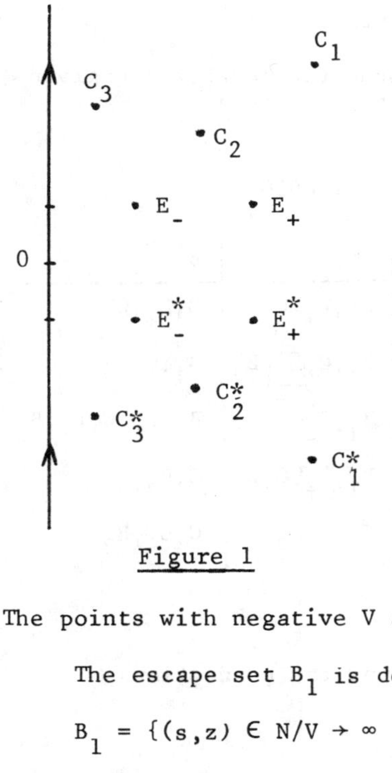

Figure 1

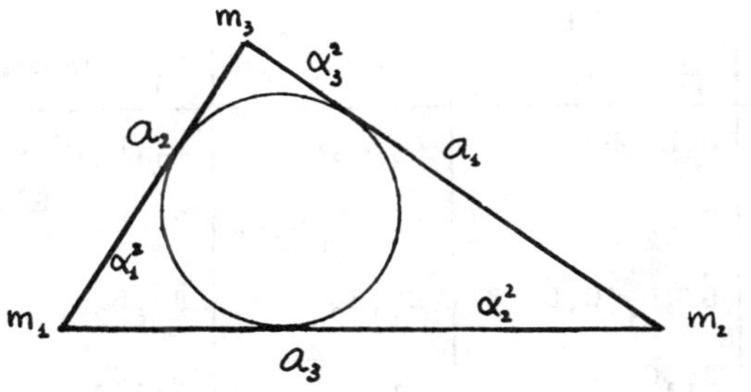

Figure 2

The points with negative V are denoted with a star.

The escape set B_1 is defined by

$$B_1 = \{(s,z) \in N/V \to \infty \quad \text{and} \quad |s_2 - s_3| \xrightarrow[t' \to \infty]{} 0\}.$$

In an analogous way B_2 and B_3 are defined. The critical points and the escape sets are the asymptotic sets for the orbits in N.

5. CONNECTIONS BETWEEN CRITICAL POINTS

To understand the problem it is important to know the dimensions of the stable W^s and unstable W^u manifolds of the critical points.

For the Lagrange and Euler points the dimensions are given in Table I.

	Lagrange W^s	Lagrange W^u	Euler W^s	Euler W^u
Collision $(V < 0)$	2	2	1	3
Ejection $(V > 0)$	2	2	3	1

Table I

We can summarize some results of connections between critical points. For the collinear problem McGehee (1974) found the connections:

(i) For all masses: $C_i \to B_j, B_k$, $i \neq j,k$, $i,j,k \in \{1,2,3\}$.

(ii) For a non-empty open set of masses: $C_i^* \to B_j, B_k$.

(iii) For exceptional masses $C_i^* \to C_i$ occurs.

For the isosceles problem Simó (1980) found the connections given in Table II, as a function of the mass parameter

$$\Omega = \frac{m_3}{m_1} \quad \text{where} \quad \Omega_1 = 0.3785 \ \ldots \text{ and } \ \Omega_2 = 2.6620\ldots$$

	$\Omega < \Omega_1$	$\Omega = \Omega_1$	$\Omega_1 < \Omega < \Omega_2$	$\Omega = \Omega_2$	$\Omega_2 < \Omega$
B_1^*	C, L_-, B_2	C, L_-, B_2	B_1, L_+, C, L_-, B_2	B_1, L_+, C	B, L_+, C
L_+^*	C, B_2	$B_1, \boxed{L_+}, C, B_2$	B_1, B_2	$B_1, C, \boxed{L_-}, B_2$	B_1, C
C^*	B_1, L_+, C, L_-, B_2	B_1, B_2	B_1, B_2	B_1, B_2	B_1, L_+, C, L_-, B_2
L_-^*	B_1, C	$B_1, C, \boxed{L_-}, B_2$	B_1, B_2	$B_1, \boxed{L_+}, C, B_2$	C, B_2
B_2^*	B_1, L_+, C	B_1, L_+, C	B_1, L_+, C	C, L_-, B_2	C, L_-, B_2

Table II

For the planar problem Moeckel (1983) found the connections

(i) $E_+, E_- \longrightarrow C_i$ $\qquad i = 1,2,3$

$E_+, E_- \longrightarrow B_i$

(ii) If we suppose $C_i^* \to E_+$, then $C_i^* \longrightarrow E_-$, $C_i^* \longrightarrow C_j$, $C_i^* \longrightarrow B_j$, $j = 1,2,3$.

6. ANOTHER FORMULATION

There is another formulation for the triple collision which is suitable for numerical studies of planar triple collision. I follow Waldvogel (1982) to describe the collision problem by means of the Murnaghan's symmetric variables (the sides, a_i, of the triangle and an ignorable angle). Then the Lemaitre's regularized variables (the square roots, α_i, of the distance between a vertex of the triangle and the tangent point of the inscribed circumference) are introduced and finally the McGehee's scaling transformation is also used. (Figure 2).

The coordinates satisfy the relation $a_i = \alpha_j^2 + \alpha_k^2$, $i,j,k \in \{1,2,3\}$.

The flow on the collision manifold is represented by six differential equations with polynomial right-hand sides

$$\frac{d\underline{\alpha}}{dt} = \frac{1}{4}B(\underline{\alpha}).\underline{\pi} - \frac{1}{4}\, a_1\, a_2\, a_3\, V.\underline{\alpha}\ ,$$

$$\frac{d\underline{\pi}}{dt} = -\frac{\partial}{\partial\underline{\alpha}} K_0\ (\underline{\alpha},\underline{\pi},\ 0),$$

where $K_0(\underline{\alpha},\underline{\pi},0) = \frac{1}{8}\,\underline{\pi}^T B(\underline{\alpha})\underline{\pi} - \sum_{j\neq k} m_j m_k\ a_j a_k$ and B is the 3×3 symmetric matrix

$$B(\underline{\alpha})=\begin{bmatrix} \frac{a_1}{m_1}\alpha^2+\frac{a_2}{m_2}\alpha_3^2+\frac{a_3}{m_3}\alpha_2^2 & -\frac{a_3}{m_3}\alpha_1\alpha_2 & -\frac{a_2}{m_3}\alpha_1\alpha_3 \\ -\frac{a_3}{m_3}\alpha_1\alpha_2 & \frac{a_2}{m_2}\alpha^2+\frac{a_3}{m_3}\alpha_1^2+\frac{a_1}{m_1}\alpha_3^2 & -\frac{a_1}{m_1}\alpha_2\alpha_3 \\ -\frac{a_2}{m_3}\alpha_1\alpha_2 & -\frac{a_1}{m_1}\alpha_2\alpha_3 & \frac{a_3}{m_3}\alpha^2+\frac{a_1}{m_1}\alpha_2^2+\frac{a_2}{m_2}\alpha_3^2 \end{bmatrix}$$

where $\alpha^2 = \alpha_1^2 + \alpha_2^2 + \alpha_3^2$.

The two first integrals are written now as

$$\sum_{j \neq k} m_j m_k (\alpha_j^2 + \alpha_k^2)^2 = m_1 + m_2 + m_3 , \qquad j,k = 1,2,3,$$

$$K_0 \ (\underline{\alpha}, \underline{\pi}, 0) = 0.$$

7. LAGRANGE POINTS CONNECTIONS

The connections between two Lagrange points having different signs of V, corresponds to intersections of two 2-dimensional manifolds in the 4-dim. manifold N. Therefore, in general these connections are not possible. I have found numerically the values of masses for which this is possible. We can represent these masses as a point of the simplex of masses, T, given by $m_1 + m_2 + m_3 = 1$. My calculations began from Simo's connections and I have found two kinds of curves which correspond to the two kinds of connections (Fig.3).

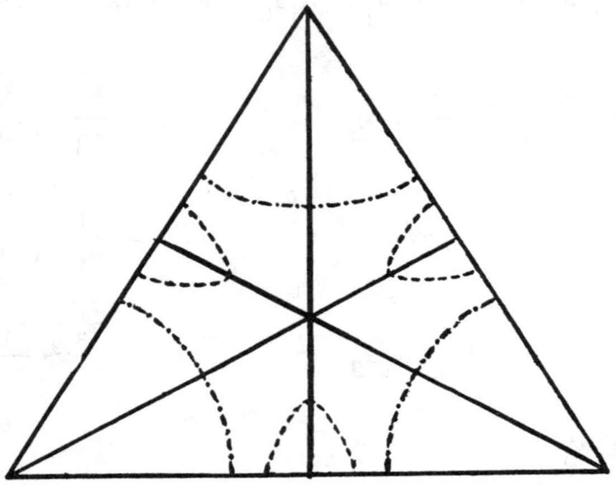

Figure 3

——— connections $E_+^* \longrightarrow E_+$, $E_-^* \longrightarrow E_-$

·—·— connections $E_+^* \longrightarrow E_-$, $E_-^* \longrightarrow E_+$

The idea of the method we use is to start with a fixed mass configuration and consider its non-rotating triple collision manifold, N.

First, we find in N the Lagrange point with negative V, E_+^*, and then study the behaviour of the orbits on the unstable manifold of E_+^*. For that we calculate a first order approximation of the unstable manifold and consider a point of a small circle around the critical point in the tangent space. Next we project this point on the unstable manifold and follow the flow until the 3-dim manifold Q defined by V = 0, in such a way that we obtain a point in Q from each point of the small original circle. This process gives us a discrete set of points on a closed curve in Q. Note that it is important to choose a good distribution of the points in the small circle in order to have almost equidistant points in the discrete set referred to above.

The calculations are simplified by the presence of a symmetry of the vector field. This is due to the following fact: The transformation $(\underline{\alpha},\underline{\pi}) \rightarrow (\underline{\alpha}, -\underline{\pi})$, together with a time reversal $t \rightarrow -t$, takes orbits into orbits. Then as a result of this symmetry, from the original curve in Q associated to E_+^*, we can obtain the curves in Q associated to $E_{+,-}$. Actually, we obtain eight curves associated to $E_{+,-}$ since these coordinates introduce four copies of each critical point.

When we have these curves, the problem is to find the two points that give us the minimum distance between the original curve and the eight copies. To have more accuracy in the calculations we take a finer partition in the original curve near these two points. In this way, we find new points that give us a smaller distance. We can repeat this refinement until the points in the original curve are so closed that we are given a good approximation of the distance between the original curve associated to E_+^* and the others, associated to $E_{+,-}$. Finally, we have obtained two minimum distance points and we consider the sign of the biggest coordinate of the vector between these two points.

Repeating the process described above, but with a little modification of the masses configuration, we will obtain a new vector between the corresponding minimum distance points. When we detect a change in the sign of the biggest coordinate of these vectors we can affirm that between the last two masses configuration there is a connecting one. This means that there exists one orbit connecting E_+^* with E_+ or E_-. In addition to this, by the symmetry, there exists one orbit connecting E_-^* with E_- or E_+ respectively.

8 FINAL GOAL AND CONJECTURE

The final goal of this study is the classification of the types of connections according to the masses and the conjecture is that the given curves are enough to make this classification. That is, the given curves are the boundaries of different open sets of the triangle of masses, and all the masses in each one of the parts give rise to non-rotating triple collision manifolds possessing the same type of connections.

REFERENCES

R. McGehee: "Triple collision in the collinear three-body problem", Invent. Math. 27 (1974), 191-227.

R. Moeckel: "Orbits near triple collision in the 3-body problem", Indiana Univ. Math. Jou., Vol.32, n.2 (1983), 221-240.

C. Simó: "Analysis of triple collision in the isosceles problem", Classical Mech. and Dynamical Syst., Marcell Dekker, New York (1980).

J. Waldvogel: "Symmetric and regularized coordinates on the plane triple collision manifold", Celestial Mechanics, 28 (1982), 69-82.

J. Waldvogel: "Stable and unstable manifolds in planar triple collision". Instabilities in Dynamical Systems (1979), 263-271.

NON-INTEGRABILITY OF HAMILTONIAN SYSTEMS BASED ON ZIGLIN'S THEOREM

Haruo Yoshida
Centre de Physique Theorique,
Ecole Polytechnique
F-91128 Palaiseau,
France

The integrable three-particle Toda lattice is described by the Hamiltonian

$$H = (1/2)(p_x^2 + p_y^2) + \exp(\sqrt{3}x+y) + \exp(-\sqrt{3}x+y) + \exp(-2y)$$

Contopoulos and Polymilis (1) made extensive numerical computations for the truncated system

$$H_n = (1/2)(p_x^2 + p_y^2) + \Sigma_{k=1,n}\{(\sqrt{3}x+y)^k + (-\sqrt{3}x+y)^k + (-2y)^k\}/(k!)$$

and observed stochastic behaviours for n = 3,5,6....,10, but not for n=4. The main result of this paper is

Theorem (5)

The truncated Toda lattice H_n (n=3,4,5,...) is always non-integrable, i.e. there cannot exist an additional analytic integral for any finite n.

The case n=3 implies non-integrability of the Hénon-Heiles system. The proof is based on Ziglin's theorem (2,3,4) using the monodromy property of the variational equation around the straight-line 'periodic' solution x=0. Non-integrability proofs of Hamiltonian systems with a homogeneous potential are given in (3,4).

REFERENCES

Contopoulos, G. and Polymilis, C.: Physica 24D (1987) 328-342.
Ziglin, S.L.: Functional Anal. Appl., 16 (1983) 181-189.
Yoshida, H. :Physica 21D (1986) 163-170.
Yoshida, H.: "A criterion for the non-existence of an additional integral in Hamiltonian systems with a homogeneous potential", Physica D (1987) in press.
Yoshida, H.: "Non-integrability of the truncated Toda lattice Hamiltonian at any order", submitted to Commun.Math. Phys. (1987).

A. E. Roy (ed.), Long-Term Dynamical Behaviour of Natural and Artificial N-Body Systems, 513.
© 1988 by Kluwer Academic Publishers.

INDEX OF NAMES

INDEX OF SUBJECTS